Business Earth Stations for Telecommunications
Walter L. Morgan and Denis Rouffet

Wireless Information Networks
Kaveh Pahlavan and Allen H. Levesque

Satellite Communications: The First Quarter Century of Service
David W. E. Rees

Fundamentals of Telecommunication Networks
Tarek N. Saadawi, Mostafa Ammar, with Ahmed El Hakeem

Meteor Burst Communications: Theory and Practice
Donald L. Schilling, Editor

Digital Communication over Fading Channels: A Unified Approach to Performance Analysis
Marvin K. Simon and Mohamed-Slim Alouini

Vector Space Projections: A Numerical Approach to Signal and Image Processing, Neural Nets, and Optics
Henry Stark and Yongyi Yang

Signaling in Telecommunication Networks
John G. van Bosse

Telecommunication Circuit Design
Patrick D. van der Puije

Worldwide Telecommunications Guide for the Business Manager
Walter H. Vignault

Digital Communication over Fading Channels

Digital Communication over Fading Channels

A Unified Approach to Performance Analysis

Marvin K. Simon
Mohamed-Slim Alouini

A Wiley-Interscience Publication
JOHN WILEY & SONS, INC.

New York • Chichester • Weinheim • Brisbane • Singapore • Toronto

This book is printed on acid-free paper. ∞

Copyright © 2000 by John Wiley & Sons, Inc. All rights reserved.

Published simultaneously in Canada.

No part of this publication may be reproduced, stored in a retrieval system or transmitted in any form or by any means, electronic, mechanical, photocopying, recording, scanning or otherwise, except as permitted under Sections 107 or 108 of the 1976 United States Copyright Act, without either the prior written permission of the Publisher, or authorization through payment of the appropriate per-copy fee to the Copyright Clearance Center, 222 Rosewood Drive, Danvers, MA 01923, (978) 750-8400, fax (978) 750-4744. Requests to the Publisher for permission should be addressed to the Permissions Department, John Wiley & Sons, Inc., 605 Third Avenue, New York, NY 10158-0012, (212) 850-6011, fax (212) 850-6008, E-mail: PERMREQ@WILEY.COM.

For ordering and customer service, call 1-800-CALL-WILEY.

Library of Congress Cataloging-in-Publication Data:

Simon, Marvin Kenneth, 1939–
   Digital communication over fading channels : a unified approach to performance analysis / Marvin K. Simon and Mohamed-Slim Alouini.
     p. cm. — (Wiley series in telecommunications and signal processing)
   Includes index.
   ISBN 0-471-31779-9 (alk. paper)
   1. Digital communications — Reliability — Mathematics. I. Alouini, Mohamed-Slim. II. Title. III. Series.
TK5103.7.S523 2000
621.382 — dc21   99-056352

Printed in the United States of America.

10 9 8 7 6 5 4 3 2 1

*Marvin K. Simon dedicates this book to his wife, Anita,
whose devotion to him and this project
never once faded during its preparation.*

*Mohamed-Slim Alouini dedicates this book
to his parents and family.*

CONTENTS

Preface   xv

PART 1 FUNDAMENTALS

Chapter 1   Introduction   3
    1.1   System Performance Measures   4
       1.1.1   Average Signal-to-Noise Ratio   4
       1.1.2   Outage Probability   5
       1.1.3   Average Bit Error Probability   6
    1.2   Conclusions   12
    References   13

Chapter 2   Fading Channel Characterization and Modeling   15
    2.1   Main Characteristics of Fading Channels   15
       2.1.1   Envelope and Phase Fluctuations   15
       2.1.2   Slow and Fast Fading   16
       2.1.3   Frequency-Flat and Frequency-Selective Fading   16
    2.2   Modeling of Flat Fading Channels   17
       2.2.1   Multipath Fading   18
       2.2.2   Log-Normal Shadowing   23
       2.2.3   Composite Multipath/Shadowing   24
       2.2.4   Combined (Time-Shared) Shadowed/Unshadowed Fading   25
    2.3   Modeling of Frequency-Selective Fading Channels   26
    References   28

Chapter 3 Types of Communication — 31
- 3.1 Ideal Coherent Detection — 31
  - 3.1.1 Multiple Amplitude-Shift-Keying or Multiple Amplitude Modulation — 33
  - 3.1.2 Quadrature Amplitude-Shift-Keying or Quadrature Amplitude Modulation — 34
  - 3.1.3 M-ary Phase-Shift-Keying — 35
  - 3.1.4 Differentially Encoded M-ary Phase-Shift-Keying — 39
  - 3.1.5 Offset QPSK or Staggered QPSK — 41
  - 3.1.6 M-ary Frequency-Shift-Keying — 43
  - 3.1.7 Minimum-Shift-Keying — 45
- 3.2 Nonideal Coherent Detection — 47
- 3.3 Noncoherent Detection — 53
- 3.4 Partially Coherent Detection — 55
  - 3.4.1 Conventional Detection: One-Symbol Observation — 55
  - 3.4.2 Multiple Symbol Detection — 57
- 3.5 Differentially Coherent Detection — 59
  - 3.5.1 M-ary Differential Phase Shift Keying — 59
  - 3.5.2 $\pi/4$-Differential QPSK — 65
- References — 65

PART 2 MATHEMATICAL TOOLS

Chapter 4 Alternative Representations of Classical Functions — 69
- 4.1 Gaussian Q-Function — 70
  - 4.1.1 One-Dimensional Case — 70
  - 4.1.2 Two-Dimensional Case — 72
- 4.2 Marcum Q-Function — 74
  - 4.2.1 First-Order Marcum Q-Function — 74
  - 4.2.2 Generalized (mth-Order) Marcum Q-Function — 81
- 4.3 Other Functions — 90
- References — 94
- Appendix 4A: Derivation of Eq. (4.2) — 95

Chapter 5 Useful Expressions for Evaluating Average Error Probability Performance — 99
- 5.1 Integrals Involving the Gaussian Q-Function — 99
  - 5.1.1 Rayleigh Fading Channel — 101

	5.1.2	Nakagami-q (Hoyt) Fading Channel	101
	5.1.3	Nakagami-n (Rice) Fading Channel	102
	5.1.4	Nakagami-m Fading Channel	102
	5.1.5	Log-Normal Shadowing Channel	104
	5.1.6	Composite Log-Normal Shadowing/Nakagami-m Fading Channel	104
5.2	Integrals Involving the Marcum Q-Function		107
	5.2.1	Rayleigh Fading Channel	108
	5.2.2	Nakagami-q (Hoyt) Fading Channel	109
	5.2.3	Nakagami-n (Rice) Fading Channel	109
	5.2.4	Nakagami-m Fading Channel	109
	5.2.5	Log-Normal Shadowing Channel	109
	5.2.6	Composite Log-Normal Shadowing/Nakagami-m Fading Channel	110
5.3	Integrals Involving the Incomplete Gamma Function		111
	5.3.1	Rayleigh Fading Channel	112
	5.3.2	Nakagami-q (Hoyt) Fading Channel	112
	5.3.3	Nakagami-n (Rice) Fading Channel	112
	5.3.4	Nakagami-m Fading Channel	113
	5.3.5	Log-Normal Shadowing Channel	114
	5.3.6	Composite Log-Normal Shadowing/Nakagami-m Fading Channel	114
5.4	Integrals Involving Other Functions		114
	5.4.1	M-PSK Error Probability Integral	114
	5.4.2	Arbitrary Two-Dimensional Signal Constellation Error Probability Integral	116
	5.4.3	Integer Powers of the Gaussian Q-Function	117
	5.4.4	Integer Powers of M-PSK Error Probability Integrals	121
References			124
Appendix 5A: Evaluation of Definite Integrals Associated with Rayleigh and Nakagami-m Fading			124

Chapter 6 New Representations of Some PDF's and CDF's for Correlative Fading Applications — **141**

6.1	Bivariate Rayleigh PDF and CDF	142
6.2	PDF and CDF for Maximum of Two Rayleigh Random Variables	146
6.3	PDF and CDF for Maximum of Two Nakagami-m Random Variables	149
References		152

PART 3 OPTIMUM RECEPTION AND PERFORMANCE EVALUATION

Chapter 7 Optimum Receivers for Fading Channels — 157

7.1 Case of Known Amplitudes, Phases, and Delays: Coherent Detection — 159

7.2 The Case of Known Phases and Delays, Unknown Amplitudes — 163
- 7.2.1 Rayleigh Fading — 163
- 7.2.2 Nakagami-m Fading — 164

7.3 Case of Known Amplitudes and Delays, Unknown Phases — 166

7.4 Case of Known Delays and Unknown Amplitudes and Phases — 168
- 7.4.1 One-Symbol Observation: Noncoherent Detection — 168
- 7.4.2 Two-Symbol Observation: Conventional Differentially Coherent Detection — 181
- 7.4.3 N-Symbol Observation: Multiple Symbol Differentially Coherent Detection — 186

7.5 Case of Unknown Amplitudes, Phases, and Delays — 188
- 7.5.1 One-Symbol Observation: Noncoherent Detection — 188
- 7.5.2 Two-Symbol Observation: Conventional Differentially Coherent Detection — 190

References — 191

Chapter 8 Performance of Single Channel Receivers — 193

8.1 Performance Over the AWGN Channel — 193
- 8.1.1 Ideal Coherent Detection — 194
- 8.1.2 Nonideal Coherent Detection — 206
- 8.1.3 Noncoherent Detection — 209
- 8.1.4 Partially Coherent Detection — 210
- 8.1.5 Differentially Coherent Detection — 213
- 8.1.6 Generic Results for Binary Signaling — 218

8.2 Performance Over Fading Channels — 219
- 8.2.1 Ideal Coherent Detection — 220
- 8.2.2 Nonideal Coherent Detection — 234
- 8.2.3 Noncoherent Detection — 239
- 8.2.4 Partially Coherent Detection — 242
- 8.2.5 Differentially Coherent Detection — 243

References — 251

Appendix 8A: Stein's Unified Analysis of the Error Probability Performance of Certain Communication Systems — 253

Chapter 9 Performance of Multichannel Receivers — 259

9.1 Diversity Combining — 260
    9.1.1 Diversity Concept — 260
    9.1.2 Mathematical Modeling — 260
    9.1.3 Brief Survey of Diversity Combining Techniques — 261
    9.1.4 Complexity–Performance Trade-offs — 264

9.2 Maximal-Ratio Combining — 265
    9.2.1 Receiver Structure — 265
    9.2.2 PDF-Based Approach — 267
    9.2.3 MGF-Based Approach — 268
    9.2.4 Bounds and Asymptotic SER Expressions — 275

9.3 Coherent Equal Gain Combining — 278
    9.3.1 Receiver Structure — 279
    9.3.2 Average Output SNR — 279
    9.3.3 Exact Error Rate Analysis — 281
    9.3.4 Approximate Error Rate Analysis — 288
    9.3.5 Asymptotic Error Rate Analysis — 289

9.4 Noncoherent Equal-Gain Combining — 290
    9.4.1 DPSK, DQPSK, and BFSK: Exact and Bounds — 290
    9.4.2 M-ary Orthogonal FSK — 304

9.5 Outage Probability Performance — 311
    9.5.1 MRC and Noncoherent EGC — 312
    9.5.2 Coherent EGC — 313
    9.5.3 Numerical Examples — 314

9.6 Impact of Fading Correlation — 316
    9.6.1 Model A: Two Correlated Branches with Nonidentical Fading — 320
    9.6.2 Model B: D Identically Distributed Branches with Constant Correlation — 323
    9.6.3 Model C: D Identically Distributed Branches with Exponential Correlation — 324
    9.6.4 Model D: D Nonidentically Distributed Branches with Arbitrary Correlation — 325
    9.6.5 Numerical Examples — 329

9.7 Selection Combining — 333
    9.7.1 MGF of Output SNR — 335

	9.7.2	Average Output SNR	336
	9.7.3	Outage Probability	338
	9.7.4	Average Probability of Error	340
9.8	Switched Diversity		348
	9.8.1	Performance of SSC over Independent Identically Distributed Branches	348
	9.8.2	Effect of Branch Unbalance	362
	9.8.3	Effect of Branch Correlation	366
9.9	Performance in the Presence of Outdated or Imperfect Channel Estimates		370
	9.9.1	Maximal-Ratio Combining	370
	9.9.2	Noncoherent EGC over Rician Fast Fading	371
	9.9.3	Selection Combining	373
	9.9.4	Switched Diversity	374
	9.9.5	Numerical Results	377
9.10	Hybrid Diversity Schemes		378
	9.10.1	Generalized Selection Combining	378
	9.10.2	Generalized Switched Diversity	403
	9.10.3	Two-Dimensional Diversity Schemes	408

References	411
Appendix 9A: Alternative Forms of the Bit Error Probability for a Decision Statistic that is a Quadratic Form of Complex Gaussian Random Variables	421
Appendix 9B: Simple Numerical Techniques for the Inversion of the Laplace Transform of Cumulative Distribution Functions	427
9B.1 Euler Summation-Based Technique	427
9B.2 Gauss–Chebyshev Quadrature-Based Technique	428
Appendix 9C: Proof of Theorem 1	430
Appendix 9D: Direct Proof of Eq. (9.331)	431
Appendix 9E: Special Definite Integrals	432

PART 4 APPLICATION IN PRACTICAL COMMUNICATION SYSTEMS

Chapter 10	Optimum Combining: A Diversity Technique for Communication Over Fading Channels in the Presence of Interference	437
	10.1 Performance of Optimum Combining Receivers	438

	10.1.1	Single Interferer, Independent Identically Distributed Fading	438
	10.1.2	Multiple Interferers, Independent Identically Distributed Fading	454
	10.1.3	Comparison with Results for MRC in the Presence of Interference	466
References			470

Chapter 11 Direct-Sequence Code-Division Multiple Access — 473

- 11.1 Single-Carrier DS-CDMA Systems — 474
  - 11.1.1 System and Channel Models — 474
  - 11.1.2 Performance Analysis — 477
- 11.2 Multicarrier DS-CDMA Systems — 479
  - 11.2.1 System and Channel Models — 480
  - 11.2.2 Performance Analysis — 483
  - 11.2.3 Numerical Examples — 489
- References — 492

PART 5 FURTHER EXTENSIONS

Chapter 12 Coded Communication Over Fading Channels — 497

- 12.1 Coherent Detection — 499
  - 12.1.1 System Model — 499
  - 12.1.2 Evaluation of Pairwise Error Probability — 502
  - 12.1.3 Transfer Function Bound on Average Bit Error Probability — 510
  - 12.1.4 Alternative Formulation of the Transfer Function Bound — 513
  - 12.1.5 Example — 514
- 12.2 Differentially Coherent Detection — 520
  - 12.2.1 System Model — 520
  - 12.2.2 Performance Evaluation — 522
  - 12.2.3 Example — 524
- 12.3 Numerical Results: Comparison of the True Upper Bounds and Union–Chernoff Bounds — 526
- References — 530
- Appendix 12A: Evaluation of a Moment Generating Function Associated with Differential Detection of M-PSK Sequences — 532

Index — 535

PREFACE

Regardless of the branch of science or engineering, theoreticians have always been enamored with the notion of expressing their results in the form of closed-form expressions. Quite often, the elegance of the closed-form solution is overshadowed by the complexity of its form and the difficulty in evaluating it numerically. In such instances, one becomes motivated to search instead for a solution that is simple in form and simple to evaluate. A further motivation is that the method used to derive these alternative simple forms should also be applicable in situations where closed-form solutions are ordinarily unobtainable. The search for and ability to find such a unified approach for problems dealing with evaluation of the performance of digital communication over generalized fading channels is what provided the impetus to write this book, the result of which represents the backbone for the material contained within its pages.

For at least four decades, researchers have studied problems of this type, and system engineers have used the theoretical and numerical results reported in the literature to guide the design of their systems. Whereas the results from the earlier years dealt mainly with simple channel models (e.g., Rayleigh or Rician multipath fading), applications in more recent years have become increasingly sophisticated, thereby requiring more complex models and improved diversity techniques. Along with the complexity of the channel model comes the complexity of the analytical solution that enables one to assess performance. With the mathematical tools that were available previously, the solutions to such problems, when possible, had to be expressed in complicated mathematical form which provided little insight into the dependence of the performance on the system parameters. Surprisingly enough, not until recently had anyone demonstrated a unified approach that not only allows previously obtained complicated results to be simplified both analytically and computationally but also permits new results to be obtained for special cases that heretofore had resisted solution in a simple form. This approach, which the authors first presented to the public in a tutorial-style article that appeared in the September 1998 issue of the *IEEE Proceedings*, has spawned a new wave of publications on the subject that, we foresee based on the variety of applications to which it has already been applied, will continue well into the new millennium. The key to the success of the approach relies on employing alternative representations of classic functions arising in the error probability analysis of digital communication systems (e.g., the Gaussian Q-function[1] and the Marcum Q-function) in such a manner that the resulting expressions for average bit or symbol error rate are in a form that is rarely more complicated than a single integral with finite limits and an integrand composed of elementary (e.g., exponential and trigonometric) functions. By virtue of replacing the conventional forms of the above-mentioned functions by their alternative representations, the integrand will contain the moment generating function (MGF) of the instantaneous fading signal-to-noise ratio (SNR), and as such, the unified approach is referred to as the *MGF-based approach*.

In dealing with application of the MGF-based approach, the coverage in this book is extremely broad, in that coherent, differentially coherent, partially coherent and noncoherent communication systems are all handled, as well as a large variety of fading channel models typical of communication links of practical interest. Both single- and multichannel reception are discussed, and in the case of the latter, a large variety of diversity types are considered. For each combination of communication (modulation/detection) type, channel fading model, and diversity type, the average bit error rate (BER) and/or symbol error rate (SER) of the system is obtained and represented by an expression that is in a form that can readily be evaluated.[2] All cases considered correspond to real practical channels, and in many instances the BER and SER expressions obtained can be evaluated numerically on a hand-held calculator.

In accomplishing the purpose set forth by the discussion above, the book focuses on developing a compendium of results that to a large extent are not readily available in standard textbooks on digital communications. Although some of these results can be found in the myriad of contributions that have been reported in the technical journal and conference literature, others are new and as yet unpublished. Indeed, aside from the fact that a significant number of the reference citations in this book are from 1999 publications, many others refer to papers that will appear in print in the new millennium. Whether or not published previously, the value of the results found in this book is that they are all colocated in a single publication with unified notation and, most important, a unified presentation framework that lends itself to simplicity of numerical evaluation. In writing this book, our intent was to spend as little space as possible duplicating material dealing with basic digital communication theory and system performance evaluation, which is well documented in many fine textbooks on the subject. Rather, this book serves to advance the material found in these books and so is of most value to those desiring to extend their knowledge

---

[1] The Gaussian Q-function has a one-to-one mapping with the complementary error function erfc x [i.e., $Q(x) = \frac{1}{2}\text{erfc}(x/\sqrt{2})$] commonly found in standard mathematical tabulations. In much of the engineering literature, however, the two functions are used interchangeably and as a matter of convenience we shall do the same in this text.

[2] The terms *bit error probability (BEP)* and *symbol error probability (SEP)* are quite often used as alternatives to *bit error rate (BER)* and *symbol error rate (SER)*. With no loss in generality, we shall employ both usages in this book.

beyond what ordinarily might be covered in the classroom. In this regard, the book should have a strong appeal to graduate students doing research in the field of digital communications over fading channels as well as to practicing engineers who are responsible for the design and performance evaluation of such systems. With regard to the latter, the book contains copious numerical evaluations that are illustrated in the form of parametric performance curves (e.g., average error probability versus average SNR). The applications chosen for the numerical illustrations correspond to real practical channels, therefore the performance curves provided will have far more than academic value. The availability of such a large collection of system performance curves in a single compilation allows the researcher or system designer to perform trade-off studies among the various communication type/fading channel/diversity combinations so as to determine the optimum choice in the face of his or her available constraints.

The book is composed of four parts, each with an express purpose. The first part contains an introduction to the subject of communication system performance evaluation followed by discussions of the various types of fading channel models and modulation/detection schemes that together form the overall system. Part 2 starts by introducing the alternative forms of the classic functions mentioned above and then proceeds to show how these forms can be used to (1) evaluate certain integrals characteristic of communication system error probability performance, and (2) find new representations for certain probability density and distribution functions typical of correlated fading applications. Part 3 is the "heart and soul" of the book, since in keeping with its title, the primary focus of this part is on performance evaluation of the various types of fading channel models and modulation/detection schemes introduced in Part 1 for both single- and multichannel (diversity) reception. Before presenting this comprehensive performance evaluation study, however, Part 3 begins by deriving the optimum receiver structures corresponding to a variety of combinations concerning the knowledge or lack thereof of the fading parameters (i.e., amplitude, phase, delay). Several of these structures might be deemed as too complex to implement in practice; nevertheless, their performances serve as benchmarks against which many suboptimum but practical structures discussed in the ensuing chapters might be compared. In Part 4, which deals with practical applications, we consider first the problem of optimum combining (diversity) in the presence of co-channel interference and then apply the unified approach to studying the performance of single- and multiple-carrier direct-sequence code-division multiple-access (DS-CDMA) systems typical of the current digital cellular wireless standard. Finally, in Part 5 we extend the theory developed in the preceding parts for uncoded communication to error-correction-coded systems.

In summary, the authors know of no other textbook currently on the market that addresses the subject of digital communication over fading channels in as comprehensive and unified a manner as is done herein. In fact, prior to the publication of this book, to the authors' best knowledge, there existed only two works (the textbook by Kennedy [1] and the reprint book by Brayer [2]) that like our book are totally dedicated to this subject, and both of them are more than a quarter of a century old. Although a number of other textbooks [3–11] devote part of their contents[3] to fading channel performance evaluation, by comparison with our book the treatment is brief and therefore incomplete. In view of the above, we believe that our book is unique in the field.

By way of acknowledgment, we wish to thank Dr. Payman Arabshahi of the Jet Propulsion Laboratory, Pasadena, CA for providing his expertise in solving a variey of problems that arose during the preparation of the electronic version of the manuscript. Mohamed-Slim Alouini would also like to express his sincere acknowledgment and gratitude to his PhD advisor Prof. Andrea J. Goldsmith of Stanford University, Palo Alto, CA for her guidance, support, and constant encouragement. Some of the material presented in Chapters 9 and 11 is the result of joint work with Prof. Goldsmith. Mohamed-Slim Alouini would also like to thank Young-Chai Ko and Yan Xin of the University of Minnesota, Minneapolis, MN for their significant contributions in some of the results presented in Chapters 9 and 7, respectively.

MARVIN K. SIMON
MOHAMED-SLIM ALOUINI

Jet Propulsion Laboratory
Pasadena, California
University of Minnesota
Minneapolis, Minnesota

REFERENCES

1. R. S. Kennedy, *Fading Dispersive Communication Channels*. New York: Wiley-Interscience, 1969.
2. K. Brayer, ed., *Data Communications via Fading Channels*. Piscataway, NJ: IEEE Press, 1975.
3. M. Schwartz, W. R. Bennett, and S. Stein, *Communication Systems and Techniques*. New York: McGraw-Hill, 1966.
4. W. C. Y. Lee, *Mobile Communications Engineering*. New York: McGraw-Hill, 1982.
5. J. Proakis, *Digital Communications*. New York: McGraw-Hill, 3rd ed., 1995 (1st and 2nd eds. in 1983, 1989, respectively).
6. M. D. Yacoub, *Foundations of Mobile Radio Engineering*. Boca Raton, FL: CRC Press, 1993.
7. W. C. Jakes, *Microwave Mobile Communication*, 2nd ed., Piscataway, NJ: IEEE Press, 1994.

---

[3] Although Reference 11 is a book that is entirely devoted to digital communication over fading channels, the focus is on error-correction coded modulations and therefore would primarily relate only to Chapter 12 of our book.

8. K. Pahlavan and A. H. Levesque, *Wireless Information Networks*. Wiley Series in Telecommunications and Signal Processing. New York: Wiley-Interscience, 1995.
9. G. L. Stüber, *Principles of Mobile Communication*. Norwell, MA: Kluwer Academic Publishers, 1996.
10. T. S. Rappaport, *Wireless Communications: Principles and Practice*. Upper Saddle River, NJ: Prentice Hall, 1996.
11. S. H. Jamali and T. Le-Ngoc, *Coded-Modulation Techniques for Fading Channels*. Norwell, MA: Kluwer Academic Publishers, 1994.

PART 1
FUNDAMENTALS

1

INTRODUCTION

As we step forward into the new millennium with wireless technologies leading the way in which we communicate, it becomes increasingly clear that the dominant consideration in the design of systems employing such technologies will be their ability to perform with adequate margin over a channel perturbed by a host of impairments not the least of which is multipath fading. This is not to imply that multipath fading channels are something new to be reckoned with, indeed they have plagued many a system designer for well over 40 years, but rather, to serve as a motivation for their ever-increasing significance in the years to come. At the same time, we do not in any way wish to diminish the importance of the fading channel scenarios that occurred well prior to the wireless revolution, since indeed many of them still exist and will continue to exist in the future. In fact, it is safe to say that whatever means are developed for dealing with the more sophisticated wireless application will no doubt also be useful for dealing with the less complicated fading environments of the past.

With the above in mind, what better opportunity is there than now to write a comprehensive book that provides simple and intuitive solutions to problems dealing with communication system performance evaluation over fading channels? Indeed, as mentioned in the preface, the primary goal of this book is to present a unified method for arriving at a set of tools that will allow the system designer to compute the performance of a host of different digital communication systems characterized by a variety of modulation/detection types and fading channel models. By set of tools we mean a compendium of analytical results that not only allow easy, yet accurate performance evaluation but at the same time provide insight into the manner in which this performance depends on the key system parameters. To emphasize what was stated above, the set of tools developed in this book are useful not only for the wireless applications that are rapidly filling our current technical journals but also to a host of others, involving satellite, terrestrial, and maritime communications.

Our repetitive use of the word performance thus far brings us to the purpose of this introductory chapter: to provide several measures of performance related to practical communication system design and to begin exploring the analytical methods by which they may be evaluated. While the deeper meaning of these measures will be truly understood only after their more formal definitions are presented in the chapters that follow, the introduction of these terms here serves to illustrate the various possibilities that exist, depending on both need and relative ease of evaluation.

1.1 SYSTEM PERFORMANCE MEASURES

1.1.1 Average Signal-to-Noise Ratio

Probably the most common and best understood performance measure characteristic of a digital communication system is *signal-to-noise ratio (SNR)*. Most often this is measured at the output of the receiver and is thus related directly to the data detection process itself. Of the several possible performance measures that exist, it is typically the easiest to evaluate and most often serves as an excellent indicator of the overall fidelity of the system. Although traditionally, the term *noise* in *signal-to-noise ratio* refers to the ever-present thermal noise at the input to the receiver, in the context of a communication system subject to fading impairment, the more appropriate performance measure is *average SNR*, where the word *average* refers to statistical averaging over the probability distribution of the fading. In simple mathematical terms, if γ denotes the instantaneous SNR [a random variable (RV)] at the receiver output, which includes the effect of fading, then

$$\bar{\gamma} \triangleq \int_0^\infty \gamma p_\gamma(\gamma) \, d\gamma \qquad (1.1)$$

is the average SNR, where $p_\gamma(\gamma)$ denotes the probability density function (PDF) of γ. To begin to get a feel for what we will shortly describe as a unified approach to performance evaluation, we first rewrite (1.1) in terms of the moment generating function (MGF) associated with γ, namely,

$$M_\gamma(s) = \int_0^\infty p_\gamma(\gamma) e^{s\gamma} \, d\gamma \qquad (1.2)$$

Taking the first derivative of (1.2) with respect to s and evaluating the result at $s = 0$, we see immediately from (1.1) that

$$\bar{\gamma} = \left. \frac{dM_\gamma(s)}{ds} \right|_{s=0} \qquad (1.3)$$

That is, the ability to evaluate the MGF of the instantaneous SNR (perhaps in closed form) allows immediate evaluation of the average SNR via a simple mathematical operation: differentiation.

To gain further insight into the power of the foregoing statement, we note that in many systems, particularly those dealing with a form of diversity (multichannel) reception known as *maximal-ratio combining (MRC)* (discussed in great detail in Chapter 9), the output SNR, γ, is expressed as a sum (combination) of the individual branch (channel) SNRs (i.e., $\gamma = \sum_{l=1}^{L} \gamma_l$, where L denotes the number of channels combined). In addition, it is often reasonable in practice to assume that the channels are independent of each other (i.e., the RVs $\gamma_l|_{l=1}^{L}$ are themselves independent). In such instances, the MGF $M_\gamma(s)$ can be expressed as the product of the MGFs associated with each channel [i.e., $M_\gamma(s) = \prod_{l=1}^{L} M_{\gamma_l}(s)$], which for a large variety of fading channel statistical models can be computed in closed form.[1] By contrast, even with the assumption of channel independence, computation of the probability density function (PDF) $p_\gamma(\gamma)$, which requires convolution of the various PDFs $p_{\gamma_l}(\gamma_l)|_{l=1}^{L}$ that characterize the L channels, can still be a monumental task. Even in the case where these individual channel PDFs are of the same functional form but are characterized by different average SNR's, $\overline{\gamma}_l$, the evaluation of $p_\gamma(\gamma)$ can still be quite tedious. Such is the power of the MGF-based approach; namely, it circumvents the need for finding the first-order PDF of the output SNR provided that one is interested in a performance measure that can be expressed in terms of the MGF. Of course, for the case of average SNR, the solution is extremely simple, namely, $\overline{\gamma} = \sum_{l=1}^{L} \overline{\gamma}_l$, regardless of whether the channels are independent or not, and in fact, one never needs to find the MGF at all. However, for other performance measures and also the average SNR of other combining statistics [e.g., the sum of an ordered set of random variables typical of generalized selection combining (GSC) (discussed in Chapter 9)], matters are not quite this simple and the points made above for justifying an MGF-based approach are, as we shall see, especially significant.

1.1.2 Outage Probability

Another standard performance criterion characteristic of diversity systems operating over fading channels is the *outage probability* denoted by P_{out} and defined as the probability that the instantaneous error probability exceeds a specified value or equivalently, the probability that the output SNR, γ, falls below a certain specified threshold, γ_{th}. Mathematically speaking,

$$P_{\text{out}} = \int_0^{\gamma_{\text{th}}} p_\gamma(\gamma)\,d\gamma \qquad (1.4)$$

which is the cumulative distribution function (CDF) of γ, namely, $P_\gamma(\gamma)$, evaluated at $\gamma = \gamma_{\text{th}}$. Since the PDF and the CDF are related by $p_\gamma(\gamma) =$

---

[1] Note that the existence of the product form for the MGF $M_\gamma(s)$ does not necessarily imply that the channels are identically distributed [i.e., each MGF $M_{\gamma_l}(s)$ is allowed to maintain its own identity independent of the others]. Furthermore, even if the channels are not assumed to be independent, the relation in (1.3) is nevertheless valid, and in many instances the MGF of the (combined) output can still be obtained in closed form.

6 INTRODUCTION

$dP_\gamma(\gamma)/d\gamma$, and since $P_\gamma(0) = 0$, the Laplace transforms of these two functions are related by[2]

$$\hat{P}_\gamma(s) = \frac{\hat{p}_\gamma(s)}{s} \tag{1.5}$$

Furthermore, since the MGF is just the Laplace transform of the PDF with argument reversed in sign [i.e., $\hat{p}_\gamma(s) = M_\gamma(-s)$], the outage probability can be found from the inverse Laplace transform of the ratio $M_\gamma(-s)/s$ evaluated at $\gamma = \gamma_{th}$, that is,

$$P_{out} = \frac{1}{2\pi j} \int_{\sigma-j\infty}^{\sigma+j\infty} \frac{M_\gamma(-s)}{s} e^{s\gamma_{th}} ds \tag{1.6}$$

where σ is chosen in the region of convergence of the integral in the complex s plane. Methods for evaluating inverse Laplace transforms have received widespread attention in the literature. (A good summary of these can be found in Ref. 1.) One such numerical technique that is particularly useful for CDFs of positive RVs (such as instantaneous SNR) is discussed in Appendix 9B and applied in Chapter 9. For our purpose here, it is sufficient to recognize once again that the evaluation of outage probability can be performed based entirely on knowledge of the MGF of the output SNR without ever having to compute its PDF.

1.1.3 Average Bit Error Probability

The third performance criterion and undoubtedly the most difficult of the three to compute is average bit error probability (BEP).[3] On the other hand, it is the one that is most revealing about the nature of the system behavior and the one most often illustrated in documents containing system performance evaluations; thus, it is of primary interest to have a method for its evaluation that reduces the degree of difficulty as much as possible.

The primary reason for the difficulty in evaluating average BEP lies in the fact that the conditional (on the fading) BEP is, in general, a nonlinear function of the instantaneous SNR, the nature of the nonlinearity being a function of the modulation/detection scheme employed by the system. For example, in the multichannel case, the average of the conditional BEP over the fading statistics is not a simple average of the per channel performance measure as was true for average SNR. Nevertheless, we shall see momentarily that an MGF-based approach is still quite useful in simplifying the analysis and in a large variety of cases allows unification under a common framework.

---

[2] The symbol "∧" above a function denotes its Laplace transform.
[3] The discussion that follows applies, in principle, equally well to average symbol error probability (SEP). The specific differences between the two are explored in detail in the chapters dealing with system performance. Furthermore, the terms *bit error rate (BER)* and *symbol error rate (SER)* are often used in the literature as alternatives to BEP and SEP. Rather than choose a preference, in this book we use these terms interchangeably.

Suppose first that the conditional BEP is of the form

$$P_b(E|\gamma) = C_1 \exp(-a_1\gamma) \tag{1.7}$$

such as would be the case for differentially coherent detection of phase-shift-keying (PSK) or noncoherent detection of orthogonal frequency-shift-keying (FSK) (see Chapter 8). Then the average BEP can be written as

$$P_b(E) \triangleq \int_0^\infty P_b(E|\gamma) p_\gamma(\gamma)\, d\gamma$$
$$= \int_0^\infty C_1 \exp(-a_1\gamma) p_\gamma(\gamma)\, d\gamma = C_1 M_\gamma(-a_1) \tag{1.8}$$

where again $M_\gamma(s)$ is the MGF of the instantaneous fading SNR and depends only on the fading channel model assumed.

Suppose next that the nonlinear functional relationship between $P_b(E|\gamma)$ and γ is such that it can be expressed as an integral whose integrand has an exponential dependence on γ in the form of (1.7), that is,[4]

$$P_b(E|\gamma) = \int_{\xi_1}^{\xi_2} C_2 h(\xi) \exp[-a_2 g(\xi)\gamma]\, d\xi \tag{1.9}$$

where for our purpose here $h(\xi)$ and $g(\xi)$ are arbitrary functions of the integration variable, and typically both ξ_1 and ξ_2 are finite (although this is not an absolute requirement for what follows).[5] Although not at all obvious at this point, suffice it to say that a relationship of the form in (1.9) can result from employing alternative forms of such classic nonlinear functions as the Gaussian Q-function and Marcum Q-function (see Chapter 4), which are characteristic of the relationship between $P_b(E|\gamma)$ and γ corresponding to, for example, coherent detection of PSK and differentially coherent detection of quadriphase-shift-keying (QPSK), respectively. Still another possibility is that the nonlinear functional relationship between $P_b(E|\gamma)$ and γ is inherently in the form of (1.9); that is, no alternative representation need be employed. An example of such occurs for the conditional symbol error probability (SEP) associated with coherent and differentially coherent detection of M-ary PSK (M-PSK) (see Chapter 8). Regardless of the particular case at hand, once again averaging (1.9) over the fading gives (after interchanging the order of integration)

$$P_b(E) = \int_0^\infty P_b(E|\gamma) p_\gamma(\gamma)\, d\gamma = \int_0^\infty \int_{\xi_1}^{\xi_2} C_2 h(\xi) \exp[-a_2 g(\xi)\gamma]\, d\xi\, p_\gamma(\gamma)\, d\gamma$$

---

[4] In the more general case, the conditional BEP might be expressed as a sum of integrals of the type in (1.9).
[5] In principle, (1.9) includes (1.7) as a special case if $h(\xi)$ is allowed to assume the form of a Dirac delta function located within the interval $\xi_1 \leq \xi \leq \xi_2$.

$$= C_2 \int_{\xi_1}^{\xi_2} h(\xi) \int_0^\infty \exp[-a_2 g(\xi)\gamma] p_\gamma(\gamma)\, d\gamma\, d\xi$$

$$= C_2 \int_{\xi_1}^{\xi_2} h(\xi) M_\gamma[-a_2 g(\xi)]\, d\xi \qquad (1.10)$$

As we shall see later in the book, integrals of the form in (1.10) can, for many special cases, be obtained in closed form. At the very worst, with rare exceptions, the resulting expression will be a single integral with finite limits and an integrand composed of elementary functions.[6] Since (1.8) and (1.10) cover a wide variety of different modulation/detection types and fading channel models, we refer to this approach for evaluating average error probability as *the unified MGF-based approach* and the associated forms of the conditional error probability as the *desired forms*. The first notion of such a unified approach was discussed in Ref. 2 and laid the groundwork for much of the material that follows in this book.

It goes without saying that not every fading channel communication problem fits the foregoing description; thus, alternative, but still simple and accurate techniques are desirable for evaluating system error probability in such circumstances. One class of problems for which a different form of MGF-based approach is possible relates to communication with symmetric binary modulations wherein the decision mechanism constitutes a comparison of a decision variable with a zero threshold. Aside from the obvious uncoded applications, the class above also includes the evaluation of pairwise error probability in error-correction-coded systems, as discussed in Chapter 12. In mathematical terms, letting $D|\gamma$ denote the decision variable,[7] the corresponding conditional BEP is of the form (assuming arbitrarily that a positive data bit was transmitted)

$$P_b(E|\gamma) = \Pr\{D|\gamma < 0\} = \int_{-\infty}^0 p_{D|\gamma}(D)\, dD = P_{D|\gamma}(0) \qquad (1.11)$$

where $p_{D|\gamma}(D)$ and $P_{D|\gamma}(D)$ are, respectively, the PDF and CDF of this variable. Aside from the fact that the decision variable $D|\gamma$ can, in general, take on both positive and negative values whereas the instantaneous fading SNR, γ, is restricted to positive values, there is a strong resemblance between the binary probability of error in (1.11) and the outage probability in (1.4). Thus, by analogy with (1.6), the conditional BEP of (1.11) can be expressed as

$$P_b(E|\gamma) = \frac{1}{2\pi j} \int_{\sigma-j\infty}^{\sigma+j\infty} \frac{M_{D|\gamma}(-s)}{s}\, ds \qquad (1.12)$$

---

[6] As we shall see in Chapter 4, the $h(\xi)$ and $g(\xi)$ that result from the alternative representations of the Gaussian and Marcum Q-functions are composed of simple trigonometric functions.

[7] The notation $D|\gamma$ is not meant to imply that the decision variable *explicitly* depends on the fading SNR. Rather, it is merely intended to indicate the dependence of this variable on the fading statistics of the channel. More about this dependence shortly.

where $M_{D|\gamma}(-s)$ now denotes the MGF of the decision variable $D|\gamma$ [i.e., the bilateral Laplace transform of $p_{D|\gamma}(D)$ with argument reversed].

To see how $M_{D|\gamma}(-s)$ might explicitly depend on γ, we now consider the subclass of problems where the conditional decision variable $D|\gamma$ corresponds to a quadratic form of independent complex Gaussian RVs (e.g., a sum of the squared magnitudes of, say, L independent complex Gaussian RVs, or equivalently, a chi-square RV with $2L$ degrees of freedom). Such a form occurs for multiple (L)-channel reception of binary modulations with differentially coherent or noncoherent detection (see Chapter 9). In this instance, the MGF $M_{D|\gamma}(s)$ happens to be exponential in γ and has the generic form

$$M_{D|\gamma}(s) = f_1(s) \exp[\gamma f_2(s)] \tag{1.13}$$

If, as before, we let $\gamma = \sum_{l=1}^{L} \gamma_l$, then substituting (1.13) into (1.12) and averaging over the fading results in the average BEP:[8]

$$P_b(E) = \frac{1}{2\pi j} \int_{\sigma-j\infty}^{\sigma+j\infty} \frac{M_D(-s)}{s} ds \tag{1.14}$$

where

$$M_D(s) \triangleq \int_0^\infty M_{D|\gamma}(s) p_\gamma(\gamma) d\gamma$$

$$= f_1(s) \int_0^\infty \exp[\gamma f_2(s)] p_\gamma(\gamma) d\gamma = f_1(s) M_\gamma(f_2(s)) \tag{1.15}$$

is the unconditional MGF of the decision variable, which also has the product form

$$M_D(s) = f_1(s) \prod_{l=1}^{L} M_{\gamma_l}(f_2(s)) \tag{1.16}$$

Finally, by virtue of the fact that the MGF of the decision variable can be expressed in terms of the MGF of the fading variable (SNR) as in (1.15) [or (1.16)], then analogous to (1.10), we are once again able to evaluate the average BEP based solely on knowledge of the latter MGF.

It is not immediately obvious how to extend the inverse Laplace transform technique discussed in Appendix 9B to CDFs of bilateral RVs; thus other methods for performing this inversion are required. A number of these, including contour integration using residues, saddle point integration, and numerical integration by Gauss–Chebyshev quadrature rules, are discussed in Refs. 3, through 6 and covered later in the book.

---

[8] The approach for computing average BEP as described by (1.13) was also described by Biglieri et al. [3] as a unified approach to computing error probabilities over fading channels.

Despite the fact that the methods dictated by (1.14) and (1.8) or (1.10) cover a wide variety of problems dealing with the performance of digital communication systems over fading channels, there are still some situations that don't lend themselves to either of these two unifying methods. An example of this is evaluation of the bit error probability performance of an M-ary noncoherent orthogonal system operating over an L-path diversity channel (see Chapter 9). However, even in this case there exists an MGF-based approach that greatly simplifies the problem and allows for a more general result [7] than that reported by Weng and Leung [8]. We now outline the method, briefly leaving the more detailed treatment to Chapter 9.

Consider an M-ary communication system where rather than comparing a single decision variable with a threshold, one decision variable $U_1|\gamma$ is compared with the remaining $M - 1$ decision variables U_m, $m = 2, 3, \ldots, M$, all of which do not depend on the fading statistics.[9] Specifically, a correct symbol decision is made if $U_1|\gamma$ is greater than U_m, $m = 2, 3, \ldots, M$. Assuming that the M decision variables are independent, then in mathematical terms, the probability of correct decision is given by

$$P_s(C|\gamma; u_1) = \Pr\{U_2 < u_1, U_3 < u_1, \ldots, U_M < u_1 | U_1|\gamma = u_1\}$$

$$= [\Pr\{U_2 < u_1 | U_1|\gamma = u_1\}]^{M-1} = \left[\int_0^{u_1} p_{U_2}(u_2)\,du_2\right]^{M-1}$$

$$= [1 - (1 - P_{U_2}(u_1))]^{M-1} \qquad (1.17)$$

Using the binomial expansion in (1.17), the conditional probability of error $P_s(E|\gamma; u_1) = 1 - P_s(C|\gamma; u_1)$ can be written as

$$P_s(E|\gamma; u_1) = \sum_{i=1}^{M-1} \binom{M-1}{i}(-1)^{i+1}[1 - P_{U_2}(u_1)]^i \stackrel{\Delta}{=} g(u_1) \qquad (1.18)$$

Averaging over u_1 and using the Fourier transform relationship between the PDF $p_{U_1|\gamma}(u_1)$ and the MGF $M_{U_1|\gamma}(j\omega)$, we obtain

$$P_s(E|\gamma) = \int_0^\infty g(u_1) p_{U_1|\gamma}(u_1)\,du_1$$

$$= \int_0^\infty \frac{1}{2\pi} \int_{-\infty}^\infty M_{U_1|\gamma}(j\omega) e^{-j\omega u_1} g(u_1)\,d\omega\,du_1 \qquad (1.19)$$

Again noting that for a noncentral chi-square RV (as is the case for $U_1|\gamma$) the conditional MGF $M_{U_1|\gamma}(j\omega)$ is of the form in (1.13), then averaging (1.19) over γ

---

[9] Again the conditional notation on γ for U_1 is not meant to imply that this decision variable is explicitly a function of the fading SNR but rather, to indicate its dependence on the fading statistics.

transforms $M_{U_1|\gamma}(j\omega)$ into $M_{U_1}(j\omega)$ of the form in (1.15), which when substituted in (1.19) and reversing the order of integration produces

$$P_s(E) = \frac{1}{2\pi} \int_{-\infty}^{\infty} f_1(j\omega) M_\gamma(f_2(j\omega)) \left[\int_0^{\infty} e^{-j\omega u_1} g(u_1) du_1 \right] d\omega \quad (1.20)$$

Finally, because the CDF $P_{U_2}(u_1)$ in (1.18) is that of a central chi-square RV with $2L$ degrees of freedom, the resulting form of $g(u_1)$ is such that the integral on u_1 in (1.20) can be obtained in closed form. Thus, as promised, what remains again is an expression for average SEP (which for M-ary orthogonal signaling can be related to average BEP by a simple scale factor) whose dependence on the fading statistics is solely through the MGF of the fading SNR.

All of the techniques considered thus far for evaluating average error probability performance rely on the ability to evaluate the MGF of the instantaneous fading SNR γ. In dealing with a form of diversity reception referred to as *equal-gain combining (EGC)* (discussed in great detail in Chapter 9), the instantaneous fading SNR at the output of the combiner takes the form $\gamma = \left[1/\sqrt{L} \sum_{l=1}^{L} \sqrt{\gamma_l} \right]^2$. In this case it is more convenient to deal with the MGF of the square root of the instantaneous fading SNR

$$x \triangleq \sqrt{\gamma} = \frac{1}{\sqrt{L}} \sum_{l=1}^{L} \sqrt{\gamma_l} = \frac{1}{\sqrt{L}} \sum_{l=1}^{L} x_l$$

since if the channels are again assumed independent, then again this MGF takes on a product form, namely, $M_x(s) = \prod_{l=1}^{L} M_{x_l}(s/\sqrt{L})$. Since the average BER can alternatively be computed from

$$P_b(E) = \int_0^{\infty} P_b(E|x) p_x(x) dx \quad (1.21)$$

then if, analogous to (1.9), $P_b(E|x)$ assumes the form

$$P_b(E|x) = \int_{\xi_1}^{\xi_2} C_2 h(\xi) \exp\left[-a_2 g(\xi) x^2 \right] d\xi \quad (1.22)$$

a variation of the procedure in (1.10) is needed to produce an expression for $P_b(E)$ in terms of the MGF of x. First, applying Parseval's theorem [9, p. 27] to (1.21) and letting $G(j\omega) = \mathcal{F}\{P_b(E|x)\}$ denote the Fourier transform of $P_b(E|x)$, then independent of the form of $P_b(E|x)$, we obtain

$$P_b(E) = \frac{1}{2\pi} \int_{-\infty}^{\infty} G(j\omega) M_x(j\omega) d\omega$$

$$= \frac{1}{\pi} \int_0^{\infty} \mathrm{Re}\{G(j\omega) M_x(j\omega)\} d\omega \quad (1.23)$$

where we have recognized that the imaginary part of the integral must be equal to zero since $P_b(E)$ is real, and that the even part of the integrand is an even function of ω. Making the change of variables $\theta = \tan^{-1} \omega$, (1.23) can be written in the form of an integral with finite limits:

$$P_b(E) = \frac{1}{\pi} \int_0^{\pi/2} \frac{1}{\cos^2 \theta} \operatorname{Re}\{G(j \tan \theta) M_x(j \tan \theta)\} d\theta$$

$$= \frac{2}{\pi} \int_0^{\pi/2} \frac{1}{\sin 2\theta} \operatorname{Re}\{\tan \theta\, G(j \tan \theta) M_x(j \tan \theta)\} d\theta \qquad (1.24)$$

Now, specifically for the form of $P_b(E|x)$ in (1.22), $G(j\omega)$ becomes

$$G(j\omega) = \int_{\xi_1}^{\xi_2} C_2 h(\xi) \int_0^\infty \exp\left[-a_2 g(\xi) x^2 + j\omega x\right] dx\, d\xi \qquad (1.25)$$

The inner integral on x can be evaluated in closed form as

$$\int_0^\infty \exp\left[-a_2 g(\xi) x^2 + j\omega x\right] dx = \frac{1}{2a_2 g(\xi)} \left\{ \sqrt{\pi a_2 g(\xi)} \exp\left[\frac{(j\omega)^2}{4a_2 g(\xi)}\right] \right.$$

$$\left. + j\omega\, {}_1F_1\left[1, \frac{3}{2}; \frac{(j\omega)^2}{4a_2 g(\xi)}\right] \right\} \qquad (1.26)$$

where ${}_1F_1(a, b; c)$ is the confluent hypergeometric function of the first kind [10, Eq. (9.210)]. Therefore, in general, evaluation of the average BER of (1.24) requires a double integration. However, for a number of specific applications [i.e., particular forms of the functions $h(\xi)$ and $g(\xi)$], the outer integral on ξ can also be evaluated in closed form; thus, in these instances, $P_b(E)$ can be obtained as a single integral with finite limits and an integrand involving the MGF of the fading. Methods of error probability evaluation based on the type of MGF approach described above have been considered in the literature [11–13] and are presented in detail in Chapter 9.

1.2 CONCLUSIONS

Without regard to the specific application or performance measure, we have briefly demonstrated in this chapter that for a wide variety of digital communication systems covering virtually all known modulation/detection techniques and practical fading channel models, there exists an MGF-based approach that simplifies the evaluation of this performance. In the biggest number of these instances, the MGF-based approach is encompassed in a unified framework which allows the development of a set of generic tools to replace the case-by-case analyses typical of previous contributions in the literature. It is the authors' hope that by the time the reader reaches the end of this book and has experienced the exhaustive set of practical circumstances where these tools are useful, he or she will fully appreciate the power behind the MGF-based approach and as such will generate for themselves an insight into finding new and exciting applications.

REFERENCES

1. J. Abate and W. Whitt, "Numerical inversion of Laplace transforms of probability distributions," *ORSA J. Comput.*, vol. 7, no. 1, 1995, pp. 36–43.
2. M. K. Simon and M.-S. Alouini, "A unified approach to the performance analysis of digital communications over generalized fading channels," *IEEE Proc.*, vol. 86, September 1998, pp. 1860–1877.
3. E. Biglieri, C. Caire, G. Taricco, and J. Ventura-Traveset, "Computing error probabilities over fading channels: a unified approach," *Eur. Trans. Telecommun.*, vol. 9, February 1998, pp. 15–25.
4. E. Biglieri, C. Caire, G. Taricco, and J. Ventura-Traveset, "Simple method for evaluating error probabilities," *Electron. Lett.*, vol. 32, February 1996, pp. 191–192.
5. J. K. Cavers and P. Ho, "Analysis of the error performance of trellis coded modulations in Rayleigh fading channels," *IEEE Trans. Commun.*, vol. 40, January 1992, pp. 74–80.
6. J. K. Cavers, J.-H. Kim and P. Ho, "Exact calculation of the union bound on performance of trellis-coded modulation in fading channels," *IEEE Trans. Commun.*, vol. 46, May 1998, pp. 576–579. Also see *Proc. IEEE, Int. Conf. Univ. Personal Commun.* (ICUPC '96), vol. 2, Cambridge, MA, September 1996, pp. 875–880.
7. M. K. Simon and M.-S. Alouini, "Bit error probability of noncoherent M-ary orthogonal modulation over generalized fading channels," *Int. J. Commun. Networks*, vol. 1, June 1999, pp. 111–117.
8. J. F. Weng and S. H. Leung, "Analysis of M-ary FSK square law combiner under Nakagami fading channels," *Electron. Lett.*, vol. 33, September 1997, pp. 1671–1673.
9. A. Papoulis, *The Fourier Integral and Its Application*. New York: McGraw-Hill, 1962.
10. I. S. Gradshteyn and I. M. Ryzhik, *Table of Integrals, Series, and Products*, 5th ed. San Diego, CA: Academic Press, 1994.
11. M.-S. Alouini and M. K. Simon, "Error rate analysis of MPSK with equal-gain combining over Nakagami fading channels," *Proc. IEEE Veh. Technol. Conf. (VTC'99)*, Houston, TX, pp. 2378–2382.
12. A. Annamalai, C. Tellambura, and V. K. Bhargava, "Exact evaluation of maximal-ratio and equal-gain diversity receivers for M-ary QAM on Nakagami fading channels," *IEEE Trans. Commun.*, vol. 47, September 1999, pp. 1335–1344.
13. A. Annamalai, C. Tellambura and V. K. Bhargava, "Unified analysis of equal-gain diversity on Rician and Nakagami fading channels," *Proc. IEEE Wireless Commun. and Networking Conf. (WCNC'99)*, New Orleans, LA, September 1999.

2

FADING CHANNEL CHARACTERIZATION AND MODELING

Radio-wave propagation through wireless channels is a complicated phenomenon characterized by various effects, such as multipath and shadowing. A precise mathematical description of this phenomenon is either unknown or too complex for tractable communications systems analyses. However, considerable efforts have been devoted to the statistical modeling and characterization of these different effects. The result is a range of relatively simple and accurate statistical models for fading channels which depend on the particular propagation environment and the underlying communication scenario.

The primary purpose of this chapter is to review briefly the principal characteristics and models for fading channels. More detailed treatment of this subject can be found in standard textbooks, such as Refs. 1,3. This chapter also introduces terminology and notation that are used throughout the book. The chapter is organized as follows. A brief qualitative description of the main characteristics of fading channels is presented in the next section. Models for frequency-flat fading channels, corresponding to narrowband transmission, are described in Section 2.2. Models for frequency-selective fading channels that characterize fading in wideband channels are described in Section 2.3.

2.1 MAIN CHARACTERISTICS OF FADING CHANNELS

2.1.1 Envelope and Phase Fluctuations

When a received signal experiences fading during transmission, both its envelope and phase fluctuate over time. For coherent modulations, the fading effects on the phase can severely degrade performance unless measures are taken to compensate for them at the receiver. Most often, analyses of systems employing such modulations assume that the phase effects due to fading are perfectly corrected at the receiver, resulting in what is referred to as ideal coherent demodulation. For noncoherent modulations, phase information is not needed at the receiver and therefore the phase variation due to fading does not affect the performance. Hence performance analyses for both ideal coherent and noncoherent modulations over fading channels requires only knowledge of the fading envelope statistics and is the case most often considered in this book. Furthermore, for slow fading (discussed next), wherein the fading is at least constant over the duration of a symbol time, the fading envelope random process can be represented by a random variable (RV) over the symbol time.

2.1.2 Slow and Fast Fading

The distinction between slow and fast fading is important for the mathematical modeling of fading channels and for the performance evaluation of communication systems operating over these channels. This notion is related to the *coherence time* T_c of the channel, which measures the period of time over which the fading process is correlated (or equivalently, the period of time after which the correlation function of two samples of the channel response taken at the same frequency but different time instants drops below a certain predetermined threshold). The coherence time is also related to the channel *Doppler spread* f_d by

$$T_c \simeq \frac{1}{f_d} \qquad (2.1)$$

The fading is said to be slow if the symbol time duration T_s is smaller than the channel's coherence time T_c; otherwise, it is considered to be fast. In slow fading a particular fade level will affect many successive symbols, which leads to burst errors, whereas in fast fading the fading decorrelates from symbol to symbol. In the latter case and when the communication receiver decisions are made based on an observation of the received signal over two or more symbol times (such as differentially coherent or coded communications), it becomes necessary to consider the variation of the fading channel from one symbol interval to the next. This is done through a range of correlation models that depend essentially on the particular propagation environment and the underlying communication scenario. These various autocorrelation models and their corresponding power spectral density are tabulated in Table 2.1, in which for convenience the variance of the fast-fading process is normalized to unity.

2.1.3 Frequency-Flat and Frequency-Selective Fading

Frequency selectivity is also an important characteristic of fading channels. If all the spectral components of the transmitted signal are affected in a similar manner, the fading is said to be *frequency nonselective* or, equivalently, *frequency flat*. This is the case for *narrowband* systems in which the transmitted signal bandwidth is much smaller than the channel's *coherence bandwidth* f_c. This

TABLE 2.1 Correlation and Spectral Properties of Various Types of Fading Processes of Practical Interest

Type of Fading Spectrum	Fading Autocorrelation, ρ	Normalized PSD
Rectangular	$\dfrac{\sin(2\pi f_d T_s)}{2\pi f_d T_s}$	$(2f_d)^{-1}, \quad |f| \leq f_d$
Gaussian	$\exp[-(\pi f_d T_s)^2]$	$\exp\left[-\left(\dfrac{f}{f_d}\right)^2\right](\sqrt{\pi}f_d)^{-1}$
Land mobile	$J_0(2\pi f_d T_s)$	$[\pi^2(f^2 - f_d^2)]^{-1/2}, \quad |f| \leq f_d$
First-order Butterworth	$\exp(-2\pi|f_d T_s|)$	$\left[\pi f_d\left(1 + \dfrac{f}{f_d}\right)^2\right]^{-1}$
Second-order Butterworth	$\exp\left(-\dfrac{\pi|f_d T_s|}{\sqrt{2}}\right)$ $\times \left(\cos\dfrac{\pi f_d T_s}{\sqrt{2}} + \sin\dfrac{\pi|f_d T_s|}{\sqrt{2}}\right)$	$\left[1 + 16\left(\dfrac{f}{f_d}\right)^4\right]^{-1}$

Source: Data from Mason [4].
[a] PSD is the power spectral density, f_d the Doppler spread, and T_s the symbol time.

bandwidth measures the frequency range over which the fading process is correlated and is defined as the frequency bandwidth over which the correlation function of two samples of the channel response taken at the same time but at different frequencies falls below a suitable value. In addition, the coherence bandwidth is related to the *maximum delay spread* $\tau_{\max}$ by

$$f_c \simeq \frac{1}{\tau_{\max}} \qquad (2.2)$$

On the other hand, if the spectral components of the transmitted signal are affected by different amplitude gains and phase shifts, the fading is said to be *frequency selective*. This applies to *wideband* systems in which the transmitted bandwidth is bigger than the channel's coherence bandwidth.

2.2 MODELING OF FLAT FADING CHANNELS

When fading affects narrowband systems, the received carrier amplitude is modulated by the fading amplitude α, where α is a RV with mean-square value $\Omega = \overline{\alpha^2}$ and probability density function (PDF) $p_\alpha(\alpha)$, which is dependent on the nature of the radio propagation environment. After passing through the fading channel, the signal is perturbed at the receiver by additive white Gaussian noise (AWGN), which is typically assumed to be statistically independent of the fading amplitude α and which is characterized by a one-sided power spectral density N_0 (W/Hz). Equivalently, the received instantaneous signal power is modulated by α^2. Thus we define the instantaneous signal-to-noise power ratio (SNR) per symbol by $\gamma = \alpha^2 E_s/N_0$ and the average SNR per symbol by $\bar{\gamma} = \Omega E_s/N_0$, where E_s is the energy per symbol.[1] In addition, the PDF of γ is obtained by introducing a change of variables in the expression for the fading PDF $p_\alpha(\alpha)$ of α, yielding

$$p_\gamma(\gamma) = \frac{p_\alpha(\sqrt{\Omega \gamma/\bar{\gamma}})}{2\sqrt{\gamma\bar{\gamma}/\Omega}}. \qquad (2.3)$$

The moment generating function (MGF) $M_\gamma(s)$ associated with the fading PDF $p_\gamma(\gamma)$ and defined by

$$M_\gamma(s) = \int_0^\infty p_\gamma(\gamma) e^{s\gamma} \, d\gamma \qquad (2.4)$$

is another important statistical characteristic of fading channels, particularly in the context of this book. In addition, the amount of fading (AF), or "fading figure," associated with the fading PDF is defined as

$$\mathrm{AF} = \frac{\mathrm{var}(\alpha^2)}{(E[\alpha^2])^2} = \frac{E[(\alpha^2 - \Omega)^2]}{\Omega^2} = \frac{E(\gamma^2) - (E[\gamma])^2}{(E[\gamma])^2} \qquad (2.5)$$

with $E[\cdot]$ denoting statistical average and $\mathrm{var}(\cdot)$ denoting variance. This figure was introduced by Charash [5, p. 29; 6] as a unified measure of the severity of the fading and is typically independent of the average fading power Ω.

We now present the various radio propagation effects involved in fading channels, their corresponding PDFs, MGFs, AFs, and their relation to physical channels. A summary of these properties is tabulated in Table 2.2.

2.2.1 Multipath Fading

Multipath fading is due to the constructive and destructive combination of randomly delayed, reflected, scattered, and diffracted signal components. This type of fading is relatively fast and is therefore responsible for the short-term signal variations. Depending on the nature of the radio propagation environment, there are different models describing the statistical behavior of the multipath fading envelope.

2.2.1.1 Rayleigh Model.
The Rayleigh distribution is frequently used to model multipath fading with no direct line-of-sight (LOS) path. In this case the channel fading amplitude α is distributed according to

$$p_\alpha(\alpha) = \frac{2\alpha}{\Omega} \exp\left(-\frac{\alpha^2}{\Omega}\right), \qquad \alpha \geq 0 \qquad (2.6)$$

---

[1] Our performance evaluation of digital communications over fading channels will generally be a function of the average SNR per symbol $\bar{\gamma}$.

TABLE 2.2 Probability Density Function (PDF) and Moment Generating Function (MGF) of the SNR per Symbol γ for Some Common Fading Channels

Type of Fading	Fading Parameter	PDF, $p_\gamma(\gamma)$	MGF, $M_\gamma(s)$
Rayleigh		$\dfrac{1}{\bar{\gamma}} \exp\left(-\dfrac{\gamma}{\bar{\gamma}}\right)$	$(1 - s\bar{\gamma})^{-1}$
Nakagami-q (Hoyt)	$0 \leq q \leq 1$	$\dfrac{(1+q^2)}{2q\bar{\gamma}} \exp\left[-\dfrac{(1+q^2)^2 \gamma}{4q^2\bar{\gamma}}\right]$ $\times I_0\left[\dfrac{(1-q^4)\gamma}{4q^2\bar{\gamma}}\right]$	$\left[1 - 2s\bar{\gamma} + \dfrac{(2s\bar{\gamma})^2 q^2}{(1+q^2)^2}\right]^{-1/2}$
Nakagami-n (Rice)	$0 \leq n$	$\dfrac{(1+n^2)e^{-n^2}}{\bar{\gamma}} \exp\left[-\dfrac{(1+n^2)\gamma}{\bar{\gamma}}\right]$ $\times I_0\left[2n\sqrt{\dfrac{(1+n^2)\gamma}{\bar{\gamma}}}\right]$	$\dfrac{(1+n^2)}{(1+n^2) - s\bar{\gamma}} \exp\left[\dfrac{n^2 s\bar{\gamma}}{(1+n^2) - s\bar{\gamma}}\right]$
Nakagami-m	$\dfrac{1}{2} \leq m$	$\dfrac{m^m \gamma^{m-1}}{\bar{\gamma}^m \Gamma(m)} \exp\left(-\dfrac{m\gamma}{\bar{\gamma}}\right)$	$\left(1 - \dfrac{s\bar{\gamma}}{m}\right)^{-m}$
Log-normal shadowing	σ	$\dfrac{4.34}{\sqrt{2\pi}\sigma\gamma} \exp\left[-\dfrac{(10\log_{10}\gamma - \mu)^2}{2\sigma^2}\right]$	$\dfrac{1}{\sqrt{\pi}} \sum_{n=1}^{N_p} H_{x_n} \exp(10^{(\sqrt{2}\sigma x_n + \mu)/10} s)$
Composite gamma/log-normal	m and $0 \leq \sigma$	$\displaystyle\int_0^\infty \dfrac{m^m \gamma^{m-1}}{w^m \Gamma(m)} \exp\left(-\dfrac{m\gamma}{w}\right)$ $\times \dfrac{\xi}{\sqrt{2\pi}\sigma w} \exp\left[-\dfrac{(10\log_{10}w - \mu)^2}{2\sigma^2}\right] dw$	$\dfrac{1}{\sqrt{\pi}} \sum_{n=1}^{N_p} H_{x_n} (1 - 10^{(\sqrt{2}\sigma x_n + \mu)/10} s/m)^{-m}$

and hence, following (2.3), the instantaneous SNR per symbol of the channel, γ, is distributed according to an exponential distribution given by

$$p_\gamma(\gamma) = \frac{1}{\overline{\gamma}} \exp\left(-\frac{\gamma}{\overline{\gamma}}\right), \qquad \gamma \geq 0 \qquad (2.7)$$

The MGF corresponding to this fading model is given by

$$M_\gamma(s) = (1 - s\overline{\gamma})^{-1} \qquad (2.8)$$

In addition, the moments associated with this fading model can be shown to be given by

$$E[\gamma^k] = \Gamma(1+k)\overline{\gamma}^k \qquad (2.9)$$

where $\Gamma(\cdot)$ is the gamma function. The Rayleigh fading model therefore has an AF equal to 1 and typically agrees very well with experimental data for mobile systems, where no LOS path exists between the transmitter and receiver antennas [3]. It also applies to the propagation of reflected and refracted paths through the troposphere [7] and ionosphere [8,9] and to ship-to-ship [10] radio links.

2.2.1.2 Nakagami-q (Hoyt) Model.

The Nakagami-q distribution, also referred to as the Hoyt distribution [11], is given in Nakagami [12, Eq. (52)] by

$$p_\alpha(\alpha) = \frac{(1+q^2)\alpha}{q\Omega} \exp\left[-\frac{(1+q^2)^2\alpha^2}{4q^2\Omega}\right] I_0\left(\frac{(1-q^4)\alpha^2}{4q^2\Omega}\right), \qquad \alpha \geq 0 \quad (2.10)$$

where $I_0(\cdot)$ is the zeroth-order modified Bessel function of the first kind, and q is the Nakagami-q fading parameter which ranges from 0 to 1. Using (2.3), it can be shown that the SNR per symbol of the channel, γ, is distributed according to

$$p_\gamma(\gamma) = \frac{1+q^2}{2q\overline{\gamma}} \exp\left[-\frac{(1+q^2)^2\gamma}{4q^2\overline{\gamma}}\right] I_0\left(\frac{(1-q^4)\gamma}{4q^2\overline{\gamma}}\right), \qquad \gamma \geq 0 \qquad (2.11)$$

It can be shown that the MGF corresponding to (2.11) is given by

$$M_\gamma(s) = \left[1 - 2s\overline{\gamma} + \frac{(2s\overline{\gamma})^2 q^2}{(1+q^2)^2}\right]^{-1/2} \qquad (2.12)$$

Also, the moments associated with this model are given by [12, Eq. (52)]

$$E(\gamma^k) = \Gamma(1+k) {}_2F_1\left(-\frac{k-1}{2}, -\frac{k}{2}; 1; \left(\frac{1-q^2}{1+q^2}\right)^2\right) \overline{\gamma}^k \qquad (2.13)$$

where $_2F_1(\cdot,\cdot;\cdot,\cdot)$ is the Gauss hypergeometric function, and the AF of the Nakagami-q distribution is therefore given by

$$\text{AF}_q = \frac{2(1+q^4)}{(1+q^2)^2}, \qquad 0 \le q \le 1 \qquad (2.14)$$

and hence ranges between 1 ($q = 1$) and 2 ($q = 0$). The Nakagami-q distribution spans the range from one-sided Gaussian fading ($q = 0$) to Rayleigh fading ($q = 1$). It is typically observed on satellite links subject to strong ionospheric scintillation [13,14]. Note that one-sided Gaussian fading corresponds to the worst-case fading or, equivalently, the largest AF for all multipath distributions considered in our analyses.

2.2.1.3 Nakagami-n (Rice) Model.

The Nakagami-n distribution is also known as the Rice distribution [15]. It is often used to model propagation paths consisting of one strong direct LOS component and many random weaker components. Here the channel fading amplitude follows the distribution [12, Eq. (50)]

$$p_\alpha(\alpha) = \frac{2(1+n^2)e^{-n^2}\alpha}{\Omega} \exp\left[-\frac{(1+n^2)\alpha^2}{\Omega}\right] I_0\left(2n\alpha\sqrt{\frac{1+n^2}{\Omega}}\right), \qquad \alpha \ge 0 \qquad (2.15)$$

where n is the Nakagami-n fading parameter which ranges from 0 to ∞ and which is related to the Rician K factor by $K = n^2$. Applying (2.3) shows that the SNR per symbol of the channel, γ, is distributed according to a noncentral chi-square distribution given by

$$p_\gamma(\gamma) = \frac{(1+n^2)e^{-n^2}}{\overline{\gamma}} \exp\left[-\frac{(1+n^2)\gamma}{\overline{\gamma}}\right] I_0\left(2n\sqrt{\frac{(1+n^2)\gamma}{\overline{\gamma}}}\right), \qquad \gamma \ge 0 \qquad (2.16)$$

It can also be shown that the MGF associated with this fading model is given by

$$M_\gamma(s) = \frac{(1+n^2)}{(1+n^2) - s\overline{\gamma}} \exp\left[\frac{n^2 s\overline{\gamma}}{(1+n^2) - s\overline{\gamma}}\right] \qquad (2.17)$$

and that the moments are given by [12, Eq. (50)]

$$E(\gamma^k) = \frac{\Gamma(1+k)}{(1+n^2)^k} {}_1F_1(-k, 1; -n^2)\overline{\gamma}^k \qquad (2.18)$$

where $_1F_1(\cdot,\cdot;\cdot)$ is the Kummer confluent hypergeometric function. The AF of the Nakagami-n distribution is given by

$$\text{AF}_n = \frac{1+2n^2}{(1+n^2)^2}, \qquad n \ge 0 \qquad (2.19)$$

and hence ranges between 0 ($n = \infty$) and 1 ($n = 0$). The Nakagami-n distribution spans the range from Rayleigh fading ($n = 0$) to no fading (constant amplitude) ($n = \infty$). This type of fading is typically observed in the first resolvable LOS paths of microcellular urban and suburban land-mobile [16], picocellular indoor [17], and factory [18] environments. It also applies to the dominant LOS path of satellite [19,20] and ship-to-ship [10] radio links.

2.2.1.4 Nakagami-m Model.
The Nakagami-m PDF is in essence a central chi-square distribution given by [12, Eq. (11)]

$$p_\alpha(\alpha) = \frac{2m^m \alpha^{2m-1}}{\Omega^m \Gamma(m)} \exp\left(-\frac{m\alpha^2}{\Omega}\right), \qquad \alpha \geq 0 \qquad (2.20)$$

where m is the Nakagami-m fading parameter which ranges from $\frac{1}{2}$ to ∞. Figure 2.1 shows the Nakagami-m PDF for $\Omega = 1$ and various values of the m parameter. Applying (2.3) shows that the SNR per symbol, γ, is distributed according to a gamma distribution given by

$$p_\gamma(\gamma) = \frac{m^m \gamma^{m-1}}{\overline{\gamma}^m \Gamma(m)} \exp\left(-\frac{m\gamma}{\overline{\gamma}}\right), \qquad \gamma \geq 0 \qquad (2.21)$$

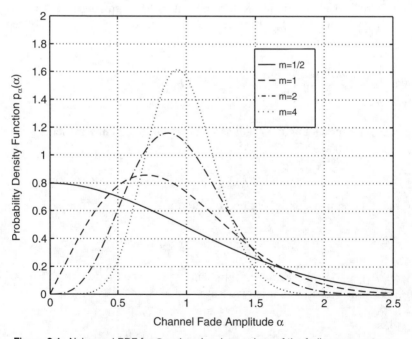

Figure 2.1. Nakagami PDF for $\Omega = 1$ and various values of the fading parameter m.

It can also be shown that the MGF is given in this case by

$$M_\gamma(s) = \left(1 - \frac{s\bar{\gamma}}{m}\right)^{-m} \quad (2.22)$$

and that the moments are given by [12, Eq. (65)]

$$E[\gamma^k] = \frac{\Gamma(m+k)}{\Gamma(m)m^k}\bar{\gamma}^k \quad (2.23)$$

which yields an AF of

$$\text{AF}_m = \frac{1}{m}, \quad m \geq \frac{1}{2} \quad (2.24)$$

Hence, the Nakagami-m distribution spans via the m parameter the widest range of AF (from 0 to 2) among all the multipath distributions considered in this book. For instance, it includes the one-sided Gaussian distribution ($m = \frac{1}{2}$) and the Rayleigh distribution ($m = 1$) as special cases. In the limit as $m \to +\infty$, the Nakagami-m fading channel converges to a nonfading AWGN channel. Furthermore, when $m < 1$, equating (2.14) and (2.24), we obtain a one-to-one mapping between the m parameter and the q parameter, allowing the Nakagami-m distribution to closely approximate the Nakagami-q (Hoyt) distribution, and this mapping is given by

$$m = \frac{(1+q^2)^2}{2(1+2q^4)}, \quad m \leq 1 \quad (2.25)$$

Similarly, when $m > 1$, equating (2.19) and (2.24) we obtain another one-to-one mapping between the m parameter and the n parameter (or, equivalently, the Rician K factor), allowing the Nakagami-m distribution to closely approximate the Nakagami-n (Rice) distribution, and this mapping is given by

$$m = \frac{(1+n^2)^2}{1+2n^2}, \quad n \geq 0$$

$$n = \sqrt{\frac{\sqrt{m^2-m}}{m-\sqrt{m^2-m}}}, \quad m \geq 1 \quad (2.26)$$

Finally, the Nakagami-m distribution often gives the best fit to land-mobile [21–23] and indoor-mobile [24] multipath propagation, as well as scintillating ionospheric radio links [9,25–28].

2.2.2 Log-Normal Shadowing

In terrestrial and satellite land-mobile systems, the link quality is also affected by slow variation of the mean signal level due to the shadowing from terrain, buildings, and trees. Communication system performance will depend on shadowing only if the radio receiver is able to average out the fast multipath fading or if an efficient microdiversity system is used to eliminate the effects of multipath. Based on empirical measurements, there is a general consensus that shadowing can be modeled by a log-normal distribution for various outdoor and indoor environments [21,29–33], in which case the path SNR per symbol γ has a PDF given by the standard log-normal expression

$$p_\gamma(\gamma) = \frac{\xi}{\sqrt{2\pi}\sigma\gamma} \exp\left[-\frac{(10\log_{10}\gamma - \mu)^2}{2\sigma^2}\right] \quad (2.27)$$

where $\xi = 10/\ln 10 = 4.3429$, and μ (dB) and σ (dB) are the mean and standard deviation of $10\log_{10}\gamma$, respectively.

The MGF associated with this slow-fading effect is given by

$$M_\gamma(s) \simeq \frac{1}{\sqrt{\pi}} \sum_{n=1}^{N_p} H_{x_n} \exp(10^{(\sqrt{2}\sigma x_n + \mu)/10} s) \quad (2.28)$$

where x_n are the zeros of the N_p-order Hermite polynomial, and H_{x_n} are the weight factors of the N_p-order Hermite polynomial and are given by Table 25.10 of Ref. 50. In addition, the moments of (2.27) are given by

$$E[\gamma^k] = \exp\left[\frac{k}{\xi}\mu + \frac{1}{2}\left(\frac{k}{\xi}\right)^2 \sigma^2\right] \quad (2.29)$$

yielding an AF of

$$\text{AF}_\sigma = \exp\left(\frac{\sigma^2}{\xi^2}\right) - 1 \quad (2.30)$$

From (2.30) the AF associated with a log-normal PDF can be arbitrarily high. However, as noted by Charash [5, p. 29], in practical situations the standard deviation of shadow fading does not exceed 9 dB [3, p. 88]. Hence, the AF of log-normal shadowing is bounded by 73. This number exceeds the maximal AF exhibited by the various multipath PDFs studied in Section 2.2.1 by several order of magnitudes.

2.2.3 Composite Multipath/Shadowing

A composite multipath/shadowed fading environment consists of multipath fading superimposed on log-normal shadowing. In this environment the receiver does not average out the envelope fading due to multipath but rather, reacts to the instantaneous composite multipath/shadowed signal [3, Sec. 2.4.2]. This is often the scenario in congested downtown areas with slow-moving pedestrians and vehicles [21,34,35]. This type of composite fading is also observed in land-mobile satellite systems subject to vegetative and/or urban shadowing [36–40]. There are two approaches and various combinations suggested in the literature for obtaining the composite distribution. Here, as an example, we present the composite gamma/log-normal PDF introduced by Ho and Stüber [35]. This PDF arises in Nakagami-m shadowed environments and is obtained by averaging the gamma distributed signal power (or, equivalently, the SNR per symbol) of (2.21) over the conditional density of the log-normally distributed mean signal power (or equivalently, the average SNR per symbol) of (2.27), giving the following channel PDF:

$$p_\gamma(\gamma) = \int_0^\infty \frac{m^m \gamma^{m-1}}{w^m \Gamma(m)} \exp\left(-\frac{m\gamma}{w}\right) \frac{\xi}{\sqrt{2\pi}\sigma w} \exp\left[-\frac{(10\log_{10} w - \mu)^2}{2\sigma^2}\right] dw \quad (2.31)$$

For the special case where the multipath is Rayleigh distributed ($m = 1$), (2.31) reduces to a composite exponential/log-normal PDF which was initially proposed by Hansen and Meno [34].

The MGF is given in this case by

$$M_\gamma(s) \simeq \frac{1}{\sqrt{\pi}} \sum_{n=1}^{N_p} H_{x_n} (1 - 10^{(\sqrt{2}\sigma x_n + \mu)/10} s/m)^{-m} \quad (2.32)$$

and the moments associated with a gamma/log-normal PDF are given by

$$E[\gamma^k] = \frac{\Gamma(m+k)}{\Gamma(m)m^k} \exp\left[\frac{k}{\xi}\mu + \frac{1}{2}\left(\frac{k}{\xi}\right)^2 \sigma^2\right] \quad (2.33)$$

and the resulting AF is given by

$$\text{AF}_{m\sigma} = \frac{1+m}{m} \exp\left(\frac{\sigma^2}{\xi^2}\right) - 1 \quad (2.34)$$

Note that when shadowing is absent ($\sigma = 0$), (2.34) reduces to (2.24), as expected. Similarly, as the fading is reduced ($m \to \infty$), (2.34) reduces to (2.30), as expected.

2.2.4 Combined (Time-Shared) Shadowed/Unshadowed Fading

From their land-mobile satellite channel characterization experiments, Lutz et al. [39] and Barts and Stutzman [41] found that the overall fading process for land-mobile satellite systems is a convex combination of unshadowed multipath fading and a composite multipath/shadowed fading. Here, as an example, we present in more detail the Lutz et al. model [39]. When no shadowing is present, the fading follows a Rice (Nakagami-n) PDF. On the other hand, when shadowing is present, it is assumed that no direct LOS path exists and the received signal power (or, equivalently, SNR per bit) is assumed to be an exponential/log-normal (Hansen–Meno) PDF [34]. The combination is characterized by the shadowing time-share factor, which is denoted by A, $0 \leq A \leq 1$; hence, the resulting combined PDF is given by

$$p_\gamma(\gamma) = (1-A)\frac{(1+K)e^{-K}}{\overline{\gamma}^u} \exp\left[-\frac{(1+K)\gamma}{\overline{\gamma}^u}\right] I_0\left[2\sqrt{\frac{K(1+K)\gamma}{\overline{\gamma}^u}}\right]$$
$$+ A \int_0^\infty \frac{1}{w} \exp\left(-\frac{\gamma}{w}\right) \frac{\xi}{\sqrt{2\pi}\sigma^s w} \exp\left[-\frac{(10\log_{10} w - \mu^s)^2}{2(\sigma^s)^2}\right] dw \quad (2.35)$$

where $\overline{\gamma}^u$ is the average SNR per symbol during the unshadowed fraction of time, and μ^s and σ^s are the average and standard deviation of $10\log_{10}\gamma$ during the shadowed fraction of time, respectively. The overall average SNR per symbol, $\overline{\gamma}$, is then given by

$$\overline{\gamma} = (1-A)\overline{\gamma}^u + A \cdot 10^{\mu^s/10 + (\ln 10)(\sigma^s)^2/200} \quad (2.36)$$

Finally, the MGF can be shown to be given by

$$M_\gamma(s) \simeq (1-A)\frac{(1+K)}{1+K-s\overline{\gamma}^u} \exp\left[\frac{Ks\overline{\gamma}^u}{(1+K)-s\overline{\gamma}^u}\right]$$
$$+ A \frac{1}{\sqrt{\pi}} \sum_{n=1}^{N_p} H_{x_n}(1 - 10^{(\sqrt{2}\sigma^s x_n + \mu^s)/10}s)^{-1} \quad (2.37)$$

2.3 MODELING OF FREQUENCY-SELECTIVE FADING CHANNELS

When wideband signals propagate through a frequency-selective channel, their spectrum is affected by the channel transfer function, resulting in a time dispersion of the waveform. This type of fading can be modeled as a linear filter characterized by the following complex-valued lowpass equivalent impulse response:

$$h(t) = \sum_{l=1}^{L_p} \alpha_l e^{-j\theta_l} \delta(t - \tau_l) \quad (2.38)$$

where $\delta(\cdot)$ is the Dirac delta function, l the channel index, and $\{\alpha_l\}_{l=1}^{L_p}$, $\{\theta_l\}_{l=1}^{L_p}$, and $\{\tau_l\}_{l=1}^{L_p}$ the random channel amplitudes, phases, and delays, respectively.

In (2.38) L_p is the number of resolvable paths (the first path being the reference path whose delay $\tau_1 = 0$) and is related to the ratio of the maximum delay spread to the symbol time. Under the slow-fading assumption, L_p is assumed to be constant over a certain period of time, and $\{\alpha_l\}_{l=1}^{L_p}$, $\{\theta_l\}_{l=1}^{L_p}$, and $\{\tau_l\}_{l=1}^{L_p}$ are all constant over a symbol interval. If the various paths of a given impulse response are generated by different scatterers, they tend to exhibit negligible correlations [33,42] and it is reasonable in that case to assume that the $\{\alpha_l\}_{l=1}^{L_p}$ are statistically independent RVs. Otherwise, the $\{\alpha_l\}_{l=1}^{L_p}$ have to be considered as correlated RVs and various fading correlation models of interest will be presented in Section 9.6.

Extending the flat fading notations, the fading amplitude α_l of the lth resolved path is assumed to be a RV whose mean-square value $\overline{\alpha_l^2}$ is denoted by Ω_l and whose PDF $p_{\alpha_l}(\alpha_l)$ can be any one of the PDFs presented above. Also as in the flat fading case, after passing through the fading channel, a wideband signal is perturbed by AWGN with a one-sided power spectral density N_0 (W/Hz). The AWGN is assumed to be independent of the fading amplitudes $\{\alpha_l\}_{l=1}^{L}$. Hence the instantaneous SNR per symbol of the lth channel is given by $\gamma_l = \alpha_l^2 E_s/N_0$, and the average SNR per symbol of the lth channel is given by $\overline{\gamma}_l = \Omega_l E_s/N_0$.

The first arriving path in the impulse response typically exhibits a lower amount of fading than subsequent paths, since it may contain the LOS path [16,23,42] Furthermore, since the specular power component typically decreases with respect to delay, the last arriving paths exhibit higher amounts of fading [23,42]. The $\{\Omega_l\}_{l=1}^{L_p}$ are related to the channel's *power delay profile* (*PDP*), which is also referred to as the *multipath intensity profile* (*MIP*) and which is typically a decreasing function of the delay. The PDP model can assume various forms, depending on whether the model is for indoor or outdoor environments and for each environment, the general propagation conditions. PDPs for indoor partitioned office buildings, indoor factory buildings with heavy machinery, high-density office buildings in urban areas, low-density residential houses in suburban areas, open rural environment, hilly or mountainous regions, and maritime environment are described in Ref. 43. For example, experimental measurements indicate that the mobile radio channel is well characterized by an exponentially decaying PDP for indoor office buildings [33] and congested urban areas [29,44]:

$$\Omega_l = \Omega_1 e^{-\tau_l/\tau_{\max}}, \qquad l = 1, 2, \ldots, L_p \qquad (2.39)$$

where Ω_1 is the average fading power corresponding to the first (reference) propagation path and $\tau_{\max}$ is the channel maximum delay spread. In the literature the delays are often assumed to be equally spaced ($\tau_{l+1} - \tau_l$ is constant and equal to the symbol time T_s) [1, Sec. 14-5-1; 45], and with this assumption, we get the equally spaced exponential profile given by

$$\Omega_l = \Omega_1 e^{-(l-1)\delta}, \qquad \delta \geq 0 \quad \text{and} \quad l = 1, 2, \ldots, L_p \qquad (2.40)$$

where the parameter δ is the *power decay factor*, which reflects the rate at which the average fading power decays. Other idealized PDP profiles reported or used in the literature include the constant (flat) [46], the flat exponential [47], the double spike [46], the Gaussian [46], the power function (polynomial) [48], and other more complicated composite profiles [49].

REFERENCES

1. J. G. Proakis, *Digital Communications*, 3rd ed. New York: McGraw-Hill, 1995.
2. T. S. Rappaport, *Wireless Communications: Principles and Practice*. Upper Saddle River, NJ: Prentice Hall, 1996.
3. G. L. Stüber, *Principles of Mobile Communications*. Norwell, MA: Kluwer Academic Publishers, 1996.
4. L. J. Mason, "Error probability evaluation of systems employing differential detection in a Rician fading environment and Gaussian noise," *IEEE Trans. Commun.*, vol. COM-35, May 1987, pp. 39–46.
5. U. Charash, "A study of multipath reception with unknown delays." Ph.D. dissertation, University of California, Berkeley, CA, January 1974.
6. U. Charash, "Reception through Nakagami fading multipath channels with random delays," *IEEE Trans. Commun.*, vol. COM-27, April 1979, pp. 657–670.
7. H. B. James and P. I. Wells, "Some tropospheric scatter propagation measurements near the radio-horizon," *Proc. IRE*, October 1955, pp. 1336–1340.
8. G. R. Sugar, "Some fading characteristics of regular VHF ionospheric propagation," *Proc. IRE*, October 1955, pp. 1432–1436.
9. S. Basu, E. M. MacKenzie, S. Basu, E. Costa, P. F. Fougere, H. C. Carlson, and H. E. Whitney, "250 MHz/GHz scintillation parameters in the equatorial, polar, and aural environments," *IEEE J. Selt. Areas Commun.*, vol. SAC-5, February 1987, pp. 102–115.
10. T. L. Staley, R. C. North, W. H. Ku, and J. R. Zeidler, "Performance of coherent MPSK on frequency selective slowly fading channels," *Proc. IEEE Veh. Technol. Conf. (VTC'96)*, Atlanta, GA, April 1996, pp. 784–788.
11. R. S. Hoyt, "Probability functions for the modulus and angle of the normal complex variate," *Bell Syst. Tech. J.*, vol. 26, April 1947, pp. 318–359.
12. M. Nakagami, "The m-distribution: a general formula of intensity distribution of rapid fading," in *Statistical Methods in Radio Wave Propagation*. Oxford: Pergamon Press, 1960, pp. 3–36.
13. B. Chytil, "The distribution of amplitude scintillation and the conversion of scintillation indices," *J. Atmos. Terr. Phys.*, vol. 29, September 1967, pp. 1175–1177.
14. K. Bischoff and B. Chytil, "A note on scintillaton indices," *Planet. Space Sci.*, vol. 17, 1969, pp. 1059–1066.
15. S. O. Rice, "Statistical properties of a sine wave plus random noise," *Bell Syst. Tech. J.*, vol. 27, January 1948, pp. 109–157.
16. K. A. Stewart, G. P. Labedz, and K. Sohrabi, "Wideband channel measurements at 900 MHz," *Proc. IEEE Veh. Technol. Conf. (VTC'95)*, Chicago, July 1995, pp. 236–240.

17. R. J. C. Bultitude, S. A. Mahmoud, and W. A. Sullivan, "A comparison of indoor radio propagation characteristics at 910 MHz and 1.75 GHz," *IEEE J. Selt. Areas Commun.*, vol. SAC-7, January 1989, pp. 20–30.
18. T. S. Rappaport and C. D. McGillem, "UHF fading in factories," *IEEE J. Selt. Areas Commun.*, vol. SAC-7, January 1989, pp. 40–48.
19. G. H. Munro, "Scintillation of radio signals from satellites," *J. Geophys. Res.*, vol. 68, April 1963.
20. P. D. Shaft, "On the relationship between scintillation index and Rician fading," *IEEE Trans. Commun.*, vol. COM-22, May 1974, pp. 731–732.
21. H. Suzuki, "A statistical model for urban multipath propagation," *IEEE Trans. Commun.*, vol. COM-25, July 1977, pp. 673–680.
22. T. Aulin, "Characteristics of a digital mobile radio channel," *IEEE Trans. Veh. Technol.*, vol. VT-30, May 1981, pp. 45–53.
23. W. R. Braun and U. Dersch, "A physical mobile radio channel model," *IEEE Trans. Veh. Technol.*, vol. VT-40, May 1991, pp. 472–482.
24. A. U. Sheikh, M. Handforth, and M. Abdi, "Indoor mobile radio channel at 946 MHz: measurements and modeling," *Proc. IEEE Veh. Technol. Conf. (VTC'93)*, Secaucus, NJ, May 1993, pp. 73–76.
25. E. J. Fremouw and H. F. Bates, "Worldwide behavior of average VHF–UHF scintillation," *Radio Sci.*, vol. 6, October 1971, pp. 863–869.
26. H. E. Whitney, J. Aarons, R. S. Allen, and D. R. Seeman, "Estimation of the cumulative probability distribution function of ionospheric scintillations," *Radio Sci.*, vol. 7, December 1972, pp. 1095–1104.
27. E. J. Fremouw, R. C. Livingston, and D. A. Miller, "On the statistics of scintillating signals," *J. Atmos. Terr. Phys.*, vol. 42, August 1980, pp. 717–731.
28. P. K. Banerjee, R. S. Dabas, and B. M. Reddy, "C-band and L-band transionospheric scintillation experiment: some results for applications to satellite radio systems," *Radio Sci.*, vol. 27, June 1992, pp. 955–969.
29. G. L. Turin, F. D. Clapp, T. L. Johnston, S. B. Fine, and D. Lavry, "A statistical model of urban multipath propagation," *IEEE Trans. Veh. Technol.*, vol. VT-21, February 1972, pp. 1–9.
30. H. Hashemi, "Simulation of the urban radio propagation channel," *IEEE Trans. Veh. Technol.*, vol. VT-28, August 1979, pp. 213–225.
31. T. S. Rappaport, S. Y. Seidel, and K. Takamizawa, "Statistical channel impulse response models for factory and open plan building radio communication system design," *IEEE Trans. Commun.*, vol. COM-39, May 1991, pp. 794–807.
32. P. Yegani and C. McGlilem, "A statistical model for the factory radio channel," *IEEE Trans. Commun.*, vol. COM-39, October 1991, pp. 1445–1454.
33. H. Hashemi, "Impulse response modeling of indoor radio propagation channels," *IEEE J. Selt. Areas Commun.*, vol. SAC-11, September 1993, pp. 967–978.
34. F. Hansen and F. I. Meno, "Mobile fading-Rayleigh and lognormal superimposed," *IEEE Trans. Veh. Technol.*, vol. VT-26, November 1977, pp. 332–335.
35. M. J. Ho and G. L. Stüber, "Co-channel interference of microcellular systems on shadowed Nakagami fading channels," *Proc. IEEE Veh. Technol. Conf. (VTC'93)*, Secaucus, NJ, May 1993, pp. 568–571.

36. C. Loo, "A statistical model for a land-mobile satellite link," *IEEE Trans. Veh. Technol.*, vol. VT-34, August 1985, pp. 122–127.
37. G. Corazza and F. Vatalaro, "A statistical model for land mobile satellite channels and its application to nongeostationary orbit systems," *IEEE Trans. Veh. Technol.*, vol. VT-43, August 1994, pp. 738–742.
38. S.-H Hwang, K.-J. Kim, J.-Y. Ahn, and K.-C. Wang, "A channel model for nongeostationary orbiting satellite system," *Proc. IEEE Veh. Technol. Conf. (VTC'97)*, Phoenix, AZ, May 1997, pp. 41–45.
39. E. Lutz, D. Cygan, M. Dippold, F. Dolainsky, and W. Papke, "The land mobile satellite communication channel: recording, statistics, and channel model," *IEEE Trans. Veh. Technol.*, vol. VT-40, May 1991, pp. 375–386.
40. M. Rice and B. Humphreys, "Statistical models for the ACTS K-band land mobile satellite channel," *Proc. IEEE Veh. Technol. Conf. (VTC'97)*, Phoenix, AZ, May 1997, pp. 46–50.
41. R. M. Barts and W. L. Stutzman, "Modeling and simulation of mobile satellite propagation," *IEEE Trans. Antennas Propagat.*, vol. AP-40, April 1992, pp. 375–382.
42. S. A. Abbas and A. U. Sheikh, "A geometric theory of Nakagami fading multipath mobile radio channel with physical interpretations," *Proc. IEEE Veh. Technol. Conf. (VTC'96)*, Atlanta, GA, April 1996, pp. 637–641.
43. D. Molkdar, "Review on radio propagation into and within buildings," *IEE Proc. H*, vol. 138, February 1991, pp. 61–73.
44. COST 207 TD(86)51-REV 3 (WG1), "Proposal on channel transfer functions to be used in GSM test late 1986," *Tech. Rep.*, Office Official Publications European Communities, September 1986.
45. T. Eng and L. B. Milstein, "Coherent DS-CDMA performance in Nakagami multipath fading," *IEEE Trans. Commun.*, vol. COM-43, February–March–April 1995, pp. 1134–1143.
46. B. Glance and L. J. Greenstein, "Frequency-selective fading effects in digital mobile radio with diversity combining," *IEEE Trans. Commun.*, vol. COM-31, September 1983, pp. 1085–1094.
47. P. F. M. Smulders and A. G. Wagemans, "Millimetre-wave biconical horn antennas for near uniform coverage in indoor picocells," *Electron. Lett.*, vol. 28, March 1992, pp. 679–681.
48. S. Ichitsubo, T. Furuno, and R. Kawasaki, "A statistical model for microcellular multipath propagation environment," in *Proc. IEEE Veh. Technol. Conf. (VTC'97)*, Phoenix, AZ, May 1997, pp. 61–66.
49. M. Wittmann, J. Marti, and T. Kürner, "Impact of the power delay profile shape on the bit error rate in mobile radio systems," *IEEE Trans. Veh. Technol.*, vol. VT-46, May 1997, pp. 329–339.
50. M. Abramowitz and I. A. Stegun, *Handbook of Mathematical Functions with Formulas, Graphs, and Mathematical Tables*. New York: Dover Publications, 1970.

3

TYPES OF COMMUNICATION

Digital modulation techniques are typically classified based on (1) the carrier attribute (e.g., phase, amplitude, frequency) that is being modulated, (2) the number of levels assigned to the modulated attribute, and (3) the degree to which the receiver extracts information about the unknown carrier phase in performing the data detection function (e.g., coherent, partially coherent, differentially coherent, noncoherent). Although most combinations of these classification categories are possible, some are more popular than others. In the simplest case, only a single carrier attribute is modulated, whereas a more sophisticated modulation scheme would allow for modulating more than one attribute (e.g., amplitude and phase), the latter affording additional degrees of freedom in satsifying the power and bandwidth requirements of the system.

Our goal in this chapter is to review the most popular digital modulation techniques (i.e., those that are most often addressed in the literature) and discuss their transmitted signal form as well as their detection over the additive white Gaussian noise (AWGN) channel. In all cases we limit our consideration to receivers that implement the maximum a posteriori (MAP) decision rule [maximum-likelihood (ML) for equiprobable signal hypotheses] and as such are optimum from the standpoint of minimizing error probability. Emphasis is placed on those modulations that might be used in applications where the channel exhibits multipath fading.

3.1 IDEAL COHERENT DETECTION

Consider a complex sinusoidal carrier, $\tilde{c}_t(t) = A_c e^{j(2\pi f_c t + \theta_c)}$, which in the simplest case is amplitude, phase, or frequency modulated by an M-level ($M = 2^m \geq 2$) digital waveform, $a(t), \theta(t)$, or $f(t)$, respectively, in accordance with the digital data to be transmitted over the channel (Fig. 3.1). The corresponding bandpass complex transmitted signal then becomes $\tilde{s}(t) = \tilde{S}(t) e^{j(2\pi f_c t + \theta_c)}$, where $\tilde{S}(t)$ is the equivalent baseband complex transmitted signal and takes on the specific forms $\tilde{S}(t) = A_c a(t)$, $\tilde{S}(t) = A_c e^{j\theta(t)}$, and $\tilde{S}(t) = A_c e^{jf(t)t}$, respectively.

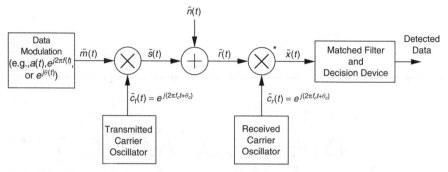

Figure 3.1. Generic complex form of transmitter and receiver for ideal coherent detection over the AWGN. (The asterisk on the multiplier denotes complex conjugate multiplication.)

When more than one attribute of the carrier is modulated (e.g., amplitude and phase), the transmitted signal would have the form $\tilde{s}(t) = A_c a(t) e^{j[2\pi f_c t + \theta_c + \theta(t)]}$. Corresponding to any of the cases above, the total received complex signal is $\tilde{r}(t) = \alpha_{ch}\tilde{s}(t) + \tilde{n}(t)$, where $\tilde{n}(t)$ is a complex white bandpass Gaussian noise process with single-sided power spectral density (PSD) N_0 (W/Hz) [i.e., $E\{\tilde{n}(t)\tilde{n}^*(t+\tau)\} = N_0\delta(t-\tau)$] and α_{ch} is the attenuation introduced by the channel. For the case of a pure AWGN channel as considered here, α_{ch} is a deterministic constant and for our purposes can be set equal to unity. For the fading channel considered later in the book, α_{ch} is a complex random variable whose statistics depend on the particular type of fading (e.g., for a Rayleigh or Rician channel, α_{ch} would be a complex Gaussian random variable).

In the case of ideal phase coherent detection (often called simply *coherent detection*), the receiver reconstructs the carrier with perfect knowledge of the phase and frequency. Thus, the receiver forms the signal[1] $\tilde{c}_r(t) = e^{j(2\pi f_c t + \theta_c)} = \tilde{c}_t(t)$ and uses this to perform a complex conjugate demodulation of the received signal (Fig. 3.1). The output of this demodulation is then $\tilde{x}(t) = \tilde{r}(t)\tilde{c}_r^*(t) = \tilde{S}(t) + \tilde{n}(t)\tilde{c}_r^*(t)$ which depending on the particular form of modulation corresponding to the three simple cases above is either $\tilde{x}(t) = A_c a(t) + \tilde{n}(t)\tilde{c}_r^*(t)$, $\tilde{x}(t) = A_c e^{j\theta(t)} + \tilde{n}(t)\tilde{c}_r^*(t)$, or $\tilde{x}(t) = A_c e^{j[2\pi f(t)t]} + \tilde{n}(t)\tilde{c}_r^*(t)$. The optimum receiver then performs matched filtering operations on $x(t)$ during each successive transmitted interval corresponding to the M possible transmitted information symbols in that interval and proceeds to make a decision based on the largest of the resulting M outputs. We now discuss a number of specific cases of the foregoing generic signal model along with the characteristics of the corresponding ideal coherent receiver.

---

[1] Again since we are considering here only the pure AWGN channel with idealized demodulation, the amplitude of the carrier reference signal is deterministic and may be normalized to unity with no loss in generality. Later when considering the fading channel, we shall see that the statistics of the fading channel must be taken into account in modeling the demodulation reference signal.

3.1.1 Multiple Amplitude-Shift-Keying or Multiple Amplitude Modulation

A *multiple amplitude-shift-keyed (M-ASK) signal* [more often referred to as *multiple amplitude modulation (M-AM)*] occurs when $a(t)$ takes on equiprobable symmetric[2] values $\alpha_i = 2i - 1 - M, i = 1, 2, \ldots, M$, in each symbol interval T_s which is related to the bit time T_b by $T_s = T_b \log_2 M$. As such, $a(t)$ is modeled as a random pulse stream, that is,

$$a(t) = \sum_{n=-\infty}^{\infty} a_n p(t - nT_s) \tag{3.1}$$

where a_n is the information (data) amplitude in the nth symbol interval $nT_s \leq t \leq (n+1)T_s$ ranging over the set of M possible values α_i as above, and $p(t)$ is a unit amplitude rectangular pulse of duration T_s seconds. The signal constellation (i.e., the locus of points of the baseband complex signal in two dimensions) is a straight line along the horizontal axis with points spaced uniformly by two units. In the nth symbol interval the transmitted complex signal is

$$\tilde{s}(t) = A_c a_n e^{j(2\pi f_c t + \theta_c)} \tag{3.2}$$

Note that because of the rectangular pulse shape, the complex baseband signal $\tilde{S}(t) = A_c a_n$ is constant in this same interval. At the receiver, after complex-conjugate demodulation by the ideal phase coherent reference $\tilde{c}_r(t) = e^{j(2\pi f_c t + \theta_c)}$, we obtain

$$\tilde{x}(t) = A_c a_n + \tilde{N}(t) \tag{3.3}$$

where $\tilde{N}(t) = \tilde{n}(t) c_r^*(t)$ is a zero-mean baseband complex Gaussian process. Passing $\tilde{x}(t)$ through M matched filters [integrate-and-dump (I&D) circuits for the assumed rectangular pulse shape of the modulation][3] results in the M outputs (Fig. 3.2a)

$$\tilde{y}_{nk} = \alpha_k a_n A_c T_s + \alpha_k \tilde{N}_n, \quad k = 1, 2, \ldots, M, \quad \tilde{N}_n = \int_{nT_s}^{(n+1)T_s} \tilde{N}(t)\, dt \tag{3.4}$$

whereupon a decision corresponding to the largest $\text{Re}\{\tilde{y}_{nk}\} = \alpha_k a_n A_c T_s + \text{Re}\{\alpha_k \tilde{N}_n\}$ is made on the transmitted amplitude. Alternatively, the amplitude scaling by the M possible levels α_i and maximum selection can be replaced by an M-level quantizer acting on the single real decision variable (see Fig. 3.2b)

$$y_n = a_n A_c T_s + N_n, \quad N_n = \text{Re}\{\tilde{N}_n\} \tag{3.5}$$

---
[2] In our discussions of AM, we consider only the case wherein the amplitude levels are distributed symmetrically around the zero level. For a discussion of asymmetric AM, see Ref. 1.
[3] As is well known, only a single matched filter is required whose output is scaled by the M possible values of α_i.

34 TYPES OF COMMUNICATION

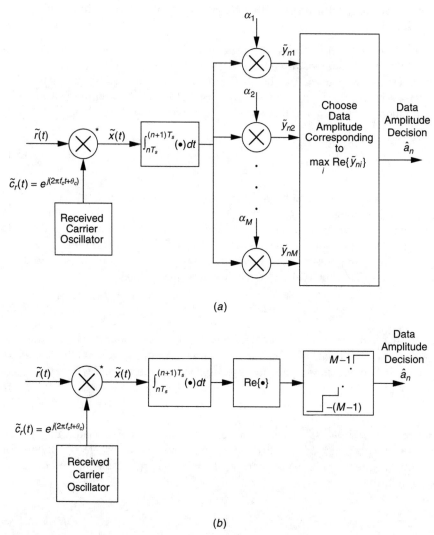

Figure 3.2. Complex forms of optimum receiver for ideal coherent detection of M-AM over the AWGN: (a) conventional maximum-likelihood form; (b) simpler decision threshold form.

3.1.2 Quadrature Amplitude-Shift-Keying or Quadrature Amplitude Modulation

A *quadrature amplitude-shift-keyed (QASK)* signal [more commonly referred to as *quadrature amplitude modulation (QAM)*] is a two-dimensional generalization of M-AM which can be viewed as a combined amplitude/phase modulation or more conveniently as a complex amplitude-modulated carrier. The signal constellation is a rectangular grid with points uniformly spaced along each axis by 2 units. Letting M still denote the number of possible transmitted waveforms, then in the nth symbol interval a QAM signal can be expressed as[4]

$$\tilde{s}(t) = A_c(a_{In} + ja_{Qn})e^{j(2\pi f_c t + \theta_c)} \quad (3.6)$$

where the information amplitudes a_{In} and a_{Qn} range independently over the sets of equiprobable values $\alpha_i = 2i - 1 - \sqrt{M}$, $i = 1, 2, \ldots, \sqrt{M}$, and $\alpha_l = 2l - 1 - \sqrt{M}$, $l = 1, 2, \ldots, \sqrt{M}$, respectively, and the I and Q subscripts denote the in-phase and quadrature channels. Here again, because of the assumed rectangular pulse shape, the complex baseband signal $\tilde{S}(t) = A_c(a_{In} + ja_{Qn})$ is constant in this same interval. At the receiver the signal is again first complex-conjugate demodulated by $\tilde{c}_r(t)$, which results in

$$\tilde{x}(t) = A_c(a_{In} + ja_{Qn}) + \tilde{N}(t) \quad (3.7)$$

Performing matched filter operations on $\tilde{x}(t)$ and recognizing the independence of the I and Q channels produces the decision variables (Fig. 3.3a)

$$y_{Ink} = \text{Re}\{\tilde{y}_k\} = \alpha_k a_{In} A_c T_s + \alpha_k N_{In}, \quad k = 1, 2, \ldots, \sqrt{M}/2,$$

$$N_{In} = \text{Re}\left\{\int_{nT_s}^{(n+1)T_s} \tilde{N}(t)\,dt\right\}$$

$$y_{Qnk} = \text{Im}\{\tilde{y}_k\} = \alpha_k a_{Qn} A_c T_s + \alpha_k N_{Qn}, \quad k = 1, 2, \ldots, \sqrt{M}/2,$$

$$N_{Qn} = \text{Im}\left\{\int_{nT_s}^{(n+1)T_s} \tilde{N}(t)\,dt\right\} \quad (3.8)$$

whereupon separate decisions corresponding to the largest y_{Ink} and y_{Qnk} are made on the I and Q components of the amplitude transmitted in the zeroth signaling (symbol) interval $0 \le t \le T_s$. Alternatively, the scaling by the M possible amplitude levels and maximum selection for the real and imaginary parts of the complex decision variable can be replaced by separate M-level quantizers acting on the single pair of I and Q decision variables

$$y_{In} = a_{In} A_c T_s + N_{In}$$

$$y_{Qn} = a_{Qn} A_c T_s + N_{Qn} \quad (3.9)$$

in which case the complex receiver of Fig. 3.3a can be redrawn in the I–Q form of Fig. 3.3b.

3.1.3 M-ary Phase-Shift-Keying

An M-ary phase-shift-keyed (M-PSK) signal occurs when $\theta(t)$ takes on equiprobable values $\beta_i = (2i-1)\pi/M$, $i = 1, 2, \ldots, M$, in each symbol interval T_s. As

---

[4] Again, one can think of the complex carrier as being modulated now by a complex random pulse stream, namely, $\tilde{a}(t) = \sum_{n=-\infty}^{\infty}(a_{In} + ja_{Qn})p(t - nT_s)$.

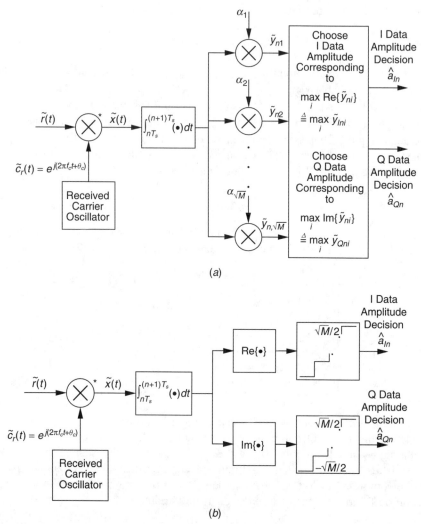

Figure 3.3. Complex forms of optimum receiver for ideal coherent detection of QAM over the AWGN: (a) conventional maximum–likelihood form; (b) simpler decision threshold form.

such, $\theta(t)$ is modeled as a random pulse stream, that is,

$$\theta(t) = \sum_{n=-\infty}^{\infty} \theta_n \, p(t - nT_s) \qquad (3.10)$$

where θ_n is the information phase in the nth symbol interval $nT_s \leq t \leq (n+1)T_s$ ranging over the set of M possible values β_i as above, and $p(t)$ is again a unit amplitude rectangular pulse of duration T_s seconds. The signal constellation is a unit circle with points uniformly spaced by $2\pi/M$ radians. Thus, the complex signal transmitted in the nth symbol interval is

$$\tilde{s}(t) = A_c e^{j(2\pi f_c t + \theta_c + \theta_n)} \tag{3.11}$$

Note again that because of the assumed rectangular pulse shape, the complex baseband signal $\tilde{S}(t) = A_c e^{j\theta_n}$ is constant in this same interval. After demodulating with the complex conjugate of $\tilde{c}_r(t)$ at the receiver, we obtain

$$\tilde{x}(t) = A_c e^{j\theta_n} + \tilde{N}(t) \tag{3.12}$$

Passing (3.12) through an I&D and then multiplying the output by $e^{-j\beta_k}$, $k = 1, 2, \ldots, M$, produces the decision variables (Fig. 3.4)

$$\tilde{y}_{nk} = A_c T_s e^{j(\theta_n - \beta_k)} + e^{-j\beta_k}\tilde{N}_n, \quad k = 1, 2, \ldots, M,$$

$$\tilde{N}_n = \int_{nT_s}^{(n+1)T_s} \tilde{N}(t)\, dt \tag{3.13}$$

from which a decision corresponding to the largest $\text{Re}\{\tilde{y}_{nk}\} = A_c T_s \cos(\theta_n - \beta_k) + \text{Re}\{e^{-j\beta_k}\tilde{N}_n\}$ is made on the information phase transmitted in the nth signaling interval.

A popular special case of M-PSK modulation is binary PSK (BPSK), which corresponds to $M = 2$. Since ideally the detection of M-PSK is independent of the location of the points around the unit circle (as long as they remain uniformly spaced by $2\pi/M$ radians), we can alternatively take as the possible values for θ_n the set $\beta_i = 2i\pi/M$, $i = 0, 1, 2, \ldots, M - 1$, which for $M = 2$ become $\beta_i = 0, \pi$. Since $e^{j0} = 1$ and $e^{j\pi} = -1$, the transmitted signal of (3.11) can be written in the form (3.2), where, in each transmission interval (now a bit interval T_b), a_n takes on the pair of equiprobable values ± 1. Thus, we observe that BPSK is the same as M-AM with $M = 2$. That is, binary amplitude and binary phase modulation are identical and are referred to as *antipodal signaling*. The receiver for BPSK is a special case of Fig. 3.4 which takes on the simpler form illustrated in Fig. 3.5 wherein the ± 1 amplitude scaling and maximum selection are replaced by a two-level quantizer (hard limiter) acting on the single real decision variable

$$y_n = a_n A_c T_b + N_n, \quad N_n = \text{Re}\{\tilde{N}_n\} \tag{3.14}$$

Another special case of M-PSK which because of its throughput efficiency (bits/second per unit of bandwidth) is quite popular is QPSK, which corresponds to $M = 4$. Here it is conventional to assume the phase set $\beta_i = \pi/4, 3\pi/4, 5\pi/4, 7\pi/4$. Projecting these information phases on the quadrature amplitude axes, we can equivalently write QPSK in the I–Q form of (3.6), where a_{In} and a_{Qn} each take on values ± 1.[5] We thus see that QPSK can also be looked upon as a special case of QAM with $M = 4$, and thus the detection of

---

[5] The actual projections of the unit circle on the I and Q coordinate axes are $1/\sqrt{2}$. However, since the carrier amplitude is arbitrary, it is convenient to rescale the carrier amplitude such that the equivalent I and Q data amplitudes take on ± 1 values.

38 TYPES OF COMMUNICATION

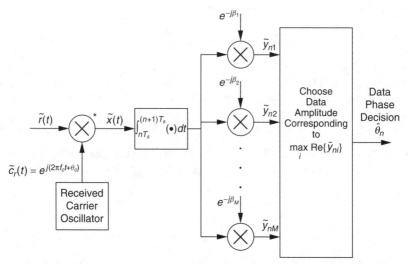

Figure 3.4. Complex form of optimum receiver for ideal coherent detection of *M*-PSK over the AWGN.

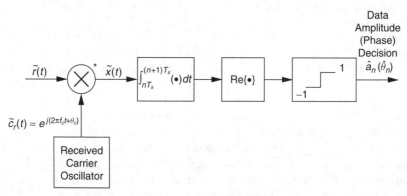

Figure 3.5. Complex form of optimum receiver for ideal coherent detection of BPSK over the AWGN.

a quadrature information phase can be obtained by combining the detections on the I and Q components of this phase. The receiver for QPSK is illustrated in Fig. 3.6 and is a two-dimensional version of that for BPSK and a special case of that for QAM. The decision variables that are input to the hard-limiting threshold devices are

$$y_{In} = \text{Re}\{\tilde{y}_n\} = a_{In} A_c T_s + N_{In}, \qquad N_{In} = \text{Re}\left\{\int_{nT_s}^{(n+1)T_s} \tilde{N}(t)\, dt\right\}$$

$$y_{Qn} = \text{Im}\{\tilde{y}_n\} = a_{Qn} A_c T_s + N_{Qn}, \qquad N_{Qn} = \text{Im}\left\{\int_{nT_s}^{(n+1)T_s} \tilde{N}(t)\, dt\right\}$$

(3.15)

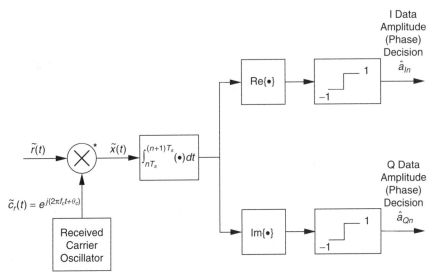

Figure 3.6. Complex form of optimum receiver for ideal coherent detection of QPSK over the AWGN.

While for M-PSK with $M = 2^m$ and m arbitrary, one can also project the information phases on the I and Q coordinates and thus make decisions on each of these multilevel amplitude signals, it should be noted that these decisions are not independent, and furthermore each pair of amplitude decisions does not necessarily render one of the transmitted phases. That is, the number of possible I–Q amplitude pairs obtained from the projections of the M possible transmitted phases exceeds M. Thus, for $M \geq 8$ it is not practical to view M-PSK in an I–Q form.

3.1.4 Differentially Encoded M-ary Phase-Shift-Keying

In an actual coherent communication system transmitting M-PSK modulation, a means must be provided at the receiver for establishing the local demodulation carrier reference signal. This means is tradionally accomplished with the aid of a suppressed carrier tracking loop [1, Chap. 2]. Such a loop for M-PSK modulation exhibits an M-fold phase ambiguity in that it can lock with equal probability at the transmitted carrier phase plus any of the M information phase values. Hence, the carrier phase used for demodulation can take on any of these same M phase values, namely, $\theta_c + \beta_i = \theta_c + 2i\pi/M$, $i = 0, 1, 2, \ldots, M - 1$. Clearly, coherent detection cannot be successful unless this M-fold phase ambiguity is resolved.

One means for resolving this ambiguity is to employ *differential phase encoding* (most often simply called *differential encoding*) at the transmitter and differential phase decoding (most often simply called *differential decoding*) at the receiver following coherent detection. That is, the information phase to be communicated is modulated on the carrier as the difference between

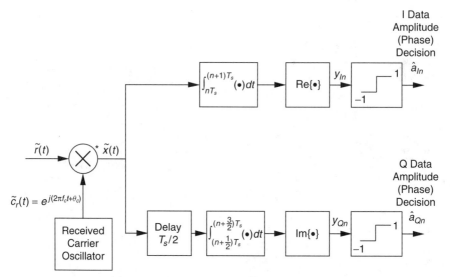

Figure 3.7. Complex form of optimum receiver for ideal coherent detection of OQPSK over the AWGN.

two adjacent transmitted phases, and the receiver takes the difference of two adjacent phase decisions to arrive at the decision on the information phase.[6] In mathematical terms, if $\Delta\theta_n$ was the information phase to be communicated in the nth transmission interval, the transmitter would first form $\theta_n = \theta_{n-1} + \Delta\theta_n$ modulo 2π (the differential encoder) and then modulate θ_n on the carrier.[7] At the receiver, successive decisions on θ_{n-1} and θ_n would be made and then differenced modulo 2π (the differential decoder) to give the decision on $\Delta\theta_n$. A block diagram of such a differentially encoded M-PSK system is illustrated in Fig. 3.7. It should be clear from this diagram that since the decision on the true information phase is obtained from the difference of two adjacent phase decisions, a performance penalty is associated with the inclusion of differential encoding/decoding in the system. The quantification of this performance penalty is discussed later in the book.

3.1.4.1 $\pi/4$-QPSK.
Depending on the set of M phases $\{\Delta\beta_i\}$ used to represent the information phase $\Delta\theta_n$ in the nth transmission interval, the actual transmitted phase θ_n in this same transmission interval can range either over the same set

---

[6] We note that this receiver (i.e., the one that makes optimum coherent decisions on two successive symbol phases and then differences these to arrive at the decision on the information phase) is suboptimum when $M > 2$ [3]. However, this receiver structure, which is the one classically used for coherent detection of differentially encoded M-PSK, can be arrived at by a suitable approximation of the likelihood function used to derive the true optimum receiver and at high SNR the difference between the two becomes mute.

[7] Note that we have shifted our notation here insofar as the information phases are concerned so as to keep the same notation for the actual transmitted phases.

$\{\beta_i\} = \{\Delta\beta_i\}$ or over another phase set. If for $M = 4$ we choose the set $\Delta\beta_i = 0, \pi/2, \pi, 3\pi/2$ to represent the information phases, then starting with an initial transmitted phase chosen from the set $\pi/4, 3\pi/4, 5\pi/4, 7\pi/4$, the subsequent transmitted phases $\{\theta_n\}$ will also range over the set $\pi/4, 3\pi/4, 5\pi/4, 7\pi/4$ in every transmission interval. This is the conventional form of differentially encoded QPSK. Now suppose instead that the set $\Delta\beta_i = \pi/4, 3\pi/4, 5\pi/4, 7\pi/4$ is used to represent the information phases $\{\Delta\theta_n\}$. Then, starting, for example, with an initial phase chosen from the set $\pi/4, 3\pi/4, 5\pi/4, 7\pi/4$, the transmitted phase in the next interval will range over the set $0, \pi/2, \pi, 3\pi/2$. In the following interval the transmitted phase will range over the set $\pi/4, 3\pi/4, 5\pi/4, 7\pi/4$, and in the interval following that one the transmitted phase will once again range over the set $0, \pi/2, \pi, 3\pi/2$. Thus we see that for this choice of phase set corresponding to the information phases $\{\Delta\theta_n\}$, the transmitted phases $\{\theta_n\}$ will alternatively range over the sets $0, \pi/2, \pi, 3\pi/2$ and $\pi/4, 3\pi/4, 5\pi/4, 7\pi/4$. Such a modulation scheme, referred to as *π/4-QPSK* [4], has an advantage relative to conventional differentially encoded QPSK as follows.

In the case of conventional differentially encoded QPSK, the maximum change in phase from transmission to transmission (which occurs when both I- and Q-channel data streams switch polarity) is π radians, which results in a complete reversal (maximum fluctuation) of the envelope of the transmitted waveform. In the case of $\pi/4$-QPSK, the maximum change in phase from transmission to transmission is $3\pi/4$ radians, which is clearly a smaller envelope fluctuation. On nonlinear transmission channels the fluctuation of the envelope is related to the regeneration of spectral sidelobes of the modulation after bandpass filtering and nonlinear amplification at the transmitter — the smaller the envelope fluctuation, the smaller the sidelobe regeneration, and vice versa. On a linear AWGN channel with ideal coherent detection, there is theoretically no advantage of $\pi/4$-QPSK over conventional differentially encoded QPSK; in fact, the two have identical error probability performance.

3.1.5 Offset QPSK or Staggered QPSK

For the same reason as using $\pi/4$-QPSK versus conventional differentially encoded QPSK on a nonlinear channel, another form of QPSK, namely, *offset QPSK (OQPSK)* [alternatively called *staggered QPSK (SQPSK)*] has become quite popular. OQPSK or SQPSK is a form of QPSK wherein the I and Q signals components are misaligned with respect to one another by half a symbol time (i.e., a bit time) interval. In mathematical terms, the complex carrier is amplitude modulated by $a_I(t) + ja_Q(t)$, where

$$a_I(t) = \sum_{n=-\infty}^{\infty} a_{In} p(t - nT_s), \qquad a_Q(t) = \sum_{n=-\infty}^{\infty} a_{Qn} p(t - nT_s - T_s/2) \quad (3.16)$$

where a_{In} and a_{Qn} are the I and Q data symbols for the nth transmission interval that take on equiprobable ± 1 values. Thus, in the nth transmission interval corresponding to the I channel, the transmitted signal has the complex form

$$\tilde{s}(t) = \begin{cases} A_c(a_{In} + ja_{Q,n-1})e^{j(2\pi f_c t + \theta_c)}, & nT_s \le t \le (n+\tfrac{1}{2})T_s \\ A_c(a_{In} + ja_{Qn})e^{j(2\pi f_c t + \theta_c)}, & (n+\tfrac{1}{2})T_s \le t \le (n+1)T_s. \end{cases} \quad (3.17)$$

Similarly, for the nth transmission interval corresponding to the Q channel, the transmitted signal has the complex form

$$\tilde{s}(t) = \begin{cases} A_c(a_{In} + ja_{Qn})e^{j(2\pi f_c t + \theta_c)}, & (n+\tfrac{1}{2})T_s \le t \le (n+1)T_s \\ A_c(a_{I,n+1} + ja_{Qn})e^{j(2\pi f_c t + \theta_c)}, & (n+1)T_s \le t \le (n+\tfrac{3}{2})T_s. \end{cases} \quad (3.18)$$

At the receiver the signal $\tilde{x}(t) = \tilde{s}(t) + \tilde{n}(t)$ is complex-conjugate demodulated by $\tilde{c}_r(t)$ and then matched filtered producing the I and Q decision variables (Fig. 3.7)

$$y_{In} = a_{In} A_c T_s + N_{In}, \quad N_{In} = \text{Re}\left\{ \int_{nT_s}^{(n+1)T_s} \tilde{N}(t)\, dt \right\}$$

$$y_{Qn} = a_{Qn} A_c T_s + N_{Qn}, \quad N_{Qn} = \text{Im}\left\{ \int_{(n+\tfrac{1}{2})T_s}^{(n+\tfrac{3}{2})T_s} \tilde{N}(t)\, dt \right\} \quad (3.19)$$

each of which is hard-limited to produce decisions on the I and Q transmitted amplitudes. Note that independent of the time offset between the I and Q channels, the decision variables of (3.19) have statistics identical to those of conventional QPSK as given by (3.15). Thus, for ideal coherent detection, QPSK and OQPSK have identical error probability performance, as will be reiterated later in the book.

Returning now to the issue of spectral sidelobe regeneration on a nonlinear channel, since the I and Q channels do not change phase at the same time instant (i.e., they are staggered by half a symbol with respect to each other), a phase change of π radians cannot occur instantaneously. Rather, if both the I and Q channels switch data polarities, the π radians that ultimately results occurs in two steps: after half a symbol the phase changes by $\pi/2$ radians, and then after the next half a symbol the phase changes by another $\pi/2$ radians. Thus we see that at any given time instant, the maximum change in phase that can occur is $\pi/2$ radians, which results in a smaller envelope fluctuation than either $\pi/4$-QPSK or conventional differentially encoded QPSK.

In summary, on a linear AWGN channel with ideal coherent detection, all three types of differentially encoded QPSK (i.e., conventional, $\pi/4$, and offset) perform identically. The differences among the three types on a linear AWGN channel occur when the carrier demodulation phase reference is not perfect (i.e., nonideal coherent detection).

3.1.6 *M-ary Frequency-Shift-Keying*

An *M-ary frequency-shift-keyed (M-FSK)* signal occurs when $f(t)$ takes on equiprobable values $\xi_i = (2i - 1 - M)\Delta f/2, i = 1, 2, \ldots, M$, in each symbol interval T_s where the frequency spacing Δf is related to the frequency modulation index h by $h = \Delta f T_s$. As such, $f(t)$ is modeled as a random pulse stream, that is,

$$f(t) = \sum_{n=-\infty}^{\infty} f_n p(t - nT_s) \qquad (3.20)$$

where f_n is the information frequency in the nth symbol interval $nT_s \leq t \leq (n+1)T_s$ ranging over the set of M possible values ξ_i as above, and $p(t)$ is again a unit amplitude rectangular pulse of duration T_s seconds. Thus the complex signal transmitted in the nth symbol interval is

$$\tilde{s}(t) = A_c e^{j[2\pi(f_c t + f_n(t - nT_s)) + \theta_c]} \qquad (3.21)$$

Note here that in contrast to the amplitude- and phase-shift-keying modulations discussed previously, the complex baseband modulation $\tilde{S}(t) = A_c e^{jf_n(t - nT_s)}$ is not constant over this same interval but rather has a sinusoidal variation. After demodulating with the complex conjugate of $\tilde{c}_r(t)$ at the receiver, we obtain

$$\tilde{x}(t) = A_c e^{j2\pi f_n(t - nT_s)} + \tilde{N}(t) \qquad (3.22)$$

Multiplying (3.22) by the set of harmonics $e^{-j2\pi \xi_k (t - nT_s)}$, $k = 1, 2, \ldots, M$, and then passing each resulting signal through an I&D produces the decision variables (Fig. 3.8)

$$\tilde{y}_{nk} = A_c \int_{nT_s}^{(n+1)T_s} e^{j2\pi (f_n - \xi_k)(t - nT_s)} dt + \tilde{N}_{nk}, \quad k = 1, 2, \ldots, M,$$

$$\tilde{N}_{nk} = \int_{nT_s}^{(n+1)T_s} e^{-j2\pi \xi_k (t - nT_s)} \tilde{N}(t) dt \qquad (3.23)$$

from which a decision corresponding to the largest $\text{Re}\{\tilde{y}_{nk}\}$ is made on the information frequency transmitted in the nth signaling interval.

For orthogonal signaling wherein the cross-correlation $\text{Re}\{\int_{nT_s}^{(n+1)T_s} \tilde{s}_k(t) \tilde{s}_l^*(t) dt\} = 0, k \neq l$, the frequency spacing is chosen such that $\Delta f = N/2T_s$ with N integer. If, for example, the transmitted frequency f_n is equal to $\xi_l = (2l - 1 - M)\Delta f/2$, then (3.23) can be expressed as

$$\tilde{y}_{nk} = A_c T_s e^{j\pi(l-k)N/2} \frac{\sin[\pi(l-k)N/2]}{\pi(l-k)N/2} + \tilde{N}_{nk}, \quad k = 1, 2, \ldots, M,$$

$$\tilde{N}_{nk} = \int_0^{T_s} e^{-j\pi(2k-1-M)Nt/2T_s} \tilde{N}(t + nT_s) dt \qquad (3.24)$$

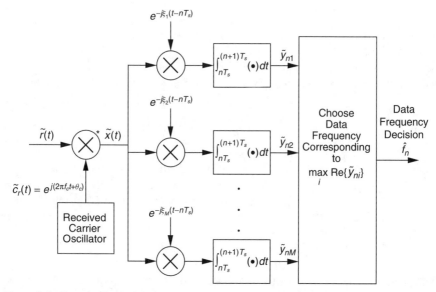

Figure 3.8. Complex form of optimum receiver for ideal coherent detection of M-FSK over the AWGN.

or, taking the real part,

$$\text{Re}\{\tilde{y}_{nk}\} = A_c T_s \frac{\sin[\pi(l-k)N]}{\pi(l-k)N} + \text{Re}\{\tilde{N}_{nk}\}, \quad k = 1, 2, \ldots, M \quad (3.25)$$

Thus we observe that for orthogonal M-FSK, only one decision variable has a nonzero mean: the one corresponding to the transmitted frequency. That is,

$$\overline{\text{Re}\{\tilde{y}_{nl}\}} = A_c T_s, \quad \overline{\text{Re}\{\tilde{y}_{nk}\}} = 0, \quad k \neq l \quad (3.26)$$

A popular special case of M-FSK modulation is binary FSK (BFSK), which corresponds to $M = 2$. In addition to orthogonal signaling (zero cross-correlation), it is possible to choose the modulation index so as to achieve the minimum cross-correlation that results in the minimum error probability (see Chapter 8). Since for arbitrary Δf we have

$$\text{Re}\left\{\int_{nT_b}^{(n+1)T_b} \tilde{s}_1(t)\tilde{s}_2^*(t)\,dt\right\} = \text{Re}\left\{A_c^2 \int_0^{T_b} e^{-j2\pi\Delta f t}\,dt\right\}$$

$$= A_c^2 T_b \frac{\sin 2\pi\Delta f T_b}{2\pi\Delta f T_b} \quad (3.27)$$

the minimum of this cross-correlation is achieved when $h = \Delta f T_b = 0.715$ [1], which results in a minimum normalized cross-correlation value

$$\rho \stackrel{\Delta}{=} \frac{\mathrm{Re}\left\{\int_{nT_b}^{(n+1)T_b} \tilde{s}_1(t)\tilde{s}_2^*(t)\,dt\right\}}{\int_{nT_b}^{(n+1)T_b} |\tilde{s}_1(t)|^2\,dt} = \frac{\mathrm{Re}\left\{\int_{nT_b}^{(n+1)T_b} \tilde{s}_1(t)\tilde{s}_2^*(t)\,dt\right\}}{\int_{nT_b}^{(n+1)T_b} |\tilde{s}_2(t)|^2\,dt}$$

$$= \frac{\sin 2\pi \Delta f T_b}{2\pi \Delta f T_b}\bigg|_{\Delta f T_b = 0.715} = -0.217 \simeq -\frac{2}{3\pi} \qquad (3.28)$$

3.1.7 Minimum-Shift-Keying

Consider a BFSK signal whose phase is maintained continuous from bit interval to bit interval, called *continuous phase frequency-shift-keying (CPFSK)* [5]. Because of this phase continuity, such a modulation has memory, and thus data bit decisions should be based on an observation longer than a single bit interval. A special case of CPFSK corresponds to a modulation index $h = \frac{1}{2}$ and is referred to as *minimum-shift-keying (MSK)* [6,7]. For this special case, the transmitted signal in the nth bit interval takes the form

$$\tilde{s}(t) = A_c e^{j[2\pi f_c t + d_n(\pi t/2T_b) + x_n]}, \qquad nT_b \le t \le (n+1)T_b \qquad (3.29)$$

where d_n is the binary (± 1) information bit and x_n is chosen to maintain the phase continuous at $t = nT_b$. Writing (3.29) in the form that characterizes the $(n-1)$st bit interval, to maintain the phase continuous at $t = nT_b$ it is straightforward to show that, assuming an initial condition $x_{-\infty} = 0$, the phase x_n satisfies the relation

$$x_n = x_{n-1} + \frac{\pi n}{2}(d_{n-1} - d_n) \qquad (3.30)$$

and thus can only take on values $(0, \pi)$ (modulo 2π). Substituting (3.30) into (3.29) and applying simple trigonometry it can be shown that MSK has an equivalent I–Q form that resembles OQPSK with, however, a pulse shape that is not rectangular. Specifically, an MSK signal has the pulse-shaped OQPSK representation

$$\tilde{s}(t) = A_c[a_I(t) + j a_Q(t)] e^{j(2\pi f_c t + \theta_c)} \qquad (3.31)$$

where $a_I(t)$ and $a_Q(t)$ are random data streams of the form in (3.16), with binary (± 1) data symbols (each of duration $T_s = 2T_b$)

$$a_{In} = \cos x_n, \qquad a_{Qn} = d_n \cos x_n = d_n a_{In} \qquad (3.32)$$

and $p(t)$ is a half sinusoid of duration T_s, that is,

$$p(t) = \begin{cases} \cos \dfrac{\pi t}{T_s}, & -\dfrac{T_s}{2} \le t \le \dfrac{T_s}{2} \\ 0, & \text{otherwise.} \end{cases} \qquad (3.33)$$

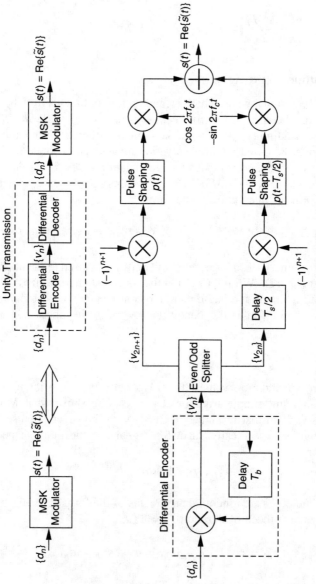

Figure 3.9. Equivalent real forms of MSK transmitters.

Note that the pulse shape for the Q data stream is

$$p\left(t - \frac{T_s}{2}\right) = \begin{cases} \sin\dfrac{\pi t}{T_s}, & 0 \le t \le T_s \\ 0, & \text{otherwise.} \end{cases} \quad (3.34)$$

There exists a direct relation between the binary data bits $\{d_n\}$ of the frequency modulation form of MSK in (3.29) and the equivalent binary data bits $\{a_{In}\}$ and $\{a_{Qn}\}$ of the I–Q form in (3.31). In particular, $\{a_{In}\}$ and $\{a_{Qn}\}$ are the odd and even bits of the differentially encoded version of $\{a_n\}$ (Fig. 3.9). That is, if $v_n = d_n v_{n-1}$ is the differentially encoded version of d_n, the equivalent I and Q data bits are given by

$$a_{In} = (-1)^{n+1} v_{2n+1}, \qquad a_{Qn} = (-1)^{n+1} v_{2n} \quad (3.35)$$

Thus, if the MSK modulation is implemented by continuous phase frequency modulating the carrier oscillator with the sequence $\{d_n\}$ and the data are to be recovered by implementing a pulse-shaped OQPSK receiver (Fig. 3.10), then following the interleaving of the I and Q decisions $\{\hat{a}_{In}\}$ and $\{\hat{a}_{Qn}\}$, one must undo the implicit differential encoding operation at the transmitter and thus employ a differential decoder to obtain the decisions on the information bits $\{d_n\}$. To get around the need for differential decoding at the receiver and the associated performance penalty (discussed in Chapter 8), one can precode the data entering the MSK modulator with a differential decoder, resulting in *precoded MSK* [1, Chap. 10]. The combination of differential decoder and MSK modulator is then identically equivalent to a pulse-shaped OQPSK modulator whose equivalent I and Q binary data bits $\{a_{In}\}$ and $\{a_{Qn}\}$ are now just the odd and even bits of $\{d_n\}$ itself (Fig. 3.11). That is, if prior to frequency modulating the carrier the information bits $\{d_n\}$ are first differentially decoded to the sequence $\{u_n\}$, where $u_n = d_n d_{n-1}$, the equivalent I and Q bits for the pulse-shaped OQPSK modulator would be

$$a_{In} = (-1)^{n+1} d_{2n+1}, \qquad a_{Qn} = (-1)^{n+1} d_{2n} \quad (3.36)$$

Thus, for precoded MSK, no differential decoder is needed at the receiver in order to recover the decisions on $\{d_n\}$ (Fig. 3.12). Since the precoder has no effect on the power spectral density of the transmitted waveform, then from a spectral point of view, MSK and pulse-shaped OQPSK are identical. Thus, from this point on, when discussing MSK modulation and demodulation, we shall assume implicitly that we are referring to precoded MSK or equivalently, pulse-shaped OQPSK.

3.2 NONIDEAL COHERENT DETECTION

In Section 3.1 we considered the ideal case of phase coherent detection wherein it was assumed that the attributes of the local carrier used to demodulate the received signal were perfectly matched to those of the transmitted carrier [i.e., $c_r(t) = c_t(t)$]. In practice, this ideal condition is never met since the local

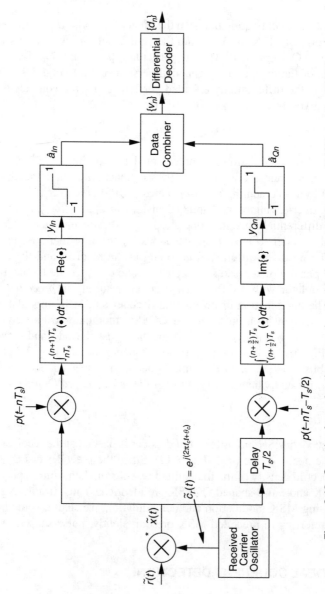

Figure 3.10. Complex form of optimum receiver for ideal coherent detection of MSK over the AWGN.

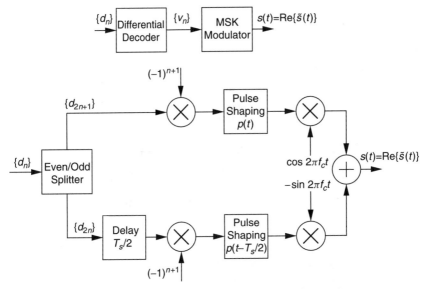

Figure 3.11. Equivalent real forms of precoded MSK transmitters.

carrier must be derived from the received signal itself, which contains the random perturbations introduced by the channel (e.g., the additive noise, fading, Doppler shift, etc.). Regardless of the manner in which the receiver creates its demodulation reference, there will result a mismatch between the phase and frequency of the received carrier and that of the locally generated carrier. Ignoring any frequency mismatch, if, as before, θ_c denotes the phase of the received carrier and $\hat{\theta}_c$ now denotes the phase of the locally generated carrier at the receiver, the phase error $\phi_c \triangleq \theta_c - \hat{\theta}_c$ would be a random variable with a specified PDF $p(\phi_c)$, which, in general, depends on the scheme used for extracting the phase estimate $\hat{\theta}_c$. We shall have more to say about the form of this PDF momentarily. For the special case of ideal phase coherent detection treated in Section 3.1, the phase error PDF was assumed to be a delta function [i.e., $p(\phi_c) = \delta(\phi_c)$].

When the nonideal carrier reference signal as above is used to demodulate the received signal, two possibilities exist with regard to the manner in which detection is subsequently performed. On the one hand, the detector can be designed assuming a perfect carrier reference (i.e., ideal coherent detection) with the nonideal nature of the demodulation reference accounted for in evaluating receiver performance. This is the case to which we direct our attention in this section. On the other hand, given the PDF of the phase error, $p(\phi_c)$, the remainder (baseband portion) of the receiver can be designed to exploit this statistical information, thereby coming up with an improved detection scheme. Such a scheme, which makes use of the available statistical information on the carrier phase error to optimize the design of the detector, is referred to as *partially coherent detection* and is discussed in Section 3.4.

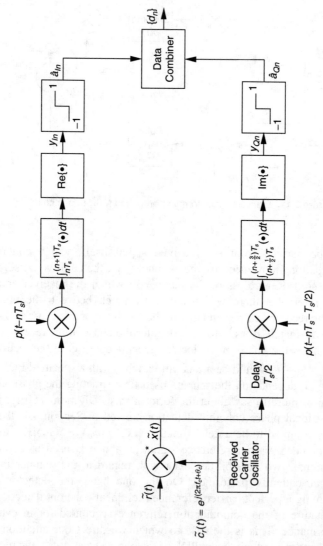

Figure 3.12. Complex form of optimum receiver for ideal coherent detection of precoded MSK (pulse-shaped OQPSK) over the AWGN.

Returning now to the manner in which the locally generated carrier is obtained at the receiver, the most common method for accomplishing this purpose is to employ a *carrier synchronization loop*[8] (e.g., a Costas loop), decision-directed loop, or form thereof [2, Chap. 2] that regenerates a carrier by continuously estimating the phase and frequency of the data-bearing received signal. Such loop structures are motivated by the MAP estimate of the carrier phase of a suppressed carrier signal and precede the data detection portion of the receiver. For a broad class of carrier reconstruction loops of the type mentioned above, the PDF of the modulo 2π-reduced phase error can be modeled as a Tikhonov distribution [10] which has the generic form[9]

$$p(\phi_c) = \frac{\exp(\rho_c \cos \phi_c)}{2\pi I_0(\rho_c)}, \qquad |\phi_c| \leq \pi \qquad (3.37)$$

with ρ_c called the loop SNR.

Another method for producing the necessary carrier synchronization at the receiver is to transmit a separate unmodulated carrier along with the data-modulated carrier and extract it at the receiver for use as the demodulation reference. Detection schemes based on such a transmitted reference are referred to as *pilot tone–aided* detection techniques and have the advantage that the method of extraction [e.g., a phase-locked loop (PLL) or narrowband filter] is not encumbered by the presence of the unknown data. On the other hand, for a given amount of total power, a portion of it must be allocated to the pilot signal and thus is not available for purposes of data detection.

In yet another method, a combination of the received signals in the previous intervals, the simplest case being just that from the previous interval, is used directly as the demodulation reference. Such detection schemes are based on observation of the received signal for more than a single symbol interval and are referred to as *differential detection*. Since these schemes in effect integrate the carrier demodulation as part of the detection operation, they are usually considered to form a class of their own, and we treat them as such in Section 3.5.

In accordance with the discussion above, the mathematical model used to define the demodulation reference signal is a complex carrier with a phase equal to the estimate of the received carrier phase [i.e., $\tilde{c}_r(t) = e^{j(2\pi f_c t + \hat{\theta}_c)}$]. Thus, for any of the complex bandpass transmitted signals $\tilde{s}(t) = A_c a(t) e^{j(2\pi f_c t + \theta_c)}$, $\tilde{s}(t) = A_c e^{j[2\pi f_c t + \theta_c + \theta(t)]}$, or $\tilde{s}(t) = A_c e^{j[2\pi (f_c + f(t))t + \theta_c]}$, the received signal after complex-conjugate demodulation becomes $\tilde{x}(t) = \tilde{S}(t) e^{j\phi_c} + \tilde{N}(t)$, which takes on the specific forms $\tilde{x}(t) = A_c a(t) e^{j\phi_c} + \tilde{N}(t)$, $\tilde{x}(t) = A_c a(t) e^{j[\theta(t) + \phi_c]} + \tilde{N}(t)$, and $\tilde{x}(t) = A_c a(t) e^{j[f(t)t + \phi_c]} + \tilde{N}(t)$, respectively, where $\tilde{N}(t) = n(t) \tilde{c}_r^*(t)$ is again a zero-mean baseband complex Gaussian process. Since ϕ_c is constant over

---

[8] Open-loop carrier synchronization techniques are also possible (see, e.g., Refs. 8 and 9), but are beyond the scope of our discussion here.
[9] The modeling of the phase error PDF for a phase-locked loop (PLL) in the form of (3.37) was also arrived at independently by Viterbi [11].

the symbol (bit) interval[10] the outputs of the matched filter for each of these types of modulation are as given in Section 3.1 [e.g., (3.4), (3.13), and (3.23), multiplied by $e^{j\phi_c}$]. As such, one can view the receiver structures for nonideal phase coherent detection as having baseband equivalents to those of ideal phase coherent detection with the addition of a phase rotation ϕ_c. Thus, if as before $\tilde{y}_{nk}, k = 1, 2, \ldots, M$, denotes the set of matched filter outputs for ideal phase coherent detection of the nth symbol, the decision variables for nonideal phase coherent detection in that same interval become $\tilde{y}_{nk} e^{j\phi_c}, k = 1, 2, \ldots, M$, where ϕ_c is distributed according to (3.37) or an appropriate variation thereof and is assumed to be independent of the $\tilde{y}_{nk}$'s. Equivalently, one can postulate a complex baseband receiver model where the kth matched filter output in the nth symbol interval is

$$\tilde{y}_{nk} = \tilde{s}_k e^{j\theta_c} + \tilde{N}_{nk} \tag{3.38}$$

which is then complex-conjugate demodulated by the complex baseband nonideal reference $\tilde{c}_r = e^{j\hat{\theta}_c}$. Here $\tilde{s}_k$ represents the signal component of the matched filter output under ideal phase-coherent conditions [i.e., the kth matched filter response to the complex baseband transmitted signal $\tilde{S}(t)$].

Another mathematical model for nonideal phase coherent detection, which is based on the complex baseband equivalent receiver above, is to treat the randomness of the phase of the demodulation reference $\tilde{c}_r = e^{j\hat{\theta}_c}$ as an equivalent AWGN source. As such, $\tilde{c}_r$ is modeled as the sum of an ideal phase coherent reference and a Gaussian random variable, that is,

$$\tilde{c}_r = \sqrt{G} A_r e^{j\theta_c} + \tilde{N}_r \tag{3.39}$$

where G is a normalized gain factor intended to reflect the SNR of the carrier synchronization technique used to produce $\hat{\theta}_c$ in the actual physical model. Although few carrier synchronizers produce a complex Gaussian reference signal, pragmatically, the mathematical nonideal reference model described by (3.39) has been demonstrated by Fitz [9,12] to be an accurate approximation of a large class of nonlinear phase estimation techniques (including the above-mentioned carrier synchronization architectures) in evaluating the average error probability performance of the system for moderate- to high-SNR applications. The advantage of the representation in (3.39) is that it affords a unified analysis akin to that suggested by Stein [13] wherein the demodulation phase reference signal and the matched filter output are both complex Gaussian processes and thus includes as a special case conventional (two-symbol observation) differential detection corresponding to $G = 1$ (see Section 3.5). This representation has a similar unifying advantage when evaluating the average error probability of such nonideal phase coherent systems in the presence of certain types of fading (see Chapter 8).

---

[10] We assume here the case where the data rate is sufficiently high relative to the carrier synchronization loop bandwidth that the phase of the demodulation reference produced by this loop is essentially constant over the duration of the data symbol.

3.3 NONCOHERENT DETECTION

In the preceding two sections it was assumed that either the carrier phase reference was provided to the receiver exactly (idealistically, by a genie), or at the very least an attempt was made to estimate it. At the other extreme, one can make the much simpler assumption that the receiver is designed not to make any attempt at estimating the carrier phase at all. Thus the local carrier used for demodulation is assumed to have an arbitrary phase which, without any loss in generality, can arbitrarily be set to zero. Detection techniques based on the absence of any knowledge of the received carrier phase are referred to as *noncoherent detection* techniques. In mathematical terms, the receiver observes the equivalent baseband signal $\tilde{R}(t) \triangleq \tilde{r}(t)e^{-j2\pi f_c t} = \tilde{S}(t)e^{j\theta_c} + \tilde{n}(t)e^{-j2\pi f_c t}$, where θ_c is unknown [and thus may be assumed to be uniformly distributed in the interval $(-\pi, \pi)$] and attempts to make a decision on $\tilde{S}(t)$.

The optimum receiver under such a scenario is well known [1] to be a structure that incorporates a form of square-law detection. Specifically, in each symbol interval the receiver first complex-conjugate demodulates the received signal with the zero-phase reference signal $c_r(t) = e^{j2\pi f_c t}$, then passes the result of this demodulation through M matched filters, one each corresponding to the transmitted baseband signals. The decision variables are then formed from the magnitudes (or equivalently, the squares of these magnitudes) of the matched filter outputs and the largest one is selected (see Fig. 3.13). In mathematical terms, the decision variables (assuming square-law detection) are given by

$$z_{nk} = |\tilde{y}_{nk}|^2 = \left| \int_{nT_s}^{(n+1)T_s} \tilde{R}(t)\tilde{S}_k^*(t)\,dt \right|^2, \qquad k = 1, 2, \ldots, M \qquad (3.40)$$

where $\tilde{S}_k(t), k = 1, 2, \ldots, M$, is the set of possible realizations of $\tilde{S}(t)$ and the decision is made in favor of the largest of the z_{nk}'s.

Suppose now that the modulation was, in fact, M-PSK and one attempted to use the receiver above for detection. Since in the absence of noise the matched filter outputs in the nth symbol interval would be given by [see Eq. (3.13), now with the addition of the unknown carrier phase θ_c] $\tilde{y}_{nk} = A_c T_s e^{j(\theta_n - \beta_k)} e^{j\theta_c}$, $k = 1, 2, \ldots, M$, the magnitudes of these outputs would all be identical and hence cannot be used for making a decision on the transmitted phase θ_n. Stated another way, since for M-PSK the information is carried in the phase of the carrier, then since the noncoherent receiver is designed to ignore this phase, it certainly cannot be used to yield a decision on it. In summary, *noncoherent detection cannot be employed with M-PSK modulation.*

Having ruled out M-PSK modulation (which would also rule out binary AM because of its equivalence with BPSK), the next most logical choice is M-FSK. Based on the results obtained in Section 3.1.6 for the matched filter outputs under

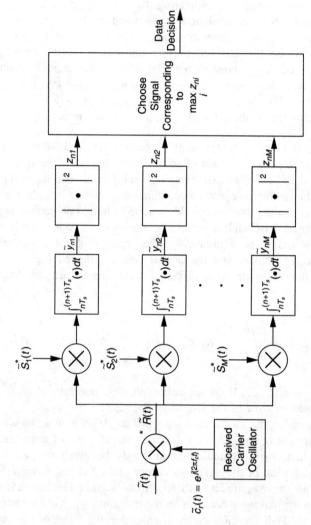

Figure 3.13. Complex form of optimum receiver for noncoherent detection over the AWGN.

ideal phase coherent conditions, we immediately write these same outputs for the noncoherent case as

$$\tilde{y}_{nk} = A_c e^{j\theta_c} \int_{nT_s}^{(n+1)T_s} e^{j2\pi(f_n-\xi_k)(t-nT_s)}\,dt + \tilde{N}_{nk}, \qquad k=1,2,\ldots,M,$$

$$\tilde{N}_{nk} = \int_{nT_s}^{(n+1)T_s} e^{-j2\pi\xi_k(t-nT_s)}\tilde{N}(t)\,dt \qquad (3.41)$$

where now $\tilde{N}(t) = \tilde{n}(t)e^{j2\pi f_c t}$. Taking the absolute value (or its square) of the $\tilde{y}_{nk}$'s in (3.41) in the absence of noise removes the unknown carrier phase but leaves the data information, which is now carried in the frequency f_n, unaltered. Thus it is feasible to use noncoherent detection with M-FSK modulation. Note, however, that the additional use of an envelope (or square-law) detector following the matched filters in the noncoherent case will result in a performance penalty relative to the coherent case, where the decision is made based on the matched filter outputs alone (see Chapter 8).

3.4 PARTIALLY COHERENT DETECTION

3.4.1 Conventional Detection: One-Symbol Observation

In Section 3.2, the assumption was made that although the true carrier demodulation was accomplished prior to data detection, the design of the detector was not in any way influenced by the randomness of the phase error statistics at the output of the demodulator (i.e., the form of the detector that is optimum for ideal phase coherent detection was still employed). When the statistics of the phase error are taken into account in the design of the detector, then based on observation of a single symbol interval, it can be shown [1,14] that the optimum detector is a linear combination of the coherent and noncoherent detectors discussed in Sections 3.1 and 3.3, respectively. In mathematical terms, the decision variables $\{z_{nk}\}$ are formed from the matched filter outputs as

$$z_{nk} = (\mathrm{Re}\{\tilde{y}_{nk}\} + \rho_c N_0/2)^2 + (\mathrm{Im}\{\tilde{y}_{nk}\})^2, \qquad k=1,2,\ldots,M \qquad (3.42)$$

or ignoring the term $(\rho_c N_0/2)^2$, which is common to all M z_{nk}'s, we have the equivalent decision variables (keeping the same notation)

$$\begin{aligned} z_{nk} &= \left(\frac{1}{N_0}\mathrm{Re}\{\tilde{y}_{nk}\}\right)^2 + \left(\frac{1}{N_0}\mathrm{Im}\{\tilde{y}_{nk}\}\right)^2 + \rho_c\left(\frac{1}{N_0}\mathrm{Re}\{\tilde{y}_{nk}\}\right) \\ &= \left|\frac{1}{N_0}\tilde{y}_{nk}\right|^2 + \rho_c\left(\frac{1}{N_0}\mathrm{Re}\{\tilde{y}_{nk}\}\right), \qquad k=1,2,\ldots,M \end{aligned} \qquad (3.43)$$

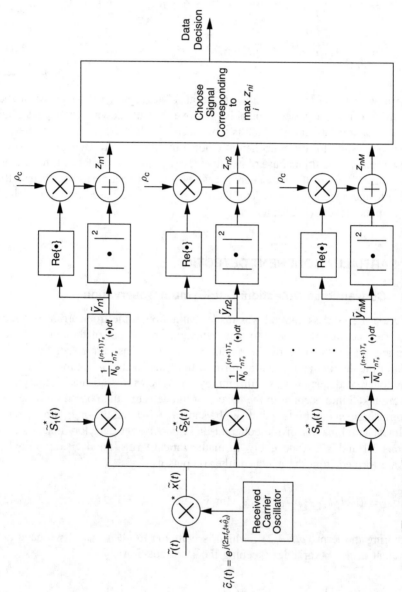

Figure 3.14. Complex form of optimum receiver for partially coherent detection over the AWGN.

where the first term is characteristic of noncoherent detection and the second term is characteristic of coherent detection. A receiver implementation based on (3.43) is illustrated in Fig. 3.14. Note that knowledge of both ρ_c and N_0 is required to implement this receiver. Such knowledge must be obtained by measurements taken on the channel and the accuracy of this knowledge will have an impact on the ultimate performance of the receiver. Since as mentioned in Section 3.3 for M-PSK modulation the first (noncoherent) term of (3.43) does not aid in the decision-making process, it can be ignored and hence the optimum partially coherent receiver of M-PSK reduces to the coherent receiver (Fig. 3.8 with a nonideal reference signal) whose performance is determined on the basis of the decision variables in Section 3.2. Regardless of the type of modulation, for $\rho_c = 0$, the receivers of Fig. 3.14 reduce to those for noncoherent detection whereas for $\rho_c = \infty$ they reduce to those for coherent detection.

3.4.2 Multiple Symbol Detection

Suppose now that we consider partially coherent detection of M-PSK based on an observation greater than a single symbol interval. If the phase error, ϕ_c, between the received carrier phase and the receiver's estimate of it is sufficiently slowly varying that it can be assumed constant over say N_s symbol intervals ($N_s \geq 2$), then an N_s-symbol observation of the received signal now contains memory, and the receiver should be able to exploit this property in arriving at an optimum design with improved performance [1, Chap. 6; 15]. As in any optimum (ML) receiver for a modulation with memory transmitted over the AWGN, the structure should employ *sequence* detection [i.e., joint (rather than symbol-by-symbol) decisions should be made on groups of N_s symbols on a block-by-block basis].

Analogous to the results in Section 3.4.1, the optimum detector based on an observation of the received signal now spanning N_s symbols, is again a linear combination of coherent and noncoherent detectors in which a set of M^{N_s} decision variables is formed from the matched filter outputs to enable selection of the most likely N_s-symbol sequence of phases. In mathematical terms, the M^{N_s} symbol-by-symbol matched filter outputs

$$\tilde{y}_{n-i,k_i} = \int_{(n-i)T_s}^{(n-i+1)T_s} \tilde{R}(t)\tilde{S}_{k_i}^*(t)\,dt, \qquad k_i = 1, 2, \ldots, M,$$
$$i = 0, 1, \ldots N_s - 1 \tag{3.44}$$

with

$$\tilde{S}_{k_i}(t) = A_c e^{j\beta_{k_i}} = A_c e^{j(2k_i-1)\pi/M}, \qquad k_i = 1, 2, \ldots, M \tag{3.45}$$

are summed over i in groups of size N_s and then used to produce the M^{N_s} decision variables

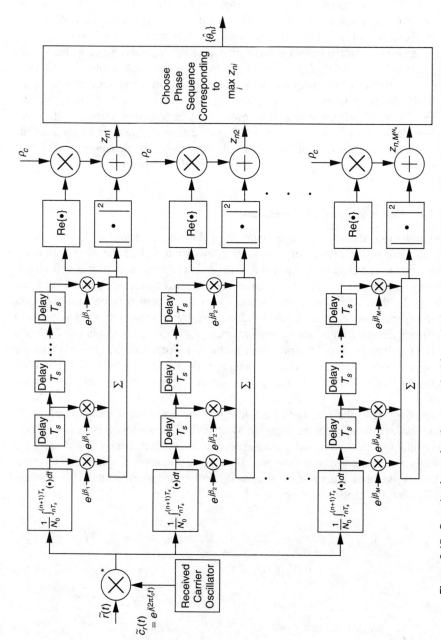

Figure 3.15. Complex form of optimum receiver for multiple symbol partially coherent detection over the AWGN.

$$z_{n\mathbf{k}} = \left(\text{Re}\left\{\sum_{i=0}^{N_s-1} \frac{1}{N_0}\tilde{y}_{n-i,k_i}\right\}\right)^2 + \left(\text{Im}\left\{\sum_{i=0}^{N_s-1} \frac{1}{N_0}\tilde{y}_{n-i,k_i}\right\}\right)^2$$

$$+ \rho_c \left(\text{Re}\left\{\sum_{i=0}^{N_s-1} \frac{1}{N_0}\tilde{y}_{n-i,k_i}\right\}\right) = \left|\sum_{i=0}^{N_s-1} \frac{1}{N_0}\tilde{y}_{n-i,k_i}\right|^2$$

$$+ \rho_c \left(\text{Re}\left\{\sum_{i=0}^{N_s-1} \frac{1}{N_0}\tilde{y}_{n-i,k_i}\right\}\right), \qquad k_i = 1, 2, \ldots, M \qquad (3.46)$$

The notation k_i in (3.44), (3.45), and (3.46) is used to indicate the fact that for each value of the transmission interval index i in the range 0 to $N_s - 1$, the transmitted signal index k can range over the set $1, 2, \ldots, M$. Also, the boldface subscript **k** on the variable z_n denotes the vector $(k_1, k_2, \ldots, k_{N_s-1})$. Finally, a decision is made on the transmitted phase sequence in the observation interval in accordance with the largest of the $z_{n\mathbf{k}}$'s. Clearly, for $N_s = 1$, (3.46) reduces to (3.43).

Note that for $N_s > 1$, the first (noncoherent) term in (3.46) in the absence of noise is not identical for all phase sequences and thus contributes to the decision-making process. This term does, however, have an associated phase ambiguity in that multiplication of each term in the sum by $e^{-j\theta_a}$ where θ_a is an arbitrary fixed phase, does not change the value of the term. Hence, based on the first term alone (i.e., for $\rho_c = 0$), the decision on the transmitted phase sequence would be ambiguous by θ_a radians, where θ_a could certainly assume the value of one of the transmitted information phases. The second term in (3.46) does not have such an associated phase ambiguity, and thus for $\rho_c \neq 0$ the decision rule would be unique. To guarantee a unique decision rule for the $\rho_c = 0$ case, one can employ differential phase encoding of the information phase symbols as discussed in Section 3.1.4. The specific details of how such differential encoding provides for a unique decision rule in this special case is discussed in Section 3.5 in connection with differential detection of M-PSK with multiple symbol observation. Figure 3.15 is an illustration of a partially coherent receiver for M-PSK based on the decision statistics of (3.46). The performance of this receiver is presented in Chapter 8.

3.5 DIFFERENTIALLY COHERENT DETECTION

3.5.1 *M*-ary Differential Phase Shift Keying

Suppose once again that one does not specifically attempt to reconstruct a local carrier at the receiver from an estimate of the received carrier phase. We saw in Section 3.3 that for an observation interval corresponding to a single transmitted symbol, the optimum noncoherent receiver could not be used to detect M-PSK modulation. Instead let us now reconsider the noncoherent detection problem assuming an observation interval greater than one symbol in duration.

This problem is akin to the partially coherent detection problem considered in the preceding section except that the memory that is introduced into the modulation now comes directly from the received carrier phase θ_c (assumed to be constant over, say, N_s symbols) rather than the phase error ϕ_c that results from its attempted estimation. As such, the maximum-likelihood solution to the problem would involve averaging the conditional likelihood function based on an N_s-symbol observation over a uniformly distributed phase (i.e., θ_c) rather than a Tikhonov-distributed phase (i.e., ϕ_c). Receivers designed according to the foregoing principles are referred to as *differential detectors* and clearly represent an extension of noncoherent reception to the case of multiple symbol observation. The term *differential* came about primarily due to the fact that in the conventional technique, a two-symbol observation is used ($N_s = 2$) and thus, as we shall see, the decision is made based on the difference between two successive matched filter outputs. However, Divsalar and Simon [16] showed that by using an observation greater than two symbols in duration, one could obtain a receiver structure that provided further improvement in performance in the limit as $N_s \to \infty$, approaching that of differentially encoded M-PSK (see Section 3.1.4). Practically speaking, it is only necessary to have N_s on the order of 3 to achieve most of the performance gain. With a little bit of thought, it should also be clear that the Tikhonov PDF of (3.37) with $\rho_\phi = 0$ becomes a uniform PDF, and thus from the above-mentioned analogy, the solution to the multiple-symbol (including $N_s = 2$) differential detection problem can be obtained directly as a special case of the results obtained for the multiple-symbol partially coherent detection problem.

3.5.1.1 Conventional Detection: Two-Symbol Observation.

We begin our discussion of differential detection of M-PSK by considering the conventional case of a two-symbol observation. Based on the discussion above, the decision variables can be obtained from the first term of (3.46) with $N_s = 2$. Substituting (3.44) together with (3.45) in this term gives

$$z_{n\mathbf{k}} = \left(\frac{1}{N_0}\right)^2 |\tilde{y}_{n,k_0} + \tilde{y}_{n-1,k_1}|^2$$

$$= \left(\frac{A_c}{N_0}\right)^2 \left| \int_{nT_s}^{(n+1)T_s} \tilde{R}(t) e^{-j\beta_{k_0}} dt + \int_{(n-1)T_s}^{nT_s} \tilde{R}(t) e^{-j\beta_{k_1}} dt \right|^2,$$

$$k_0, k_1 = 1, 2, \ldots, M \quad (3.47)$$

where β_{k_0} represents the assumed value for the information phase θ_0 transmitted in the nth symbol interval and β_{k_1} represents the assumed value for the information phase θ_{-1} transmitted in the $(n-1)$st symbol interval. As mentioned above, multiplying each of the two matched filter outputs in (3.47) by $e^{-j\theta_a}$ with θ_a arbitrary does not change the decision variables. To resolve this phase ambiguity we employ differential phase encoding at the transmitter as discussed in Section 3.1.4. In particular, the transmitted information phases, now denoted by $\{\Delta\theta_n\}$, are first converted (differentially encoded) to the set of phases $\{\theta_n\}$ in accordance with the relation

$$\theta_n = \theta_{n-1} + \Delta\theta_n \quad \text{modulo } 2\pi \tag{3.48}$$

where β_{k_0} and β_{k_1} in (3.47) now represent the assumed values for the differentially encoded phases in the nth and $(n-1)$st symbol intervals, respectively. Note that for θ_n and θ_{n-1} to both range over the set $\beta_k = (2k-1)\pi/M, k = 1, 2, \ldots, M$, we must now restrict the information phase $\Delta\theta_n$ to range over the set $\Delta\beta_k = 2k\pi/M, k = 0, 1, 2, \ldots, M-1$. If we now choose the arbitrary phase equal to the negative of the information phase in the $(n-1)$st interval (i.e., $\theta_a = -\beta_{k_0}$), then multiplying each matched output term in (3.47) by $e^{-j\theta_a} = e^{j\beta_{k_0}}$, we can rewrite (3.47) as [ignoring the $(A_c/N_0)^2$ scaling term]

$$\begin{aligned} z_{nk} &= \left| \int_{nT_s}^{(n+1)T_s} \tilde{R}(t)\, dt + \int_{(n-1)T_s}^{nT_s} \tilde{R}(t) e^{-j(\beta_{k_1}-\beta_{k_0})}\, dt \right|^2 \\ &= \left| \int_{nT_s}^{(n+1)T_s} \tilde{R}(t)\, dt + \int_{(n-1)T_s}^{nT_s} \tilde{R}(t) e^{-j\Delta\beta_k}\, dt \right|^2, \\ & k = 0, 1, \ldots, M-1 \end{aligned} \tag{3.49}$$

Choosing the largest of the z_{nk}'s in (3.49) then directly gives an unambiguous decision on the information phase $\Delta\theta_n$. Expanding the squared magnitude in (3.49) as

$$\begin{aligned} & \left| \int_{nT_s}^{(n+1)T_s} \tilde{R}(t)\, dt + \int_{(n-1)T_s}^{nT_s} \tilde{R}(t) e^{-j\Delta\beta_k}\, dt \right|^2 \\ &= \left| \int_{nT_s}^{(n+1)T_s} \tilde{R}(t)\, dt \right|^2 + \left| \int_{(n-1)T_s}^{nT_s} \tilde{R}(t) e^{-j\Delta\beta_k}\, dt \right|^2 \\ &\quad + 2\operatorname{Re}\left\{ \left(\int_{nT_s}^{(n+1)T_s} \tilde{R}(t)\, dt \right)^* \left(\int_{(n-1)T_s}^{nT_s} \tilde{R}(t) e^{-j\Delta\beta_k}\, dt \right) \right\} \end{aligned} \tag{3.50}$$

and noting that the first two terms of (3.50) are independent of the decision index k, an equivalent decision rule is to choose the largest of

$$\begin{aligned} z_{nk} &= \operatorname{Re}\left\{ \left(\int_{nT_s}^{(n+1)T_s} \tilde{R}(t)\, dt \right)^* \left(\int_{(n-1)T_s}^{nT_s} \tilde{R}(t) e^{-j\Delta\beta_k}\, dt \right) \right\} \\ &= \operatorname{Re}\left\{ e^{-j\Delta\beta_k} \left(\int_{nT_s}^{(n+1)T_s} \tilde{R}(t)\, dt \right)^* \left(\int_{(n-1)T_s}^{nT_s} \tilde{R}(t)\, dt \right) \right\}, \\ & k = 0, 1, \ldots, M-1 \end{aligned} \tag{3.51}$$

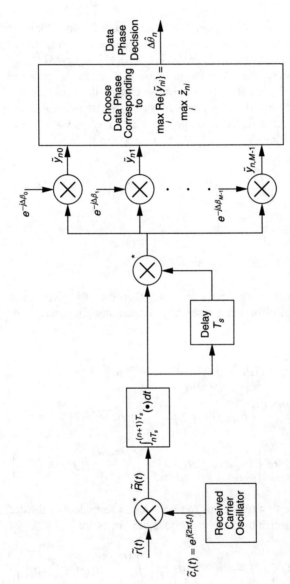

Figure 3.16. Complex form of optimum receiver for conventional (two-symbol observation) differentially coherent detection of M-PSK over the AWGN.

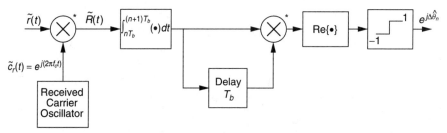

Figure 3.17. Complex form of optimum receiver for conventional (two-symbol observation) differentially coherent detection of DPSK over the AWGN.

A receiver that implements this decision rule is illustrated in Fig. 3.16 and is the optimum receiver under the constraint of a two-symbol observation. For binary DPSK, the decision rule simplifies to

$$e^{j\Delta\hat{\theta}_n} = \text{sgn}\left[\text{Re}\left\{\left(\int_{nT_b}^{(n+1)T_b}\tilde{R}(t)\,dt\right)^*\left(\int_{(n-1)T_b}^{nT_b}\tilde{R}(t)\,dt\right)\right\}\right] \quad (3.52)$$

and is implemented by the receiver illustrated in Fig. 3.17. Note that the structure of the receiver in Fig. 3.16 and its special case in Fig. 3.17 is such that the previous matched filter output acts as the effective baseband demodulation reference for the current matched filter output. In this context the differentially coherent receiver behaves like the nonideal coherent receiver discussed in Section 3.2 with a reference signal as in (3.39) having a gain $G = 1$ and an additive noise independent of that associated with the received signal.

3.5.1.2 Multiple-Symbol Detection.
Analogous to what was true for partially coherent detection, the performance of the differentially coherent detection system can be improved by optimally designing the receiver based on an observation of the received signal for more than two symbol intervals [16]. The appropriate decision variables are now obtained from the first term of (3.46) with $N_s > 2$. Once again using differential phase encoding to resolve the phase ambiguity inherent in this term — in particular, setting the arbitrary phase $\theta_a = -\theta_{n-N_s+1}$ and using the differential encoding algorithm of (3.48) — we obtain analogous to (3.49) the decision variables

$$z_{n\mathbf{k}} = \left|\int_{nT_s}^{(n+1)T_s}\tilde{R}(t)\,dt + \int_{(n-1)T_s}^{nT_s}\tilde{R}(t)e^{-j\Delta\beta_{k_1}}\,dt + \cdots \right.$$
$$\left. + \int_{(n-N_s+1)T_s}^{(n-N_s+2)T_s}\tilde{R}(t)e^{-j\Delta\beta_{k_{N_s-1}}}\,dt\right|^2,$$
$$k_i = 0, 1, \ldots, M-1, \quad i = 1, 2, \ldots, N_s - 1 \quad (3.53)$$

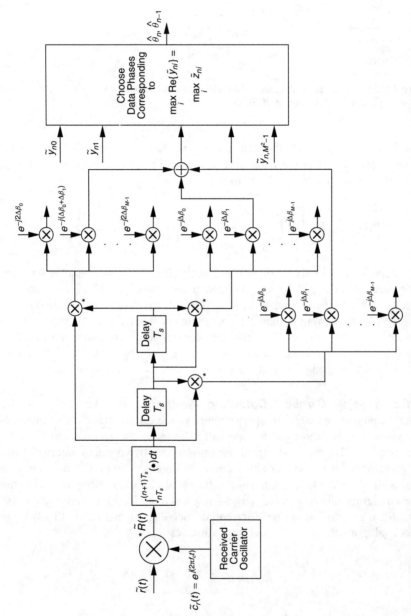

Figure 3.18. Complex form of optimum receiver for three-symbol differentially coherent detection of *M*-PSK over the AWGN.

from which a decision on the information sequence $\Delta\theta_{n-N_s+2}, \Delta\theta_{n-N_s+3}, \ldots,$ $\Delta\theta_{n-1}, \Delta\theta_n$ is made corresponding to the largest of the $z_{n\mathbf{k}}$'s. Note that an N_s-symbol observation results in a simultaneous decision on $N_s - 1$ information phase symbols. The squared magnitude in (3.53) can be expanded analogous to (3.50) to simplify the decision rule. For example, for $N_s = 3$ the decision rule is to choose the pair of information phases $\Delta\theta_{n-1}, \Delta\theta_n$ corresponding to the maximum over k_1 and k_2 of

$$\begin{aligned} z_{n\mathbf{k}} = \mathrm{Re} \Bigg\{ & e^{-j\Delta\beta_{k_1}} \left(\int_{nT_s}^{(n+1)T_s} \tilde{R}(t)\, dt \right)^* \left(\int_{(n-1)T_s}^{nT_s} \tilde{R}(t)\, dt \right) \\ & + e^{-j\Delta\beta_{k_2}} \left(\int_{(n-1)T_s}^{nT_s} \tilde{R}(t)\, dt \right)^* \left(\int_{(n-2)T_s}^{(n-1)T_s} \tilde{R}(t)\, dt \right) \\ & + e^{-j(\Delta\beta_{k_1}+\Delta\beta_{k_2})} \left(\int_{nT_s}^{(n+1)T_s} \tilde{R}(t)\, dt \right)^* \left(\int_{(n-2)T_s}^{(n-1)T_s} \tilde{R}(t)\, dt \right) \Bigg\}, \\ & k_1, k_2 = 0, 1, \ldots, M-1 \end{aligned} \quad (3.54)$$

A receiver that implements this decision rule is illustrated in Fig. 3.18.

We conclude this section by mentioning that although it appears that the complexity of the receiver implementation grows exponentially with the observation block size N_s [1, Sec. 7.2.3], Mackenthun [17] has developed algorithms for implementing multiple symbol differential detection of M-PSK that considerably reduce this complexity, thus making it a feasible alternative to coherent detection of differentially encoded M-PSK. These algorithms and their complexity in terms of the number of operations per N_s-symbol block being processed are also discussed in Ref. 1.

3.5.2 $\pi/4$-Differential QPSK

The $\pi/4$-QPSK introduced in Section 3.1.4.1 in combination with coherent detection as a means of reducing the regeneration of spectral sidelobes in bandpass filtered/nonlinear systems can also be used for the same purpose when combined with differential detection. The resulting scheme, called *$\pi/4$-differential QPSK ($\pi/4$-DQPSK)*, behaves quite similar to ordinary differential detection of QPSK as discussed in Section 3.5.1, with the following exception. Since the set of phases $\{\Delta\beta_k\}$ used to represent the information phases $\{\Delta\theta_n\}$ is now $\Delta\beta_k = (2k-1)\pi/4, k = 1, 2, 3, 4$, this set must be used in place of the set $\Delta\beta_k = k\pi/4, k = 0, 1, 2, 3$, in the phase comparison portion of Fig. 3.16.

REFERENCES

1. M. K. Simon, S. M. Hinedi, and W. C. Lindsey, *Digital Communication Techniques: Signal Design and Detection*. Upper Saddle River, NJ: Prentice Hall, 1995.

2. W. C. Lindsey and M. K. Simon, *Telecommunication Systems Engineering*. Upper Saddle River, NJ: Prentice Hall, 1973.
3. M. K. Simon and D. Divsalar, "On the optimality of classical coherent receivers of differentially encoded *M*-PSK," *IEEE Commun. Letters.*, vol. 1, May 1997, pp. 67–70.
4. P. A. Baker, "Phase-modulation data sets for serial transmission at 2000 and 2400 bits per second, part I," *AIEE Trans. Commun. Electron.*, July 1962.
5. J. B. Anderson, T. Aulin, and C.-E. Sundberg, *Digital Phase Modulation*. New York: Plenum Press, 1986.
6. M. L. Doelz, and E. T. Heald, "Minimum-shift data communication system," U.S. patent 2,977,417, March 28, 1961.
7. S. Pasupathy, "Minimum shift keying: a spectrally efficient modulation," *IEEE Commun.*, vol. 17, July 1979, pp. 14–22.
8. A. J. Viterbi and A. M. Viterbi, "Nonlinear estimation of PSK modulation carrier phase with application to burst digital transmission," *IEEE Trans. Inf. Theory*, vol, IT-32, July 1983, pp. 543–551.
9. M. P. Fitz, "Open loop techniques for carrier synchronization," Ph.D. dissertation, University of Southern California, Los Angeles, June 1989.
10. V. I. Tikhonov, "The effect of noise on phase-locked oscillator operation," *Autom. Remote Control*, vol. 20, 1959, pp. 1160–1168. Transated from *Autom. Telemek.*, Akademya Nauk, SSSR, vol. 20, September 1959.
11. A. J. Viterbi, "Phase-locked loop dynamics in the presence of noise by Fokker–Planck techniques," *Proc. IEEE*, vol. 51, December 1963, pp. 1737–1753.
12. M. P. Fitz, "Further results in the unified analysis of digital communication systems," *IEEE Trans. Commun.*, vol. 40, March 1992, pp. 521–532.
13. S. Stein, "Unified analysis of certain coherent and noncoherent binary communication systems," *IEEE Trans. Inf. Theory*, vol. IT-10, January 1964, pp. 43–51.
14. A. J. Viterbi, "Optimum detection and signal selection for partially coherent binary communication," *IEEE Trans. Inf. Theory*, vol. IT-11, April 1965, pp. 239–246.
15. M. K. Simon and D. Divsalar, "Multiple symbol partially coherent detection of MPSK," *IEEE Trans. Commun.*, vol. 42, February/March/April 1994, pp. 430–439.
16. D. Divsalar and M. K. Simon, "Multiple-symbol differential detection of MPSK," *IEEE Trans. Commun.*, vol. 38, March 1990, pp. 300–308.
17. K. M. Mackenthun, Jr., "A fast algorithm for multiple-symbol differential detection of MPSK," *IEEE Trans. Commun.*, vol. 42, February/March/April 1994, pp. 1471–1474.

PART 2

MATHEMATICAL TOOLS

4

ALTERNATIVE REPRESENTATIONS OF CLASSICAL FUNCTIONS

Having characterized and classified the various types of fading channels and modulation/detection combinations that can be communicated over these channels, the next logical consideration is evaluation of the average error probability performance of the receivers of such signals. Before moving on in the next part of the book to a description of these receivers and the details of their performance on the generalized fading channel, we divert our attention to developing a set of mathematical tools that will unify and greatly simplify these evaluations. The key to such a unified approach is the development of alternative representations of two classical mathematical functions (i.e., the Gaussian Q-function and the Marcum Q-function) that characterize the error probability performance of digital signals communicated over the AWGN channel in a form that is analytically more desirable for the fading channel. The specific nature and properties of this desired form will become clear shortly. For the moment, suffice it to say that the canonical forms of the Gaussian and Marcum Q-functions that have been around for many decades and to this day still dominate the literature dealing with error performance evaluation have an intrinsic value in their own right with respect to their relation to well-known probability distributions. What we aim to show, however, is that aside from this intrinsic value, these canonical forms suffer a major disadvantage in situations where the argument(s) of the functions depend on random parameters that require further statistical averaging. Such is the case when evaluating average error probability on the fading channel as well as on many other channels with random disturbances. Herein lies the most significant value of the alternative representations of these functions: namely, their ability to enable simple and in many cases closed-form evaluation of such statistical averages.

4.1 GAUSSIAN Q-FUNCTION

4.1.1 One-Dimensional Case

The one-dimensional Gaussian Q-function (often referred to as the Gaussian probability integral), $Q(x)$, is defined as the complement (with respect to unity) of the cumulative distribution function (CDF) corresponding to the normalized (zero mean, unit variance) Gaussian random variable (RV) X. The canonical representation of this function is in the form of a semi-infinite integral of the corresponding probability density function (PDF), namely,

$$Q(x) = \int_x^\infty \frac{1}{\sqrt{2\pi}} \exp\left(-\frac{y^2}{2}\right) dy \qquad (4.1)$$

In principle, the representation of (4.1) suffers from two disadvantages. From a computational standpoint, this relation requires truncation of the upper infinite limit when using numerical integral evaluation or algorithmic techniques. More important, however, the presence of the argument of the function as the lower limit of the integral poses analytical difficulties when this argument depends on other random parameters that ultimately require statistical averaging over their probability distributions. For the pure AWGN channel, only the first of the two disadvantages comes into play which ordinarily poses little difficulty and therefore accounts for the popularity of this form of the Gaussian Q-function in the performance evaluation literature. However, for channels perturbed by other disturbances, in particular the fading channel, the second disadvantage plays an important role since, as we shall see later, the argument of the Q-function depends, among other parameters, on the random fading amplitudes of the various received signal components. Thus, to evaluate the *average* error probability in the presence of fading, one must average the Q-function over the fading amplitude distributions. It is primarily this second disadvantage, namely, the inability to average analytically over one or more random variables when they appear in the lower limit of an integral, that serves as the primary motivation for seeking alternative representations of this and similar functions. Clearly, then, what would be more desirable in such evaluations would be to have a form for $Q(x)$ wherein the argument of the function is in neither the upper nor the lower limit of the integral and furthermore, appears in the integrand as the argument of an elementary function (e.g., an exponential). Still more desirable would be a form wherein the argument-independent limits are finite. In what follows, any function that has the two properties above will be said to be in the *desired form*.

A number of years ago, Craig [1] cleverly showed that evaluation of the average probability of error for the two-dimensional AWGN channel could be considerably simplified by choosing the origin of coordinates for each decision region as that defined by the *signal* vector as opposed to using a fixed coordinate system origin for all decision regions derived from the *received* vector. This shift in vector space coordinate systems allowed the integrand of the two-dimensional integral describing the conditional (on the transmitted signal) probability of error to be independent of the transmitted signal. A by-product of Craig's work was a definite integral form for the Gaussian Q-function, which was in the desired form.[1]

In particular, $Q(x)$ of (4.1) could also now be defined (but only for $x \geq 0$) by

$$Q(x) = \frac{1}{\pi} \int_0^{\pi/2} \exp\left(-\frac{x^2}{2\sin^2\theta}\right) d\theta \qquad (4.2)$$

The form in (4.2) is not readily obtainable by a change of variables directly in (4.1). However, by first extending (4.1) to two dimensions (x and y) where one of the dimensions (y) is integrated over the half plane, a change of variables from rectangular to polar coordinates readily produces (4.2). Furthermore, (4.2) can be obtained directly by a straightforward change of variables of a standard known integral involving $Q(x)$, in particular [5, Eq. (3.363.2)]. Both of these techniques for arriving at (4.2) are described in Appendix 4A. Yet another derivation of (4.2) is given in Ref. 6 and is based on the fact that since the product of two independent random variables, one of which is a Rayleigh and the other a sinusoidal random process with random phase, is a Gaussian random variable, determining the CDF of this product variable is equivalent to evaluating the Gaussian Q-function.

Based on our previous discussion, it is clear that $Q(x)$ of (4.2) is in the desired form, that is, in addition to the advantage of having finite integration limits independent of the argument of the function, x, it has the further advantage that *the integrand now has a Gaussian form with respect to x!* We shall see in Chapter 5 that this exponential dependence of the integrand on the argument of the Q-function will play a very important role in simplifying the evaluation of performance results for coherent communication over generalized fading channels. Before exploiting this property of (4.2) in great detail, however, we wish to give further insight into the alternative definition of the Gaussian Q-function with regard to how it relates to the well-known Chernoff bound.

Note that the maximum of the integrand in (4.2) occurs when $\theta = \pi/2$ [i.e., the integrand achieves its maximum value, namely, $\exp(-x^2/2)$, at the upper limit]. Thus, replacing the integrand by its maximum value, we immediately get the well-known upper bound on $Q(x)$, namely, $Q(x) \leq \frac{1}{2}\exp(-x^2/2)$, which is the Chernoff bound. As we shall see on many occasions later in the book, the advantage of this observation is that the form of $Q(x)$ in (4.2) allows manipulations akin to those afforded by the Chernoff bound *but without the necessity of invoking a bound!* In principle, one simply operates on the integrand in the same fashion as if the Q-function had been replaced by the Chernoff bound, and then at the end performs a single integration over the variable θ. For

---

[1] This form of the Gaussian Q-function was earlier implied in the work of Pawula et al. [2] and Weinstein [3]. The earliest reference to this form of the Gaussian Q-function found by the authors appeared in a classified report (which has since become unclassified) by Nuttall [4]. The relation given there is actually for the complementary error function, which is related to the Gaussian Q-function by $\text{erfc}(x) = 2Q(\sqrt{2}x)$.

4.1.2 Two-Dimensional Case

The normalized two-dimensional Gaussian probability integral is defined by

$$Q(x_1, y_1; \rho) = \frac{1}{2\pi\sqrt{1-\rho^2}} \int_{x_1}^{\infty} \int_{y_1}^{\infty} \exp\left[-\frac{x^2 + y^2 - 2\rho xy}{2(1-\rho^2)}\right] dx\, dy \quad (4.3)$$

Rewriting (4.3) as

$$Q(x_1, y_1; \rho) = \frac{1}{2\pi\sqrt{1-\rho^2}}$$
$$\times \int_0^{\infty} \int_0^{\infty} \exp\left[-\frac{(x+x_1)^2 + (y+y_1)^2 - 2\rho(x+x_1)(y+y_1)}{2(1-\rho^2)}\right] dx\, dy \quad (4.4)$$

we see that we can interpret this double integral as the probability that a signal vector $\mathbf{s} = (x + x_1)$ received in correlated unit variance Gaussian noise falls in the upper right quadrant of the (x, y) plane. Defining

$$\overline{S} = \sqrt{x_1^2 + y_1^2}, \qquad \phi_s = \tan^{-1}\frac{y_1}{x_1} \quad (4.5)$$

then using the geometry of Fig. 4.1, it is straightforward to show that $Q(x_1, y_1; \rho)$ can be expressed as

$$Q(x_1, y_1; \rho) = \frac{1}{2\pi} \int_0^{\pi/2 - \phi_s} \frac{\sqrt{1-\rho^2}}{1 - \rho \sin 2\theta} \exp\left[-\frac{\overline{S}^2}{2} \frac{1 - \rho \sin 2\theta}{(1-\rho^2)} \frac{\cos^2 \phi_s}{\sin^2 \theta}\right] d\theta$$
$$+ \frac{1}{2\pi} \int_0^{\phi_s} \frac{\sqrt{1-\rho^2}}{1 - \rho \sin 2\theta} \exp\left[-\frac{\overline{S}^2}{2} \frac{1 - \rho \sin 2\theta}{(1-\rho^2)} \frac{\sin^2 \phi_s}{\sin^2 \theta}\right] d\theta \quad (4.6)$$

which using (4.6) simplifies still further to

$$Q(x_1, y_1; \rho) = \frac{1}{2\pi} \int_0^{\pi/2 - \tan^{-1} y_1/x_1} \frac{\sqrt{1-\rho^2}}{1 - \rho \sin 2\theta} \exp\left[-\frac{x_1^2}{2} \frac{1 - \rho \sin 2\theta}{(1-\rho^2)\sin^2 \theta}\right] d\theta$$
$$+ \frac{1}{2\pi} \int_0^{\tan^{-1} y_1/x_1} \frac{\sqrt{1-\rho^2}}{1 - \rho \sin 2\theta} \exp\left[-\frac{y_1^2}{2} \frac{1 - \rho \sin 2\theta}{(1-\rho^2)\sin^2 \theta}\right] d\theta$$
$$\quad (4.7)$$

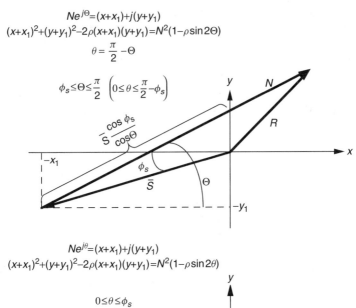

Figure 4.1. Geometry for (4.6).

For the special case of $\rho = 0$, (4.7) simplifies to

$$Q(x_1, y_1; 0) = Q(x_1)Q(y_1)$$
$$= \frac{1}{2\pi} \int_0^{\pi/2 - \tan^{-1} y_1/x_1} \exp\left(-\frac{x_1^2}{2\sin^2\theta}\right) d\theta$$
$$+ \frac{1}{2\pi} \int_0^{\tan^{-1} y_1/x_1} \exp\left(-\frac{y_1^2}{2\sin^2\theta}\right) d\theta \qquad (4.8)$$

In addition, when $x_1 = y_1 = x$, we have

$$Q(x, x; 0) = Q^2(x) = \frac{1}{\pi} \int_0^{\pi/4} \exp\left(-\frac{x^2}{2\sin^2\theta}\right) d\theta \qquad (4.9)$$

which is a single-integral form for the square of the Gaussian Q-function.[2] The form of the result in (4.9) can also be obtained directly from (4.1) by squaring the latter, rewriting it as a double integral of a two-dimensional Gaussian PDF, and then converting from rectangular to polar coordinates (see Appendix 4A). Comparing (4.9) with (4.2), we see that to compute the *square* of the one-dimensional Gaussian probability integral, one integrates the same integrand but only over the first half of the domain.

4.2 MARCUM Q-FUNCTION

Motivated by the form of the alternative Gaussian Q-function in (4.2), one questions whether a similar form is possible for the generalized Marcum Q-function [8], which as we shall see in later chapters is common in performance results for communication problems dealing with partially coherent, differentially coherent, and noncoherent detection. We now present the steps leading up to this desirable form and then show how it offers the same advantages as the alternative representation of the Gaussian Q-function. For simplicity of the presentation, we shall first demonstrate the approach for the first-order ($m = 1$) Marcum Q-function and then generalize to the mth-order function, where in general m can be noninteger as well as integer. The derivations and specific forms that will be derived can be found in Ref. 9, with similar derivations and forms found in Ref. 10.

4.2.1 First-Order Marcum Q-Function

The first-order Marcum Q-function, $Q_1(s, \sqrt{y})$, is defined as the complement (with respect to unity) of the CDF corresponding to the normalized noncentral chi-square random variable, $Y = \sum_{k=1}^{2} X_k^2$, whose canonical representation is in the form of a semi-infinite integral of the corresponding probability density function (PDF), namely,[3]

$$Q_1(s, \sqrt{y}) = \int_{\sqrt{y}}^{\infty} x \exp\left(-\frac{x^2 + s^2}{2}\right) I_0(sx)\,dx \qquad (4.10)$$

where s^2 is referred to as the *noncentrality parameter*. Also, for simplicity of notation, we shall replace the arguments s and $\sqrt{y}$ in (4.10) by α and β,

---

[2] This result can also be obtained from Lebedev [7, Chap 2, Prob. 6] after making the change of variables $\theta = \pi/2 - \tan^{-1} t$.
[3] It is common in the literature to omit the "1" subscript on the Marcum Q-function when referring to the first-order function. For the purpose of clarity and distinction from the generalized (mth-order) Marcum Q-function to be introduced shortly, we shall maintain the subscript notation.

respectively, in which case (4.10) is rewritten in the more common form[4]

$$Q_1(\alpha, \beta) = \int_\beta^\infty x \exp\left(-\frac{x^2 + \alpha^2}{2}\right) I_0(\alpha x)\, dx \qquad (4.11)$$

Using integration by parts, it has also been shown [12,13] that the first-order Marcum Q-function has the series form

$$Q_1(\alpha, \beta) = \exp\left(-\frac{\alpha^2 + \beta^2}{2}\right) \sum_{k=0}^\infty \left(\frac{\alpha}{\beta}\right)^k I_k(\alpha\beta)$$

$$= \exp\left[-\frac{\beta^2}{2}(1 + \zeta^2)\right] \sum_{k=0}^\infty \zeta^k I_k(\beta^2 \zeta) \qquad (4.12)$$

where $\zeta \triangleq \alpha/\beta$. The reason for introducing the parameter ζ to represent the ratio of the arguments of the Marcum Q-function is in the same sense that the definition in (4.10) has one argument that represents the true argument of the function (i.e., $\sqrt{y}$), whereas the second argument (i.e., s) is a parameter. More insight into the significance of ζ in the digital communications application and its dependence on the modulation/detection form is given in Chapter 5. Suffice it to say at the moment that in terms of the analogy with Craig's result, we are attempting to express the Marcum Q-function as an integral with finite limits and an integrand that is a Gaussian function of β.

---

[4] It is interesting to note that the complement (with respect to unity) of the first-order Marcum Q-function can be looked upon as a special case of the incomplete Toronto function [11, pp. 227–228], which finds its roots in the radar literature and is defined by

$$T_B(m, n, r) = 2r^{n-m+1} e^{-r^2} \int_0^B t^{m-n} e^{-t^2} I_n(2rt)\, dt.$$

In particular, we have

$$T_{\beta/\sqrt{2}}\left(1, 0, \frac{\alpha}{\sqrt{2}}\right) = 1 - Q_1(\alpha, \beta).$$

Furthermore, as $\beta \to \infty$, $Q_1(\alpha, \beta)$ can be related to the Gaussian Q-function as follows. Using the asymptotic (for large argument) form of the zero-order modified Bessel function of the first kind, we get [4, Eq. (A-27)]

$$Q_1(\alpha, \beta) \simeq \int_\beta^\infty x \exp\left(-\frac{x^2 + \alpha^2}{2}\right) \frac{\exp(\alpha x)}{\sqrt{2\pi\alpha x}}\, dx$$

$$\simeq \sqrt{\frac{\beta}{\alpha}} \frac{1}{\sqrt{2\pi}} \int_\beta^\infty \exp\left[-\frac{(x-\alpha)^2}{2}\right] dx = \sqrt{\frac{\beta}{\alpha}} Q(\beta - \alpha)$$

The modified Bessel function of kth order can be expressed as the integral [5, Eqs. (8.406.3) and (8.411.1)]

$$I_k(z) = \frac{1}{2\pi} \int_{-\pi}^{\pi} (-je^{-j\theta})^k e^{-z\sin\theta} d\theta \tag{4.13}$$

where $j = \sqrt{-1}$ and it is clear that the imaginary part of the right-hand side of (4.13) must be equal to zero [since $I_k(z)$ is a real function of the real argument z]. Although (4.13) is not restricted to values of ζ less than unity, to arrive at the alternative representation of the Marcum Q-function it will be convenient to make this assumption. (Shortly we shall give an alternative series form from which an alternative representation can be derived for the case where the ratio α/β is greater than unity.) Thus, assuming in (4.13) that $0 \leq \zeta < 1$, after substitution in (4.12) we obtain

$$Q_1(\alpha, \beta) = \exp\left[-\frac{\beta^2}{2}(1+\zeta^2)\right] \frac{1}{2\pi} \int_{-\pi}^{\pi} \sum_{k=0}^{\infty} [\zeta(-je^{-j\theta})]^k e^{-\beta^2 \zeta \sin\theta} d\theta$$

$$= \exp\left[-\frac{\beta^2}{2}(1+\zeta^2)\right] \frac{1}{2\pi} \int_{-\pi}^{\pi} \frac{1}{1+\zeta(je^{-j\theta})} e^{-\beta^2 \zeta \sin\theta} d\theta \tag{4.14}$$

Simplifying the complex factor of the integrand as

$$\frac{1}{1+\zeta(je^{-j\theta})} = \frac{1}{1+\zeta(\sin\theta + j\cos\theta)} = \frac{1+\zeta(\sin\theta - j\cos\theta)}{(1+\zeta\sin\theta)^2 + (\zeta\cos\theta)^2}$$

$$= \frac{1+\zeta(\sin\theta - j\cos\theta)}{1+2\zeta\sin\theta + \zeta^2} \tag{4.15}$$

and recognizing again that the imaginary part of (4.15) must result in a zero integral [since $Q_1(\alpha, \beta)$ is real], substituting (4.15) into (4.14) gives the final result

$$Q_1(\alpha, \beta) = Q_1(\beta\zeta, \beta) = \frac{1}{2\pi} \int_{-\pi}^{\pi} \frac{1+\zeta\sin\theta}{1+2\zeta\sin\theta + \zeta^2}$$
$$\times \exp\left[-\frac{\beta^2}{2}(1+2\zeta\sin\theta + \zeta^2)\right] d\theta, \quad \beta > \alpha \geq 0 \quad (0 \leq \zeta < 1) \tag{4.16}$$

which is in the desired form of a single integral with finite limits and an integrand that is bounded and well behaved over the interval $-\pi \leq \theta \leq \pi$ and is Gaussian in the argument β.

We observe from (4.16) that ζ is restricted to be less than unity (i.e., $\alpha \neq \beta$). The reason for this stems from the closed form used for the geometric series in (4.14), which, strictly speaking, is valid only when $\zeta < 1$. This special case, which has limited interest in communication performance applications, has been evaluated [14, Eq. (A-3-2)] and has the closed-form result

$$Q_1(\alpha, \alpha) = \frac{1 + \exp(-\alpha^2)I_0(\alpha^2)}{2} \qquad (4.17)$$

For the case $\alpha > \beta \geq 0$, the appropriate series form is [12,13][5]

$$Q_1(\alpha, \beta) = 1 - \exp\left(-\frac{\alpha^2 + \beta^2}{2}\right) \sum_{k=1}^{\infty} \left(\frac{\beta}{\alpha}\right)^k I_k(\alpha\beta)$$

$$= 1 - \exp\left[-\frac{\alpha^2}{2}(1 + \zeta^2)\right] \sum_{k=1}^{\infty} \zeta^k I_k(\alpha^2\zeta) \qquad (4.18)$$

whereupon an analogous development to that leading up to (4.16) would yield the result[6]

$$Q_1(\alpha, \beta) = Q_1(\alpha, \alpha\zeta) = 1 + \frac{1}{2\pi} \int_{-\pi}^{\pi} \frac{\zeta^2 + \zeta \sin\theta}{1 + 2\zeta \sin\theta + \zeta^2}$$

$$\times \exp\left[-\frac{\alpha^2}{2}(1 + 2\zeta \sin\theta + \zeta^2)\right] d\theta, \qquad \alpha > \beta \geq 0 \quad (0 \leq \zeta < 1) \qquad (4.19)$$

where now $\zeta \stackrel{\Delta}{=} \beta/\alpha < 1$. Once again the expression in (4.19) is a single integral with finite limits and an integrand that is bounded and well behaved over the interval $-\pi \leq \theta \leq \pi$ and is Gaussian in one of the arguments, in this case, α. Aside from its analytical desirability in the applications discussed in later chapters, the form of (4.16) and (4.19) is also computationally desirable relative to other methods suggested previously by Parl [16] and Cantrell and Ojha [17] for numerical evaluation of the Marcum Q-function.

The results in (4.16) and (4.19) can be put in a form with a more reduced integration interval. In particular, using the symmetry properties of the trigonometric functions over the intervals $(-\pi, 0)$ and $(0, \pi)$, we obtain the alternative forms

$$Q_1(\alpha, \beta) = Q_1(\beta\zeta, \beta) = \frac{1}{\pi} \int_0^{\pi} \frac{1 \pm \zeta \cos\theta}{1 \pm 2\zeta \cos\theta + \zeta^2}$$

$$\times \exp\left[-\frac{\beta^2}{2}(1 \pm 2\zeta \cos\theta + \zeta^2)\right] d\theta, \qquad \beta > \alpha \geq 0 \quad (0 \leq \zeta < 1) \qquad (4.20)$$

---

[5] We note that (4.18) is valid even if $\alpha < \beta$, but for our purpose the series form given in (4.12) is more convenient for this case.

[6] At first glance it might appear from (4.19) that the Marcum Q-function can exceed unity. However, the integral in (4.19) is always less than or equal to zero. It should also be noted that the results in (4.16) and (4.19) can also be obtained from the work of Pawula [15] dealing with the relation between the Rice Ie-function and the Marcum Q-function. In particular, equating Eqs. (2a) and (2c) of Ref. 15 and using the integral representation of the zero-order Bessel function obtained from (4.13) with $k = 0$ in the latter of the two equations, one can, with an appropriate change of variables, arrive at (4.16) and (4.19).

and

$$Q_1(\alpha, \beta) = Q_1(\alpha, \alpha\zeta) = 1 + \frac{1}{\pi} \int_0^\pi \frac{\zeta^2 \pm \zeta \cos\theta}{1 \pm 2\zeta \cos\theta + \zeta^2}$$

$$\times \exp\left[-\frac{\alpha^2}{2}(1 \pm 2\zeta \cos\theta + \zeta^2)\right] d\theta, \qquad \alpha > \beta \geq 0 \quad (0 \leq \zeta < 1)$$

(4.21)

Since, as we shall soon see, for the generalized (mth-order) Marcum Q-function the reduced integration interval form is considerably more complex than the form between symmetrical $(-\pi, \pi)$ limits, we shall tend to use (4.16) and (4.19) when dealing with the applications.

As a simple check on the validity of (4.16) and (4.19), we examine the limiting cases $Q_1(0, \beta)$ and $Q_1(\alpha, 0)$. Letting $\zeta = 0$ in (4.16), we immediately have the well-known result

$$Q_1(0, \beta) = \exp\left(-\frac{\beta^2}{2}\right) \tag{4.22}$$

Similarly, letting $\zeta = 0$ in (4.19) gives

$$Q_1(\alpha, 0) = 1 \tag{4.23}$$

Simple upper and lower bounds on $Q_1(\alpha, \beta)$ can be obtained in the same manner that the Chernoff bound on the Gaussian Q-function was obtained from (4.2). In particular, for $\beta > \alpha \geq 0$, we observe that the maximum and minimum of the integrand in (4.16) occurs for $\theta = -\pi/2$ and $\theta = \pi/2$, respectively. Thus, replacing the integrand by its maximum and minimum values leads to the upper and lower "Chernoff-type" bounds

$$\frac{1}{1+\zeta} \exp\left[-\frac{\beta^2(1+\zeta)^2}{2}\right] \leq Q_1(\beta\zeta, \beta) \leq \frac{1}{1-\zeta} \exp\left[-\frac{\beta^2(1-\zeta)^2}{2}\right] \quad (4.24a)$$

or equivalently,

$$\frac{\beta}{\beta+\alpha} \exp\left[-\frac{(\beta+\alpha)^2}{2}\right] \leq Q_1(\alpha, \beta) \leq \frac{\beta}{\beta-\alpha} \exp\left[-\frac{(\beta-\alpha)^2}{2}\right] \quad (4.24b)$$

which, in view of (4.22), are asymptotically tight as $\alpha \to 0$.

For $\alpha > \beta \geq 0$, the integrand in (4.19) has a minimum at $\theta = -\pi/2$ and a maximum at $\theta = \pi/2$. Since the maximum of the integrand, $[\zeta/(1+\zeta)]\exp[-\alpha^2(1+\zeta)^2/2]$, is always positive, the upper bound obtained by replacing the integrand by this value would exceed unity and hence be useless. On the other hand, the minimum of the integrand, $-[\zeta/(1-\zeta)]\exp[-\alpha^2(1-\zeta)^2/2]$ is always negative. Hence a lower Chernoff-type bound on $Q_1(\alpha, \beta)$ is given by[7]

$$1 - \frac{\zeta}{1-\zeta} \exp\left[-\frac{\alpha^2(1-\zeta)^2}{2}\right] \leq Q_1(\alpha, \alpha\zeta) \tag{4.25a}$$

or equivalently,

$$1 - \frac{\alpha}{\alpha-\beta} \exp\left[-\frac{(\alpha-\beta)^2}{2}\right] \leq Q_1(\alpha, \beta) \tag{4.25b}$$

Another alternative and in some sense simpler form of the first-order Marcum Q-function was recently disclosed in Ref. 18. This form dispenses with the trigonometric factor that precedes the exponential in the integrands of (4.16) and (4.19) in favor of the sum of two purely exponential integrands each still having the desired dependence on β or α as appropriate. In particular, with a change in notation suitable to that used previously in this chapter, the results obtained in Ref. 18 can be expressed as follows:

$$Q_1(\alpha, \beta) = Q_1(\beta\zeta, \beta) = \frac{1}{4\pi} \int_{-\pi}^{\pi} \left\{ \exp\left[-\frac{\beta^2}{2}(1 + 2\zeta \sin\theta + \zeta^2)\right] \right.$$
$$\left. + \exp\left[-\frac{\beta^2}{2}\left(\frac{(1-\zeta^2)^2}{1 + 2\zeta \sin\theta + \zeta^2}\right)\right] \right\} d\theta, \quad \beta \geq \alpha \geq 0 \quad (0 \leq \zeta \leq 1) \tag{4.26}$$

$$Q_1(\alpha, \beta) = Q_1(\alpha, \alpha\zeta) = 1 + \frac{1}{4\pi} \int_{-\pi}^{\pi} \left\{ \exp\left[-\frac{\alpha^2}{2}(1 + 2\zeta \sin\theta + \zeta^2)\right] \right.$$
$$\left. - \exp\left[-\frac{\alpha^2}{2}\left(\frac{(1-\zeta^2)^2}{1 + 2\zeta \sin\theta + \zeta^2}\right)\right] \right\} d\theta, \quad \alpha \geq \beta \geq 0 \quad (0 \leq \zeta \leq 1) \tag{4.27}$$

or equivalently, in the reduced forms analogous to (4.20) and (4.21):

$$Q_1(\alpha, \beta) = Q_1(\beta\zeta, \beta) = \frac{1}{2\pi} \int_0^{\pi} \left\{ \exp\left[-\frac{\beta^2}{2}(1 \pm 2\zeta \cos\theta + \zeta^2)\right] \right.$$
$$\left. + \exp\left[-\frac{\beta^2}{2}\left(\frac{(1-\zeta^2)^2}{1 \pm 2\zeta \cos\theta + \zeta^2}\right)\right] \right\} d\theta, \quad \beta \geq \alpha \geq 0 \quad (0 \leq \zeta \leq 1) \tag{4.28}$$

$$Q_1(\alpha, \beta) = Q_1(\alpha, \alpha\zeta) = 1 + \frac{1}{2\pi} \int_0^{\pi} \left\{ \exp\left[-\frac{\alpha^2}{2}(1 \pm 2\zeta \cos\theta + \zeta^2)\right] \right.$$
$$\left. - \exp\left[-\frac{\alpha^2}{2}\left(\frac{(1-\zeta^2)^2}{1 \pm 2\zeta \cos\theta + \zeta^2}\right)\right] \right\} d\theta, \quad \alpha \geq \beta \geq 0 \quad (0 \leq \zeta \leq 1) \tag{4.29}$$

[7] Clearly, since $Q_1(\alpha, \beta)$ can never be negative, the lower bound of (4.25a) or (4.25b) is only useful for values of the arguments that result in a nonnegative value.

Since the first exponential integrand in each of (4.26) through (4.29) is identical to the exponential integrand in the corresponding equations (4.16), (4.19), (4.20), and (4.21), we can look upon the second exponential in the integrands of the former group of equations as compensating for the lack of the trigonometric multiplying factor in the integrands of the latter equation group.

The forms of the Marcum Q-function in (4.26) and (4.27) [or (4.28) and (4.29)] immediately allow obtaining tighter upper and lower bounds of this function than those in (4.24) and (4.25). In particular, once again recognizing that for $\beta > \alpha \geq 0$ the maximum and minimum of the first exponential integrand in (4.26) occurs for $\theta = -\pi/2$ and $\theta = \pi/2$, respectively, and vice versa for the second exponential integrand, we immediately obtain[8]

$$\exp\left[-\frac{\beta^2(1+\zeta)^2}{2}\right] \leq Q_1(\beta\zeta, \beta) \leq \exp\left[-\frac{\beta^2(1-\zeta)^2}{2}\right] \quad (4.30a)$$

or equivalently,

$$\exp\left[-\frac{(\beta+\alpha)^2}{2}\right] \leq Q_1(\alpha, \beta) \leq \exp\left[-\frac{(\beta-\alpha)^2}{2}\right] \quad (4.30b)$$

Making a similar recognition in (4.27), then for $\alpha > \beta \geq 0$ we obtain the lower bound

$$1 - \frac{1}{2}\left\{\exp\left[-\frac{\alpha^2(1-\zeta)^2}{2}\right] - \exp\left[-\frac{\alpha^2(1+\zeta)^2}{2}\right]\right\} \leq Q_1(\alpha, \alpha\zeta) \quad (4.31a)$$

or equivalently,[9]

$$1 - \frac{1}{2}\left\{\exp\left[-\frac{(\alpha-\beta)^2}{2}\right] - \exp\left[-\frac{(\alpha+\beta)^2}{2}\right]\right\} \leq Q_1(\alpha, \beta) \quad (4.31b)$$

---

[8] It has been pointed out to the authors by W. F. McGee of Ottawa, Canada that the same tighter bounds can be obtained from (4.16) by upper and lower bounding only the exponential factor in the integrand (thus making it independent of the integration variable θ) and then recognizing that the integral of the remaining factor of the integrand can be obtained in closed form and evaluates to unity. We point out to the reader that this procedure of only upper and lower bounding the exponential is valid when the remaining factor is positive over the entire domain of the integral as is the case in (4.16).

[9] Note that the upper bound in this case would become

$$Q_1(\alpha, \beta) \leq 1 + \frac{1}{2}\left\{\exp\left[-\frac{(\alpha-\beta)^2}{2}\right] - \exp\left[-\frac{(\alpha+\beta)^2}{2}\right]\right\}$$

which exceeds unity and is thus not useful.

We note that the bounds in (4.31a) and (4.31b) cannot be obtained directly from (4.19) by lower bounding the exponential in the integrand since the factor that precedes it is not positive over the entire domain of the integral.

Before concluding this section, we alert the reader to the inclusion of the endpoint $\alpha = \beta$ ($\zeta = 1$) in the alternative representations of (4.26) through (4.29), all of which yield the value of $Q_1(\alpha, \alpha)$ in (4.17). This is in contrast to the alternative representation pairs (4.16), (4.19) or (4.20), (4.21), which yield different limits as α approaches β (ζ approaches 1) from the left and right, respectively. The reason for these different left and right limits [the arithmetic average of which does in fact produce the result in (4.17)] is again tied to the fact that these representations rely on the convergence of a geometric series which, strictly speaking, is not convergent at the point $\zeta = 1$. On the other hand, the derivation of the representations in (4.26) through (4.29) is based on a different approach [18] and as such are continuous across the point $\zeta = 1$. Thus, even in the neighborhood of $\zeta = 1$, one would anticipate better behavior from these representations.

4.2.2 Generalized (*m*th-Order) Marcum Q-Function

The generalized Marcum Q-function is defined analogous to (4.10) by

$$Q_m(s, \sqrt{y}) = \frac{1}{s^{m-1}} \int_{\sqrt{y}}^{\infty} x^m \exp\left(-\frac{x^2 + s^2}{2}\right) I_{m-1}(sx)\, dx \qquad (4.32)$$

or, equivalently,[10]

$$Q_m(\alpha, \beta) = \frac{1}{\alpha^{M-1}} \int_{\beta}^{\infty} x^m \exp\left(-\frac{x^2 + \alpha^2}{2}\right) I_{m-1}(\alpha x)\, dx \qquad (4.33)$$

---

[10] The complement of the generalized Marcum Q-function can also be viewed as a special case of the incomplete Toronto function. In particular,

$$T_{\beta/\sqrt{2}}\left(2m - 1, m - 1, \frac{\alpha}{\sqrt{2}}\right) = 1 - Q_m(\alpha, \beta)$$

Furthermore, as $\beta \to \infty$, $Q_m(\alpha, \beta)$ can be related to the Gaussian Q-function in the same manner as was done for the first-order Marcum Q-function. Specifically, since the asymptotic (for large argument) form of the kth-order modified Bessel function of the first kind is independent of the order, then

$$Q_m(\alpha, \beta) \simeq \int_{\beta}^{\infty} x \left(\frac{x}{\alpha}\right)^{m-1} \exp\left(-\frac{x^2 + \alpha^2}{2}\right) \frac{\exp(\alpha x)}{\sqrt{2\pi \alpha x}}\, dx$$

$$\simeq \left(\frac{\beta}{\alpha}\right)^{m-1/2} \frac{1}{\sqrt{2\pi}} \int_{\beta}^{\infty} \exp\left[-\frac{(x - \alpha)^2}{2}\right] dx$$

$$= \left(\frac{\beta}{\alpha}\right)^{m-1/2} Q(\beta - \alpha)$$

where for m integer, the canonical form in (4.32) has the significance of being the complement (with respect to unity) of the CDF corresponding to the normalized noncentral chi-square random variable, $Y = \sum_{k=1}^{m+1} X_k^2$. It would be desirable to obtain integral forms analogous to (4.16) and (4.19) to represent the generalized Marcum Q-function regardless of whether m is integer or noninteger. Unfortunately, this has been shown to be possible only for the case of m integer, at least in the sense of an exact representation [9,10]. As we shall see from the derivation of these forms, however, the ones derived for m integer are also applicable in an approximate sense to the case of m noninteger in certain regions of the function's arguments. Thus, we begin by proceeding with an approach analogous to that taken in arriving at (4.16) and (4.19) without restricting m to be integer, applying this restriction only when it becomes necessary. The details are as follows.

Applying integration by parts to (4.33) with $u = x^{m-1} I_{m-1}(\alpha x)$ and $dv = x \exp[-(x^2 + \alpha^2)/2] dx$ and using the Bessel function recursion relation $I_{m-1}(x) - I_{m+1}(x) = (2m/x) I_m(x)$ [19, Eq. (9.6.26)], it is straightforward to show that the generalized Marcum Q-function satisfies the recursion relation

$$Q_m(\alpha, \beta) = \frac{\beta}{\alpha} \exp\left(-\frac{\alpha^2 + \beta^2}{2}\right) I_{m-1}(\alpha\beta) + Q_{m-1}(\alpha, \beta) \quad (4.34)$$

Recognizing that regardless of the values of α and β, $Q_{-\infty}(\alpha, \beta) = 0$ and $Q_\infty(\alpha, \beta) = 1$, then iterating (4.34) in both the forward and backward directions gives the series forms

$$Q_m(\alpha, \beta) = \exp\left(-\frac{\alpha^2 + \beta^2}{2}\right) \sum_{r=1-m}^{\infty} \left(\frac{\alpha}{\beta}\right)^r I_{-r}(\alpha\beta) \quad (4.35)$$

and

$$Q_m(\alpha, \beta) = 1 - \exp\left(-\frac{\alpha^2 + \beta^2}{2}\right) \sum_{r=m}^{\infty} \left(\frac{\beta}{\alpha}\right)^r I_r(\alpha\beta) \quad (4.36)$$

Note that when m is integer, the values of the summation index r are also integer, and since in this case $I_{-r}(x) = I_r(x)$, we can rewrite (4.35) as

$$Q_m(\alpha, \beta) = \exp\left(-\frac{\alpha^2 + \beta^2}{2}\right) \sum_{r=1-m}^{\infty} \left(\frac{\alpha}{\beta}\right)^r I_r(\alpha\beta) \quad (4.37)$$

Equations (4.36) and (4.37) are the series forms of the generalized Marcum Q-function that are found in the literature and apply when m is integer. When m is noninteger, the values of the summation index r are also noninteger, and since in this case $I_{-r}(x) \neq I_r(x)$, then (4.37) is no longer valid; instead one must use

(4.35). Note that (4.36) is valid for m integer or m noninteger and together with (4.37) reduce to (4.18) and (4.12), respectively, for $m = 1$.

Although the discussion above appears to make a mute point, it is important in the approach taken in Ref. 9 since certain trigonometric manipulations applied there when deriving the alternative representation of the Marcum Q-function from the series representation hold only for m integer. Despite this fact, however, if the $I_r(x)$ function could still be represented exactly by the integral $I_r(x) = (1/2\pi) \int_{-\pi}^{\pi} (-je^{-j\theta})^r e^{-x\sin\theta} d\theta$ [which is the same as (4.13) with r substituted for k], then even though the summation indices in (4.36) and (4.37) are noninteger, adjacent values are separated by unity and the same geometric series manipulations could be performed as were done previously for the first-order Marcum Q-function. Unfortunately, however, the integral representation of $I_r(x)$ above is approximately valid only when its argument x is large irrespective of the value of r, and thus the steps that follow and the results that ensue are only approximate when m, the order of the Marcum Q-function, is noninteger. In what follows, however, we shall proceed as though this integral representation is exact (which it is for r integer, or equivalently, m integer) with the understanding that the final integral representations obtained for the mth-order Marcum Q-function will be exact for m integer and approximate (for large values of the argument β or α as appropriate) for m noninteger.

As discussed previously with regard to the application of the alternative representation, it is convenient to introduce the parameter $\zeta < 1$ to represent the ratio of the smaller to the larger of the two variables of the Marcum Q-function. We can therefore rewrite (4.35) and (4.36) as

$$Q_m(\beta\zeta, \beta) = \exp\left[-\frac{\beta^2}{2}(1+\zeta^2)\right] \sum_{r=1-m}^{\infty} \zeta^r I_{-r}(\beta^2\zeta), \quad 0^+ \leq \zeta \stackrel{\Delta}{=} \alpha/\beta < 1 \quad (4.38)$$

$$Q_m(\alpha, \alpha\zeta) = 1 - \exp\left[-\frac{\alpha^2}{2}(1+\zeta^2)\right] \sum_{r=m}^{\infty} \zeta^r I_r(\alpha^2\zeta), \quad 0 \leq \zeta \stackrel{\Delta}{=} \beta/\alpha < 1 \quad (4.39)$$

Letting $N < m < N+1$ (i.e., N is the largest integer less than or equal to m), substituting the integral form of the modified Bessel function in (4.38) gives

$$Q_m(\beta\zeta, \beta) = \exp\left[-\frac{\beta^2}{2}(1+\zeta^2)\right] \frac{1}{2\pi} \int_{-\pi}^{\pi} \sum_{r=1-m}^{\infty} \zeta^r (-je^{-j\theta})^{-r} e^{-\beta^2\zeta\sin\theta} d\theta$$

$$= \exp\left[-\frac{\beta^2}{2}(1+\zeta^2)\right] \frac{1}{2\pi} \int_{-\pi}^{\pi} \left[\sum_{r=1-m}^{N-m} (e^{j(\theta+\pi/2)}\zeta)^r \right.$$

$$\left. + \sum_{r=N-m+1}^{\infty} (e^{j(\theta+\pi/2)}\zeta)^r \right] e^{-\beta^2\zeta\sin\theta} d\theta \quad (4.40)$$

Recognizing as mentioned above that the sums in (4.40) are still geometric series despite the fact that the summation index r does not take on integer values, we obtain

$$Q_m(\beta\zeta, \beta) = \exp\left[-\frac{\beta^2}{2}(1+\zeta^2)\right]\frac{1}{2\pi}$$

$$\times \int_{-\pi}^{\pi}\left[\zeta^{-(m-1)}e^{-j(m-1)(\theta+\pi/2)}\frac{1-\zeta^N e^{jN(\theta+\pi/2)}}{1-\zeta e^{j(\theta+\pi/2)}}\right.$$

$$\left.+ \zeta^{N+1-m}e^{j(N+1-m)(\theta+\pi/2)}\frac{1}{1-\zeta e^{j(\theta+\pi/2)}}\right]e^{-\beta^2\zeta\sin\theta}d\theta \quad (4.41)$$

Since $Q_m(\alpha, \beta)$ is a real function of its arguments, then taking the real part of the right hand side of (4.41) and simplifying results in the desired expression

$$Q_m(\beta\zeta, \beta) = \frac{1}{2\pi}\int_{-\pi}^{\pi}\frac{\zeta^{-(m-1)}\{\cos[(m-1)(\theta+\pi/2)] - \zeta\cos[m(\theta+\pi/2)]\}}{1+2\zeta\sin\theta+\zeta^2}$$

$$\times \exp\left[-\frac{\beta^2}{2}(1+2\zeta\sin\theta+\zeta^2)\right]d\theta, \quad 0^+ \leq \zeta = \alpha/\beta < 1 \quad (4.42)$$

Note that the limit of $Q_m(\beta\zeta, \beta)$ as $\zeta \to 0$ is difficult to evaluate directly from the form in (4.42), which explains the restriction on its region of validity. However, this limit can be evaluated starting with the integral form of (4.33) and using the small argument form of the modified Bessel function, that is,

$$I_\nu(z) \simeq \frac{(z/2)^\nu}{\Gamma(\nu+1)} \quad (4.43)$$

When this is done, the following results:

$$Q_m(0, \beta) = \frac{\Gamma(m, \beta^2/2)}{\Gamma(m)} \quad (4.44)$$

where $\Gamma(\alpha, x)$ is the complementary Gauss incomplete gamma function [5, Eq. (8.350.2)]. Using a particular integral representation of $\Gamma(\alpha, x)$ [20, Eq. (11.10)], then after some changes of variables, $Q_m(0, \beta)$ can be put in the desired form,

$$Q_m(0, \beta) = \frac{\beta^{2m}}{2^{m-1}\Gamma(m)}\int_0^{\pi/2}\frac{\cos\theta}{(\sin\theta)^{1+2m}}\exp\left(-\frac{\beta^2}{2\sin^2\theta}\right)d\theta \quad (4.45)$$

For m integer, the gamma function can be evaluated in closed form [5, Eq. (8.352.2)] and (4.44) reduces to

$$Q_m(0, \beta) = \sum_{n=0}^{m-1} \exp\left(-\frac{\beta^2}{2}\right) \frac{(\beta^2/2)^n}{n!} \tag{4.46}$$

which is a special case of another form of the Marcum Q-function proposed by Dillard [21], namely,

$$Q_m(\alpha, \beta) = \sum_{n=0}^{\infty} \exp\left(-\frac{\alpha^2}{2}\right) \frac{(\alpha^2/2)^n}{n!} \sum_{k=0}^{n+m-1} \exp\left(-\frac{\beta^2}{2}\right) \frac{(\beta^2/2)^k}{k!} \tag{4.47}$$

In a similar fashion, substituting the integral form of the modified Bessel function in (4.39) gives

$$Q_m(\alpha, \alpha\zeta) = 1 - \exp\left[-\frac{\alpha^2}{2}(1+\zeta^2)\right] \frac{1}{2\pi} \int_{-\pi}^{\pi} \sum_{r=m}^{\infty} \zeta^r (-je^{-j\theta})^{-r} e^{-\beta^2 \zeta \sin\theta} d\theta$$

$$= 1 - \exp\left[-\frac{\beta^2}{2}(1+\zeta^2)\right] \frac{1}{2\pi} \int_{-\pi}^{\pi} \sum_{r=m}^{\infty} (e^{j(\theta+\pi/2)}\zeta)^r e^{-\beta^2 \zeta \sin\theta} d\theta$$

$$\tag{4.48}$$

where upon recognizing the sum as a geometric series, we get

$$Q_m(\alpha, \alpha\zeta) = 1 - \exp\left[-\frac{\alpha^2}{2}(1+\zeta^2)\right] \frac{1}{2\pi}$$

$$\times \int_{-\pi}^{\pi} \left[\zeta^{m+1} e^{j(m+1)(\theta+\pi/2)} \frac{1}{1-\zeta e^{j(\theta+\pi/2)}}\right] e^{-\beta^2 \zeta \sin\theta} d\theta \tag{4.49}$$

Finally, taking the real part of the right-hand side of (4.49) and simplifying gives the complementary expression to (4.42), namely,

$$Q_m(\alpha, \alpha\zeta) = 1 - \frac{1}{2\pi} \int_{-\pi}^{\pi} \frac{\zeta^m \{\cos[m(\theta+\pi/2)] - \zeta \cos[(m-1)(\theta+\pi/2)]\}}{1 + 2\zeta \sin\theta + \zeta^2}$$

$$\times \exp\left[-\frac{\alpha^2}{2}(1 + 2\zeta \sin\theta + \zeta^2)\right] d\theta, \qquad 0 \le \zeta = \beta/\alpha < 1 \tag{4.50}$$

For m integer, (4.42) and (4.50) simplify slightly to

$$Q_m(\beta\zeta, \beta) = \frac{1}{2\pi} \int_{-\pi}^{\pi} \frac{(-1)^{(m-1)/2} \zeta^{-(m-1)}[\cos(m-1)\theta + \zeta \sin m\theta]}{1 + 2\zeta \sin\theta + \zeta^2}$$

$$\times \exp\left[-\frac{\beta^2}{2}(1 + 2\zeta \sin\theta + \zeta^2)\right] d\theta, \qquad 0^+ < \zeta = \alpha/\beta < 1, \quad m \text{ odd}$$

$$\tag{4.51}$$

$$Q_m(\beta\zeta, \beta) = \frac{1}{2\pi} \int_{-\pi}^{\pi} \frac{(-1)^{m/2}\zeta^{-(m-1)}[\sin(m-1)\theta - \zeta\cos m\theta]}{1 + 2\zeta\sin\theta + \zeta^2}$$

$$\times \exp\left[-\frac{\beta^2}{2}(1 + 2\zeta\sin\theta + \zeta^2)\right] d\theta, \quad 0^+ < \zeta = \alpha/\beta < 1, \quad m \text{ even}$$

$$Q_m(\alpha, \zeta\alpha) = 1 + \frac{1}{2\pi} \int_{-\pi}^{\pi} \frac{(-1)^{(m-1)/2}\zeta^m[\sin m\theta + \zeta\cos(m-1)\theta]}{1 + 2\zeta\sin\theta + \zeta^2}$$

$$\times \exp\left[-\frac{\alpha^2}{2}(1 + 2\zeta\sin\theta + \zeta^2)\right] d\theta, \quad 0 \le \zeta = \beta/\alpha < 1, \quad m \text{ odd}$$
(4.52)

$$Q_m(\alpha, \zeta\alpha) = \frac{1}{2\pi} \int_{-\pi}^{\pi} \frac{(-1)^{m/2}\zeta^m[\cos m\theta - \zeta\sin(m-1)\theta]}{1 + 2\zeta\sin\theta + \zeta^2}$$

$$\times \exp\left[-\frac{\alpha^2}{2}(1 + 2\zeta\sin\theta + \zeta^2)\right] d\theta, \quad 0 \le \zeta = \beta/\alpha < 1, \quad m \text{ even}$$

which are the forms reported by Simon [9, Eqs. (8) and (10)]. Finally, the limit of (4.50) as $\zeta \to 0$ is easily seen to be $Q_m(\alpha, 0) = 1$, which is in agreement with the similar result in (4.23) for the first-order Marcum Q-function.

As before, we observe from (4.42) and (4.50) that ζ is restricted to be less than unity (i.e., $\alpha \ne \beta$) for the reason mentioned previously relative to the alternative representations of the first-order Marcum Q-function. For m integer, this special case has the closed-form result [10]

$$Q_m(\alpha, \alpha) = \frac{1}{2} + \exp(-\alpha^2)\left[\frac{I_0(\alpha^2)}{2} + \sum_{k=1}^{m-1} I_k(\alpha^2)\right]$$
(4.53)

For m noninteger, the authors have been unable to arrive at an approximate closed-form result.

Finally, we note that the approach taken in Ref. 18 for arriving at the alternative forms for the first-order Marcum Q-function given in (4.26) through (4.29) unfortunately does not produce an equivalent simplification in the case of the mth-order Marcum Q-function. Similarily, upper and lower bounds on the mth-order Marcum Q-function are not readily obtainable by upper and lower bounding the exponential in the integrands of (4.42) and (4.50) since the first factor of these integrands is not positive over the domain of the integral. Thus, throughout the remainder of the book, unless the forms in (4.26) through (4.29) produce a specific analytical advantage, we shall tend to use the alternative forms of the first-order Marcum Q-function function given in (4.16) and (4.19) because of their synergy with the equivalent forms in (4.42) and (4.50) for the mth-order Marcum Q-function.

Despite the fact that upper and lower bounds on the mth-order Marcum Q-function are not readily obtainable from (4.42) and (4.50), it is nevertheless possible [22] for m integer to obtain such bounds by using the upper and lower bounds on the first-order Marcum Q-function given in (4.30a) and (4.30b) together with the recursive relation of (4.34).[11] In particular, (4.34) can first be rewritten as

$$Q_m(\alpha, \beta) = \exp\left(-\frac{\alpha^2 + \beta^2}{2}\right) \sum_{n=1}^{m-1} \left(\frac{\beta}{\alpha}\right)^n I_n(\alpha\beta) + Q_1(\alpha, \beta) \quad (4.54)$$

Now expressing $I_n(z)$ in its integral form analogous to (4.13), that is,

$$I_n(z) = \frac{1}{\pi} \int_0^\pi e^{z \cos\theta} \cos n\theta \, d\theta \quad (4.55)$$

and recognizing that the exponential part of the integrand has maximum and minimum values of e^z and e^{-z}, respectively, then because of the n-fold periodicity of $\cos n\theta$ and the equally spaced (by π/n) regions where $\cos n\theta$ is alternately positive and negative within the interval $0 \le \theta \le \pi$, we can upper bound $I_n(z)$ by[12]

$$I_n(z) \le \frac{n}{2} \left(e^z \frac{1}{\pi} \int_0^{\pi/2n} \cos n\theta \, d\theta + e^{-z} \frac{1}{\pi} \int_{\pi/2n}^{3\pi/2n} \cos n\theta \, d\theta + e^z \frac{1}{\pi} \int_{3\pi/2n}^{2\pi/n} \cos n\theta \, d\theta \right)$$

$$= \frac{e^z - e^{-z}}{\pi}, \quad z \ge 0 \quad (4.56)$$

which is independent of n for $n \ge 1$. This allows the series in (4.54) to be summed as a geometric series that has a closed-form result. Finally, using (4.56) in (4.54) together with the upper bound on $Q_1(\alpha, \beta)$ for $0^+ \le \zeta = \alpha/\beta < 1$ as given by (4.30b), we obtain after some manipulation

$$Q_m(\alpha, \beta) \le \exp\left[-\frac{(\beta - \alpha)^2}{2}\right] + \frac{1}{\pi} \left\{ \exp\left[-\frac{(\beta - \alpha)^2}{2}\right] \right.$$

$$\left. - \exp\left[-\frac{(\beta + \alpha)^2}{2}\right] \right\} \left(\frac{\beta}{\alpha}\right)^{m-1} \left[\frac{1 - (\alpha/\beta)^{m-1}}{1 - \alpha/\beta}\right] \quad (4.57a)$$

or equivalently,

$$Q_m(\beta\zeta, \beta) \le \exp\left[-\frac{\beta^2(1-\zeta)^2}{2}\right] + \frac{1}{\pi} \left\{ \exp\left[-\frac{\beta^2(1-\zeta)^2}{2}\right] \right.$$

$$\left. - \exp\left[-\frac{\beta^2(1+\zeta)^2}{2}\right] \right\} \frac{1}{\zeta^{m-1}} \left(\frac{1 - \zeta^{m-1}}{1 - \zeta}\right) \quad (4.57b)$$

---
[11] We emphasize that we are again looking for simple (exponential-type) bounds recognizing that although these may not be the tightest bounds achievable over all ranges of their arguments, relative to others previously reported in the literature [23], they are particularly useful in the context of evaluating error probability performance over fading channels.
[12] Note that (4.56) is valid for n odd as well as n even.

The first term of (4.57a) or (4.57b) represents the upper bound on the first-order Marcum Q-function, and thus, as would be expected, for $m = 1$ the remaining terms in these equations evaluate to zero.

To obtain the lower bound on $Q_m(\alpha, \beta)$ for $0^+ \leq \zeta = \alpha/\beta < 1$, we can again use the lower bound on $Q_1(\alpha, \beta)$ as given by (4.30b) in (4.54); however, the procedure used to obtain the upper bound on $I_n(z)$ that led to (4.56) would now yield the lower bound

$$I_n(z) \geq \frac{e^{-z} - e^z}{\pi} \tag{4.58}$$

which for $z \geq 0$ is always less than or equal to zero and therefore not useful relative to the simpler lower bound $I_n(z) \geq 0, n \geq 1$. Thus, to get a useful lower bound on $I_n(z)$, we must employ an alternative form of its integral definition, namely [19, Eq. (9.6.18)]

$$I_n(z) = \frac{(z/2)^n}{\sqrt{\pi}\Gamma(n + \frac{1}{2})} \int_0^\pi e^{z \cos \theta} \sin^{2n} \theta \, d\theta \tag{4.59}$$

Once again replacing the exponential factor of the integrand by its minimum value, e^{-z}, we obtain the lower bound

$$I_n(z) \geq \frac{(z/2)^n}{\sqrt{\pi}\Gamma(n + \frac{1}{2})} e^{-z} \int_0^\pi \sin^{2n} \theta \, d\theta \tag{4.60}$$

which using [5, Eqs. (3.621.3) and (8.339.2)] yields

$$I_n(z) \geq \frac{z^n}{(2n)!!} e^{-z} \tag{4.61}$$

Finally, substituting (4.61) in (4.54) and using the lower bound on $Q_1(\alpha, \beta)$ as given by (4.30b) results after some simplification in

$$\exp\left[-\frac{(\beta + \alpha)^2}{2}\right] \sum_{n=0}^{m-1} \frac{(\beta^2/2)^n}{n!} \leq Q_m(\alpha, \beta), \quad 0 \leq \alpha < \beta \tag{4.62}$$

or equivalently,

$$\exp\left[-\frac{\beta^2(1 + \zeta)^2}{2}\right] \sum_{n=0}^{m-1} \frac{(\beta^2/2)^n}{n!} \leq Q_m(\beta\zeta, \beta), \quad 0 \leq \zeta = \alpha/\beta < 1 \tag{4.63}$$

Again the first term (corresponding to $n = 0$) is the lower bound on the first-order Marcum Q-function, and as would be expected, for $m = 1$ there are no other terms in the sum. Also, for $\zeta = 0$, (4.63) becomes equal to the exact result for $Q_m(0, \beta)$ as given by (4.46). Thus one would anticipate that the lower bound would be asymptotically tight for small values of ζ.

For the parameter range $0 \leq \zeta = \beta/\alpha < 1$, we can obtain a lower bound on $Q_m(\alpha, \beta)$ by using the lower bound on the first-order Marcum Q-function as in (4.31b) together with the lower bound on $I_n(z)$ as given by (4.61), which results in

$$1 - \frac{1}{2}\left\{\exp\left[-\frac{(\alpha-\beta)^2}{2}\right] - \exp\left[-\frac{(\alpha+\beta)^2}{2}\right]\right\}$$
$$+ \exp\left[-\frac{(\alpha+\beta)^2}{2}\right]\sum_{n=1}^{m-1}\frac{(\beta^2/2)^n}{n!} \leq Q_m(\alpha, \beta) \qquad (4.64a)$$

or equivalently,

$$1 - \frac{1}{2}\left\{\exp\left[-\frac{\alpha^2(1-\zeta)^2}{2}\right] - \exp\left[-\frac{\alpha^2(1+\zeta)^2}{2}\right]\right\}$$
$$+ \exp\left[-\frac{\alpha^2(1-\zeta)^2}{2}\right]\sum_{n=1}^{m-1}\frac{(\alpha^2\zeta^2/2)^n}{n!} \leq Q_m(\alpha, \alpha\zeta) \qquad (4.64b)$$

Figures 4.2, 4.3, and 4.4 are plots of $Q_1(\alpha, \beta)$, $Q_2(\alpha, \beta)$, and $Q_4(\alpha, \beta)$ versus β together with their upper and lower bounds, as determined from (4.57a) and (4.62) for values of $\alpha = 1, 5$, and 10, respectively. Also illustrated are Chernoff-type upper and lower bounds derived from Ref. 23.[13] We observe that as anticipated the upper bound of (4.57a), corresponding to $\beta > \alpha$ is asymptotically tight, whereas for the same region, the lower bound as given by (4.62) is quite loose and gets looser as α/β increases. Fortunately (we shall see why in later chapters), the reverse is true for the lower bound of (4.64a), corresponding to the region $\alpha > \beta$ (i.e., it is always extremely tight). In the case of (4.64a), the lower bound was examined both with and without the additional term involving the summation, the latter being equivalent to (4.31b). Over the range of values considered, the numerical results that take into account the presence of the extra series term are indistinguishable (when plotted) from those without it. Hence we can conclude that this series term can be dropped without losing tightness on the overall result. This observation will be important in the application discussions that follow in later chapters.

---

[13] It is to be noted that whereas these upper and lower bounds of Ref. 23 are of interest on their own, their regions of validity do not share a common boundary in the α versus β plane, thus prohibiting their use in evaluating upper bounds on expressions containing the difference of two Marcum Q-functions with reversed arguments [i.e., $Q_m(\alpha, \beta) - Q_m(\beta, \alpha)$]. We shall see later in the book that expressions of this type are characteristic of many types of error probability evaluations over fading channels, and thus upper bounding such error probabilities requires an upper bound on the first Q-function and a lower bound on the second, with a boundary between their regions of validity given by $\alpha = \beta$. The bounds presented in this chapter clearly satisfy this requirement, and thus with regard to the primary subject matter of this book, they are the only bounds of interest.

90 ALTERNATIVE REPRESENTATIONS OF CLASSICAL FUNCTIONS

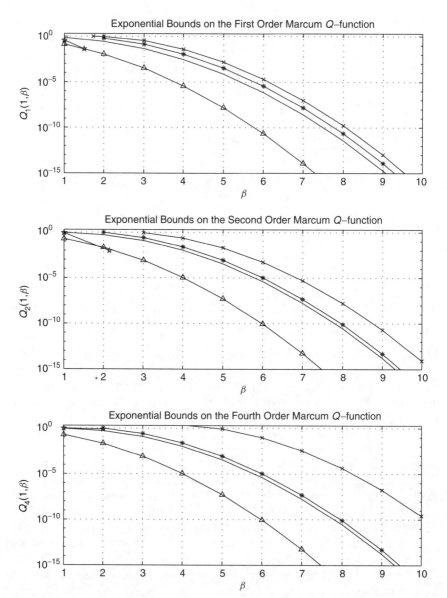

Figure 4.2. Plots of $Q_1(1, \beta)$, $Q_2(1, \beta)$, $Q_4(1, \beta)$, and their bounds versus β: —, Exact; ∗, upper bound (4.57a); ×, Chernoff upper bound from Ref. 23; ☆, Chernoff lower bound from Ref. 23; △, lower bound of (4.62).

4.3 OTHER FUNCTIONS

Before going on to discuss how these alternative representations of the Gaussian and Marcum Q-functions allow for unification and simplification of the evaluation of average error probability performance of digital communication

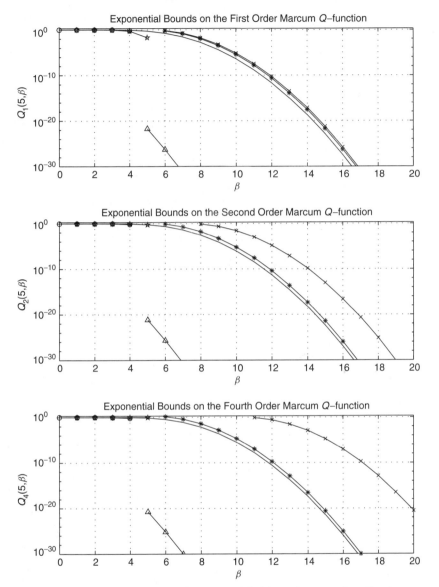

Figure 4.3. Plots of $Q_1(5, \beta)$, $Q_2(5, \beta)$, $Q_4(5, \beta)$, and their bounds versus β. –, Exact; ∗, upper bound of (4.57a); ×, Chernoff upper bound from Ref. 23; ☆, Chernoff lower bound from Ref. 23; △, lower bound of (4.62).

over generalized fading channels, we consider alternative representations of yet two other functions that can be derived from the results above and are also of interest in characterizing this performance.

One function that occurs in the error probability analysis of conventional noncoherent communication systems and also in certain differentially and

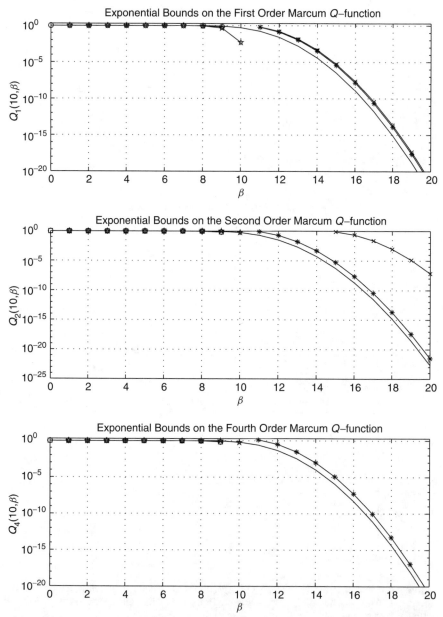

Figure 4.4. Plots of $Q_1(10,\beta)$, $Q_2(10,\beta)$, $Q_4(10,\beta)$, and their bounds versus β. −, Exact; ∗, upper bound of (4.57a); ×, Chernoff upper bound from Ref. 23; ☆, Chernoff lower bound from Ref. 23; △ lower bound of (4.62). Note that the lower bound given by (4.62) and the Chernoff upper bound from Ref. 23 ($m = 4$) are out of the range of the plot.

partially coherent communication systems is $\exp[-(\alpha^2+\beta^2)/2]I_0(\alpha\beta)$, where typically, $\beta > \alpha \geq 0$. Once again defining $\zeta = \alpha/\beta < 1$ and using (4.12), we get a form analogous to (4.16), namely,

$$\exp\left(-\frac{\alpha^2+\beta^2}{2}\right)I_0(\alpha\beta) = \frac{1}{2\pi}\int_{-\pi}^{\pi}\exp\left[-\frac{\beta^2}{2}(1+2\zeta\sin\theta+\zeta^2)\right]d\theta \quad (4.65)$$

A second function that is particularly useful in simplifying the error probability analysis of conventional differentially coherent communication modulations (i.e., M-DPSK) transmitted on the AWGN and fading channels and again has the desirable properties of finite integration limits and a Gaussian integrand was developed by Pawula et al. [2] in the general context of studying the distribution of the phase between two random vectors. In particular, for the M-DPSK application, consider the geometry of Fig. 4.5, where $\mathbf{s}_1 = Ae^{j\phi_1}$ and $\mathbf{s}_2 = Ae^{j\phi_2}$ represent the signal vectors transmitted in successive symbol intervals and $\mathbf{V}_1 = R_1 e^{j\theta_1}$ and $\mathbf{V}_2 = R_2 e^{j\theta_2}$ are the corresponding noisy observations. The components of the zero-mean Gaussian noise vectors that produce $\mathbf{V}_1$ from $\mathbf{s}_1$ and $\mathbf{V}_2$ from $\mathbf{s}_2$ each have variance σ^2 and are uncorrelated. Denoting the angle between the signal vectors by $\Delta\Phi = (\phi_2 - \phi_1)$ modulo 2π and the corresponding angle between the noisy observation vectors by $\psi = (\theta_2 - \theta_1)$ modulo 2π, Pawula et al. [2] defined the function

$$F(\psi) = \frac{\sin(\Delta\phi - \psi)}{4\pi}\int_{-\pi/2}^{\pi/2}\frac{1}{1-\cos(\Delta\phi-\psi)\cos t}$$
$$\times \exp\left\{-\frac{A^2}{2\sigma^2}[1-\cos(\Delta\phi-\psi)\cos t]\right\}dt \quad (4.66)$$

which like a probability distribution function is monotonically increasing in the interval $-\pi \leq \psi \leq \pi$ except for a jump discontinuity at $\psi = \Delta\Phi$, where

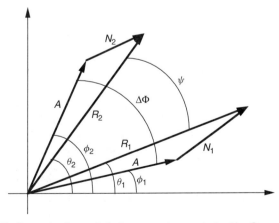

Figure 4.5. Geometry for angle between vectors perturbed by Gaussian noise.

$F(\Delta\Phi^-) - F(\Delta\Phi^+) = -1$. For evaluating the symbol error probability of M-DPSK conditioned on a fixed amplitude A, the special case of $\Delta\Phi = 0$ is of interest since the symmetry of the problem allows one arbitrarily to assume transmission of a zero information phase (i.e., successive transmission of two identical signal vectors). For this case, (4.66) simplifies to

$$F(\psi) = -\frac{\sin\psi}{4\pi} \int_{-\pi/2}^{\pi/2} \frac{1}{1 - \cos\psi \cos t} \exp\left[-\frac{A^2}{2\sigma^2}(1 - \cos\psi \cos t)\right] dt \quad (4.67)$$

Once again notice the similarity in form of (4.66) and (4.67) with the representations of the Gaussian and Marcum Q-functions in (4.2) and (4.16), respectively.

Using the approach taken in Ref. 18 to arrive at the alternative forms of the first-order Marcum Q-function in (4.26) through (4.29), a somewhat simpler form of (4.67) can be obtained as

$$F(\psi) = -\frac{1}{4\pi} \int_{-(\pi-\psi)}^{\pi-\psi} \exp\left(-\frac{A^2}{2\sigma^2} \frac{\sin^2\psi}{1 + \cos\psi \cos t}\right) dt \quad (4.68)$$

Here the trigonometric factor in the integrand of (4.67) is replaced by a different integrand for the exponential as well as integration limits that depend on the argument of the function.

REFERENCES

1. J. W. Craig, "A new, simple and exact result for calculating the probability of error for two-dimensional signal constellations," *IEEE MILCOM'91 Conf. Rec.*, Boston, pp. 25.5.1–25.5.5.
2. R. F. Pawula, S. O. Rice, and J. H. Roberts, "Distribution of the phase angle between two vectors perturbed by Gaussian noise," *IEEE Trans. Commun.*, vol. 30, August 1982, pp. 1828–1841.
3. F. S. Weinstein, "Simplified relationships for the probability distribution of the phase of a sine wave in narrow-band normal noise," *IEEE Trans. Inf. Theory*, vol. IT-20, September 1974, pp. 658–661.
4. A. Nuttall, Some Integrals Involving the Q-Function, Tech. Rep. 4297, Naval Underwater Systems Center, New London, CT, April 17, 1972.
5. I. S. Gradshteyn and I. M. Ryzhik, *Table of Integrals, Series, and Products*, 5th ed. San Diego, CA: Academic Press, 1994.
6. K. Lever, "New derivation of Craig's formula for the Gaussian probability function," *Electron. Lett.*, vol. 34, September 1998, pp. 1821–1822.
7. N. N. Lebedev, *Special Functions and Their Applications*, translated from Russian and edited by R. A. Silverman. New York: Dover Publications, 1972.
8. J. I. Marcum, *Table of Q Functions*, U.S. Air Force Project RAND Research Memorandum M-339, ASTIA Document AD 1165451, Rand Corporation, Santa Monica, CA, January 1, 1950.

9. M. K. Simon, "A new twist on the Marcum Q-function and its application," *IEEE Commun. Lett.*, vol. 2, February 1998, pp. 39–41.
10. C. W. Helstrom, *Elements of Signal Detection and Estimation*, Upper Saddle River, NJ: Prentice Hall, 1995.
11. J. I. Marcum and P. Swerling, "Studies of target detection by pulsed radar," *IEEE Trans. Inf. Theory*, vol. IT-6, April 1960.
12. C. W. Helstrom, *Statistical Theory of Signal Detection*. New York: Pergamon Press, 1960.
13. J. Proakis, *Digital Communications*, 3rd ed. New York: McGraw-Hill, 1995.
14. M. Schwartz, W. R. Bennett, and S. Stein, *Communication Systems and Techniques*. New York: McGraw-Hill, 1966.
15. R. F. Pawula, "Relations between the Rice Ie-function and Marcum Q-function with applications to error rate calculations," *Electron. Lett.*, vol. 31, September 28, 1995, pp. 1717–1719.
16. S. Parl, "A new method of calculating the generalized Q-function," *IEEE Trans. Inf. Theory*, vol. IT-26, January 1980, pp. 121–124.
17. P. E. Cantrell and A. K. Ojha, "Comparison of generalized Q-function algorithms," *IEEE Trans. Inf. Theory*, vol. IT-33, July 1987, pp. 591–596.
18. R. F. Pawula, "A new formula for MDPSK symbol error probability," *IEEE Commun. Lett.*, vol. 2, October 1998, pp. 271–272.
19. M. Abramowitz and I. A. Stegun, *Handbook of Mathematical Functions with Formulas, Graphs, and Mathematical Tables*, 9th ed. New York: Dover Press, 1972.
20. N. M. Temme, *Special Functions: An Introduction to Classical Functions of Mathematical Physics*, New York: Wiley, 1996.
21. G. M. Dillard, "Recursive computation of the generalized Q-function," *IEEE Trans. Aerosp. Electron. Syst.*, vol. AES-9, July 1973, pp. 614–615.
22. M. K. Simon and M.-S. Alouini, "Exponential-type bounds on the generalized Marcum Q-function with application to fading channel error probability analysis," *IEEE Trans. Commun.*, March 2000, pp. 359–366.
23. S. S. Rappaport, "Computing approximations for the generalized Q function and its complement," *IEEE Trans. Inf. Theory*, July 1971, pp. 497–498.

APPENDIX 4A: DERIVATION OF EQ. (4.2)

In this appendix we present two proofs of the alternative form of the Gaussian Q-function given in Eq. (4.2). (A third proof can be obtained by applying the asymptotic relation between the Marcum and Gaussian Q-functions as given in footnote 4 of this chapter to the closed form of the integral in Nuttall [4, Eq. (74)] in the limit as b approaches unity.)

Consider the integral in Gradshteyn and Ryzhik [5, Eq. (3.363.2)], namely,

$$\int_u^\infty \frac{e^{-\mu x}}{x\sqrt{x-u}}\,dx = \frac{\pi}{\sqrt{u}}\,\text{erfc}(\sqrt{u\mu}) \tag{4A.1}$$

Multiplying both sides of (4A.1) by $\frac{1}{2}e^{\mu u}$ and then letting $u = y^2$ gives

$$\frac{1}{2}\int_{y^2}^{\infty} \frac{e^{-\mu x}e^{\mu y^2}}{x\sqrt{x-y^2}}\,dx = \frac{\pi}{2y}e^{\mu y^2}\,\text{erfc}(y\sqrt{\mu}) \qquad (4\text{A}.2)$$

Now let $u = x - y^2$ in (4A.2). Then

$$\frac{1}{2}\int_{0}^{\infty} \frac{e^{-\mu u}}{(u+y^2)\sqrt{u}}\,du = \frac{\pi}{2y}e^{\mu y^2}\,\text{erfc}(y\sqrt{\mu}) \qquad (4\text{A}.3)$$

Next, let $u = t^2$, and $du = 2t\,dt = 2\sqrt{u}\,dt$. Thus (4A.3) becomes

$$\int_{0}^{\infty} \frac{e^{-\mu t^2}}{t^2+y^2}\,dt = \frac{\pi}{2y}e^{\mu y^2}\,\text{erfc}(y\sqrt{\mu}) \qquad (4\text{A}.4)$$

This intermediate form of the desired result appears as Eq. (3.466.1) in Ref. 5 and also as Eq. (7.4.11) in Ref. 19. In addition, Pawula et al. [2, Eq. (34)] used it to derive their expression [2, Eq. (71)] for the average symbol error probability of M-PSK. The reason for mentioning this here is that Pawula et al. point out clearly that for $M = 2$, [2, Eq. (71)] reduces to the well-known result for binary PSK, which is expressed strictly in terms of the Gaussian Q-function. Since for $M = 2$, [2, Eq. (71)] becomes the representation of Craig [1, Eq. (9)], as given here in (4.2), it is worthy of note that as early as 1982, Pawula recognized the existence of this alternative representation. We now proceed with the final steps to arrive at (4.2).

Let $y = 1$ and $\mu = z^2$ in (4A.4), which results in

$$\frac{2}{\pi}\int_{0}^{\infty} \frac{e^{-z^2(t^2+1)}}{t^2+1}\,dt = \text{erfc}(z) \qquad (4\text{A}.5)$$

Finally, let $\sin^2\theta = (t^2+1)^{-1}$, $\cos^2\theta = t^2(t^2+1)^{-1}$, and $dt = -(t^2+1)\,d\theta$, in which case (4A.5) becomes the desired result

$$\frac{2}{\pi}\int_{0}^{\pi/2} \exp\left(-\frac{z^2}{\sin^2\theta}\right) d\theta = \text{erfc}(z) \qquad (4\text{A}.6)$$

or equivalently, letting $z = x/\sqrt{2}$,

$$\frac{1}{\pi}\int_{0}^{\pi/2} \exp\left(-\frac{x^2}{2\sin^2\theta}\right) d\theta = Q(x) \qquad (4\text{A}.7)$$

APPENDIX 4A: DERIVATION OF EQ. (4.2)

Another neat method of arriving at (4.2) is to start by extending the definition in (4.1) (with some name changes in the variables) to two dimensions, namely,

$$Q(z) = 2 \int_0^\infty \frac{1}{\sqrt{2\pi}} \exp\left(-\frac{y^2}{2}\right) dy \overbrace{\int_z^\infty \frac{1}{\sqrt{2\pi}} \exp\left(-\frac{x^2}{2}\right) dx}^{=1}$$

$$= \frac{1}{\pi} \int_z^\infty \int_0^\infty \exp\left(-\frac{x^2 + y^2}{2}\right) dy\, dx \qquad (4A.8)$$

Now make the change of variables from rectangular to polar coordinates, that is,

$$x = r \cos \phi$$
$$y = r \sin \phi$$
$$dx\, dy = r\, dr\, d\phi \qquad (4A.9)$$

Thus

$$Q(z) = \frac{1}{\pi} \int_0^{\pi/2} \int_{z/\cos\phi}^\infty r \exp\left(-\frac{r^2}{2}\right) dr\, d\phi$$

$$= \frac{1}{\pi} \int_0^{\pi/2} \exp\left(-\frac{z^2}{2\cos^2\phi}\right) d\phi \qquad (4A.10)$$

Finally, letting $x = z$ and $\theta = \pi/2 - \phi$, we obtain (4.2).

The advantage of this proof over the former is that it can readily be extended to arrive at (4.9) for $Q^2(z)$ as follows. Once again, start by extending the definition to two dimensions, namely,

$$Q^2(z) = \int_z^\infty \frac{1}{\sqrt{2\pi}} \exp\left(-\frac{y^2}{2}\right) dy \int_z^\infty \frac{1}{\sqrt{2\pi}} \exp\left(-\frac{x^2}{2}\right) dx$$

$$= \frac{1}{2\pi} \int_z^\infty \int_z^\infty \exp\left(-\frac{x^2 + y^2}{2}\right) dy\, dx \qquad (4A.11)$$

Making the same change of variables as in (4A.9) and dividing the rectangular region of integration into two triangular parts gives

$$Q^2(z) = \frac{1}{2\pi} \int_0^{\pi/4} \int_{z/\sin\phi}^\infty r \exp\left(-\frac{r^2}{2}\right) dr\, d\phi + \frac{1}{2\pi} \int_{\pi/4}^{\pi/2} \int_{z/\cos\phi}^\infty r \exp\left(-\frac{r^2}{2}\right) dr\, d\phi$$

$$= \frac{1}{2\pi} \int_0^{\pi/4} \exp\left(-\frac{z^2}{2\sin^2\phi}\right) d\phi + \frac{1}{2\pi} \int_{\pi/4}^{\pi/2} \exp\left(-\frac{z^2}{2\cos^2\phi}\right) d\phi \qquad (4A.12)$$

Letting $x = z$ and also $\theta = \pi/2 - \phi$ in the second integral, then combining the two terms, we obtain (4.9).

5

USEFUL EXPRESSIONS FOR EVALUATING AVERAGE ERROR PROBABILITY PERFORMANCE

As alluded to in Chapter 4, the alternative representations of the Gaussian and Marcum Q-functions in the desired form are the key mathematical tools in unifying evaluation of the average error probability performance of digital communication systems over the generalized fading channel. Before going on to present the specific details of such performances in the remaining parts of the book, we digress in this chapter to derive a set of expressions which can be looked upon as additional mathematical tools that will prove to be particularly useful in carrying out these evaluations. Each of these expressions will consist of an integral of the product of the Gaussian or Marcum Q-function and an instantaneous SNR per bit PDF that is characteristic of the fading channels discussed in Chapter 2 and will be specified either in closed form, as a single integral with finite limits and an integrand composed of elementary (e.g., trigonometric and exponential) functions, or as a single integral with finite limits and an integrand consisting of a Gauss–Hermite quadrature integral [1, Eq. (25.4.46)]. Since, as we shall see later, a great deal of commonality exists among the performances of various modulation/detection schemes over a given channel type, it will be convenient to have these expressions at one's disposal rather than have to rederive them in each instance. It is for this reason that we have elected to include a mathematical chapter of this type prior to discussing the practical applications of such tools.

5.1 INTEGRALS INVOLVING THE GAUSSIAN Q-FUNCTION

When characterizing the performance of coherent digital communications, the generic form of the expression for the error probability involves the Gaussian Q-function (and occasionally, the square of the Gaussian Q-function) with an argument proportional to the square root of the instantaneous SNR of the

100 USEFUL EXPRESSIONS FOR EVALUATING AVERAGE ERROR PROBABILITY PERFORMANCE

received signal. In the case of communication over a slow-fading channel, the instantaneous SNR per bit, γ, is a time-invariant random variable with a PDF, $p_\gamma(\gamma)$, defined by the type of fading discussed in Chapter 2. To compute the average error probability[1] one must evaluate an integral whose integrand consists of the product of the above-mentioned Gaussian Q-function and fading PDF, that is,[2]

$$I = \int_0^\infty Q(a\sqrt{\gamma}) p_\gamma(\gamma)\, d\gamma \tag{5.1}$$

where a is a constant that depends on the specific modulation/detection combination. If one were to use the classical definition of the Gaussian Q-function of (4.1) in (5.1) then, in general, evaluation of (5.1) is difficult because of the presence of $\sqrt{\gamma}$ in the lower limit of the Gaussian Q-function integral. If, instead, we were to use the desired form of the Gaussian Q-function of (4.2) in (5.1), the result would be

$$\begin{aligned} I &= \int_0^\infty \frac{1}{\pi} \int_0^{\pi/2} \exp\left(-\frac{a^2 \gamma}{2 \sin^2 \theta}\right) d\theta \, p_\gamma(\gamma)\, d\gamma \\ &= \frac{1}{\pi} \int_0^{\pi/2} \left[\int_0^\infty \exp\left(-\frac{a^2 \gamma}{2 \sin^2 \theta}\right) p_\gamma(\gamma)\, d\gamma \right] d\theta \end{aligned} \tag{5.2}$$

where the inner integral (in brackets) is in the form of a Laplace transform with respect to the variable γ. Since the moment generating function (MGF)[3] of γ [i.e., $M_\gamma(s) \triangleq \int_0^\infty e^{s\gamma} p_\gamma(\gamma)\, d\gamma$] is the Laplace transform of $p_\gamma(\gamma)$ with the exponent

---

[1] In this chapter we do not distinguish between bit and character (symbol) error probability.
[2] This is the simplest form of integral required to evaluate average error probability performance and is characteristic of single-channel reception, which we discuss in great detail in Chapter 8. More complicated (e.g., multidimensional) forms of integrals are required to evaluate the performance of multichannel reception (see Chapter 9). However, in a large majority of cases, the new representation of the Gaussian Q-function allows these to be partitioned into a product of single-dimensional integrals of the type in (5.1). Thus it is sufficient at this point to consider only integrals of this type.
[3] For a real nonnegative continuous random variable X, most textbooks dealing with probability define the *moment generating function* by $M_X(t) = E\{e^{tX}\} = \int_0^\infty e^{tx} p_X(x)\, dx$, where t is a real variable. Based on this definition the nth moment of X would then be obtained from

$$E\{X^n\} = \left. \frac{d^n}{dt^n} M_X(t) \right|_{t=0}$$

Since our interest is primarily in the transform property of the moment generating function rather than on its ability to generate the moments of the random variable, for convenience of notation we replace the real variable t with the complex variable s, in which case the Laplace transform of the PDF is given by $M_X(-s) = \int_0^\infty e^{-sx} p_X(x)\, dx$. Also, if s is purely imaginary (i.e., $s = j\omega$) one obtains the *characteristic function*, namely,

$$\psi_X(\omega) = E\{e^{j\omega X}\} = \int_0^\infty e^{j\omega x} p_X(x)\, dx = M_X(j\omega)$$

reversed in sign, (5.2) can be rewritten as

$$I = \frac{1}{\pi} \int_0^{\pi/2} M_\gamma\left(-\frac{a^2}{2\sin^2\theta}\right) d\theta \tag{5.3}$$

Since tables of Laplace transforms are readily available, the desired form of the Gaussian Q-function therefore allows evaluation of I in the simplest possible way, in most cases resulting in a single integral on θ (when the Laplace transform is available in closed form). In the remainder of this section, we evaluate I of (5.3) for the variety of fading channel PDF's derived in Chapter 2.

5.1.1 Rayleigh Fading Channel

The simplest fading channel from the standpoint of analytical characterization is the Rayleigh channel, whose instantaneous SNR per bit PDF is given by [see (2.7)]

$$p_\gamma(\gamma) = \frac{1}{\bar{\gamma}} \exp\left(-\frac{\gamma}{\bar{\gamma}}\right), \qquad \gamma \geq 0 \tag{5.4}$$

where $\bar{\gamma}$ is the average SNR per bit. The Laplace transform of the Rayleigh PDF can be evaluated in closed form with the result [2, Eq. (17)]

$$M_\gamma(-s) = \frac{1}{1+s\bar{\gamma}}, \qquad s > 0 \tag{5.5}$$

Substituting (5.5) into (5.3) gives

$$I \triangleq I_r(a, \bar{\gamma}) = \frac{1}{\pi} \int_0^{\pi/2} \left(1 + \frac{a^2\bar{\gamma}}{2\sin^2\theta}\right)^{-1} d\theta = \frac{1}{2}\left(1 - \sqrt{\frac{a^2\bar{\gamma}/2}{1+a^2\bar{\gamma}/2}}\right) \tag{5.6}$$

5.1.2 Nakagami-q (Hoyt) Fading Channel

For the Nakagami-q (Hoyt) distribution with instantaneous SNR per bit PDF given by [see (2.11)]

$$p_\gamma(\gamma) = \frac{1+q^2}{2q\bar{\gamma}} \exp\left[-\frac{(1+q^2)^2\gamma}{4q^2\bar{\gamma}}\right] I_0\left[\frac{(1-q^4)\gamma}{4q^2\bar{\gamma}}\right], \qquad \gamma \geq 0 \tag{5.7}$$

with Laplace transform [2, Eq. (109)]

$$M_\gamma(-s) = \left[1 + 2s\bar{\gamma} + \frac{4q^2s^2\bar{\gamma}^2}{(1+q^2)^2}\right]^{-1/2}, \qquad s > 0 \tag{5.8}$$

the integral in (5.3) evaluates to

$$I \triangleq I_q(a, q, \overline{\gamma}) = \frac{1}{\pi} \int_0^{\pi/2} \left[1 + \frac{a^2}{\sin^2 \theta} \overline{\gamma} + \frac{q^2 a^2 \overline{\gamma}^2}{(1+q^2)^2 \sin^4 \theta} \right]^{-1/2} d\theta \quad (5.9)$$

5.1.3 Nakagami-n (Rice) Fading Channel

For the Nakagami-n (Rice) distribution with instantaneous SNR per bit PDF given by [see (2.16)]

$$p_\gamma(\gamma) = \frac{(1+n^2)e^{-n^2}}{\overline{\gamma}} \exp\left[-\frac{(1+n^2)\gamma}{\overline{\gamma}} \right] I_0 \left[2n \sqrt{\frac{(1+n^2)\gamma}{\overline{\gamma}}} \right], \quad \gamma \geq 0 \quad (5.10)$$

with Laplace transform[4]

$$M_\gamma(-s) = \frac{1+n^2}{1+n^2+s\overline{\gamma}} \exp\left(-\frac{n^2 s \overline{\gamma}}{1+n^2+s\overline{\gamma}} \right), \quad s > 0 \quad (5.11)$$

the integral in (5.3) evaluates to

$$I \triangleq I_n(a, n, \overline{\gamma})$$
$$= \frac{1}{\pi} \int_0^{\pi/2} \frac{(1+n^2)\sin^2 \theta}{(1+n^2)\sin^2 \theta + a^2 \overline{\gamma}/2} \exp\left[-\frac{n^2 a^2 \overline{\gamma}/2}{(1+n^2)\sin^2 \theta + a^2 \overline{\gamma}/2} \right] d\theta,$$
$$s > 0 \quad (5.12)$$

To obtain the desired result for the Rician fading channel, we merely substitute $n^2 = K$ in (5.12), which results in

$$I \triangleq I_n(a, K, \overline{\gamma})$$
$$= \frac{1}{\pi} \int_0^{\pi/2} \frac{(1+K)\sin^2 \theta}{(1+K)\sin^2 \theta + a^2 \overline{\gamma}/2} \exp\left[-\frac{K a^2 \overline{\gamma}/2}{(1+K)\sin^2 \theta + a^2 \overline{\gamma}/2} \right] d\theta,$$
$$s > 0 \quad (5.13)$$

5.1.4 Nakagami-m Fading Channel

For the Nakagami-m distribution with instantaneous SNR per bit PDF given by [see (2.21)]

$$p_\gamma(\gamma) = \frac{m^m \gamma^{m-1}}{\overline{\gamma}^m \Gamma(m)} \exp\left(-\frac{m\gamma}{\overline{\gamma}} \right), \quad \gamma \geq 0 \quad (5.14)$$

---
[4] This particular Laplace transform is not tabulated directly in Ref. 2 but can be evaluated from a definite integral in the same reference, in particular, Eq. (6.631.4).

with Laplace transform [2, Eq. (3)]

$$M_\gamma(-s) = \left(1 + \frac{s\bar{\gamma}}{m}\right)^{-m}, \qquad s > 0 \qquad (5.15)$$

the integral in (5.3) evaluates to

$$I \triangleq I_m(a, m, \bar{\gamma}) = \frac{1}{\pi} \int_0^{\pi/2} \left(1 + \frac{a^2\bar{\gamma}}{2m\sin^2\theta}\right)^{-m} d\theta \qquad (5.16)$$

which can be evaluated in closed form using the definite integral derived in Appendix 5A, namely,[5]

$$\frac{1}{\pi} \int_0^{\pi/2} \left(1 + \frac{c}{\sin^2\theta}\right)^{-m} d\theta$$

$$= \begin{cases} \dfrac{1}{2}\left[1 - \mu^2(c)\sum_{k=0}^{m-1}\binom{2k}{k}\left(\dfrac{1-\mu(c)}{4}\right)^k\right], & \mu(c) \triangleq \sqrt{\dfrac{c}{1+c}}, \\ \qquad m \text{ integer} & (5.17a) \\ \\ \dfrac{1}{2\sqrt{\pi}}\dfrac{\sqrt{c}}{(1+c)^{m+1/2}}\dfrac{\Gamma(m+\frac{1}{2})}{\Gamma(m+1)}\,_2F_1\left(1, m+\dfrac{1}{2}; m+1; \dfrac{1}{1+c}\right), \\ \qquad m \text{ noninteger} & (5.17b) \end{cases}$$

where $_2F_1(\cdot, \cdot; \cdot; \cdot)$ is the Gauss hypergeometric function [1, Eq. (15.1.1)]. Thus, using (5.17) in (5.16) gives

$I_m(a, m, \bar{\gamma})$

$$= \begin{cases} \dfrac{1}{2}\left[1 - \mu\left(\dfrac{a^2\bar{\gamma}}{2m}\right)\sum_{k=0}^{m-1}\binom{2k}{k}\left(\dfrac{1-\mu^2(a^2\bar{\gamma}/2m)}{4}\right)^k\right], \\ \mu\left(\dfrac{a^2\bar{\gamma}}{2m}\right) \triangleq \sqrt{\dfrac{a^2\bar{\gamma}/2}{m + a^2\bar{\gamma}/2}}, \quad m \text{ integer} & (5.18a) \\ \\ \dfrac{1}{2\sqrt{\pi}}\dfrac{\sqrt{a^2\bar{\gamma}/2m}}{(1+a^2\bar{\gamma}/2m)^{m+1/2}}\dfrac{\Gamma(m+\frac{1}{2})}{\Gamma(m+1)}\,_2F_1\left(1, m+\dfrac{1}{2}; m+1; \dfrac{m}{m+a^2\bar{\gamma}/2}\right), \\ \qquad m \text{ noninteger} & (5.18b) \end{cases}$$

Note that for $m = 1$, (5.18a) reduces to the result for the Rayleigh case as given by (5.6).

---

[5] This definite integral appears not to be available in standard integral tables such as Ref. 2.

5.1.5 Log-Normal Shadowing Channel

For the log-normal shadowing distribution with instantaneous SNR per bit PDF given by [see (2.25)]

$$p_\gamma(\gamma) = \frac{10/\ln 10}{\sqrt{2\pi\sigma^2}\gamma} \exp\left[-\frac{(10\log_{10}\gamma - \mu)^2}{2\sigma^2}\right], \quad \gamma \geq 0$$

$$\mu(\text{in dB}) = 10\log_{10}\overline{\gamma}$$

$$\sigma(\text{in dB}) = \text{logarithmic standard deviation of shadowing} \quad (5.19)$$

the Laplace transform cannot be obtained in closed form. Instead, we substitute (5.19) into (5.2) directly and then make a change of variables, namely, $x = (10\log_{10}\gamma - \mu)/\sqrt{2}\sigma$, which results in

$$I \triangleq I_{\ln}(a, \mu, \sigma)$$

$$= \frac{1}{\pi} \int_0^{\pi/2} \left[\frac{1}{\sqrt{\pi}} \int_{-\infty}^{\infty} \exp\left(-\frac{a^2}{2\sin^2\theta} \cdot 10^{(x\sqrt{2}\sigma+\mu)/10}\right) e^{-x^2} dx\right] d\theta \quad (5.20)$$

The inner integral can be efficiently computed using a Gauss–Hermite quadrature integration [1, Eq. (25.4.46)], that is,

$$\frac{1}{\sqrt{\pi}} \int_{-\infty}^{\infty} \exp\left(-\frac{a^2}{2\sin^2\theta} \cdot 10^{(x\sqrt{2}\sigma+\mu)/10}\right) e^{-x^2} dx$$

$$= \frac{1}{\sqrt{\pi}} \sum_{i=1}^{n} w_i \exp\left(-\frac{a^2}{2\sin^2\theta} \cdot 10^{(x_i\sqrt{2}\sigma+\mu)/10}\right) \quad (5.21)$$

where $\{x_i\}$, $i = 1, 2, \ldots, n$, are the zeros of the nth-order Hermite polynomial $He_n(x)$ and $\{w_i\}$, $i = 1, 2, \ldots, n$, are weight factors tabulated in Table 25.10 of Ref. 1 for values of n from 2 to 20. Since the x_i's and w_i's are independent of θ, substituting (5.21) in (5.20) and making use of the desired form of the Gaussian Q-function as given in (4.2), we get

$$I_{\ln}(a, \mu, \sigma) = \frac{1}{\sqrt{\pi}} \sum_{i=1}^{n} w_i Q\left(a\sqrt{10^{(x_i\sqrt{2}\sigma+\mu)/10}}\right) \quad (5.22)$$

where the value of n is chosen depending on the desired degree of accuracy.

5.1.6 Composite Log-Normal Shadowing/Nakagami-m Fading Channel

The class of composite shadowing–fading channels is discussed in Section 2.2.3. A popular example of this class that is characteristic of congested downtown areas with a large number of slow-moving pedestrians and vehicles is the composite log-normal shadowing/Nakagami-m fading channel. For this channel, $p_\gamma(\gamma)$ is obtained by averaging the instantaneous Nakagami-m fading average power (treated now as a random variable) over the conditional PDF of the log-normal shadowing, which from (5.14) and (5.19) results in the composite gamma/log-normal PDF

$$p_\gamma(\gamma) = \int_0^\infty \frac{m^m \gamma^{m-1}}{\Omega^m \Gamma(m)} \exp\left(-\frac{m\gamma}{\Omega}\right)$$
$$\times \left\{\frac{10/\ln 10}{\sqrt{2\pi\sigma^2}\Omega} \exp\left[-\frac{(10\log_{10}\Omega - \mu)^2}{2\sigma^2}\right]\right\} d\Omega, \quad \gamma \geq 0 \quad (5.23)$$

Since the Laplace transform of the Nakagami-m fading portion of (5.23) is known in closed form [see (5.15)], the Laplace transform of the composite PDF in (5.23) can be obtained as the single integral

$$M_\gamma(-s) = \int_0^\infty \left(1 + \frac{s\Omega}{m}\right)^{-m}$$
$$\times \left\{\frac{10/\ln 10}{\sqrt{2\pi\sigma^2}\Omega} \exp\left[-\frac{(10\log_{10}\Omega - \mu)^2}{2\sigma^2}\right]\right\} d\Omega, \quad s > 0 \quad (5.24)$$

Substituting (5.24) into (5.2) and then making a change of variables, namely, $x = (10\log_{10}\Omega - \mu)/\sqrt{2}\sigma$, results in

$$I \triangleq I_{g/\ln}(a, \mu, \sigma, m)$$
$$= \frac{1}{\pi} \int_0^{\pi/2} \left[\frac{1}{\sqrt{\pi}} \int_{-\infty}^\infty \left(1 + \frac{a^2}{2m\sin^2\theta} 10^{(x\sqrt{2}\sigma+\mu)/10}\right)^{-m} e^{-x^2} dx\right] d\theta \quad (5.25)$$

Once again the inner integral can be computed efficiently using a Gauss–Hermite quadrature integration [1, Eq. (25.4.46)], that is,

$$\frac{1}{\sqrt{\pi}} \int_{-\infty}^\infty \left(1 + \frac{a^2}{2m\sin^2\theta} \cdot 10^{(x\sqrt{2}\sigma+\mu)/10}\right)^{-m} e^{-x^2} dx$$
$$= \frac{1}{\sqrt{\pi}} \sum_{i=1}^n w_i \left(1 + \frac{a^2}{2m\sin^2\theta} \cdot 10^{(x_i\sqrt{2}\sigma+\mu)/10}\right)^{-m} \quad (5.26)$$

Since, as mentioned previously, the x_i's and w_i's are independent of θ, then substituting (5.26) in (5.25) and making use of the closed-form integral in (5.17a), we get

$$I_{g/\ln}(a, \mu, \sigma, m) = \frac{1}{2\sqrt{\pi}} \sum_{i=1}^n w_i \left[1 - \mu(c_i) \sum_{k=0}^{m-1} \binom{2k}{k} \left(\frac{1 - \mu^2(c_i)}{4}\right)^k\right],$$

$$\mu(c_i) \triangleq \sqrt{\frac{c_i}{1+c_i}}, \quad c_i \triangleq \frac{a^2}{2m} \cdot 10^{(x_i\sqrt{2}\sigma+\mu)/10} \quad (5.27)$$

Before moving on to a consideration of integrals involving the Marcum Q-function, we give brief attention to integrals involving the square of the Gaussian Q-function, since these will be found useful when we discuss evaluating average symbol error probability of coherently detected square QAM over generalized fading channels. Analogous to (5.1), then, it is of interest to evaluate

$$I = \int_0^\infty Q^2(a\sqrt{\gamma}) p_\gamma(\gamma)\, d\gamma \tag{5.28}$$

for the various fading channel PDFs. Using the classical definition of the Gaussian Q-function, such integrals would be extremely difficult to obtain in closed form since $Q^2(a\sqrt{\gamma})$ would be written as a double integral each of which has $\sqrt{\gamma}$ in its lower limit. However, in view of the similarity between the desired forms of the Gaussian Q-function and the square of the Gaussian Q-function [compare (4.2) and (4.9)], in principal it becomes a simple matter to evaluate I of (5.28) — in particular, one merely need replace the $\pi/2$ upper limit in the integration on θ in the evaluations of I of (5.1) with $\pi/4$ to arrive at the desired results. Although this may seem like a simple generalization, depending on the channel, the foregoing replacement of the upper limit can lead to closed-form expressions that are significantly more complicated. For the Rayleigh fading channel, the analogous result to (5.6) is straightforward in view of the fact that the indefinite integral form of this equation has a closed-form result [see (5A.11) in Appendix 5A]. Thus, using (5A.13), we arrive at

$$I \stackrel{\Delta}{=} I_r^{(2)}(a, \bar{\gamma}) = \frac{1}{\pi} \int_0^{\pi/4} \left(1 + \frac{a^2 \bar{\gamma}}{2 \sin^2 \theta}\right)^{-1} d\theta$$

$$= \frac{1}{4}\left[1 - \sqrt{\frac{a^2 \bar{\gamma}/2}{1 + a^2 \bar{\gamma}/2}} \left(\frac{4}{\pi} \tan^{-1} \sqrt{\frac{1 + a^2 \bar{\gamma}/2}{a^2 \bar{\gamma}/2}}\right)\right] \tag{5.29}$$

For the Nakagami-m channel with m integer, the result is considerably more complex than (5.18a). However, using (5A.17) with $M = 4$, we obtain

$$I \stackrel{\Delta}{=} I_m^{(2)}(a, m, \bar{\gamma})$$

$$= \frac{1}{\pi} \int_0^{\pi/4} \left(1 + \frac{a^2 \bar{\gamma}}{2m \sin^2 \theta}\right)^{-m} d\theta$$

$$= \frac{1}{4} - \frac{1}{\pi}\alpha \left\{ \left(\frac{\pi}{2} - \tan^{-1}\alpha\right) \sum_{k=0}^{m-1} \binom{2k}{k} \frac{1}{[4(1+c)]^k} \right.$$

$$\left. - \sin(\tan^{-1}\alpha) \sum_{k=1}^{m-1} \sum_{i=1}^{k} \frac{T_{ik}}{(1+c)^k} [\cos(\tan^{-1}\alpha)]^{2(k-i)+1} \right\} \tag{5.30}$$

where

$$c = \frac{a^2\bar{\gamma}}{2m},$$

$$\alpha \triangleq \mu = \sqrt{\frac{c}{1+c}} = \sqrt{\frac{a^2\bar{\gamma}/2}{m + a^2\bar{\gamma}/2}} \tag{5.31}$$

and

$$T_{ik} \triangleq \frac{\binom{2k}{k}}{\binom{2(k-i)}{k-i} 4^i[2(k-i)+1]} \tag{5.32}$$

5.2 INTEGRALS INVOLVING THE MARCUM Q-FUNCTION

When characterizing the performance of differentially coherent and noncoherent digital communications, the generic form of the expression for the error probability typically involves the generalized Marcum Q-function, both of whose arguments are proportional to the square root of the instantaneous SNR of the received signal. To compute the average error probability over a slow-fading channel, one must evaluate an integral whose integrand consists of the product of the above-mentioned Marcum Q-function and the PDF of the instantaneous SNR per bit. Thus, analogous to (5.1), we wish to investigate integrals having the generic form

$$I = \int_0^\infty Q_l(a\sqrt{\gamma}, b\sqrt{\gamma}) p_\gamma(\gamma) \, d\gamma \tag{5.33}$$

where a and b are constants that depend on the specific modulation/detection combination, l the order of the Marcum Q-function, and $p_\gamma(\gamma)$ again depends on the type of fading, as discussed in Chapter 2. As was true for the Gaussian Q-function, if one were to use the classical definition of the Marcum Q-function given by Eq. (4.33) in (5.33), then, in general, evaluation of (5.33) is difficult because of the presence of $\sqrt{\gamma}$ in the lower limit of the Marcum Q-function integral. If, instead, we were to use the desired form of the Marcum Q-function of (4.42) or (4.50) in (5.33), the result of this substitution would be

$$I = \frac{1}{2\pi} \int_{-\pi}^{\pi} \frac{\zeta^{-(l-1)}\{\cos[(l-1)(\theta + \pi/2)] - \zeta \cos[l(\theta + \pi/2)]\}}{1 + 2\zeta \sin\theta + \zeta^2}$$

$$\times \left\{ \int_0^\infty \exp\left[-\frac{b^2\gamma}{2}(1 + 2\zeta \sin\theta + \zeta^2)\right] p_\gamma(\gamma) \, d\gamma \right\} d\theta,$$

$$0^+ \leq \zeta = a/b < 1 \tag{5.34}$$

or

$$I = 1 - \frac{1}{2\pi}\int_{-\pi}^{\pi}\frac{\zeta^l\{\cos[l(\theta+\pi/2)] - \zeta\cos[(l-1)(\theta+\pi/2)]\}}{1+2\zeta\sin\theta+\zeta^2}$$

$$\times\left\{\int_0^{\infty}\exp\left[-\frac{a^2\gamma}{2}(1+2\zeta\sin\theta+\zeta^2)\right]p_\gamma(\gamma)\,d\gamma\right\}d\theta,$$

$$0^+ \le \zeta = b/a < 1 \qquad (5.35)$$

where the inner integral is again in the form of a Laplace transform with respect to the variable γ. That is, if, as in Section 5.1, $M_\gamma(s) \triangleq \int_0^{\infty} e^{s\gamma}p_\gamma(\gamma)\,d\gamma$ denotes the MGF of γ, (5.34) and (5.35) can be rewritten as

$$I = \frac{1}{2\pi}\int_{-\pi}^{\pi}\frac{\zeta^{-(l-1)}\{\cos[(l-1)(\theta+\pi/2)] - \zeta\cos[l(\theta+\pi/2)]\}}{1+2\zeta\sin\theta+\zeta^2}$$

$$\times M_\gamma\left[-\frac{b^2}{2}(1+2\zeta\sin\theta+\zeta^2)\right]d\theta, \qquad 0^+ \le \zeta = a/b < 1 \qquad (5.36)$$

or

$$I = 1 - \frac{1}{2\pi}\int_{-\pi}^{\pi}\frac{\zeta^l\{\cos[l(\theta+\pi/2)] - \zeta\cos[(l-1)(\theta+\pi/2)]\}}{1+2\zeta\sin\theta+\zeta^2}$$

$$\times M_\gamma\left[-\frac{a^2}{2}(1+2\zeta\sin\theta+\zeta^2)\right]d\theta, \qquad 0^+ \le \zeta = b/a < 1 \qquad (5.37)$$

In the remainder of this section, we evaluate I of (5.36) for the variety of fading channel PDFs derived in Chapter 2, where, for simplicity of notation, we introduce the functions

$$g(\theta;\zeta) \triangleq 1 + 2\zeta\sin\theta + \zeta^2$$

$$h(\theta;\zeta,l) \triangleq \zeta^{-(l-1)}\left\{\cos\left[(l-1)\left(\theta+\frac{\pi}{2}\right)\right] - \zeta\cos\left[l\left(\theta+\frac{\pi}{2}\right)\right]\right\} \qquad (5.38)$$

Also, the corresponding results for I of (5.37) can then be obtained by inspection.

5.2.1 Rayleigh Fading Channel

For the Rayleigh channel with a Laplace transform of the instantaneous SNR per bit PDF given by (5.5), the integral I of (5.36) [or equivalently, (5.33) for $a < b$] evaluates to

$$I \triangleq J_r(b,\zeta,\bar{\gamma},l) = \frac{1}{2\pi}\int_{-\pi}^{\pi}\frac{h(\theta;\zeta,l)}{g(\theta;\zeta)}\left[1+\frac{b^2\bar{\gamma}}{2}g(\theta;\zeta)\right]^{-1}d\theta \qquad (5.39)$$

5.2.2 Nakagami-q (Hoyt) Fading Channel

For the Nakagami-q (Hoyt) distribution with a Laplace transform of the instantaneous SNR per bit PDF given by (5.7), the integral I of (5.36) evaluates to

$$I \overset{\Delta}{=} J_q(b, \zeta, q, \overline{\gamma}, l)$$
$$= \frac{1}{2\pi} \int_{-\pi}^{\pi} \frac{h(\theta; \zeta, l)}{g(\theta; \zeta)} \left[1 + b^2 \overline{\gamma} g(\theta; \zeta) + \frac{q^2 b^4 \overline{\gamma}^2 g^2(\theta; \zeta)}{(1+q^2)^2} \right]^{-1/2} d\theta \quad (5.40)$$

5.2.3 Nakagami-n (Rice) Fading Channel

For the Nakagami-n (Rice) distribution with a Laplace transform of the instantaneous SNR per bit PDF given by (5.11), the integral I of (5.36) evaluates to

$$I \overset{\Delta}{=} J_n(b, \zeta, n, \overline{\gamma}, l) = \frac{1}{2\pi} \int_{-\pi}^{\pi} \frac{h(\theta; \zeta, l)}{g(\theta; \zeta)}$$
$$\times \left[\frac{1+n^2}{1+n^2+(b^2\overline{\gamma}/2)g(\theta;\zeta)} \exp\left(-\frac{n^2(b^2\overline{\gamma}/2)g(\theta;\zeta)}{1+n^2+(b^2\overline{\gamma}/2)g(\theta;\zeta)} \right) \right] d\theta \quad (5.41)$$

or equivalently, in terms of the Rician parameter

$$I \overset{\Delta}{=} J_n(b, \zeta, K, \overline{\gamma}, l) = \frac{1}{2\pi} \int_{-\pi}^{\pi} \frac{h(\theta; \zeta, l)}{g(\theta; \zeta)}$$
$$\times \left[\frac{1+K}{1+K+(b^2\overline{\gamma}/2)g(\theta;\zeta)} \exp\left(-\frac{(Kb^2\overline{\gamma}/2)g(\theta;\zeta)}{1+K+(b^2\overline{\gamma}/2)g(\theta;\zeta)} \right) \right] d\theta \quad (5.42)$$

5.2.4 Nakagami-m Fading Channel

For the Nakagami-m distribution with a Laplace transform of the instantaneous SNR per bit PDF given by (5.15), the integral I of (5.36) evaluates to

$$I \overset{\Delta}{=} J_m(b, \zeta, m, \overline{\gamma}, l) = \frac{1}{2\pi} \int_{-\pi}^{\pi} \frac{h(\theta; \zeta, l)}{g(\theta; \zeta)} \left(1 + \frac{b^2\overline{\gamma}}{2m} g(\theta;\zeta) \right)^{-m} d\theta \quad (5.43)$$

which reduces to (5.39) for the Rayleigh ($m = 1$) case.

5.2.5 Log-Normal Shadowing Channel

As discussed in Section 5.1.5, the Laplace transform of the instantaneous SNR per bit PDF for the log-normal shadowing distribution cannot be obtained in closed form. Thus, we proceed as before and substitute (5.19) directly into (5.34) and then make a change of variables, namely, $x = (10 \log_{10} \gamma - \mu)/\sqrt{2}\sigma$, which results in

$$I \triangleq J_{\ln}(b, \zeta, \mu, \sigma, l) = \frac{1}{2\pi} \int_{-\pi}^{\pi} \frac{h(\theta; \zeta, l)}{g(\theta; \zeta)}$$

$$\times \left[\frac{1}{\sqrt{\pi}} \int_{-\infty}^{\infty} \exp\left(-\frac{b^2 g(\theta; \zeta)}{2} \cdot 10^{(x\sqrt{2}\sigma+\mu)/10}\right) e^{-x^2} dx \right] d\theta \quad (5.44)$$

The inner integral can be efficiently computed using a Gauss–Hermite quadrature integration [1, Eq. (25.4.46)], that is,

$$\frac{1}{\sqrt{\pi}} \int_{-\infty}^{\infty} \exp\left(-\frac{b^2 g(\theta; \zeta)}{2} \cdot 10^{(x\sqrt{2}\sigma+\mu)/10}\right) e^{-x^2} dx$$

$$= \frac{1}{\sqrt{\pi}} \sum_{i=1}^{n} w_i \exp\left(-\frac{b^2 g(\theta; \zeta)}{2} \cdot 10^{(x_i\sqrt{2}\sigma+\mu)/10}\right) \quad (5.45)$$

Substituting (5.45) into (5.44) and making use of the desired from of the generalized Marcum Q-function as given in (4.42), we get

$$J_{\ln}(b, \zeta, \mu, \sigma, l) = \frac{1}{\sqrt{\pi}} \sum_{i=1}^{n} w_i Q_l \left(b\zeta \sqrt{10^{(x_i\sqrt{2}\sigma+\mu)/10}}, b\sqrt{10^{(x_i\sqrt{2}\sigma+\mu)/10}}\right) \quad (5.46)$$

5.2.6 Composite Log-Normal Shadowing/Nakagami-m Fading Channel

Finally, we consider the composite log-normal shadowing/Nakagami-m fading channel treated in Section 5.1.6. For this channel, we again make use of the single integral form of the Laplace transform of $p_\gamma(\gamma)$ as given in (5.24), which upon substitution into (5.36) together with the change of variables $x = (10 \log_{10} \Omega - \mu)/\sqrt{2}\sigma$ results in

$$I \triangleq J_{g/\ln}(b, \zeta, \mu, \sigma, m, l)$$

$$= \frac{1}{2\pi} \int_{-\pi}^{\pi} \frac{h(\theta; \zeta, l)}{g(\theta; \zeta)} \left[\frac{1}{\sqrt{\pi}} \int_{-\infty}^{\infty} \left(1 + \frac{b^2 g(\theta; \zeta)}{2m} \cdot 10^{(x\sqrt{2}\sigma+\mu)/10}\right)^{-m} e^{-x^2} dx \right] d\theta \quad (5.47)$$

Once again the inner integral can be computed efficiently using a Gauss–Hermite quadrature integration [1, Eq. (25.4.46)], that is,

$$\frac{1}{\sqrt{\pi}} \int_{-\infty}^{\infty} \left(1 + \frac{b^2 g(\theta; \zeta)}{2m} \cdot 10^{(x\sqrt{2}\sigma+\mu)/10}\right)^{-m} e^{-x^2} dx$$

$$= \frac{1}{\sqrt{\pi}} \sum_{i=1}^{n} w_i \left(1 + \frac{b^2 g(\theta; \zeta)}{2m} \cdot 10^{(x_i\sqrt{2}\sigma+\mu)/10}\right)^{-m} \quad (5.48)$$

Substituting (5.48) in (5.47) and making use of the closed-form integral in (5.17), we get

$$J_{g/\ln}(b, \zeta, \mu, \sigma, m, l) = \frac{1}{\sqrt{\pi}} \sum_{i=1}^{n} w_i \left[\frac{1}{2\pi} \int_{-\pi}^{\pi} \frac{h(\theta; \zeta, l)}{g(\theta; \zeta)} \right.$$
$$\left. \times \left(1 + \frac{b^2 g(\theta; \zeta)}{2m} \cdot 10^{(x_i \sqrt{2}\sigma + \mu)/10} \right)^{-m} d\theta \right] \quad (5.49)$$

Unfortunately, because a closed-form result was not obtainable for (5.43), we cannot similarly obtain a closed-form result for (5.49).

5.3 INTEGRALS INVOLVING THE INCOMPLETE GAMMA FUNCTION

In the preceding section, we considered integrals involving the Marcum Q-function $Q_m(\alpha, \beta)$, $0 < \alpha < \beta$, where the desired form of this function as given by (4.42) was used to simplify the evaluations. A special case of the Marcum Q-function corresponding to its first argument equal to zero is expressible as a ratio of complementary Gauss incomplete gamma functions [see Eq. (4.44)]. As we shall see in Chapter 8, integrals involving such a ratio are appropriate to the unification of the error probability performance of coherent, differentially coherent, and noncoherent binary PSK and FSK systems over generalized fading channels. However, since the desired form of the Marcum Q-function of (4.42) requires that the first argument be greater than zero, the specific results derived in Section 5.2 cannot be used in this instance. Fortunately, however, the special case $Q_m(0, \beta)$ can be put in a separate desired form[6] as given by (4.45). In this section we derive the analogous results to those in Section 5.2 using this special desired form of $Q_m(0, \beta)$.

Based on the discussion above, then, we are interested in evaluating

$$I = \int_0^\infty Q_l(0, b\sqrt{\gamma}) p_\gamma(\gamma) \, d\gamma = \int_0^\infty \frac{\Gamma(l, b^2\gamma/2)}{\Gamma(l)} p_\gamma(\gamma) \, d\gamma \quad (5.50)$$

for the various characterizations of $p_\gamma(\gamma)$ or substituting the form of (4.45) in (5.50), we are equivalently interested in evaluating

$$I = \int_0^\infty \frac{(b\sqrt{\gamma})^{2l}}{2^{l-1}\Gamma(l)} \int_0^{\pi/2} \frac{\cos\theta}{(\sin\theta)^{1+2l}} \exp\left(-\frac{b^2\gamma}{2\sin^2\theta}\right) d\theta p_\gamma(\gamma) \, d\gamma \quad (5.51)$$

---

[6] The desired form of the integral for $Q_m(0, \beta)$ is slightly less desirable than that for $Q_m(\alpha, \beta)$, $0 < \alpha < \beta$, in that the integrand contains a term β^{2m} in addition to the usual Gaussian dependence on β. Nevertheless, it is still useful in carrying out integrals involving the statistics of the fading channel by using Laplace transform manipulations.

Reversing the order of integration and grouping together like variables, we can rewrite (5.51) as

$$I = \frac{b^{2l}}{2^{l-1}\Gamma(l)} \int_0^{\pi/2} \frac{\cos\theta}{(\sin\theta)^{1+2l}} \int_0^\infty \gamma^l \exp\left(-\frac{b^2\gamma}{2\sin^2\theta}\right) p_\gamma(\gamma)\,d\gamma\,d\theta \quad (5.52)$$

where the integral on γ is in the form of a Laplace transform that is similar to but slightly more complicated than the MGF of γ.

5.3.1 Rayleigh Fading Channel

Substituting (5.4) in (5.52) and making use of Eq. (3.381.4) of Ref. 2, we obtain

$$I \triangleq J_r(b, \bar{\gamma}, l) = 2l \left(\frac{b^2\bar{\gamma}}{2}\right)^l \int_0^{\pi/2} \frac{\cos\theta}{(\sin\theta)^{1+2l}} \left(1 + \frac{b^2\bar{\gamma}}{2\sin^2\theta}\right)^{-l-1} d\theta \quad (5.53)$$

Making the change of variables $t = (1 + b^2\bar{\gamma}/2\sin^2\theta)^{-1}$, after some manipulation we arrive at the equivalent compact result

$$J_r(b, \bar{\gamma}, l) = l \int_0^{(1+b^2\bar{\gamma}/2)^{-1}} (1-t)^{l-1} dt = l B_{(1+b^2\bar{\gamma}/2)^{-1}}(1, l) \quad (5.54)$$

where

$$B_x(p, q) \triangleq \int_0^x t^{p-1}(1-t)^{q-1} dt \quad (5.55)$$

is the incomplete beta function [2, Eq. (8.391)].

5.3.2 Nakagami-q (Hoyt) Fading Channel

Substituting (5.7) in (5.52) and making use of the Laplace transform found in Erdelyi et al. [3, Eq. (8)], recognizing the relation between the associated Legendre function and the Gaussian hypergeometric function [2, Eq. (8.771.1)], we obtain

$$I \triangleq J_q(b, q, \bar{\gamma}, l) = l \left(\frac{b^2\bar{\gamma}}{2}\right)^l \left(\frac{1+q^2}{q}\right) \int_0^{\pi/2} \frac{\cos\theta}{(\sin\theta)^{1+2l}}$$

$$\times \left[\left(\frac{b^2\bar{\gamma}}{2\sin^2\theta} + \frac{(1+q^2)^2}{4q^2}\right)^2 - \left(\frac{1-q^4}{4q^2}\right)^2\right]^{-[(l+1)/2]}$$

$$\times {}_2F_1\left(-l, l+1; 1; \frac{1}{2} - \frac{1}{2} \frac{\frac{b^2\bar{\gamma}}{2\sin^2\theta} + \frac{(1+q^2)^2}{4q^2}}{\sqrt{\left(\frac{b^2\bar{\gamma}}{2\sin^2\theta} + \frac{(1+q^2)^2}{4q^2}\right)^2 - \left(\frac{1-q^4}{4q^2}\right)^2}}\right) d\theta$$

(5.56)

5.3.3 Nakagami-n (Rice) Fading Channel

Substituting (5.11) in (5.52) and making use of the Laplace transform found in Endelyi et al. [3, Eq. (20)], then recognizing the relation between the Whittaker function and the confluent hypergeometric function [2, Eq. (9.220.2)], we obtain

$$I \triangleq J_n(b, n, \bar{\gamma}, l)$$

$$= 2l \left(\frac{b^2\bar{\gamma}}{2}\right)^l (1+n^2)e^{-n^2} \int_0^{\pi/2} \frac{\cos\theta}{(\sin\theta)^{1+2l}} \left(1+n^2+\frac{b^2\bar{\gamma}}{2\sin^2\theta}\right)^{-l-1}$$

$$\times {}_1F_1\left(1+l, 1; \frac{n^2(1+n^2)}{1+n^2+b^2\bar{\gamma}/2\sin^2\theta}\right) d\theta \qquad (5.57)$$

or equivalently in terms of the Rician parameter,

$$I \triangleq J_n(b, K, \bar{\gamma}, l)$$

$$= 2l \left(\frac{b^2\bar{\gamma}}{2}\right)^l (1+K)e^{-K} \int_0^{\pi/2} \frac{\cos\theta}{(\sin\theta)^{1+2l}} \left(1+K+\frac{b^2\bar{\gamma}}{2\sin^2\theta}\right)^{-l-1}$$

$$\times {}_1F_1\left(1+l; 1; \frac{K(1+K)}{1+K+b^2\bar{\gamma}/2\sin^2\theta}\right) d\theta \qquad (5.58)$$

where ${}_1F_1(\cdot;\cdot;\cdot)$ is the confluent hypergeometric function [2, Sec. 9.20].

5.3.4 Nakagami-m Fading Channel

Substituting (5.15) in (5.52) and making use of Eq. (3.381.4) of Ref. 2, we obtain

$$I \triangleq J_m(b, m, \bar{\gamma}, l) = \frac{2}{B(m, l)} \left(\frac{b^2\bar{\gamma}}{2m}\right)^l \int_0^{\pi/2} \frac{\cos\theta}{(\sin\theta)^{1+2l}} \left(1+\frac{b^2\bar{\gamma}}{2m\sin^2\theta}\right)^{-l-m} d\theta \qquad (5.59)$$

where

$$B(m, l) = B(l, m) \triangleq \frac{\Gamma(m)\Gamma(l)}{\Gamma(m+l)} \qquad (5.60)$$

is the beta function [2, Eq. (8.384.1)]. Making the change of variables $t = (1+b^2\bar{\gamma}/2m\sin^2\theta)^{-1}$, then after some manipulation we arrive at the equivalent compact result

$$J_m(b, \bar{\gamma}, l) = \frac{1}{B(m, l)} \int_0^{(1+b^2\bar{\gamma}/2m)^{-1}} t^{m-1}(1-t)^{l-1} dt = \frac{B_{(1+b^2\bar{\gamma}/2m)^{-1}}(m, l)}{B(m, l)} \qquad (5.61)$$

or in terms of the incomplete beta function ratio [2, Eq. (8.392)],

$$I_x(p, q) \triangleq \frac{B_x(p, q)}{B(p, q)} \tag{5.62}$$

the still simpler form

$$J_m(b, \overline{\gamma}, l) = I_{(1+b^2\overline{\gamma}/2m)^{-1}}(m, l) \tag{5.63}$$

For the Rayleigh ($m = 1$) case, (5.61) clearly reduces to (5.54) since $B(1, l) = l^{-1}$.

5.3.5 Log-Normal Shadowing Channel

Substituting the PDF of (5.19) into (5.52) and making the change of variables, $x = (10 \log_{10} \gamma - \mu)/\sqrt{2}\sigma$ results after much simplification in

$$I \triangleq J_{\ln}(b, \mu, \sigma, l) = \frac{1}{\sqrt{\pi}\Gamma(l)} \sum_{i=1}^{n} \Gamma\left(l, \frac{b^2}{2} \cdot 10^{(x_i\sqrt{2}\sigma+\mu)/10}\right) \tag{5.64}$$

where again $\{x_i\}$, $i = 1, 2, \ldots, n$, are the zeros of the nth-order Hermite polynomial $He_n(x)$, as discussed in Section 5.1.5.

5.3.6 Composite Log-Normal Shadowing/Nakagami-*m* Fading Channel

Finally, for the composite log-normal shadowing/Nakagami-*m* fading channel treated in Section 5.1.6, we substitute the PDF of (5.23) into (5.52) together with the change of variables $x = (10 \log_{10} \Omega - \mu)/\sqrt{2}\sigma$, resulting in

$$I \triangleq J_{g/\ln}(b, \mu, \sigma, m, l) = \frac{1}{\sqrt{\pi}} \sum_{i=1}^{n} w_i I_{[1+(b^2/2m)\cdot 10^{(x_i\sqrt{2}\sigma+\mu)/10}]^{-1}}(m, l) \tag{5.65}$$

where now in addition $\{w_i\}$, $i = 1, 2, \ldots, n$, are the Gauss–quadrature weights as discussed in Section 5.1.5.

5.4 INTEGRALS INVOLVING OTHER FUNCTIONS

When studying the error probability performance of certain modulation schemes over generalized fading channels, we shall have reason to evaluate integrals involving special functions other than the three considered previously in this chapter. In this section we consider integrals involving two such special functions corresponding to well-known modulation schemes.

5.4.1 *M*-PSK Error Probability Integral

When studying the average error probability performance of M-PSK over generalized fading channels, we shall have reason to evaluate integrals of the form

$$K = \int_0^\infty \frac{1}{\pi} \int_0^{(M-1)\pi/M} \exp\left(-\frac{a^2\gamma}{2\sin^2\theta}\right) d\theta \, p_\gamma(\gamma) \, d\gamma$$

$$= \frac{1}{\pi} \int_0^{(M-1)\pi/M} \left[\int_0^\infty \exp\left(-\frac{a^2\gamma}{2\sin^2\theta}\right) p_\gamma(\gamma) \, d\gamma\right] d\theta \quad (5.66)$$

where specifically $a^2 = 2\sin^2\pi/M$. The integral in (5.66) is a generalization of the one in (5.2) in the sense that the latter is a special case of the form corresponding to $M = 2$. Thus (5.66) follows directly from (5.3) and is given by

$$K = \frac{1}{\pi} \int_0^{(M-1)\pi/M} M_\gamma\left(-\frac{a^2}{2\sin^2\theta}\right) d\theta \quad (5.67)$$

Although this may seem like a simple generalization, unfortunately the replacement of the $\pi/2$ upper limit in (5.3) by $(M-1)\pi/M$ results wherever possible in closed-form expressions for (5.67) that, in general, are significantly more complicated. Without further ado, we present the results for the evaluation of (5.67) corresponding to the various types of fading channels, where closed-form results can be obtained. The results corresponding to the remainder of the fading channels can be obtained by the same upper limit replacement as mentioned above in the corresponding expressions of Section 5.1.

5.4.1.1 Rayleigh Fading Channel.
Substituting (5.5) in (5.67) and making use of (5A.15), we obtain

$$K \triangleq K_r(a, \bar{\gamma}, M) = \frac{1}{\pi} \int_0^{(M-1)\pi/M} \left(1 + \frac{a^2\bar{\gamma}}{2\sin^2\theta}\right)^{-1} d\theta$$

$$= \frac{M-1}{M}\left\{1 - \sqrt{\frac{a^2\bar{\gamma}/2}{1+a^2\bar{\gamma}/2}} \frac{M}{(M-1)\pi}\left[\frac{\pi}{2} + \tan^{-1}\left(\sqrt{\frac{a^2\bar{\gamma}/2}{1+a^2\bar{\gamma}/2}} \cot\frac{\pi}{M}\right)\right]\right\}$$

$$(5.68)$$

which reduces to (5.6) when $M = 2$.

5.4.1.2 Nakagami-m Fading Channel.
Here we need to substitute the Laplace transform of (5.15) into (5.67). After this is done, then making use of (5A.17), we obtain

116 USEFUL EXPRESSIONS FOR EVALUATING AVERAGE ERROR PROBABILITY PERFORMANCE

$$K \triangleq K_m(a, \overline{\gamma}, m, M)$$

$$= \frac{1}{\pi} \int_0^{(M-1)\pi/M} \left(1 + \frac{a^2\overline{\gamma}}{2m\sin^2\theta}\right)^{-m} d\theta$$

$$= \frac{M-1}{M} - \frac{1}{\pi}\sqrt{\frac{a^2\overline{\gamma}/2m}{1+a^2\overline{\gamma}/2m}} \left\{ \left(\frac{\pi}{2} + \tan^{-1}\alpha\right) \sum_{k=0}^{m-1} \binom{2k}{k} \frac{1}{[4(1+a^2\overline{\gamma}/2m)]^k} \right.$$

$$\left. + \sin(\tan^{-1}\alpha) \sum_{k=1}^{m-1}\sum_{i=1}^{k} \frac{T_{ik}}{(1+a^2\overline{\gamma}/2m)^k} [\cos(\tan^{-1}\alpha)]^{2(k-i)+1} \right\} \quad (5.69)$$

where

$$\alpha \triangleq \sqrt{\frac{a^2\overline{\gamma}/2m}{1+a^2\overline{\gamma}/2m}} \cot\frac{\pi}{M} \quad (5.70)$$

and T_{ik} is again given by (5.32).

5.4.2 Arbitrary Two-Dimensional Signal Constellation Error Probability Integral

As a generalization of QAM, Craig [4] showed that the evaluation of the average error probability performance of an arbitrary two-dimensional (2-D) signal constellation with polygon-shaped decision regions over the AWGN channel can be expressed as a summation of integrals of the form[7]

$$P_i = \frac{1}{2\pi} \int_0^{\theta_i} \exp\left[-\frac{a_i^2 \sin^2\psi_i}{2\sin^2(\theta+\psi_i)}\right] d\theta \quad (5.71)$$

where a_i^2 is a signal-to-noise ratio parameter associated with the ith signal in the set and θ_i and ψ_i are angles associated with the correct decision region corresponding to that signal. Thus, when studying the average error probability performance of these 2-D signal constellations over generalized fading channels, we shall have reason to evaluate integrals of the form

$$L = \int_0^\infty \frac{1}{2\pi} \int_0^{\theta_i} \exp\left[-\frac{a_i^2 \gamma \sin^2\psi_i}{2\sin^2(\theta+\psi_i)}\right] d\theta \, p_\gamma(\gamma) \, d\gamma$$

$$= \frac{1}{2\pi} \int_0^{\theta_i} \left\{ \int_0^\infty \exp\left[-\frac{a_i^2 \gamma \sin^2\psi_i}{2\sin^2(\theta+\psi_i)}\right] p_\gamma(\gamma) \, d\gamma \right\} d\theta \quad (5.72)$$

---

[7] Equation (5.71) appears as Eq. (13) in Ref. 4 but with an error of a factor of $\frac{1}{2}$ [i.e., the factor $1/\pi$ that premultiplies the integral there should be $1/2\pi$, as shown in (5.71)].

By comparison with (5.66), we observe that (5.72) can be expressed in the form of (5.67), namely,

$$L = \frac{1}{2\pi} \int_0^{\theta_i} M_\gamma \left[-\frac{a_i^2 \sin^2 \psi_i}{2 \sin^2(\theta + \psi_i)} \right] d\theta \tag{5.73}$$

where again $M_\gamma(s)$ is the MGF of γ. Evaluation of the Laplace transform integrand in (5.73) for the various types of fading channels follows exactly along the lines of the previous results and hence is not repeated here. Unfortunately, however, for arbitrary θ_i it is not always possible now to obtain closed-form expressions for L even when the integrand is obtainable in closed form. However, for the Rayleigh channel, using (5.5) for $M_\gamma(-s)$ and the indefinite form of the integral in (5A.11), it is straightforward to obtain the following closed-form solution:

$$L \triangleq L_r(a_i, \overline{\gamma}, \theta_i, \psi_i)$$

$$= \frac{1}{4} - \frac{c_i}{2\pi} \sqrt{\frac{1}{c_i(1+c_i)}} \tan^{-1}\left(\sqrt{\frac{1+c_i}{c_i}} \tan \theta \right) \Bigg|_{-\psi_i}^{\theta_i - \psi_i}$$

$$= \frac{1}{4} - \frac{c_i}{2\pi} \sqrt{\frac{1}{c_i(1+c_i)}} \left\{ \tan^{-1}\left[\sqrt{\frac{1+c_i}{c_i}} \tan(\theta_i - \psi_i) \right] \right.$$

$$\left. + \tan^{-1}\left(\sqrt{\frac{1+c_i}{c_i}} \tan \psi_i \right) \right\} \tag{5.74}$$

where

$$c_i \triangleq \frac{a_i^2 \overline{\gamma}}{2} \sin^2 \psi_i \tag{5.75}$$

For Nakagami-m fading, using the Laplace transform in (5.15), we obtain

$$L \triangleq L_m(a_i, \overline{\gamma}, m, \theta_i, \psi_i)$$

$$= \frac{1}{2\pi} \left[\int_0^{\theta_i - \psi_i} \left(\frac{\sin^2 \phi}{\sin^2 \phi + c_i/m} \right)^m d\phi + \int_0^{\psi_i} \left(\frac{\sin^2 \phi}{\sin^2 \phi + c_i/m} \right)^m d\phi \right] \tag{5.76}$$

with c_i still as defined in (5.75). If, depending on the signal constellation, θ_i and $\theta_i - \psi_i$ both turn out to be either in the form $(M-1)\pi/M$ or π/M for $M = 2^m$, m integer, the closed-form results of (5A.16) and (5A.21) can be used to obtain (5.76) in closed form. Otherwise, the single-integral form of (5.76) must be used.

The results for the other fading channel types will, in general, be expressed as a single integral with finite limits $(0, \theta_i)$ in accordance with (5.73) and the various closed-form expressions previously obtained for $M_\gamma(-s)$.

5.4.3 Integer Powers of the Gaussian Q-Function

Associated with the study of the average error probability performance of coherent communication systems using differentially encoded QPSK and M-ary orthogonal signals in the presence of slow fading, we shall have need to evaluate integrals of the form

$$I_k \overset{\Delta}{=} \int_0^\infty Q^k(a\sqrt{\gamma}) p_\gamma(\gamma) \, d\gamma \qquad (5.77)$$

where k is assumed to be integer. In general, for arbitrary integer values of k, I_k cannot be obtained in the desired form. However, certain special cases, namely, $k = 1, 2, 3, 4$, do exist either in closed form or in the form of a single integral with finite limits and an integrand composed of elementary functions. For $k = 1$, the results were presented in Section 5.1. The specific results corresponding to $k = 2, 3, 4$ for Rayleigh and Nakagami-m fading are presented in what follows.

5.4.3.1 Rayleigh Fading Channel.
To evaluate (5.77) for $k = 2$, we substitute the alternative form of $Q^2(x)$ of (4.9) into this equation, resulting in

$$I_2 = \frac{1}{\pi} \int_0^{\pi/4} M_\gamma\left(-\frac{a^2}{2\sin^2\theta}\right) d\theta \qquad (5.78)$$

which is identical to (5.3) except that the upper limit is now $\pi/4$ rather than $\pi/2$. Using (5.5) for $M_\gamma(-s)$, (5.78) becomes [analogous to (5.6)]

$$I_2 \overset{\Delta}{=} I_{2,r}(a, \bar{\gamma}) = \frac{1}{\pi} \int_0^{\pi/4} \left(1 + \frac{a^2 \bar{\gamma}}{2\sin^2\theta}\right)^{-1} d\theta$$

$$= \frac{1}{\pi} \int_0^{\pi/4} \frac{\sin^2\theta}{\sin^2\theta + a^2\bar{\gamma}/2} \, d\theta \qquad (5.79)$$

The integral in (5.79) is evaluated in closed form in Appendix 5A. In particular, using (5A.13), we obtain

$$I_{2,r}(a, \bar{\gamma}) = \frac{1}{4}\left[1 - \sqrt{\frac{c}{1+c}}\left(\frac{4}{\pi} \tan^{-1}\sqrt{\frac{1+c}{c}}\right)\right], \quad c \overset{\Delta}{=} \frac{a^2 \bar{\gamma}}{2} \qquad (5.80)$$

For $k = 3$, an expression for $Q^3(x)$ in the form of (4.2) and (4.9) has not been found. Nevertheless, it is still possible to evaluate I_3 in the single-integral form referred to above. In particular, writing $Q^3(x)$ as the product $Q(x)Q^2(x)$ and using (4.2) and (4.9) in (5.77), the following sequence of steps occurs.

$$I_3 \overset{\Delta}{=} I_{3,r}(a, \bar{\gamma}) = \frac{1}{\pi} \int_0^{\pi/4} \frac{1}{\pi} \int_0^{\pi/2} \int_0^\infty \exp\left[\frac{a^2\gamma}{2}\left(\frac{1}{\sin^2\theta} + \frac{1}{\sin^2\phi}\right)\right] p_\gamma(\gamma) \, d\gamma \, d\theta \, d\phi$$

$$= \frac{1}{\pi} \int_0^{\pi/4} \frac{1}{\pi} \int_0^{\pi/2} P_\gamma\left[\frac{a^2}{2}\left(\frac{1}{\sin^2\theta} + \frac{1}{\sin^2\phi}\right)\right] d\theta \, d\phi$$

$$= \frac{1}{\pi} \int_0^{\pi/4} \frac{1}{\pi} \int_0^{\pi/2} \left(1 + \frac{a^2\bar{\gamma}}{2\sin^2\theta} + \frac{a^2\bar{\gamma}}{2\sin^2\phi}\right)^{-1} d\theta\, d\phi$$

$$= \frac{1}{\pi} \int_0^{\pi/4} \frac{2}{a^2\bar{\gamma}} c(\phi) \left[\frac{1}{\pi} \int_0^{\pi/2} \frac{\sin^2\theta}{\sin^2\theta + c(\phi)} d\theta\right] d\phi,$$

$$c(\phi) \triangleq \frac{a^2\bar{\gamma}}{2}\left(\frac{\sin^2\phi}{\sin^2\phi + a^2\bar{\gamma}/2}\right) \tag{5.81}$$

Using the closed-form result from (5A.9) for the inner integral (in brackets), we get the desired result

$$I_{3,r}(a, \bar{\gamma}) = (a^2\bar{\gamma})^{-1} \frac{1}{\pi} \int_0^{\pi/4} c(\phi) \left[1 - \sqrt{\frac{c(\phi)}{1 + c(\phi)}}\right] d\phi \tag{5.82}$$

It is also possible to obtain a single-integral form for I_4 by writing $Q^4(x)$ as the product $Q^2(x)Q^2(x)$ and then using (4.9) twice in (5.77) followed by the closed-form expression in (5.80) to evaluate the inner integral. The steps leading to the result parallel those in (5.81) and produce

$$I_4 \triangleq I_{4,r}(a, \bar{\gamma}) = \frac{1}{\pi} \int_0^{\pi/4} \frac{2}{a^2\bar{\gamma}} c(\phi) \left[\frac{1}{\pi} \int_0^{\pi/4} \frac{\sin^2\theta}{\sin^2\theta + c(\phi)} d\theta\right] d\phi \tag{5.83}$$

Finally, using Eq. (5A.13) for the integral in brackets in (5.83) produces the desired result:

$$I_{4,r}(a, \bar{\gamma}) = \left(\frac{a^2\bar{\gamma}}{2}\right)^{-1} \frac{1}{\pi} \int_0^{\pi/4} \frac{1}{4}$$

$$\times \left\{1 - \sqrt{\frac{c(\phi)}{1 + c(\phi)}} \left[\frac{4}{\pi} \tan^{-1}\left(\sqrt{\frac{1 + c(\phi)}{c(\phi)}}\right)\right]\right\} d\phi \tag{5.84}$$

5.4.3.2 Nakagami-m Fading Channel.
Following the same procedure as for the Rayleigh fading channel, we can evaluate (5.77) for the Nakagami-m fading channel as follows. For $k = 2$, we again start with (5.78) but now use (5.15) for $M_\gamma(-s)$, which produces [analogous to (5.16)]

$$I_2 \triangleq I_{2,m}(a, m, \bar{\gamma}) = \frac{1}{\pi} \int_0^{\pi/4} \left(1 + \frac{a^2\bar{\gamma}}{2\sin^2\theta}\right)^{-m} d\theta$$

$$= \frac{1}{\pi} \int_0^{\pi/4} \left(\frac{\sin^2\theta}{\sin^2\theta + a^2\bar{\gamma}/2}\right)^m d\theta \tag{5.85}$$

120 USEFUL EXPRESSIONS FOR EVALUATING AVERAGE ERROR PROBABILITY PERFORMANCE

The integral in (5.85) is evaluated in closed form in Appendix 5A. In particular, using (5A.21), we obtain (for m integer)

$$I_{2,m}(a, m, \bar{\gamma}) = \frac{1}{4} - \frac{1}{\pi}\sqrt{\frac{c}{1+c}} \left\{ \left(\frac{\pi}{2} - \tan^{-1}\sqrt{\frac{c}{1+c}} \right) \sum_{k=0}^{m-1} \binom{2k}{k} \right.$$

$$\times \frac{1}{[4(1+c)]^k} - \sin\left(\tan^{-1}\sqrt{\frac{c}{1+c}}\right) \sum_{k=1}^{m-1}\sum_{i=1}^{k} \frac{T_{ik}}{(1+c)^k}$$

$$\left. \times \left[\cos\left(\tan^{-1}\sqrt{\frac{c}{1+c}}\right)\right]^{2(k-i)+1} \right\} \qquad (5.86)$$

where c is defined in (5.80) and T_{ik} in (5.32). For $k = 3$, the steps analogous to (5.81) are as follows:

$$I_3 \triangleq I_{3,m}(a, m, \bar{\gamma})$$

$$= \frac{1}{\pi}\int_0^{\pi/4} \frac{1}{\pi}\int_0^{\pi/2} M_\gamma \left[\frac{a^2}{2}\left(\frac{1}{\sin^2\theta} + \frac{1}{\sin^2\phi}\right) \right] d\theta\, d\phi$$

$$= \frac{1}{\pi}\int_0^{\pi/4} \frac{1}{\pi}\int_0^{\pi/2} \left(1 + \frac{a^2\bar{\gamma}}{2m\sin^2\theta} + \frac{a^2\bar{\gamma}}{2m\sin^2\phi}\right)^{-m} d\theta\, d\phi$$

$$= \frac{1}{\pi}\int_0^{\pi/4} \left(\frac{2}{a^2\bar{\gamma}}c(\phi)\right)^m \left[\frac{1}{\pi}\int_0^{\pi/2} \left(\frac{\sin^2\theta}{\sin^2\theta + c(\phi)}\right)^m d\theta\right] d\phi \qquad (5.87)$$

where $c(\phi)$ is still as defined in (5.81). Using the closed-form result in (5A.4b) we obtain the desired result as

$$I_{3,m}(a, m, \bar{\gamma}) = \frac{1}{\pi}\int_0^{\pi/4} \left(\frac{2}{a^2\bar{\gamma}}c(\phi)\right)^m \left[\frac{1 - \mu(c(\phi))}{2}\right]^m$$

$$\times \sum_{k=0}^{m-1} \binom{m-1+k}{k} \left[\frac{1 + \mu(c(\phi))}{2}\right]^k d\phi \qquad (5.88)$$

where [see (5A.4a)]

$$\mu(c) \triangleq \sqrt{\frac{c}{1+c}} \qquad (5.89)$$

Finally, for $k = 4$ we get

$$I_4 \triangleq I_{4,m}(a, m, \bar{\gamma})$$

$$= \frac{1}{\pi}\int_0^{\pi/4} \frac{1}{\pi}\int_0^{\pi/4} M_\gamma\left[\frac{a^2}{2}\left(\frac{1}{\sin^2\theta} + \frac{1}{\sin^2\phi}\right)\right] d\theta\, d\phi$$

$$= \frac{1}{\pi} \int_0^{\pi/4} \frac{1}{\pi} \int_0^{\pi/4} \left(1 + \frac{a^2 \bar{\gamma}}{2m \sin^2 \theta} + \frac{a^2 \bar{\gamma}}{2m \sin^2 \phi}\right)^{-m} d\theta \, d\phi$$

$$= \frac{1}{\pi} \int_0^{\pi/4} \left(\frac{2}{a^2 \bar{\gamma}} c(\phi)\right)^m \left[\frac{1}{\pi} \int_0^{\pi/4} \left(\frac{\sin^2 \theta}{\sin^2 \theta + c(\phi)}\right)^m d\theta\right] d\phi \quad (5.90)$$

whereupon using (5.86) for the term in brackets with c replaced by $c(\phi)$, we get

$I_{4,m}(a, m, \bar{\gamma})$

$$= \frac{1}{\pi} \int_0^{\pi/4} \left(\frac{2}{a^2 \bar{\gamma}} c(\phi)\right)^m \left[\frac{1}{4} - \frac{1}{\pi} \sqrt{\frac{c(\phi)}{1+c(\phi)}} \left\{\left(\frac{\pi}{2} - \tan^{-1} \sqrt{\frac{c(\phi)}{1+c(\phi)}}\right)\right.\right.$$

$$\times \sum_{k=0}^{m-1} \binom{2k}{k} \frac{1}{[4(1+c(\phi))]^k} - \sin\left(\tan^{-1} \sqrt{\frac{c(\phi)}{1+c(\phi)}}\right)$$

$$\left.\left. \times \sum_{k=1}^{m-1} \sum_{i=1}^{k} \frac{T_{ik}}{[1+c(\phi)]^k} \left[\cos\left(\tan^{-1} \sqrt{\frac{c(\phi)}{1+c(\phi)}}\right)\right]^{2(k-i)+1}\right\}\right] d\phi \quad (5.91)$$

Although an equation like (5.91) gives the appearance of being complex, we remind the reader that we have accomplished our goal, namely, to express the result in a form no more complicated than a single integral with finite limits and an integrand containing elementary (in this case, pure trigonometric) functions.

5.4.4 Integer Powers of M-PSK Error Probability Integrals

Associated with the study of the average error probability performance of coherently detected differentially encoded M-PSK in the presence of slow fading, we shall have need to evaluate integrals of the form

$$K_2 \triangleq \int_0^\infty \left[\frac{1}{\pi} \int_0^{(M-1)\pi/M} \exp\left(-\frac{a^2 \gamma}{2 \sin^2 \theta}\right) d\theta\right]^2 p_\gamma(\gamma) \, d\gamma \quad (5.92)$$

and

$$L_2(\theta_{u1}, \theta_{u2}, a_1, a_2) \triangleq \int_0^\infty \left[\frac{1}{\pi} \int_0^{\theta_{u1}} \exp\left(-\frac{a_1^2 \gamma}{2 \sin^2 \theta}\right) d\theta\right]$$

$$\times \left[\frac{1}{\pi} \int_0^{\theta_{u2}} \exp\left(-\frac{a_2^2 \gamma}{2 \sin^2 \theta}\right) d\theta\right] p_\gamma(\gamma) \, d\gamma \quad (5.93)$$

where, as was the case in Section 5.4.1, $a^2 = 2\sin^2\pi/M$, and now, in addition, a_1^2 and a_2^2 assume the possible values $2\sin^2(2k\pm 1)\pi/M$, $k = 0, 1, 2, \ldots, M-1$, and θ_{u1} and θ_{u2} assume the possible values $\pi[1-(2k\pm 1)/M]$. While (5.92) can be evaluated in the desired form for both Rayleigh and Nakagami-m fading, unfortunately, (5.93) can be obtained in such a form only for the Rayleigh case. Thus we shall only present the results for this single fading case.

5.4.4.1 Rayleigh Fading Channel. Since (5.92) can be viewed as a special case of (5.93) corresponding to $a_1^2 = a_2^2 = a^2$ and $\theta_{u1} = \theta_{u2} = (M-1)\pi/M$, we shall consider only the generic form in (5.93), where $a_1^2, a_2^2, \theta_{u1}$, and θ_{u2} are allowed to be completely arbitrary. Following steps analogous to those in (5.81), we proceed as follows:

$$\begin{aligned}
L_2(\theta_{u1}, \theta_{u2}, a_1, a_2) \\
&= \frac{1}{\pi}\int_0^{\theta_{u1}} \frac{1}{\pi}\int_0^{\theta_{u2}} M_\gamma\left[-\frac{1}{2}\left(\frac{a_1^2}{\sin^2\theta} + \frac{a_2^2}{\sin^2\phi}\right)\right] d\theta\, d\phi \\
&= \frac{1}{\pi}\int_0^{\theta_{u1}} \frac{1}{\pi}\int_0^{\theta_{u2}} \left(1 + \frac{a_1^2\overline{\gamma}}{2\sin^2\theta} + \frac{a_2^2\overline{\gamma}}{2\sin^2\phi}\right)^{-1} d\theta\, d\phi \\
&= \frac{1}{\pi}\int_0^{\theta_{u1}} \left(\frac{2}{a_1^2\overline{\gamma}}c_{12}(\phi)\right)\left[\frac{1}{\pi}\int_0^{\theta_{u2}} \frac{\sin^2\theta}{\sin^2\theta + c_{12}(\phi)}d\theta\right] d\phi \quad (5.94)
\end{aligned}$$

where $c_{12}(\phi)$ is defined analogous to $c(\phi)$ in (5.81) as

$$c_{12}(\phi) \triangleq \frac{a_1^2\overline{\gamma}}{2}\left(\frac{\sin^2\phi}{\sin^2\phi + a_2^2\overline{\gamma}/2}\right) \quad (5.95)$$

Rewriting the integral in brackets as

$$\int_0^{\theta_{u2}} \frac{\sin^2\theta}{\sin^2\theta + c_{12}(\phi)}d\theta = \theta_{u2} - \int_0^{\theta_{u2}} \frac{c_{12}(\phi)}{\sin^2\theta + c_{12}(\phi)}d\theta \quad (5.96)$$

then making use of Eq. (2.562.1) of Ref. 2, we obtain

$$\int_0^{\theta_{u2}} \frac{\sin^2\theta}{\sin^2\theta + c_{12}(\phi)}d\theta \\ = \frac{1}{\pi}\left[\theta_{u2} - \sqrt{\frac{c_{12}(\phi)}{1+c_{12}(\phi)}}\tan^{-1}\left(\sqrt{\frac{1+c_{12}(\phi)}{c_{12}(\phi)}}\tan\theta_{u2}\right)\right] \quad (5.97)$$

and hence

$$L_2(\theta_{u1}, \theta_{u2}, a_1, a_2) = \left(\frac{1}{\pi}\right)^2 \frac{2}{a_1^2 \gamma} \int_0^{\theta_{u1}} c_{12}(\phi)$$

$$\times \left[\theta_{u2} - \sqrt{\frac{c_{12}(\phi)}{1+c_{12}(\phi)}} \tan^{-1}\left(\sqrt{\frac{1+c_{12}(\phi)}{c_{12}(\phi)}} \tan \theta_{u2}\right)\right] d\phi$$

(5.98)

Since as mentioned above, $K_2 = L_2((M-1)\pi/M, (M-1)\pi/M, a, a)$, this special case evaluates as

$$K_2 = \left(\frac{1}{\pi}\right)^2 \frac{2}{a^2 \gamma} \int_0^{(M-1)\pi/M} c(\phi) \left[\frac{(M-1)\pi}{M} - \sqrt{\frac{c(\phi)}{1+c(\phi)}}\right.$$

$$\left. \times \tan^{-1}\left(\sqrt{\frac{1+c(\phi)}{c(\phi)}} \tan \frac{(M-1)\pi}{M}\right)\right] d\phi \qquad (5.99)$$

where $c(\phi)$ is as defined in (5.81). The other special cases that will be of interest in later chapters dealing with differentially encoded, coherently detected M-PSK are $L_2(\theta_+, \theta_+, a_+, a_+)$, $L_2(\theta_-, \theta_-, a_-, a_-)$, and $L_2(\theta_+, \theta_-, a_+, a_-)$, where $\theta_\pm \triangleq \pi[1 - (2k \pm 1)/M]$, $a_\pm^2 \triangleq 2\sin^2(2k \pm 1)\pi/M$, $k = 0, 1, 2, \ldots, M-1$. These special cases of (5.98) evaluate as

$$L_2(\theta_\pm, \theta_\pm, a_\pm, a_\pm)$$

$$= \left(\frac{1}{\pi}\right)^2 \left(\frac{2}{a_\pm^2 \gamma}\right) \int_0^{\pi(1-(2k\pm 1)/M)} c_\pm(\phi) \left[\pi\left(1 - \frac{2k \pm 1}{M}\right) - \sqrt{\frac{c_\pm(\phi)}{1+c_\pm(\phi)}}\right.$$

$$\left. \times \tan^{-1}\left\{\sqrt{\frac{1+c_\pm(\phi)}{c_\pm(\phi)}} \tan\left[\pi\left(1 - \frac{2k \pm 1}{M}\right)\right]\right\}\right] d\phi \qquad (5.100)$$

and

$$L_2(\theta_+, \theta_-, a_+, a_-)$$

$$= \left(\frac{1}{\pi}\right)^2 \left(\frac{2}{a_+^2 \gamma}\right) \int_0^{\pi(1-(2k+1)/M)} c_{+-}(\phi) \left[\pi\left(1 - \frac{2k-1}{M}\right) - \sqrt{\frac{c_{+-}(\phi)}{1+c_{+-}(\phi)}}\right.$$

$$\left. \times \tan^{-1}\left\{\sqrt{\frac{1+c_{+-}(\phi)}{c_{+-}(\phi)}} \tan\left[\pi\left(1 - \frac{2k-1}{M}\right)\right]\right\}\right] d\phi \qquad (5.101)$$

where

$$c_{\pm}(\phi) \triangleq \frac{a_{\pm}^2 \bar{\gamma}}{2}\left(\frac{\sin^2 \phi}{\sin^2 \phi + a_{\pm}^2 \bar{\gamma}/2}\right), \quad c_{+-}(\phi) \triangleq \frac{a_{+}^2 \bar{\gamma}}{2}\left(\frac{\sin^2 \phi}{\sin^2 \phi + a_{-}^2 \bar{\gamma}/2}\right) \quad (5.102)$$

REFERENCES

1. M. Abramowitz and I. A. Stegun, *Handbook of Mathematical Functions with Formulas, Graphs, and Mathematical Tables*, 9th ed. New York: Dover Press, 1972.
2. I. S. Gradshteyn and I. M. Ryzhik, *Table of Integrals, Series, and Products*, 5th ed. San Diego, CA: Academic Press, 1994.
3. A. Erdelyi, W. Magnus, F. Oberhettinger, and F. G. Tricomi, *Table of Integral Transforms*, vol. 1, New York: McGraw-Hill, 1954.
4. J. W. Craig, "A new, simple and exact result for calculating the probability of error for two-dimensional signal constellations," *IEEE MILCOM'91 Conf. Rec.*, Boston, pp. 25.5.1–25.5.5.
5. T. Eng and L. B. Milstein, "Coherent DS-CDMA performance in Nakagami multipath fading," *IEEE Trans. Commun.*, vol. 43, February/March/April 1995, pp. 1134–1143.
6. J. Proakis, *Digital Communications*, 3rd ed. New York: McGraw-Hill, 1995.
7. S. Chennakeshu and J. B. Anderson, "Error rates for Rayleigh fading multichannel reception of MPSK signals," *IEEE Trans. Commun.*, vol. 43, February/March/April 1995, pp. 338–346.
8. J. Edwards, *A Treatise on the Integral Calculus*, Vol. II. London: Macmillan, 1922.
9. E. Villier, "Performance analysis of optimum combining with multiple interferers in flat Rayleigh fading," *IEEE Trans. Commun.*, vol. 47, October 1999, pp. 1503–1510.

APPENDIX 5A: EVALUATION OF DEFINITE INTEGRALS ASSOCIATED WITH RAYLEIGH AND NAKAGAMI-m FADING

1. $\dfrac{1}{\pi}\displaystyle\int_0^{\pi/2}\left(\dfrac{\sin^2\theta}{\sin^2\theta + c}\right)^m d\theta$

We wish to consider evaluating the integral

$$I_m = \frac{1}{\pi}\int_0^{\pi/2}\left(\frac{\sin^2\theta}{\sin^2\theta + c}\right)^m d\theta \quad (5A.1)$$

for m both integer and noninteger. To do this we shall make an equivalence with another definite integral for which closed-form results have been reported in the literature. In particular, it has been shown [5, Eq. (A8)] that the integral

$$J_m(a,b) \triangleq \frac{a^m}{\Gamma(m)}\int_0^\infty e^{-at}t^{m-1}Q(\sqrt{bt})\,dt, \qquad n \geq 0 \quad (5A.2)$$

BOUND PRINTED MATTER

Orlando, FL 32816-2430
P.O. Box 16430
4000 Central Florida Blvd
Library - Interlibrary Loan
University of Central Florida

Ship To:

...Loan Department
...Community College Library
...Clearlake Road
...oa, FL 32922

Via: Library Rate

NeedBy: 10/28/02

Borrower: FTU
ILL: 981491z
Req Date: 9/13/02
OCLC #: 42780405
Patron: OKEKE, OBIECHINA DEPT: EG STA
Author: Simon, Marvin Kenneth, 1939-
Title: Digital communication over fading channe
Article:
Vol:
Date:
Verified: OCLC [Format: Book]
Maxcost: 30.00IFM
Due Date:

No:
Pages:

Lending Notes:
Bor Notes: ...

APPENDIX 5A: EVALUATION OF DEFINITE INTEGRALS

has the closed-form result

$$J_m(a,b) \triangleq J_m(c) = \frac{\sqrt{c/\pi}}{2(1+c)^{m+1/2}} \frac{\Gamma\left(m+\frac{1}{2}\right)}{\Gamma(m+1)} {}_2F_1\left(1, m+\frac{1}{2}; m+1; \frac{1}{1+c}\right),$$

$$c \triangleq \frac{b}{2a} \qquad m \text{ noninteger} \qquad (5A.3)$$

When m is restricted to positive integer values, it has been further shown [5, Eq. (A13)] that (5A.3) simplifies to

$$J_m(a,b) \triangleq J_m(c) = \frac{1}{2}\left[1 - \mu(c) \sum_{k=0}^{m-1} \binom{2k}{k} \left(\frac{1-\mu^2(c)}{4}\right)^k\right],$$

$$\mu(c) \triangleq \sqrt{\frac{c}{1+c}} \qquad m \text{ integer} \qquad (5A.4a)$$

which was also obtained previously by Proakis [6, Eq. (14-4-15)] in the form

$$J_m(c) = \left(\frac{1-\mu(c)}{2}\right)^m \sum_{k=0}^{m-1} \binom{m-1+k}{k} \left(\frac{1+\mu(c)}{2}\right)^k, \qquad m \text{ integer}$$

$$(5A.4b)$$

Using the alternative representation of the Gaussian Q-function as given in Eq. (4.2) in (5A.2) gives

$$J_m(a,b) = \frac{a^m}{\Gamma(m)} \int_0^\infty e^{-at} t^{m-1} \left(\frac{1}{\pi} \int_0^{\pi/2} e^{-b t/2 \sin^2 \theta} d\theta\right) dt$$

$$= \frac{a^m}{\pi \Gamma(m)} \int_0^{\pi/2} \int_0^\infty t^{m-1} e^{-(a+b/2 \sin^2 \theta)t} dt\, d\theta \qquad (5A.5)$$

The inner integral on t can be expressed in terms of the integral definition of the gamma function, namely [1, Eq. (6.1.1)],

$$\Gamma(m) = \alpha^m \int_0^\infty t^{m-1} e^{-\alpha t} dt \qquad (5A.6)$$

Thus, using (5A.6) in (5A.5), we obtain

$$J_m(a,b) = \frac{a^m}{\pi \Gamma(m)} \int_0^{\pi/2} \frac{\Gamma(m)}{(a+b/2 \sin^2 \theta)^m} d\theta = \frac{1}{\pi} \int_0^{\pi/2} \frac{1}{(1+b/2a \sin^2 \theta)^m} d\theta$$

$$(5A.7)$$

Finally, letting $c = b/2a$, we can rewrite (5A.7) as

$$J_m(a,b) \triangleq J_m(c) = \frac{1}{\pi} \int_0^{\pi/2} \left(\frac{\sin^2 \theta}{\sin^2 \theta + c}\right)^m d\theta \qquad (5A.8)$$

which is identical with I_m of (5A.1). Thus, equating (5A.8) with (5A.3) and (5A.4) establishes the desired results for m noninteger and m integer, respectively.

One final note is to observe from (5A.4) that $J_1(c) = [1 - \mu(c)]/2$. Thus, a special case of (5A.8) that is of interest on Rayleigh channels is

$$\frac{1}{\pi} \int_0^{\pi/2} \frac{\sin^2 \theta}{\sin^2 \theta + c} d\theta = \frac{1}{2} \left(1 - \sqrt{\frac{c}{1+c}} \right) \tag{5A.9}$$

which could also be obtained directly as follows:

$$\frac{1}{\pi} \int_0^{\pi/2} \frac{\sin^2 \theta}{\sin^2 \theta + c} d\theta = \frac{1}{\pi} \int_0^{\pi/2} \left(1 - \frac{c}{\sin^2 \theta + c} \right) d\theta$$

$$= \frac{1}{2} - \frac{1}{\pi} \int_0^{\pi/2} \frac{c}{\sin^2 \theta + c} d\theta \tag{5A.10}$$

Making use of the definite integral in Eq. (2.562.1) of Ref. 2, we arrive at

$$\frac{1}{\pi} \int_0^{\pi/2} \frac{\sin^2 \theta}{\sin^2 \theta + c} d\theta = \frac{1}{2} - \frac{c}{\pi} \sqrt{\frac{1}{c(1+c)}} \tan^{-1} \left(\sqrt{\frac{1+c}{c}} \tan \theta \right) \Bigg|_0^{\pi/2}$$

$$= \frac{1}{2} \left(1 - \sqrt{\frac{c}{1+c}} \right) \triangleq P(c) \tag{5A.11}$$

The reason for including this alternative derivation is that it is useful in deriving closed-form results for two other integrals of interest related to evaluating the performance of QAM and M-PSK over Rayleigh channels. In particular, for QAM we will have a need to evaluate

$$\frac{1}{\pi} \int_0^{\pi/4} \frac{\sin^2 \theta}{\sin^2 \theta + c} d\theta = \frac{1}{\pi} \int_0^{\pi/4} \left(1 - \frac{c}{\sin^2 \theta + c} \right) d\theta$$

$$= \frac{1}{4} - \frac{1}{\pi} \int_0^{\pi/4} \frac{c}{\sin^2 \theta + c} d\theta \tag{5A.12}$$

Making use of the same indefinite integral as used in (5A.11) we immediately arrive at the desired result, namely,

$$\frac{1}{\pi} \int_0^{\pi/4} \frac{\sin^2 \theta}{\sin^2 \theta + c} d\theta = \frac{1}{4} - \frac{c}{\pi} \sqrt{\frac{1}{c(1+c)}} \tan^{-1} \left(\sqrt{\frac{1+c}{c}} \tan \theta \right) \Bigg|_0^{\pi/4}$$

$$= \frac{1}{4} \left[1 - \sqrt{\frac{c}{1+c}} \left(\frac{4}{\pi} \tan^{-1} \sqrt{\frac{1+c}{c}} \right) \right] \tag{5A.13}$$

2. $\dfrac{1}{\pi} \displaystyle\int_0^{(M-1)\pi/M} \dfrac{\sin^2\theta}{\sin^2\theta + c}\, d\theta$

For M-PSK, we will have a need to evaluate

$$\frac{1}{\pi} \int_0^{(M-1)\pi/M} \frac{\sin^2\theta}{\sin^2\theta + c}\, d\theta = \frac{M-1}{M} - \frac{1}{\pi} \int_0^{(M-1)\pi/M} \frac{c}{\sin^2\theta + c}\, d\theta \tag{5A.14}$$

Making use of the same indefinite integral as used in (5A.11) we immediately arrive at the desired result, namely,

$$\frac{1}{\pi} \int_0^{(M-1)\pi/M} \frac{\sin^2\theta}{\sin^2\theta + c}\, d\theta$$

$$= \left(\frac{M-1}{M}\right) \left[1 - \sqrt{\frac{c}{1+c}} \frac{M}{(M-1)\pi} \tan^{-1}\left(\sqrt{\frac{1+c}{c}} \tan\frac{(M-1)\pi}{M}\right)\right]$$

$$= \left(\frac{M-1}{M}\right) \left\{1 - \sqrt{\frac{c}{1+c}} \frac{M}{(M-1)\pi} \left[\frac{\pi}{2} + \tan^{-1}\left(\sqrt{\frac{c}{1+c}} \cot\frac{\pi}{M}\right)\right]\right\} \tag{5A.15}$$

3. $\dfrac{1}{\pi} \displaystyle\int_0^{(M-1)\pi/M} \left(\dfrac{\sin^2\theta}{\sin^2\theta + c}\right)^m d\theta$

For evaluation of symbol error probability corresponding to single-channel reception of M-PSK on Nakagami-m fading channels and also for multichannel reception of M-PSK on Rayleigh fading channels, we shall have need to evaluate

$$K_m = \frac{1}{\pi} \int_0^{(M-1)\pi/M} \left(\frac{\sin^2\theta}{\sin^2\theta + c}\right)^m d\theta \tag{5A.16}$$

Using a result [7, Eq. (21)] for the symbol error probability performance of M-PSK over a Rayleigh channel with multichannel reception, it is straightforward to show that for m integer,

$$\frac{1}{\pi} \int_0^{(M-1)\pi/M} \left(\frac{\sin^2\theta}{\sin^2\theta + c}\right)^m d\theta$$

$$= \frac{M-1}{M} - \frac{1}{\pi}\sqrt{\frac{c}{1+c}} \left\{\left(\frac{\pi}{2} + \tan^{-1}\alpha\right) \sum_{k=0}^{m-1} \binom{2k}{k} \frac{1}{[4(1+c)]^k}\right.$$

$$\left. + \sin(\tan^{-1}\alpha) \sum_{k=1}^{m-1}\sum_{i=1}^{k} \frac{T_{ik}}{(1+c)^k} [\cos(\tan^{-1}\alpha)]^{2(k-i)+1}\right\} \tag{5A.17}$$

where

$$\alpha \triangleq \sqrt{\frac{c}{1+c}} \cot \frac{\pi}{M} \tag{5A.18}$$

and

$$T_{ik} \triangleq \frac{\binom{2k}{k}}{\binom{2(k-i)}{k-i} 4^i [2(k-i)+1]} \tag{5A.19}$$

For $m = 1$, (5A.17) reduces to (5A.15).

4. $\displaystyle \frac{1}{\pi} \int_0^{\pi/4} \left(\frac{\sin^2 \theta}{\sin^2 \theta + c} \right)^m d\theta$

For evaluation of symbol error probability corresponding to single-channel reception of QAM on Nakagami-m fading channels and also for multichannel reception of QAM on Rayleigh fading channels, we shall have need to evaluate

$$L_m = \frac{1}{\pi} \int_0^{\pi/4} \left(\frac{\sin^2 \theta}{\sin^2 \theta + c} \right)^m d\theta \tag{5A.20}$$

Using a result [7, Eq. (18)] with $\theta_U = (M+1)\pi/M$ and $\theta_L = (M-1)\pi/M$, it is straightforward to show that for m integer,

$$\begin{aligned}
\frac{1}{\pi} &\int_0^{\pi/M} \left(\frac{\sin^2 \theta}{\sin^2 \theta + c} \right)^m d\theta \\
&= \frac{1}{M} - \frac{1}{\pi} \sqrt{\frac{c}{1+c}} \Bigg\{ \left(\frac{\pi}{2} - \tan^{-1} \alpha\right) \sum_{k=0}^{m-1} \binom{2k}{k} \frac{1}{[4(1+c)]^k} \\
&\quad - \sin(\tan^{-1} \alpha) \sum_{k=1}^{m-1} \sum_{i=1}^{k} \frac{T_{ik}}{(1+c)^k} [\cos(\tan^{-1} \alpha)]^{2(k-i)+1} \Bigg\}
\end{aligned} \tag{5A.21}$$

where α and T_{ik} are as evaluated in (5A.18) and (5A.19), respectively. Letting $M = 4$ in (5A.21) whereupon $\alpha = \sqrt{c/(1+c)}$ gives the desired result in (5A.20).

Finally, for exact evaluation of bit error probability corresponding to single-channel reception of M-PSK on Nakagami-m fading channels and also for multichannel reception of M-PSK on Rayleigh fading channels, we shall have need to evaluate integrals of the form in (5A.17) or (5A.21) but with upper limits given by $\pi[1 - (2k \pm 1)/M]$ for $k = 1, 2, \ldots, M-1$. What is needed to evaluate the bit error probabilities above is the difference of specific pairs of

APPENDIX 5A: EVALUATION OF DEFINITE INTEGRALS   129

these integrals which can be related to the generic closed-form result given by Eq. (18) of Ref. 7. Specifically, it can be shown that

$I_m(\theta_U, \theta_L; K)$

$$= \frac{1}{2\pi} \int_0^{\pi-\theta_L} \left(\frac{\sin^2\theta}{\sin^2\theta + \mu_L^2}\right)^m d\theta - \frac{1}{2\pi} \int_0^{\pi-\theta_U} \left(\frac{\sin^2\theta}{\sin^2\theta + \mu_U^2}\right)^m d\theta$$

$$= \frac{\theta_U - \theta_L}{2\pi} + \frac{1}{2\pi}\beta_U \left\{ \left(\frac{\pi}{2} + \tan^{-1}\alpha_U\right) \sum_{k=0}^{m-1} \binom{2k}{k} \frac{1}{[4(1+\mu_U^2)]^k} \right.$$

$$\left. + \sin(\tan^{-1}\alpha_U) \sum_{k=1}^{m-1} \sum_{i=1}^{k} \frac{T_{ik}}{(1+\mu_U^2)^k}[\cos(\tan^{-1}\alpha_U)]^{2(k-i)+1} \right\}$$

$$- \frac{1}{2\pi}\beta_L \left\{ \left(\frac{\pi}{2} + \tan^{-1}\alpha_L\right) \sum_{k=0}^{m-1} \binom{2k}{k} \frac{1}{[4(1+\mu_L^2)]^k} \right.$$

$$\left. + \sin(\tan^{-1}\alpha_L) \sum_{k=1}^{m-1} \sum_{i=1}^{k} \frac{T_{ik}}{(1+\mu_L^2)^k}[\cos(\tan^{-1}\alpha_L)]^{2(k-i)+1} \right\} \quad (5A.22)$$

where

$$\mu_L \triangleq \sqrt{\frac{K}{m}}\sin\theta_L, \quad \beta_L \triangleq \frac{\mu_L}{\sqrt{1+\mu_L^2}}, \quad \alpha_L \triangleq \beta_L \cot\theta_L$$

$$\mu_U \triangleq \sqrt{\frac{K}{m}}\sin\theta_U, \quad \beta_U \triangleq \frac{\mu_U}{\sqrt{1+\mu_U^2}}, \quad \alpha_U \triangleq \beta_U \cot\theta_U \quad (5A.23)$$

with K a constant. Our interest will be in the case where $\theta_U = (2k+1)\pi/M$, $\theta_L = (2k-1)\pi/M$ and K is related to signal-to-noise ratio. Alternatively, for $\theta_U = (M+1)\pi/M$, $\theta_L = (M-1)\pi/M$, then $\mu_L = -\mu_U \triangleq \sqrt{c}$, $\beta_L = -\beta_U = \sqrt{c/(1+c^2)}$, and $\alpha_L = \alpha_U \triangleq \alpha$, in which case (5A.22) simplifies immediately to (5A.21).

5. $\displaystyle \frac{1}{\pi}\int_0^\phi \left(\frac{\sin^2\theta}{\sin^2\theta + c}\right)^m d\theta$

Interestingly enough, a closed-form expression for the integral in (5A.16) or (5A.21) with arbitrary upper limit, say ϕ, can be obtained from (5A.22). In particular, setting $\theta_L = \pi - \phi$ and $\theta_U = \pi$, whereupon the second integral in (5A.22) disappears, we arrive at the result

$$I_m(\phi; c) = \frac{1}{\pi} \int_0^\phi \left(\frac{\sin^2\theta}{\sin^2\theta + c}\right)^m d\theta = \frac{\phi}{\pi} - \frac{1}{\pi}\beta\left\{\left(\frac{\pi}{2} + \tan^{-1}\alpha\right)\right.$$

$$\times \sum_{k=0}^{m-1} \binom{2k}{k} \frac{1}{[4(1+c)]^k} + \sin(\tan^{-1}\alpha)$$

$$\left.\times \sum_{k=1}^{m-1} \sum_{i=1}^{k} \frac{T_{ik}}{(1+c)^k} [\cos(\tan^{-1}\alpha)]^{2(k-i)+1}\right\}, \quad -\pi \le \phi \le \pi \quad (5A.24)$$

where

$$\beta \triangleq \sqrt{\frac{c}{1+c}} \operatorname{sgn} \phi, \qquad \alpha \triangleq -\beta \cot \phi \tag{5A.25}$$

Clearly, (5A.24) reduces to (5A.16) and (5A.21) when $\phi = (M-1)\pi/M$ and $\phi = \pi/M$, respectively.

Another closed form for the integral in (5A.24) has been suggested to the authors by R. F. Pawula, which is readily derived using a clever change of variables due to Euler and Legendre [8, p. 316]. Although this alternative closed form is quite similar in structure to (5A.24) and therefore does not offer a significant computational advantage, it is nevertheless worth documenting because of the elegance associated with its derivation and the simplicity with which the final result is obtained relative to that employed in arriving at (5A.24).

To begin, we first employ simple trigonometry to convert the integral to a slightly different form as follows:

$$I_m(\phi; c) = \frac{1}{\pi} \int_0^\phi \left(\frac{\sin^2\theta}{\sin^2\theta + c}\right)^m d\theta = \frac{1}{\pi} \int_0^\phi \left(\frac{1 - \cos 2\theta}{1 + 2c - \cos 2\theta}\right)^m d\theta$$

$$= \frac{1}{2\pi(1+2c)^m} \int_0^{2\phi} \left(\frac{1 - \cos\xi}{1 - d\cos\xi}\right)^m d\xi \tag{5A.26}$$

where $d \triangleq 1/(1+2c)$. Next, employing the Euler–Legendre change of variables

$$1 - d\cos\xi = \frac{1 - d^2}{1 + d\cos x}, \qquad d\xi = \frac{\sqrt{1-d^2}}{1 + d\cos x} dx \tag{5A.27}$$

then after some algebraic and trigonometric manipulation, we obtain the form

$$I_m(\phi; c) = \frac{d\sqrt{c}}{2^m \pi (1+c)^{m-1/2}} \int_0^{x_{\max}} \frac{(1-\cos x)^m}{1 + d\cos x} dx \tag{5A.28}$$

where

$$\tan x_{\max} = \frac{\sqrt{1-d^2} \sin 2\phi}{\cos 2\phi - d} = \frac{2\sqrt{c(1+c)} \sin 2\phi}{(1+2c)\cos 2\phi - 1} \tag{5A.29}$$

APPENDIX 5A: EVALUATION OF DEFINITE INTEGRALS   131

Finally, letting $x = 2t$ and taking care to assure that $x_{\max}$ as derived from (5A.29) is intepreted in the four-quadrant arctangent sense, we get the simpler integral form

$$I_m(\phi; c) = \frac{\sqrt{c}}{\pi(1+c)^{m-1/2}} \int_0^T \frac{\sin^{2m} t}{c + \cos^2 t} dt \qquad (5A.30)$$

where

$$T = \frac{x_{\max}}{2} = \frac{1}{2} \tan^{-1} \frac{N}{D} + \frac{\pi}{2} \left[1 - \left(\frac{1 + \operatorname{sgn} D}{2} \right) \operatorname{sgn} N \right] \qquad (5A.31)$$

with

$$N = 2\sqrt{c(1+c)} \sin 2\phi, \qquad D = (1 + 2c) \cos 2\phi - 1 \qquad (5A.32)$$

The integral form of (5A.30) is valid for m integer as well as m noninteger but is restricted to values of ϕ [the upper limit in the integral of (5A.26)] between zero and π. Later, after obtaining the desired closed-form result, we will show how to remove this restriction.

To obtain the closed form of (5A.30), we use the well-known geometric series $\sum_{k=0}^{m-1} x^k = (1 - x^m)/(1 - x)$ to rewrite this equation as

$$I_m(\phi; c) = \frac{1}{\pi} \sqrt{\frac{c}{1+c}} \int_0^T \frac{1 - (1 - a^{2m} \sin^{2m} t)}{1 - a^2 \sin^2 t} dt$$

$$= \frac{1}{\pi} \sqrt{\frac{c}{1+c}} \int_0^T \frac{1}{1 - a^2 \sin^2 t} dt - \frac{1}{\pi} \sqrt{\frac{c}{1+c}} \sum_{k=0}^{m-1} a^{2k} \int_0^T \sin^{2k} t \, dt$$

(5A.33)

where $a^2 \triangleq 1/(1+c)$. The first term is the original integral when $m = 0$ and thus from (5A.26) must be equal to ϕ/π. The second integral is available in Eq. (2.513.1) of Ref. 2, namely,

$$\int_0^T \sin^{2k} t \, dt = \frac{T}{2^{2k}} \binom{2k}{k} + \frac{(-1)^k}{2^{2k-1}} \sum_{j=0}^{k-1} (-1)^j \binom{2k}{j} \frac{\sin[(2k-2j)T]}{2k - 2j}$$

(5A.34)

Combining these two results and simplifying gives the alternative closed-form result

$$I_m(\phi; c) = \frac{\phi}{\pi} - \frac{T}{\pi} \sqrt{\frac{c}{1+c}} \sum_{k=0}^{m-1} \binom{2k}{k} \frac{1}{[4(1+c)]^k}$$

$$- \frac{2}{\pi} \sqrt{\frac{c}{1+c}} \sum_{k=0}^{m-1} \sum_{j=0}^{k-1} \binom{2k}{j} \frac{(-1)^{j+k}}{[4(1+c)]^k} \frac{\sin[(2k-2j)T]}{2k - 2j},$$

$$0 \leq \phi \leq \pi \qquad (5A.35)$$

To extend this result so as to apply for upper integration limits in the region $\pi \leq \phi \leq 2\pi$, we proceed as follows. First we partition the integral in (5A.26) as

$$I_m(\phi; c) = \frac{1}{\pi} \int_0^\phi \left(\frac{\sin^2 \theta}{c + \sin^2 \theta}\right)^m d\theta$$

$$= \frac{1}{\pi} \int_0^\pi \left(\frac{\sin^2 \theta}{c + \sin^2 \theta}\right)^m d\theta + \frac{1}{\pi} \int_\pi^\phi \left(\frac{\sin^2 \theta}{c + \sin^2 \theta}\right)^m d\theta \quad (5A.36)$$

In the second integral make the change of variables $\theta' = \theta - \pi$. Then

$$I_m(\phi; c) = \frac{1}{\pi} \int_0^\pi \left(\frac{\sin^2 \theta}{c + \sin^2 \theta}\right)^m d\theta + \frac{1}{\pi} \int_0^{\phi-\pi} \left(\frac{\sin^2 \theta'}{c + \sin^2 \theta'}\right)^m d\theta' \quad (5A.37)$$

The second integral in (5A.37) can be evaluated using (5A.35) with ϕ replaced by $\phi - \pi$. For the first integral we have to first evaluate T in the limit when $\phi = \pi$ and then use (5A.35). Since ϕ approaches π from below, it is straightforward to show that the first term of (5A.31) will be zero and the second term will approach π. Thus, $\lim_{\phi \to \pi} T = \pi$. Using this value of T in (5A.35), the double sum evaluates to zero and hence the first integral above becomes

$$\frac{1}{\pi} \int_0^\pi \left(\frac{\sin^2 \theta}{c + \sin^2 \theta}\right)^m d\theta = 1 - \sqrt{\frac{c}{1+c}} \sum_{k=0}^{m-1} \binom{2k}{k} \frac{1}{[4(1+c)]^k} \quad (5A.38)$$

Thus, when $\pi \leq \phi \leq 2\pi$, the final result can be written as

$$I_m(\phi; c) = 1 - \sqrt{\frac{c}{1+c}} \sum_{k=0}^{m-1} \binom{2k}{k} \frac{1}{[4(1+c)]^k}$$

$$+ \frac{\phi - \pi}{\pi} - \frac{T'}{\pi} \sqrt{\frac{c}{1+c}} \sum_{k=0}^{m-1} \binom{2k}{k} \frac{1}{[4(1+c)]^k}$$

$$- \frac{2}{\pi} \sqrt{\frac{c}{1+c}} \sum_{k=0}^{m-1} \sum_{j=0}^{k-1} \binom{2k}{j} \frac{(-1)^{j+k}}{[4(1+c)]^k} \frac{\sin[(2k-2j)T']}{2k-2j} \quad (5A.39)$$

where T' is T evaluated with ϕ replaced by $\phi - \pi$. However, because of the periodicity of T with respect to the 2ϕ process, we have $T' = T$. Thus, the final result is

$$I_m(\phi; c) = \frac{\phi}{\pi} - \left(1 + \frac{T}{\pi}\right) \sqrt{\frac{c}{1+c}} \sum_{k=0}^{m-1} \binom{2k}{k} \frac{1}{[4(1+c)]^k}$$

$$- \frac{2}{\pi} \sqrt{\frac{c}{1+c}} \sum_{k=0}^{m-1} \sum_{j=0}^{k-1} \binom{2k}{j} \frac{(-1)^{j+k}}{[4(1+c)]^k} \frac{\sin[(2k-2j)T]}{2k-2j},$$

$$\pi \leq \phi \leq 2\pi \quad (5A.40)$$

APPENDIX 5A: EVALUATION OF DEFINITE INTEGRALS   133

or combining this with (5A.35)

$$I_m(\phi; c) = \frac{\phi}{\pi} - \left(\frac{1 + \text{sgn}(\phi - \pi)}{2} + \frac{T}{\pi}\right)\sqrt{\frac{c}{1+c}} \sum_{k=0}^{m-1} \binom{2k}{k} \frac{1}{[4(1+c)]^k}$$

$$- \frac{2}{\pi}\sqrt{\frac{c}{1+c}} \sum_{k=0}^{m-1}\sum_{j=0}^{k-1} \binom{2k}{j} \frac{(-1)^{j+k}}{[4(1+c)]^k} \frac{\sin[(2k-2j)T]}{2k-2j},$$

$$0 \leq \phi \leq 2\pi \qquad (5A.41)$$

6. $\displaystyle \frac{1}{\pi}\int_0^\phi \left(\frac{\sin^2\theta}{\sin^2\theta + c_1}\right)^m \left(\frac{\sin^2\theta}{\sin^2\theta + c_2}\right) d\theta$

In the study of generalized diversity selection combining to be discussed in Chapter 9, we shall have need to evaluate an extension of the integral in (5A.24), namely,

$$I_m(\phi; c_1, c_2) = \frac{1}{\pi}\int_0^\phi \left(\frac{\sin^2\theta}{\sin^2\theta + c_1}\right)^m \left(\frac{\sin^2\theta}{\sin^2\theta + c_2}\right) d\theta \qquad (5A.42)$$

where, in general $c_1 \neq c_2$. Since a closed form for such an integral cannot be obtained from the results of Ref. 7 nor for that matter from any other reported contributions, we turn once again to the method suggested by Pawula for arriving at the alternative closed form for $I_m(\phi; c)$ given in (5A.35), but instead apply it now to (5A.42). In particular, following steps analogous to (5A.26) through (5A.30), it is straightforward to show that

$$I_m(\phi; c_1, c_2) = \frac{\sqrt{c_1}(1-d_1)d_2}{\pi(1+c_1)^{m-1/2}(d_1-d_2)} \int_0^{T_1} \left(\frac{\sin^{2(m+1)} t}{c_1 + \cos^2 t}\right) \left(\frac{1}{D + \cos^2 t}\right) dt \qquad (5A.43)$$

where, as before, $d_i \triangleq 1/(1+2c_i)$, $i = 1, 2$ and now also

$$D \triangleq \frac{1 - d_1 d_2 - d_1 + d_2}{2(d_1 - d_2)} \qquad (5A.44)$$

In addition, T_1 corresponds to T of (5A.31) with c replaced by c_1.

Now using the same geometric series manipulation as in (5A.33), we can rewrite (5A.43) as

$$I_m(\phi; c_1, c_2) = \frac{\sqrt{c_1(1+c_1)}(1-d_1)d_2 b_1^2}{\pi(d_1 - d_2)} \int_0^{T_1} \frac{1 - (1 - a_1^{2(m+1)} \sin^{2(m+1)} t)}{(1 - a_1^2 \sin^2 t)(1 - b_1^2 \sin^2 t)} dt \qquad (5A.45)$$

where, as before, $a_1^2 \triangleq 1/(1+c_1)$ and now, in addition,

$$b_1^2 \triangleq \frac{1}{1+D} = \frac{2(d_1 - d_2)}{1 + d_1 - d_2 - d_1 d_2} = \frac{c_2 - c_1}{c_2(1+c_1)} \quad (5A.46)$$

Expanding the integrand of (5A.45) into a partial fraction expansion and evaluating the fractional coefficient in front of the integral purely in terms of c_1 and c_2, we obtain, after considerable algebraic simplification,

$$I_m(\phi; c_1, c_2) = \frac{1}{\pi}\sqrt{\frac{c_1}{1+c_1}} \int_0^{T_1} \frac{1-(1-a_1^{2m}\sin^{2m} t)}{1-a_1^2 \sin^2 t} dt$$

$$- \frac{1}{\pi}\sqrt{\frac{c_1}{1+c_1}} \left(\frac{c_2}{c_2-c_1}\right)^m \int_0^{T_1} \frac{1-(1-b_1^{2m}\sin^{2m} t)}{1-b_1^2 \sin^2 t} dt \quad (5A.47)$$

Comparing the first term of (5A.47) with (5A.33), we see immediately that

$$I_m(\phi; c_1, c_2) = I_m(\phi; c_1) - \frac{1}{\pi}\sqrt{\frac{c_1}{1+c_1}} \left(\frac{c_2}{c_2-c_1}\right)^m \int_0^{T_1} \frac{1-(1-b_1^{2m}\sin^{2m} t)}{1-b_1^2 \sin^2 t} dt \quad (5A.48)$$

which indicates that the second term in (5A.48) accounts for the additional factor in the integrand of $I_m(\phi; c_1, c_2)$ that is not present in the integrand of $I_m(\phi; c_1)$.

Since for $c_1 = c_2$, we have from (5A.46) that $b_1^2 = 0$, then writing the second term of (5A.48) as

$$\frac{1}{\pi}\sqrt{\frac{c_1}{1+c_1}} \left(\frac{c_2}{c_2-c_1}\right)^m \int_0^{T_1} \frac{b_1^{2m} \sin^{2m} t}{1-b_1^2 \sin^2 t} dt$$

$$= \frac{1}{\pi}\sqrt{\frac{c_1}{1+c_1}} \frac{1}{(1+c_1)^m} \int_0^{T_1} \frac{\sin^{2m} t}{1-b_1^2 \sin^2 t} dt$$

$$= \frac{1}{\pi}\sqrt{\frac{c_1}{1+c_1}} \frac{1}{(1+c_1)^m} \int_0^{T_1} \sin^{2m} t \, dt \quad (5A.49)$$

and using (5A.34), we obtain

$$\frac{1}{\pi}\sqrt{\frac{c_1}{1+c_1}} \left(\frac{c_2}{c_2-c_1}\right)^m \int_0^{T_1} \frac{b_1^{2m} \sin^{2m} t}{1-b_1^2 \sin^2 t} dt$$

$$= \frac{T_1}{\pi}\sqrt{\frac{c_1}{1+c_1}} \binom{2m}{m} \frac{1}{[4(1+c_1)]^m}$$

$$- \frac{2}{\pi}\sqrt{\frac{c_1}{1+c_1}} \sum_{j=0}^{m-1} \binom{2m}{j} \frac{(-1)^{j+m}}{[4(1+c_1)]^m} \frac{\sin[(2m-2j)T_1]}{2m-2j} \quad (5A.50)$$

APPENDIX 5A: EVALUATION OF DEFINITE INTEGRALS **135**

Substituting (5A.50) into (5A.48) and recognizing the form of $I_m(\phi; c)$ in (5A.35), we immediately see that for $c_1 = c_2$,

$$I_m(\phi; c_1, c_1) = I_{m+1}(\phi; c_1) \tag{5A.51}$$

as it should from the definition of $I_m(\phi; c_1, c_2)$ in (5A.42).

For the case $c_1 \neq c_2$, we return to the form in (5A.48) and analogous to (5A.33) partition it into two integrals, that is,

$$I_m(\phi; c_1, c_2) = I_m(\phi; c_1) - \left(\frac{c_2}{c_2 - c_1}\right)^m \left[\frac{1}{\pi}\sqrt{\frac{c_1}{1+c_1}} \int_0^{T_1} \frac{1}{1 - b_1^2 \sin^2 t} dt \right.$$

$$\left. - \frac{1}{\pi}\sqrt{\frac{c_1}{1+c_1}} \sum_{k=0}^{m-1} b_1^{2k} \int_0^{T_1} \sin^{2k} t \, dt \right] \tag{5A.52}$$

The first integral in (5A.52) can be evaluated by first noting from (5A.47) that

$$I_0(\phi; c_1, c_2) = \frac{1}{\pi} \int_0^\phi \frac{\sin^2 \theta}{\sin^2 \theta + c_2} d\theta = I_1(\phi; c_2)$$

$$= \frac{1}{\pi}\sqrt{\frac{c_1}{1+c_1}} \int_0^{T_1} \frac{1}{1 - a_1^2 \sin^2 t} dt - \frac{1}{\pi}\sqrt{\frac{c_1}{1+c_1}} \int_0^{T_1} \frac{1}{1 - b_1^2 \sin^2 t} dt$$

$$= \frac{\phi}{\pi} - \frac{1}{\pi}\sqrt{\frac{c_1}{1+c_1}} \int_0^{T_1} \frac{1}{1 - b_1^2 \sin^2 t} dt \tag{5A.53}$$

Evaluating $I_1(\phi; c_2)$ from (5A.35) as

$$I_1(\phi; c_2) = \frac{\phi}{\pi} - \frac{T_2}{\pi}\sqrt{\frac{c_2}{1+c_2}} \tag{5A.54}$$

where T_2 now corresponds to T of (5A.31) with c replaced by c_2, then combining (5A.53) and (5A.54), we get

$$\frac{1}{\pi}\sqrt{\frac{c_1}{1+c_1}} \int_0^{T_1} \frac{1}{1 - b_1^2 \sin^2 t} dt = \frac{T_2}{\pi}\sqrt{\frac{c_2}{1+c_2}} \tag{5A.55}$$

The second integral of (5A.52) is evaluated as before using (5A.34). Without further ado we present the desired closed-form result for $I_m(\phi; c_1, c_2)$, which is

$$I_m(\phi; c_1, c_2) = I_m(\phi; c_1) - \frac{T_2}{\pi}\sqrt{\frac{c_2}{1+c_2}} \left(\frac{c_2}{c_2 - c_1}\right)^m$$

$$+ \frac{T_1}{\pi}\sqrt{\frac{c_1}{1+c_1}} \sum_{k=0}^{m-1} \left(\frac{c_2}{c_2 - c_1}\right)^{m-k} \binom{2k}{k} \frac{1}{[4(1+c_1)]^k}$$

$$+ \frac{2}{\pi}\sqrt{\frac{c_1}{1+c_1}} \sum_{k=0}^{m-1}\sum_{j=0}^{k-1} \left(\frac{c_2}{c_2-c_1}\right)^{m-k} \binom{2k}{j}$$

$$\times \frac{(-1)^{j+k}}{[4(1+c_1)]^k} \frac{\sin[(2k-2j)T_1]}{2k-2j}, \qquad 0 \le \phi \le \pi \qquad (5A.56)$$

To extend the range of coverage of the upper integration limit from $0 \le \phi \le \pi$ to $0 \le \phi \le 2\pi$, we proceed as before and arrive at the final desired result:

$$I_m(\phi; c_1, c_2)$$

$$= I_m(\phi; c_1) - \left(\frac{1+\mathrm{sgn}(\phi-\pi)}{2} + \frac{T_2}{\pi}\right)\sqrt{\frac{c_2}{1+c_2}} \left(\frac{c_2}{c_2-c_1}\right)^m$$

$$+ \left(\frac{1+\mathrm{sgn}(\phi-\pi)}{2} + \frac{T_1}{\pi}\right)\sqrt{\frac{c_1}{1+c_1}} \sum_{k=0}^{m-1} \left(\frac{c_2}{c_2-c_1}\right)^{m-k} \binom{2k}{k}$$

$$\times \frac{1}{[4(1+c_1)]^k} + \frac{2}{\pi}\sqrt{\frac{c_1}{1+c_1}} \sum_{k=0}^{m-1}\sum_{j=0}^{k-1} \left(\frac{c_2}{c_2-c_1}\right)^{m-k} \binom{2k}{j}$$

$$\times \frac{(-1)^{j+k}}{[4(1+c_1)]^k} \frac{\sin[(2k-2j)T_1]}{2k-2j}, \qquad 0 \le \phi \le 2\pi \qquad (5A.57)$$

where now $I_m(\phi; c_1)$ is evaluated from (5A.41).

7. $\dfrac{1}{\pi}\displaystyle\int_0^{\pi/2} \left(\dfrac{\sin^2\theta}{\sin^2\theta + c_1}\right)^{m_1} \left(\dfrac{\sin^2\theta}{\sin^2\theta + c_2}\right)^{m_2} d\theta$

An extension of the preceding integral wherein each of the two factors in the integrand is raised to an arbitrary power is of interest in the study of diversity (optimum) combining in the presence of interference (see Chapter 10 for a complete discussion of this topic). Unfortunately, it appears difficult to apply the previous derivation approaches to obtain a result for the most generic form of this integral, where the powers are not necessarily restricted to be integer and the upper limit of the integral is arbitrary. However, for the case where the upper limit is equal to $\pi/2$ and the powers are restricted to be integer, which is of interest in evaluating the average error probability performance of PSK with optimum combining over a Rayleigh fading channel, making an association with a closed-form result obtained by Villier [9], we present (without derivation) the following result:

$$\frac{1}{\pi}\int_0^{\pi/2}\left(\frac{\sin^2\theta}{\sin^2\theta+c_1}\right)^{m_1}\left(\frac{\sin^2\theta}{\sin^2\theta+c_2}\right)^{m_2}d\theta$$

$$=\frac{(c_1/c_2)^{m_2-1}}{2(1-c_1/c_2)^{m_1+m_2-1}}\left[\sum_{k=0}^{m_2-1}\left(\frac{c_2}{c_1}-1\right)^k B_k I_k(c_2)\right.$$

$$\left.-\frac{c_1}{c_2}\sum_{k=0}^{m_1-1}\left(1-\frac{c_1}{c_2}\right)^k C_k I_k(c_1)\right] \tag{5A.58}$$

where[1]

$$B_k \triangleq \frac{A_k}{\binom{m_1+m_2-1}{k}},\qquad C_k \triangleq \sum_{n=0}^{m_2-1}\frac{\binom{k}{n}}{\binom{m_1+m_2-1}{n}}A_n,$$

$$A_k \triangleq (-1)^{m_2-1+k}\frac{\binom{m_2-1}{k}}{(m_2-1)!}\prod_{\substack{n=1\\n\ne k+1}}^{m_2}(m_1+m_2-n) \tag{5A.59}$$

and

$$I_k(c)=1-\sqrt{\frac{c}{1+c}}\left[1+\sum_{n=1}^k\frac{(2n-1)!!}{n!2^n(1+c)^n}\right] \tag{5A.60}$$

with the double factorial notation denoting the product of only odd integers from 1 to $2k-1$. It is straightforward (although requiring some tedious manipulations) to show that (5A.58) reduces to (5A.56) when $m_1=m$ and $m_2=1$. Also, by symmetry it can be shown that (5A.58) reduces to (5A.56) with c_1 and c_2 switched when $m_1=1$ and $m_2=m$.

8. $\dfrac{1}{\pi}\displaystyle\int_0^\phi \dfrac{\sin^{2m}\theta}{c+\sin^2\theta}d\theta$

Yet another integral that arises in the study of generalized diversity selection combining to be discussed in Chapter 9 is

$$J_m(\phi;c)\triangleq\frac{1}{\pi}\int_0^\phi\frac{\sin^{2m}\theta}{c+\sin^2\theta}d\theta \tag{5A.61}$$

---

[1] Note that by convention, $\binom{k}{n}=0$ for $n>k$. Also, for $m_2=1$, by convention the product $\prod_{\substack{n=1\\n\ne k+1}}^{m_2}(m_1+m_2-n)=1$ and the only nonzero-valued coefficients are $A_0=B_0=C_k=1$. For $m_2>1$, the coefficients A_k, B_k, and C_k clearly depend on both m_1 and m_2.

This integral is similar in form to (5A.30) and can be evaluated by following an approach analogous to that used in arriving at the closed form in (5A.35). The procedure is as follows. Let $a^2 = 1/c$. Then

$$J_m(\phi; c) = \frac{1}{\pi} \frac{a^2}{a^{2m}} \int_0^\phi \frac{a^{2m} \sin^{2m} \theta}{1 + a^2 \sin^2 \theta} d\theta$$

$$= \frac{1}{\pi a^{2(m-1)}} \int_0^\phi \frac{1 - (1 - a^{2m} \sin^{2m} \theta)}{1 + a^2 \sin^2 \theta} d\theta$$

$$= \frac{1}{\pi a^{2(m-1)}} \left[\int_0^\phi \frac{1}{1 + a^2 \sin^2 \theta} d\theta - \int_0^\phi \frac{1 - a^{2m} \sin^{2m} \theta}{1 + a^2 \sin^2 \theta} d\theta \right] \quad (5A.62)$$

For l odd, $\sum_{i=0}^l (-1)^i x^i = (1 - x^{l+1})/(1 + x)$. Thus, letting $x = a^2 \sin^2 \phi$, then for m even we get

$$J_m(\phi; c) = \frac{1}{\pi a^{2(m-1)}} \left[\int_0^\phi \frac{1}{1 + a^2 \sin^2 \theta} d\theta - \sum_{i=0}^{m-1} (-1)^i a^{2i} \int_0^\phi \sin^{2i} \theta \, d\theta \right] \quad (5A.63)$$

Finally, using Gradshteyn and Ryzhik [2, Eq. (2.562)] to evaluate the first integral, that is,

$$\int_0^\phi \frac{1}{1 + a^2 \sin^2 \theta} d\theta = \frac{1}{\sqrt{1 + a^2}} \tan^{-1} \left(\sqrt{1 + a^2} \tan \phi \right) \quad (5A.64)$$

and (5A.34) for the second integral, we arrive at the desired result (for m even)

$$J_m(\phi; c) = \frac{c^{m-1}}{\pi} \left\{ \sqrt{\frac{c}{1+c}} \tan^{-1} \left(\sqrt{\frac{1+c}{c}} \tan \phi \right) - \sum_{i=0}^{m-1} (-1)^i \frac{1}{c^i} \right.$$
$$\left. \times \left[\frac{\phi}{2^{2i}} \binom{2i}{i} + \frac{(-1)^i}{2^{2i-1}} \sum_{j=0}^{i-1} (-1)^j \binom{2i}{j} \frac{\sin[(2i - 2j)\phi]}{2i - 2j} \right] \right\} \quad (5A.65)$$

For m odd we slightly change the procedure. First rewriting (5A.62) as

$$J_m(\phi; c) = \frac{1}{\pi a^{2(m-1)}} \int_0^\phi \frac{-1 + (1 + a^{2m} \sin^{2m} \theta)}{1 + a^2 \sin^2 \theta} d\theta$$

$$= \frac{1}{\pi a^{2(m-1)}} \left[-\int_0^\phi \frac{1}{1 + a^2 \sin^2 \theta} d\theta + \int_0^\phi \frac{1 + a^{2m} \sin^{2m} \theta}{1 + a^2 \sin^2 \theta} d\theta \right]$$

$$(5A.66)$$

then noting that for l even, $\sum_{i=0}^{l}(-1)^i x^i = (1+x^{l+1})/(1+x)$, we obtain

$$J_m(\phi; c) = \frac{1}{\pi a^{2(m-1)}} \left[-\int_0^\phi \frac{1}{1+a^2 \sin^2 \theta} d\theta + \sum_{i=0}^{m-1} (-1)^i a^{2i} \int_0^\phi \sin^{2i} \theta \, d\theta \right]$$
(5A.67)

which is the negative of (5A.63). Thus, for arbitrary integer m, we have

$$J_m(\phi; c) = (-1)^m \frac{c^{m-1}}{\pi} \left\{ \sqrt{\frac{c}{1+c}} \tan^{-1}\left(\sqrt{\frac{1+c}{c}} \tan \phi\right) - \sum_{i=0}^{m-1} (-1)^i \frac{1}{c^i} \right.$$

$$\left. \times \left[\frac{\phi}{2^{2i}} \binom{2i}{i} + \frac{(-1)^i}{2^{2i-1}} \sum_{j=0}^{i-1} (-1)^j \binom{2i}{j} \frac{\sin[(2i-2j)\phi]}{2i-2j} \right] \right\}$$
(5A.68)

A special case of interest is when $\phi = \pi/2$, in which case (5A.68) simplifies to

$$J_m(\pi/2; c) = (-1)^m \frac{c^{m-1}}{2} \left[\sqrt{\frac{c}{1+c}} - \sum_{i=0}^{m-1} (-1)^i \frac{1}{2^{2i} c^i} \binom{2i}{i} \right] \quad (5A.69)$$

which reduces to (5A.9) when $m = 1$, as it should.

6

NEW REPRESENTATIONS OF SOME PDF's AND CDF's FOR CORRELATIVE FADING APPLICATIONS

Later in the book we shall have reason to study the performance of digital communication systems over correlative fading channels. Such channels occur, for example, in small terminals equipped with space antenna diversity where the antenna spacing is insufficient to provide independent fading among the various signal paths. In such instances, the received signal will consist of two or more replicas of the transmitted signal with fading amplitudes that are correlated random variables. To assess the performance of receivers of such signals, it is therefore necessary to study the joint statistics of correlated random variables with probability distributions characterized by the various fading channel models of Chapter 2.

One important application of the above scenario pertains to a system wherein the channel is assumed to be modeled by two paths and the receiver thus implements a diversity combiner with two branches. Evaluation of the performance of such a *dual diversity combining receiver* (discussed in great detail in Chapter 9) requires, in general, knowledge of the two-dimensional (bivariate) fading amplitude PDF and CDF. For the specific case of *selection combining (SC)* [1, Sec. 10-4], the combiner chooses the branch with the highest signal-to-noise ratio (or equivalently, with the strongest signal assuming equal noise power among the branches) and outputs this signal to the threshold decision device. To evaluate performance in this instance, it is sufficient to obtain the one-dimensional PDF and CDF of the SC output, which is tantamount to finding the PDF and CDF of the maximum of two correlated fading random variables. The SC output CDF is used to evaluate outage probability (the probability that neither SC input exceeds the detection threshold, or equivalently, the probability that the SC output falls below this threshold), while the SC output PDF is used to evaluate average error probability.

In what follows we focus on the Rayleigh and Nakagami-m fading channels since they are the most commonly used in digital communication system analyses and, as discussed previously, are typical of many wireless environments.

6.1 BIVARIATE RAYLEIGH PDF AND CDF

From a purely mathematical standpoint, the bivariate Rayleigh and Nakagami-m distributions can be viewed as the joint statistics of the envelopes, R_1 and R_2, of two correlated chi-square random variables of degree 2 and $2m$, respectively. Specifically, the bivariate Nakagami-m PDF is given by [1, Eq. (126); 3, Eq. (1)]

$$p_{R_1,R_2}(r_1, \Omega_1; r_2, \Omega_2 | m, \rho)$$

$$= \frac{4m^{m+1}(r_1 r_2)^m}{\Gamma(m)\Omega_1\Omega_2(1-\rho)(\sqrt{\Omega_1\Omega_2\rho})^{m-1}} \exp\left[-\frac{m}{1-\rho}\left(\frac{r_1^2}{\Omega_1} + \frac{r_2^2}{\Omega_2}\right)\right]$$

$$\times I_{m-1}\left(\frac{2m\sqrt{\rho}r_1 r_2}{\sqrt{\Omega_1\Omega_2}(1-\rho)}\right), \qquad r_1, r_2 \geq 0 \qquad (6.1)$$

where $\Omega_i = \overline{r_i^2}, i = 1, 2$ and $\rho = \mathrm{cov}(r_1^2, r_2^2)/\sqrt{\mathrm{var}(r_1^2)\,\mathrm{var}(r_2^2)}$ is the correlation coefficient ($0 \leq \rho < 1$). The special case of the bivariate Rayleigh PDF is given by [2, Eq. (122); 4, Eq. (3.7-13)]

$$p_{R_1,R_2}(r_1, \Omega_1; r_2, \Omega_2 | \rho) = \frac{4 r_1 r_2}{\Omega_1\Omega_2(1-\rho)} \exp\left[-\frac{1}{1-\rho}\left(\frac{r_1^2}{\Omega_1} + \frac{r_2^2}{\Omega_2}\right)\right]$$

$$\times I_0\left(\frac{2\sqrt{\rho}r_1 r_2}{(1-\rho)\sqrt{\Omega_1\Omega_2}}\right), \qquad r_1, r_2 \geq 0 \qquad (6.2)$$

Tan and Beaulieu [3] were successful in finding infinite series representations of the CDFs corresponding to (6.1) and (6.2), in particular,

$$P_{R_1,R_2}(r_1, \Omega_1; r_2, \Omega_2 | m, \rho)$$

$$= \frac{(1-\rho)^m}{\Gamma(m)} \sum_{k=0}^{\infty} \rho^k \frac{\gamma(m+k, mr_1^2/\Omega_1(1-\rho))\gamma(m+k, mr_2^2/\Omega_2(1-\rho))}{k!\Gamma(m+k)}$$

$$(6.3)$$

where $\gamma(\alpha, x) \triangleq \int_0^x e^{-t} t^{\alpha-1} dt$, $\mathrm{Re}\{\alpha\} > 0$ is the incomplete gamma function [5, Eq. (6.5.2)] and

$$P_{R_1,R_2}(r_1, \Omega_1; r_2, \Omega_2 | \rho)$$

$$= (1-\rho) \sum_{k=0}^{\infty} \rho^k P\left(k+1, \frac{r_1^2}{\Omega_1(1-\rho)}\right) P\left(k+1, \frac{r_2^2}{\Omega_2(1-\rho)}\right) \qquad (6.4)$$

where $P(\alpha, x) = (1/\Gamma(\alpha)) \int_0^x e^{-t} t^{\alpha-1} \, dt$, $\text{Re}\{\alpha\} > 0$ is another common form of the incomplete gamma function [5, Eq. (6.5.3)]. Although (6.3) and (6.4) appear to have a simple structure, they have the drawback that because they are infinite series of the product of pairs of integrals, their computation requires truncation of the series. Bounds on the error resulting from this truncation along with empirical results for indicating the rate of convergence and tightness of the ensuing bounds, are discussed in Ref. 3. Tan and Beaulieu [3] go further to point out that the complementary Rayleigh bivariate CDF (and thus also the Rayleigh bivariate CDF itself) had previously been expressed in terms of the Marcum Q-function [1, App. A], that is,

$$P_{R_1, R_2}(r_1, \Omega_1; r_2, \Omega_2 | \rho)$$
$$= 1 - \Pr\{R_1 > r_1\} - \Pr\{R_2 > r_2\} + \Pr\{R_1 > r_1, R_2 > r_2\}$$
$$= 1 - \exp\left(-\frac{r_1^2}{\Omega_1}\right) Q_1\left(\sqrt{\frac{2}{1-\rho}} \frac{r_2}{\sqrt{\Omega_2}}, \sqrt{\frac{2\rho}{1-\rho}} \frac{r_1}{\sqrt{\Omega_1}}\right)$$
$$- \exp\left(-\frac{r_2^2}{\Omega_2}\right) \left[1 - Q_1\left(\sqrt{\frac{2\rho}{1-\rho}} \frac{r_2}{\sqrt{\Omega_2}}, \sqrt{\frac{2}{1-\rho}} \frac{r_1}{\sqrt{\Omega_1}}\right)\right] \quad (6.5)$$

Although Tan and Beaulieu [3] abandoned this result because of the lack of availability of the Marcum Q-function in standard distributions of such mathematical software packages as Maple V, MATLAB, and Mathematica, Simon and Alouini [6] recognized the value of (6.5) in terms of the desired form of the Marcum Q-function as described by (4.16) and (4.19). Indeed, as we shall soon see, this desired form of the Marcum Q-function allows the bivariate Rayleigh CDF to be similarly expressed as a single integral with finite limits and an integrand that includes a type of bivariate Gaussian PDF. This resulting form is simple, exact, and requires no special function evaluations (i.e., the integrand is entirely composed of elementary functions such as exponentials and trigonometrics).

Since the Marcum Q-function as represented by (4.16) and (4.19) depends on the relative values of its arguments, we must consider its use in (6.5) separately for different regions of the arguments r_1 and r_2. For simplicity of notation, we shall also introduce the normalized (by the square root of the average power) envelope random variables $Y_i \triangleq r_i/\sqrt{\Omega_i}$, $i = 1, 2$.

Consider first the region of r_1 and r_2 such that

$$\sqrt{\frac{2}{\Omega_2(1-\rho)}} r_2 < \sqrt{\frac{2\rho}{\Omega_1(1-\rho)}} r_1$$

or, equivalently, $Y_2 < \sqrt{\rho} Y_1$ which corresponds to the first argument being less than the second argument in the first Marcum Q-function in (6.5). Since in this region we would also have $\sqrt{\rho}Y_2 < Y_1$, then in the second Marcum Q-function in (6.5), the first argument is also less than the second argument. As such, we now substitute (4.16) in both of these two terms. After much simplification, one arrives at the desired result, namely,

$$P_{R_1,R_2}(r_1, \Omega_1; r_2, \Omega_2|\rho) = 1 - \exp(-Y_2^2)$$
$$+ \frac{1}{2\pi} \int_{-\pi}^{\pi} \exp\left(-\frac{Y_1^2 + Y_2^2 + 2\sqrt{\rho}Y_1Y_2\sin\theta}{1-\rho}\right)$$
$$\times \left[\frac{(1-\rho^2)Y_1^2Y_2^2 + \sqrt{\rho}(1-\rho)Y_1Y_2(Y_1^2+Y_2^2)\sin\theta}{(\rho Y_1^2 + 2\sqrt{\rho}Y_1Y_2\sin\theta + Y_2^2) \times (Y_1^2 + 2\sqrt{\rho}Y_1Y_2\sin\theta + \rho Y_2^2)}\right] d\theta$$
(6.6)

The complement of the region just considered is where $Y_2 > \sqrt{\rho}Y_1$ or equivalently, $\sqrt{\rho}Y_2 > \rho Y_1$. Here, however, we can have either $\sqrt{\rho}Y_2 > Y_1$ or $\rho Y_1 < \sqrt{\rho}Y_2 < Y_1$. Thus, two separate subcases must be considered. For the first subcase where $\sqrt{\rho}Y_2 > Y_1$, we would certainly also have $\sqrt{\rho}Y_2 > \rho Y_1$ and thus for both Marcum Q-function terms in (6.5), the second argument is greater than the first argument. Thus, substituting (4.19) in both of these terms, we obtain after much simplification the identical result of (6.6) except that the second term, namely, $\exp(-Y_2^2)$, now becomes $\exp(-Y_1^2)$. Finally, for the second subcase where $\rho Y_1 < \sqrt{\rho}Y_2 < Y_1$, once again (6.6) is appropriate with, however, the second term, $\exp(-Y_2^2)$, now replaced by $\exp(-Y_1^2) + \exp(-Y_2^2)$.

What remains is to evaluate the bivariate Rayleigh CDF at the endpoints between the regions where one must make use of the relation in (4.17). When this is done, the following results are obtained for the second term in (6.6). When $Y_2 = \sqrt{\rho}Y_1$, use $\frac{1}{2}\exp(-Y_1^2) + \exp(-\rho Y_1^2)$ and when $Y_1 = \sqrt{\rho}Y_2$ use $\frac{1}{2}\exp(-Y_2^2) + \exp(-\rho Y_2^2)$. Summarizing, the bivariate Rayleigh can be expressed in the form of a single integral with finite limits and an integrand composed of elementary functions as follows:

$$P_{R_1,R_2}(r_1, \Omega_1; r_2, \Omega_2|\rho) = 1 - g(Y_1, Y_2|\rho)$$
$$+ \frac{1}{2\pi} \int_{-\pi}^{\pi} \exp\left(-\frac{Y_1^2 + Y_2^2 + 2\sqrt{\rho}Y_1Y_2\sin\theta}{1-\rho}\right)$$
$$\times \left[\frac{(1-\rho^2)Y_1^2Y_2^2 + \sqrt{\rho}(1-\rho)Y_1Y_2(Y_1^2+Y_2^2)\sin\theta}{(\rho Y_1^2 + 2\sqrt{\rho}Y_1Y_2\sin\theta + Y_2^2) \times (Y_1^2 + 2\sqrt{\rho}Y_1Y_2\sin\theta + \rho Y_2^2)}\right] d\theta,$$
$$Y_i \triangleq r_i/\sqrt{\Omega_i}$$
(6.7)

where

$$g(Y_1, Y_2|\rho) = \begin{cases} \exp(-Y_2^2), & 0 \leq Y_2 < \sqrt{\rho}Y_1 \\ \frac{1}{2}\exp(-Y_1^2) + \exp(-\rho Y_1^2), & Y_2 = \sqrt{\rho}Y_1 \\ \exp(-Y_1^2) + \exp(-Y_2^2), & \sqrt{\rho}Y_1 < Y_2 < Y_1/\sqrt{\rho} \\ \frac{1}{2}\exp(-Y_2^2) + \exp(-\rho Y_2^2), & Y_2 = Y_1/\sqrt{\rho} \\ \exp(-Y_1^2), & Y_1/\sqrt{\rho} < Y_2 \end{cases} \quad (6.8)$$

At first glance, one might conclude from (6.8) that the bivariate CDF as given by (6.7) is discontinuous at the boundaries $Y_2 = \sqrt{\rho}Y_1$ and $Y_2 = Y_1/\sqrt{\rho}$. Clearly, this cannot be true since the Marcum Q-function itself is continuous over the entire range of both of its arguments and thus from the form in (6.5), the CDF must also be continuous over these same ranges. The explanation for this apparent discontinuity is that the integral portion of (6.7) is also discontinuous at these same boundaries but in such a way as to compensate completely for the discontinuities in $g(Y_1, Y_2|\rho)$ and thus produce a CDF that is continuous for all positive Y_1 and Y_2.

The bivariate Rayleigh CDF of (6.7) has been evaluated numerically using Mathematica and compared with the double-integral representation [3, Eqs. (1) and (2)], the infinite series representation [3, Eq. (4)] and (6.5) using direct evaluation of the Marcum Q-function. Both the infinite sum and the proposed integral representation have a significant speed-up factor compared to the other two methods (double-integral approach and the one where Marcum-Q is evaluated numerically). Furthermore, the proposed approach always gives the exact result (up to the precision/accuracy allowed by the platform), whereas the infinite series representations (when programmed with the available Mathematica routines and setting the upper limit to infinity as allowed by Mathematica) loses its accuracy for high values of ρ such as 0.8 and 0.9 and a truncation of the series is required.[1] Note that the number of terms for the truncation must be determined for each set of values of r_1, r_2 and ρ. Tan and Beaulieu [3] derived a bound on the error resulting from truncation of the infinite series but reported that this bound becomes loose as ρ approaches 1, which we have verified is the case.

An alternative simple form of the bivariate Rayleigh CDF can be obtained by substituting the representations of the first-order Marcum Q-function of (4.26) and (4.27) in (6.5). When this is done, then after considerable algebraic manipulation the following result is obtained:

$$P_{R_1,R_2}(r_1, \Omega_1; r_2, \Omega_2|\rho) = 1 - g(Y_1, Y_2|\rho) + \text{sgn}(Y_2 - \sqrt{\rho}Y_1)I(Y_1, Y_2|\rho) \\ + \text{sgn}(Y_1 - \sqrt{\rho}Y_2)I(Y_2, Y_1|\rho) \quad (6.9)$$

---

[1] Note that the infinite series representation itself converges to the correct result for all values of ρ between zero and one. It is the limitation of the numerical evaluation of this series caused by the software used to make this evaluation that results in the loss of accuracy for large ρ.

where analogous to (6.8),

$$g(Y_1, Y_2|\rho) = \begin{cases} \exp(-Y_2^2), & 0 \le Y_2 < \sqrt{\rho}Y_1 \\ \exp(-Y_1^2) + \exp(-Y_2^2), & \sqrt{\rho}Y_1 \le Y_2 < Y_1/\sqrt{\rho} \\ \exp(-Y_1^2), & Y_1/\sqrt{\rho} \le Y_2 \end{cases} \quad (6.10)$$

and

$$I(Y_1, Y_2|\rho) = \frac{1}{4\pi} \int_{-\pi}^{\pi} \exp\left\{-\left[Y_1^2 + \frac{1}{1-\rho}\frac{(\rho Y_1^2 - Y_2^2)^2}{Y_2^2 + 2\sqrt{\rho}Y_1 Y_2 \sin\theta + \rho Y_1^2}\right]\right\} \quad (6.11)$$

Note that the compensation for the discontinuities in $g(Y_1, Y_2|\rho)$ at the boundaries $Y_2 = \sqrt{\rho}Y_1$ and $Y_2 = Y_1/\sqrt{\rho}$ is now immediately obvious from the form of the last two terms in (6.9). Moreover, the values of the CDF at these endpoints are given as

$$P_{R_1, R_2}(r_1, \Omega_1; r_2, \Omega_2|\rho) = 1 - \frac{1}{2}\exp(-Y_1^2) - \exp(-Y_2^2)$$

$$+ \frac{1}{4\pi}\int_{-\pi}^{\pi} \exp\left\{-\left[Y_2^2 + \frac{1}{1-\rho}\frac{Y_1^2(1-\rho^2)^2}{1+2\rho\sin\theta+\rho^2}\right]\right\},$$

$$Y_2 = \sqrt{\rho}Y_1 \quad (6.12)$$

and

$$P_{R_1, R_2}(r_1, \Omega_1; r_2, \Omega_2|\rho) = 1 - \frac{1}{2}\exp(-Y_2^2) - \exp(-Y_1^2)$$

$$+ \frac{1}{4\pi}\int_{-\pi}^{\pi} \exp\left\{-\left[Y_1^2 + \frac{1}{1-\rho}\frac{Y_2^2(1-\rho^2)^2}{1+2\rho\sin\theta+\rho^2}\right]\right\},$$

$$Y_1 = \sqrt{\rho}Y_2 \quad (6.13)$$

One might anticipate that the bivariate Nakagami-m CDF could be expressed in a form analogous to (6.5), depending instead on the mth-order Marcum Q-function. If this were possible, then using the desired form of the generalized Marcum Q-function as in (4.42) and (4.50), one could also express the bivariate Nakagami-m CDF in the desired form. Unfortunately, to the author's knowledge an expression analogous to (6.5) has not been reported in the literature and the author's have themselves been unable to arrive at one.

6.2 PDF AND CDF FOR MAXIMUM OF TWO RAYLEIGH RANDOM VARIABLES

In this section we consider the distributions of the random variable $R = \max(R_1, R_2)$, where R_1 and R_2 are correlated Rayleigh random variables with joint PDF as in (6.2). As mentioned previously, the random variable R characterizes the output of an SC whose inputs are R_1 and R_2. Since $\Pr\{R \le R^*\} = \Pr\{R_1 \le$

$R^*, R_2 \leq R^*\}$, the CDF of R is obtained immediately from the joint CDF of R_1, R_2 by equating its two arguments. Since we are ultimately interested in the PDF of the instantaneous SNR per bit,[2] $\gamma \triangleq r^2 E_b/N_0$ with mean $\overline{\gamma} = \overline{r^2} E_b/N_0 = \Omega E_b/N_0$, it is convenient for the Rayleigh case to start by renormalizing the bivariate CDF of (6.7).[3] Thus, noting that $Y_i^2 \triangleq r_i^2/\Omega_i = \gamma_i/\overline{\gamma}_i$, $i = 1, 2$, the joint CDF of γ_1 and γ_2 is given by

$$P_{\gamma_1,\gamma_2}(\gamma_1, \overline{\gamma}_1; \gamma_2, \overline{\gamma}_2 | \rho) = 1 - G(H(\gamma_1, \overline{\gamma}_1), H(\gamma_2, \overline{\gamma}_2) | \rho)$$

$$+ \frac{1}{2\pi} \int_{-\pi}^{\pi} \exp\left[-\frac{\frac{\gamma_1}{\overline{\gamma}_1} + \frac{\gamma_2}{\overline{\gamma}_2} + 2\sqrt{\rho\left(\frac{\gamma_1}{\overline{\gamma}_1}\right)\left(\frac{\gamma_2}{\overline{\gamma}_2}\right)} \sin\theta}{1-\rho}\right]$$

$$\times \left[\frac{(1-\rho^2)\left(\frac{\gamma_1}{\overline{\gamma}_1}\right)\left(\frac{\gamma_2}{\overline{\gamma}_2}\right) + \sqrt{\rho(1-\rho)}\sqrt{\left(\frac{\gamma_1}{\overline{\gamma}_1}\right)\left(\frac{\gamma_2}{\overline{\gamma}_2}\right)}\left(\frac{\gamma_1}{\overline{\gamma}_1} + \frac{\gamma_2}{\overline{\gamma}_2}\right) \sin\theta}{\left(\rho\frac{\gamma_1}{\overline{\gamma}_1} + 2\sqrt{\rho\left(\frac{\gamma_1}{\overline{\gamma}_1}\right)\left(\frac{\gamma_2}{\overline{\gamma}_2}\right)} \sin\theta + \frac{\gamma_2}{\overline{\gamma}_2}\right)}\right.$$

$$\left.\times \left(\frac{\gamma_1}{\overline{\gamma}_1} + 2\sqrt{\rho\left(\frac{\gamma_1}{\overline{\gamma}_1}\right)\left(\frac{\gamma_2}{\overline{\gamma}_2}\right)} \sin\theta + \rho\frac{\gamma_2}{\overline{\gamma}_2}\right)\right] d\theta$$

(6.14)

where

$$G(H(\gamma_1, \overline{\gamma}_1), H(\gamma_2, \overline{\gamma}_2) | \rho) = \begin{cases} H(\gamma_2, \overline{\gamma}_2), & 0 \leq \frac{\gamma_2}{\overline{\gamma}_2} < \rho \frac{\gamma_1}{\overline{\gamma}_1} \\ \frac{1}{2}H(\gamma_1, \overline{\gamma}_1) + H(\gamma_2, \overline{\gamma}_2), & \frac{\gamma_2}{\overline{\gamma}_2} = \rho \frac{\gamma_1}{\overline{\gamma}_1} \\ H(\gamma_1, \overline{\gamma}_1) + H(\gamma_2, \overline{\gamma}_2), & \rho \frac{\gamma_1}{\overline{\gamma}_1} < \frac{\gamma_2}{\overline{\gamma}_2} < \frac{1}{\rho}\frac{\gamma_1}{\overline{\gamma}_1} \\ \frac{1}{2}H(\gamma_2, \overline{\gamma}_2) + H(\gamma_1, \overline{\gamma}_1), & \frac{\gamma_2}{\overline{\gamma}_2} = \frac{1}{\rho}\frac{\gamma_1}{\overline{\gamma}_1} \\ H(\gamma_1, \overline{\gamma}_1), & \frac{1}{\rho}\frac{\gamma_1}{\overline{\gamma}_1} < \frac{\gamma_2}{\overline{\gamma}_2} \end{cases}$$

(6.15)

---

[2] As in Chapter 5, we do not distinguish between bit and character instantaneous SNR. Thus, the results derived here apply equally to the instantaneous SNR per symbol when relating to digital communication systems with modulations that are higher order than binary.

[3] We shall use the form of the CDF in (6.7) rather than that in (6.9) because of its synergy with the corresponding results for correlated Nakagami-m RVs discussed in the next section.

with $H(\gamma_i, \overline{\gamma}_i) = \exp(-\gamma_i/\overline{\gamma}_i)$, $i = 1, 2$. Defining the instantaneous SNR per bit at the SC output by $\gamma = \max(\gamma_1, \gamma_2)$, the CDF of γ, namely $P_\gamma(\gamma)$, is obtained immediately by substituting $\gamma_1 = \gamma_2 = \gamma$ in (6.14), that is,[4]

$$P_\gamma(\gamma) = 1 - G(H(\gamma, \overline{\gamma}_1), H(\gamma, \overline{\gamma}_2)|\rho) + \frac{1}{2\pi} \int_{-\pi}^{\pi} \exp\left[-\gamma \frac{\overline{\gamma}_1 + \overline{\gamma}_2 + 2\sqrt{\rho\overline{\gamma}_1\overline{\gamma}_2}\sin\theta}{\overline{\gamma}_1\overline{\gamma}_2(1-\rho)}\right]$$

$$\times \left[\frac{(1-\rho^2)\overline{\gamma}_1\overline{\gamma}_2 + \sqrt{\rho}(1-\rho)\sqrt{\overline{\gamma}_1\overline{\gamma}_2}(\overline{\gamma}_1 + \overline{\gamma}_2)\sin\theta}{(\rho\overline{\gamma}_2 + 2\sqrt{\rho\overline{\gamma}_1\overline{\gamma}_2}\sin\theta + \overline{\gamma}_1)(\overline{\gamma}_2 + 2\sqrt{\rho\overline{\gamma}_1\overline{\gamma}_2}\sin\theta + \rho\overline{\gamma}_1)}\right] d\theta$$

$$\stackrel{\triangle}{=} 1 - G(\exp(-\gamma/\overline{\gamma}_1), \exp(-\gamma/\overline{\gamma}_2)|\rho) + \frac{1}{2\pi}\int_{-\pi}^{\pi} \exp[-\gamma h_1(\theta|\rho)]h_2(\theta|\rho)\,d\theta \quad (6.16)$$

where

$$h_1(\theta|\rho) \stackrel{\triangle}{=} \frac{\overline{\gamma}_1 + \overline{\gamma}_2 + 2\sqrt{\rho\overline{\gamma}_1\overline{\gamma}_2}\sin\theta}{\overline{\gamma}_1\overline{\gamma}_2(1-\rho)}$$

$$h_2(\theta|\rho) \stackrel{\triangle}{=} \frac{(1-\rho^2)\overline{\gamma}_1\overline{\gamma}_2 + \sqrt{\rho}(1-\rho)\sqrt{\overline{\gamma}_1\overline{\gamma}_2}(\overline{\gamma}_1 + \overline{\gamma}_2)\sin\theta}{(\rho\overline{\gamma}_2 + 2\sqrt{\rho\overline{\gamma}_1\overline{\gamma}_2}\sin\theta + \overline{\gamma}_1)(\overline{\gamma}_2 + 2\sqrt{\rho\overline{\gamma}_1\overline{\gamma}_2}\sin\theta + \rho\overline{\gamma}_1)} \quad (6.17)$$

To obtain the PDF of γ, we differentiate (6.16). Since the dependence γ in (6.16) is purely exponential, it is a simple matter to arrive at the result, namely,

$$p_\gamma(\gamma) = -G'(H(\gamma, \overline{\gamma}_1), H(\gamma, \overline{\gamma}_2)|\rho)$$

$$+ \frac{1}{2\pi}\int_{-\pi}^{\pi}[-h_1(\theta|\rho)]\exp[-\gamma h_1(\theta|\rho)]h_2(\theta|\rho)\,d\theta$$

$$= G(-H'(\gamma, \overline{\gamma}_1), -H'(\gamma, \overline{\gamma}_2)|\rho)$$

$$+ \frac{1}{2\pi}\int_{-\pi}^{\pi}[-h_1(\theta|\rho)]\exp[-\gamma h_1(\theta|\rho)]h_2(\theta|\rho)\,d\theta \quad (6.18)$$

where the prime denotes differentiation with respect to γ and thus $-H'(\gamma, \overline{\gamma}_i) = (1/\overline{\gamma}_i)\exp(-\gamma/\overline{\gamma}_i)$, $i = 1, 2$. Note that the dependence of $p_\gamma(\gamma)$ on γ is also purely exponential and as such resembles the behavior of the instantaneous SNR per bit corresponding to a single Rayleigh RV, namely, $p_\gamma(\gamma) = (1/\overline{\gamma})\exp(-\gamma/\overline{\gamma})$. Because of this similarity, it is possible to draw an analogy with results for the average error probability performance of single-channel (no diversity) digital modulations transmitted over a Rayleigh fading channel (see

---

[4] Note that when $\gamma_1 = \gamma_2 = \gamma$, as will be the case for the SC output PDF and CDF, the five regions of validity for $G(\bullet, \bullet|\rho)$ of (6.15) are independent of γ and become (1) $0 \leq \overline{\gamma}_1 < \rho\overline{\gamma}_2$, (2) $\overline{\gamma}_1 = \rho\overline{\gamma}_2$, (3) $\rho\overline{\gamma}_2 < \overline{\gamma}_1 < \overline{\gamma}_2/\rho$, (4) $\overline{\gamma}_1 = \overline{\gamma}_2/\rho$, and (5) $\overline{\gamma}_2/\rho < \overline{\gamma}_1$.

Chapter 8) which make use of the integrals developed in Sections 5.1.1 and 5.2.1 based on the desired forms of the Gaussian and Marcum Q-functions. However, because of the additional integration on θ required by the second term in (6.18), the functional form of the results will be somewhat different.

6.3 PDF AND CDF FOR MAXIMUM OF TWO NAKAGAMI-m RANDOM VARIABLES

As mentioned in Section 6.1, the alternative representation of the Marcum Q-function discussed in Chapter 4 is not helpful in simplifying the bivariate Nakagami-m CDF in the form of a single integral with finite limits as was possible for the Rayleigh case; thus, the method used to arrive at the CDF and PDF of the SC output in Section 6.2 cannot be used here. Fortunately, however, Fedele et al. [7] were able to arrive directly at an expression for the SC output PDF in terms of the mth-order Marcum Q-function directly from the defining expression for the bivariate Nakagami-m CDF as in (6.3). Using the alternative representation of the generalized Marcum Q-function given in (4.42) and (4.50), Simon and Alouini [8] were then able to simplify the expression for the SC output PDF and working backward (i.e., integrating rather than differentiating), obtain the SC output CDF. Following this approach, one never needs to find the joint CDF of the SC input. While it is true that the results from Ref. 7 could also be used to obtain directly the PDF and CDF of the SC output for Rayleigh fading by considering the special case of the Nakagami-m distribution corresponding to $m = 1$, the method used in Section 6.1 for solving the Rayleigh case allows for additional simplifications of the resulting expressions for outage probability and average error probability, as will be demonstrated later in the book.

The PDF of the SC output $R = \max(R_1, R_2)$ can be found directly from the bivariate Nakagami-m PDF of R_1 and R_2 as

$$p_R(r) = \frac{d}{dr} \int_0^r \int_0^r p_{R_1,R_2}(r_1, \Omega_1; r_2, \Omega_2 | m, \rho) \, dr_1 \, dr_2 \qquad (6.19)$$

Substituting (6.3) in (6.19) results in [7, Eq. (20)]

$$p_R(r) = \frac{2m^m r^{2m-1}}{\Gamma(m)\Omega_1^m} \exp\left(-\frac{mr^2}{\Omega_1}\right) \left[1 - Q_m\left(\sqrt{\frac{2m\rho}{(1-\rho)\Omega_1}} r, \sqrt{\frac{2m\rho}{(1-\rho)\Omega_2}} r\right)\right]$$
$$+ \frac{2m^m r^{2m-1}}{\Gamma(m)\Omega_2^m} \exp\left(-\frac{mr^2}{\Omega_2}\right) \left[1 - Q_m\left(\sqrt{\frac{2m\rho}{(1-\rho)\Omega_2}} r, \sqrt{\frac{2m\rho}{(1-\rho)\Omega_1}} r\right)\right],$$
$$r \geq 0 \qquad (6.20)$$

which when rewritten in terms of the instantaneous SC output instantaneous SNR per bit, γ, becomes

$$p_\gamma(\gamma) = \frac{m^m}{\Gamma(m)\overline{\gamma}_1} \left(\frac{\gamma}{\overline{\gamma}_1}\right)^{m-1} \exp\left(-\frac{m\gamma}{\overline{\gamma}_1}\right)$$

$$\times \left[1 - Q_m\left(\sqrt{\frac{2m\rho}{1-\rho}\left(\frac{\gamma}{\overline{\gamma}_1}\right)}, \sqrt{\frac{2m\rho}{1-\rho}\left(\frac{\gamma}{\overline{\gamma}_2}\right)}\right)\right]$$

$$+ \frac{m^m}{\Gamma(m)\overline{\gamma}_2} \left(\frac{\gamma}{\overline{\gamma}_2}\right)^{m-1} \exp\left(-\frac{m\gamma}{\overline{\gamma}_2}\right)$$

$$\times \left[1 - Q_m\left(\sqrt{\frac{2m\rho}{1-\rho}\left(\frac{\gamma}{\overline{\gamma}_2}\right)}, \sqrt{\frac{2m\rho}{1-\rho}\left(\frac{\gamma}{\overline{\gamma}_1}\right)}\right)\right], \qquad \gamma \geq 0 \qquad (6.21)$$

Applying the alternative representation of the generalized Marcum Q-function to (6.21), then analogous to (6.18) we obtain for $\rho \neq 0$[5]:

$$p_\gamma(\gamma) = G(-H'(\gamma, \overline{\gamma}_1, m), -H'(\gamma, \overline{\gamma}_2, m)|\rho)$$

$$- \frac{m^m}{\Gamma(m)} \frac{1}{2\pi} \int_{-\pi}^{\pi} \gamma^{m-1} \exp[-m\gamma h_1(\theta|\rho)]h(\theta|\rho)\, d\theta, \qquad \gamma > 0 \qquad (6.22)$$

with

$$-H'(\gamma, \overline{\gamma}_i, m) \triangleq \frac{m^m}{\Gamma(m)\overline{\gamma}_i} \left(\frac{\gamma}{\overline{\gamma}_i}\right)^{m-1} \exp\left(-\frac{m\gamma}{\overline{\gamma}_i}\right), \qquad i = 1, 2 \qquad (6.23)$$

Also, $h_1(\theta|\rho)$ is still given by (6.17), which is independent of m and

$$h(\theta|\rho) \triangleq \frac{1}{\overline{\gamma}_1^m} \left(\frac{\overline{\gamma}_1}{\rho\overline{\gamma}_2}\right)^{(m-1)/2}$$

$$\times \left[\frac{-\overline{\gamma}_1 \cos[(m-1)(\theta+\pi/2)] + \sqrt{\rho\overline{\gamma}_1\overline{\gamma}_2}\cos[m(\theta+\pi/2)]}{\rho\overline{\gamma}_2 + 2\sqrt{\rho\overline{\gamma}_1\overline{\gamma}_2}\sin\theta + \overline{\gamma}_1}\right]$$

$$+ \frac{1}{\overline{\gamma}_2^m}\left(\frac{\rho\overline{\gamma}_1}{\overline{\gamma}_2}\right)^{-(m-1)/2}$$

$$\times \left[\frac{\overline{\gamma}_2 \cos[(m-1)(\theta+\pi/2)] - \sqrt{\rho\overline{\gamma}_1\overline{\gamma}_2}\cos[m(\theta+\pi/2)]}{\overline{\gamma}_2 + 2\sqrt{\rho\overline{\gamma}_1\overline{\gamma}_2}\sin\theta + \rho\overline{\gamma}_1}\right] \qquad (6.24)$$

---

[5] Note that the alternative representation of the generalized Marcum Q-function ($m \neq 1$) is valid only for $\rho \neq 0$.

Note that $m = 1$, (6.24) simplifies to

$$h(\theta|\rho) = \frac{1}{\overline{\gamma}_1}\left[\frac{\overline{\gamma}_1 + \sqrt{\rho\overline{\gamma}_1\overline{\gamma}_2}\sin\theta}{\rho\overline{\gamma}_2 + 2\sqrt{\rho\overline{\gamma}_1\overline{\gamma}_2}\sin\theta + \overline{\gamma}_1}\right] + \frac{1}{\overline{\gamma}_2}\left[\frac{\overline{\gamma}_2 + \sqrt{\rho\overline{\gamma}_1\overline{\gamma}_2}\sin\theta}{\overline{\gamma}_2 + 2\sqrt{\rho\overline{\gamma}_1\overline{\gamma}_2}\sin\theta + \rho\overline{\gamma}_1}\right] \quad (6.25)$$

which can be shown to be equal to $h_1(\theta|\rho)h_2(\theta|\rho)$ with $h_2(\theta|\rho)$ obtained from (6.17). Thus, also noting that $-H'(\gamma, \overline{\gamma}_i, 1) = (1/\overline{\gamma}_i)\exp(-\gamma/\overline{\gamma}_i)$, $i = 1, 2$, the PDF of (6.24) reduces to (6.18), as it should.

Note here that the dependence on γ of $p_\gamma(\gamma)$ in (6.22) resembles the behavior of the instantaneous SNR per bit corresponding to a single Rayleigh RV, namely, $p_\gamma(\gamma) = [m^m \gamma^{m-1}/\overline{\gamma}^m \Gamma(m)]\exp(-m\gamma/\overline{\gamma})$. Because of this similarity, it is possible to draw an analogy with results for the average error probability performance of single-channel (no diversity) digital modulations transmitted over a Nakagami-m fading channel (see Chapter 8) which make use of the integrals developed in Sections 5.1.4 and 5.2.4 based on the desired forms of the Gaussian and Marcum Q-functions. However, because of the additional integration on θ required by the second term in (6.18), the functional form of the results will be somewhat different.

The CDF of the SC output can now be found directly by integration of (6.22) with the result (for $\rho \neq 0$)

$$P_\gamma(\gamma) = G(-H(\gamma, \overline{\gamma}_1, m), -H(\gamma, \overline{\gamma}_2, m)|\rho)$$
$$- \frac{m^m}{\Gamma(m)}\frac{1}{2\pi}\int_{-\pi}^{\pi}\left\{\int_0^\gamma y^{m-1}\exp[-myh_1(\theta|\rho)]\,dy\right\}h(\theta|\rho)\,d\theta$$
$$= G(-H(\gamma, \overline{\gamma}_1, m), -H(\gamma, \overline{\gamma}_2, m)|\rho)$$
$$- \frac{m^m}{\Gamma(m)}\frac{1}{2\pi}\int_{-\pi}^{\pi}(h_1(\theta|\rho))^{-m}[-H(\gamma, h_1^{-1}(\theta|\rho), m)]h(\theta|\rho)\,d\theta \quad (6.26)$$

where now

$$-H(\gamma, \overline{\gamma}_i, m) \triangleq \int_0^\gamma -H'(y, \overline{\gamma}_i, m)\,dy$$
$$= 1 - \exp\left(-\frac{m\gamma}{\overline{\gamma}_i}\right)\sum_{k=0}^{m-1}\frac{(m\gamma/\overline{\gamma}_i)^k}{k!}, \quad i = 1, 2 \quad (6.27)$$

For $\rho = 0$, the PDF γ can be obtained from Fedele et al. [7, Eq. (20)], which after some changes of variables becomes

$$p_\gamma(\gamma) = \frac{m^m}{\Gamma(m)\overline{\gamma}_1}\left(\frac{\gamma}{\overline{\gamma}_1}\right)^{m-1}\exp\left(-\frac{m\gamma}{\overline{\gamma}_1}\right)\left[1 - \frac{\Gamma(m, m\gamma/\overline{\gamma}_2)}{\Gamma(m)}\right]$$
$$+ \frac{m^m}{\Gamma(m)\overline{\gamma}_2}\left(\frac{\gamma}{\overline{\gamma}_2}\right)^{m-1}\exp\left(-\frac{m\gamma}{\overline{\gamma}_2}\right)\left[1 - \frac{\Gamma(m, m\gamma/\overline{\gamma}_1)}{\Gamma(m)}\right], \quad \gamma \geq 0 \quad (6.28)$$

where $\Gamma(m, x) = \int_x^\infty e^{-t} t^{m-1} dt$ is the complementary incomplete gamma function [5, Eq. (6.5.3)]. For m integer $\Gamma(m, x)$ has a closed-form expression [9, Eq. (8.352.2)] and (6.28) simplifies to

$$p_\gamma(\gamma) = \frac{m^m}{(m-1)!\overline{\gamma}_1} \left(\frac{\gamma}{\overline{\gamma}_1}\right)^{m-1} \exp\left(-\frac{m\gamma}{\overline{\gamma}_1}\right) [-H(\gamma, \overline{\gamma}_2, m)]$$
$$+ \frac{m^m}{(m-1)!\overline{\gamma}_2} \left(\frac{\gamma}{\overline{\gamma}_2}\right)^{m-1} \exp\left(-\frac{m\gamma}{\overline{\gamma}_2}\right) [-H(\gamma, \overline{\gamma}_1, m)], \quad \gamma \geq 0$$
(6.29)

The corresponding CDFs are obtained by integration of (6.28) and (6.29) between 0 and γ. For m noninteger, integration of (6.28) does not produce a closed-form result, whereas for m integer, integration of (6.29) results in

$$P_\gamma(\gamma) = -H(\gamma, \overline{\gamma}_1, m) - H(\gamma, \overline{\gamma}_2, m)$$
$$- \sum_{n=0}^{m-1} \frac{(n+m-1)!}{n!(m-1)!} \frac{(\overline{\gamma}_1)^n (\overline{\gamma}_2)^m + (\overline{\gamma}_1)^m (\overline{\gamma}_2)^n}{(\overline{\gamma}_1 + \overline{\gamma}_2)^{n+m}} \left[-H_n\left(\gamma, \frac{\overline{\gamma}_1 \overline{\gamma}_2}{\overline{\gamma}_1 + \overline{\gamma}_2}, m\right)\right]$$
(6.30)

where analogous to (6.27),

$$-H_n(\gamma, \overline{\gamma}, m) \triangleq 1 - \exp\left(-\frac{m\gamma}{\overline{\gamma}}\right) \sum_{k=0}^{m+n-1} \frac{(m\gamma/\overline{\gamma})^k}{k!}$$
(6.31)

Note that $-H_0(\gamma, \overline{\gamma}, m)$ is equal $-H(\gamma, \overline{\gamma}, m)$ of (6.27).

REFERENCES

1. M. Schwartz, W. R. Bennett, and S. Stein, *Communication Systems and Techniques.* New York: McGraw-Hill, 1966.
2. M. Nakagami, "The m-distribution: A general formula of intensity distribution of rapid fading," in *Statistical Methods in Radio Wave Propagation.* Oxford: Pergamon Press, 1960, pp. 3–36.
3. C. C. Tan and N. C. Beaulieu, "Infinite series representation of the bivariate Rayleigh and Nakagami-m distributions," *IEEE Trans. Commun.*, vol. 45, no. 10, October 1997, pp. 1159–1161.
4. S. O. Rice, "Mathematical analysis of random noise," *Bell Syst. Tech. J.*, vol. 23, 1944, pp. 282–332; vol. 24, 1945, pp. 46–156.
5. M. Abramowitz and I. A. Stegun, *Handbook of Mathematical Functions with Formulas, Graphs, and Mathematical Tables*, 9th ed. New York: Dover Press, 1972.
6. M. K. Simon and M.-S. Alouini, "A simple single integral representation of the bivariate Rayleigh distribution," *IEEE Commun. Lett.*, vol. 2, no. 5, May 1998, pp. 128–130.

7. G. Fedele, I. Izzo, and M. Tanda, "Dual diversity reception of M-ary DPSK signals over Nakagami fading channels," *IEEE International Symp. Personal, Indoor, and Mobile Radio Commun.*, Toronto, Ontario, Canada, September 1995, pp. 1195–1201.
8. M. K. Simon and M.-S. Alouini, "A unified performance analysis of digital communication with dual selection combining diversity over correlated Rayleigh and Nakagami-m fading channels," *IEEE Trans. Commun.*, vol. 47, no. 1, January 1999, pp. 33–43. Also presented in part in the GLOBECOM '98 Conference Record, Sydney, Australia, November 8–12, 1998.
9. I. S. Gradshteyn and I. M. Ryzhik, *Table of Integrals, Series, and Products*, 5th ed. San Diego, CA: Academic Press, 1994.

PART 3

OPTIMUM RECEPTION AND PERFORMANCE EVALUATION

7

OPTIMUM RECEIVERS FOR FADING CHANNELS

As far back as the 1950s, researchers and communication engineers recognized the need for investigating the form of receivers that would provide optimum detection of digital modulations transmitted over a channel composed of a combination of AWGN and multiplicative fading. For the most part, most of these contributions dealt with only the simplest of modulation/detection schemes and fading channels (i.e., BPSK with coherent detection and Rayleigh or Rician fading). In some instances, the work pertained to single-channel reception, while in others multichannel reception was considered. Our goal in this chapter is to present the work of the past under a unified framework based on the maximum-likelihood approach and also to consider a larger number of situations corresponding to more sophisticated modulations, detection schemes, and fading channels. In addition, we treat a variety of combinations of channel state knowledge relating to the amplitude, phase, and delay parameter vectors associated with the fading channels. In many instances, implementation of the optimum structure may not be simple or even feasible and thus a suboptimum solution is preferable and is discussed. Also, evaluating the error probability performance of these optimum receivers may not always be possible to accomplish using the analytical tools discussed previously in this book or anywhere else for that matter. Nevertheless, it is of interest to determine in each case the optimum receiver since it serves as a benchmark against which to measure the suboptimum structure, which is simpler both to implement and to analyze.

We begin our discussion by reviewing the mathematical models for the transmitted signal and generalized fading channel as introduced in previous chapters. In particular, consider that during a symbol period of T_s seconds the transmitter sends the real bandpass signal[1]

---

[1] Without any loss in generality, we shall assume that the carrier phase, θ_c, is arbitrarily set equal to zero since the various paths that compose the channel will each introduce their own random phase into the transmission.

158 OPTIMUM RECEIVERS FOR FADING CHANNELS

$$s_k(t) = \text{Re}\{\tilde{s}_k(t)\} = \text{Re}\{\tilde{S}_k(t)e^{j2\pi f_c t}\} \tag{7.1}$$

where $\tilde{s}_k(t)$ is the kth complex bandpass signal and $\tilde{S}_k(t)$ is the corresponding kth complex baseband signal chosen from the set of M equiprobable message waveforms representing the transmitted information. At this point, we do not restrict the signal set $\{\tilde{S}_k(t)\}$ in any way (e.g., we do not require that the signals have equal energy), and thus we are able to handle all of the various modulation types discussed in Chapter 3.

The signal of (7.1) is transmitted over the generalized fading channel which is characterized by L_p independent paths, each of which is a slowly varying channel which attenuates, delays, and phase shifts the signal and adds an AWGN noise source. Thus the received signal is a set of noisy replicas of the transmitted signal, that is,[2]

$$\begin{aligned} r_l(t) &= \text{Re}\{\alpha_l \tilde{s}_k(t-\tau_l)e^{j\theta_l} + \tilde{n}_l(t)\} \\ &= \text{Re}\{\alpha_l \tilde{S}_k(t-\tau_l)e^{j(2\pi f_c t + \theta_l)} + \tilde{N}_l(t)e^{j2\pi f_c t}\} \\ &= \text{Re}\{\tilde{r}_l(t)\} = \text{Re}\{\tilde{R}_l(t)e^{j2\pi f_c t}\}, \quad l = 1, 2, \ldots, L_p \end{aligned} \tag{7.2}$$

where $\{\tilde{N}_l(t)\}_{l=1}^{L_p}$ is a set of statistically independent[3] complex AWGN processes each with PSD $2N_l$ watts/Hz. The sets $\{\alpha_l\}_{l=1}^{L_p}$, $\{\theta_l\}_{l=1}^{L_p}$, and $\{\tau_l\}_{l=1}^{L_p}$ are the random channel amplitudes, phases, and delays, respectively, which because of the slow-fading assumption, are assumed to be constant over the transmission (symbol) interval T_s. Also, without loss of generality, we take the first channel to be the reference channel whose delay $\tau_1 = 0$ and assume further that the delays are ordered (i.e., $\tau_1 < \tau_2 < \cdots < \tau_{L_p}$).

The optimum receiver computes the set of a posteriori probabilities $p(s_k(t)|\{r_l(t)\}_{l=1}^{L_p})$, $k = 1, 2, \ldots, M$, and chooses as its decision that message whose signal $s_k(t)$ corresponds to the largest of these probabilities.[4] Since the messages (signals) are assumed to be equiprobable, then by Bayes rule, the equivalent decision rule is to choose $s_k(t)$ corresponding to the largest of the conditional probabilities (likelihoods) $p(\{r_l(t)\}_{l=1}^{L_p}|s_k(t))$, $k = 1, 2, \ldots, M$, which is the maximum-likelihood (ML) decision rule. Using the law of conditional

---

[2] In deriving the various optimum receiver configurations, we assume a "one-shot" approach (i.e., a single transmission), wherein intersymbol interference (ISI) that would be produced by the presence of the path delays on continuous transmission is ignored.

[3] It should be noted that Turin [1] originally considered optimal diversity reception for the more general case where the link noises (as well as the link fades) could be mutually correlated; however, the noises and fades were statistically independent. Later, however, Turin [2] restricted his considerations to link noises that were white Gaussian and statistically independent. (The link fades, however, were still allowed to be correlated—statistically independent and exponentially correlated fades were considered as special cases.)

[4] The receiver is assumed to be time-synchronized to the transmitted signal (i.e., it knows the time epoch of the beginning of the transmission).

probability, each of these conditional probabilities can be expressed as[5]

$$\iiint p(\{r_l(t)\}_{l=1}^{L_p}|s_k(t), \{\alpha_l\}_{l=1}^{L_p}, \{\theta_l\}_{l=1}^{L_p}, \{\tau_l\}_{l=1}^{L_p})$$
$$\times p(\{\alpha_l\}_{l=1}^{L_p}, \{\theta_l\}_{l=1}^{L_p}, \{\tau_l\}_{l=1}^{L_p}) d\{\alpha_l\}_{l=1}^{L_p} d\{\theta_l\}_{l=1}^{L_p} d\{\tau_l\}_{l=1}^{L_p} \quad (7.3)$$

and as such depends on the degree of knowledge [amount of *channel state information (CSI)*] available on the parameter sets $\{\alpha_l\}_{l=1}^{L_p}$, $\{\theta_l\}_{l=1}^{L_p}$, and $\{\tau_l\}_{l=1}^{L_p}$. For instance, if any of the three parameter sets are assumed to be known (e.g., through channel measurement), the statistical averages on that set of parameters need not be performed. In the limiting case (to be considered shortly) where all parameters are assumed to be known to the receiver, none of the statistical averages in (7.3) need be performed, and hence the ML decision rule simplifies to choosing the largest of $p(\{r_l(t)\}_{l=1}^{L_p}|s_k(t), \{\alpha_l\}_{l=1}^{L_p}, \{\theta_l\}_{l=1}^{L_p}, \{\tau_l\}_{l=1}^{L_p})$, $k = 1, 2, \ldots, M$.

Receivers that make use of CSI have been termed *self-adaptive* [3] in that the estimates of the system parameters are utilized to adjust the decision structure, thereby improving system performance by adaptation to slowly varying channel changes. We start our detailed discussion of optimum receivers with the most general case of all parameters known since the decision rule is independent of the statistics of the channel parameters and leads to a well-known classic structure whose performance is better than all others that are based on less than complete parameter knowledge. Also, since detection schemes are typically classified based on the degree of knowledge related to the phase(s) of the received signal, ideal coherent detection implying perfect knowledge falls into this category.

7.1 CASE OF KNOWN AMPLITUDES, PHASES, AND DELAYS: COHERENT DETECTION

Conditioned on perfect knowledge of the the amplitudes, phases, and delays, the conditional probability $p(\{r_l(t)\}_{l=1}^{L_p}|s_k(t), \{\alpha_l\}_{l=1}^{L_p}, \{\theta_l\}_{l=1}^{L_p}, \{\tau_l\}_{l=1}^{L_p})$ is a joint Gaussian PDF which because of the independence assumption on the additive noise components can be written as

$$p(\{r_l(t)\}_{l=1}^{L_p}|s_k(t), \{\alpha_l\}_{l=1}^{L_p}, \{\theta_l\}_{l=1}^{L_p}, \{\tau_l\}_{l=1}^{L_p})$$
$$= \prod_{l=1}^{L_p} K_l \exp\left[-\frac{1}{2N_l}\int_{\tau_l}^{T_s+\tau_l}\left|\tilde{r}_l(t) - \alpha_l \tilde{s}_k(t-\tau_l)e^{j\theta_l}\right|^2 dt\right]$$
$$= \prod_{l=1}^{L_p} K_l \exp\left[-\frac{1}{2N_l}\int_{\tau_l}^{T_s+\tau_l}\left|\tilde{R}_l(t) - \alpha_l \tilde{S}_k(t-\tau_l)e^{j\theta_l}\right|^2 dt\right] \quad (7.4)$$

---
[5] Each integral in (7.3) is, in fact, an L_p-fold integral.

where K_l is an integration constant. Substituting (7.2) into (7.4) and simplifying yields

$$p\left(\{r_l(t)\}_{l=1}^{L_p} \big| s_k(t), \{\alpha_l\}_{l=1}^{L_p}, \{\theta_l\}_{l=1}^{L_p}, \{\tau_l\}_{l=1}^{L_p}\right)$$

$$= K \prod_{l=1}^{L_p} \exp\left[\text{Re}\left\{\frac{\alpha_l}{N_l} e^{-j\theta_l} y_{kl}(\tau_l)\right\} - \frac{\alpha_l^2 E_k}{N_l}\right]$$

$$= K \exp\left[\sum_{l=1}^{L_p} \text{Re}\left\{\frac{\alpha_l}{N_l} e^{-j\theta_l} y_{kl}(\tau_l)\right\} - \sum_{l=1}^{L_p} \frac{\alpha_l^2 E_k}{N_l}\right] \quad (7.5)$$

where

$$y_{kl}(\tau_l) \triangleq \int_{\tau_l}^{T_s + \tau_l} \tilde{R}_l(t) \tilde{S}_k^*(t - \tau_l) \, dt = \int_0^{T_s} \tilde{R}_l(t + \tau_l) \tilde{S}_k^*(t) \, dt \quad (7.6)$$

is the complex cross-correlation of the lth received signal and the kth signal waveform and

$$E_k = \frac{1}{2} \int_0^{T_s} |\tilde{S}_k(t)|^2 \, dt = \frac{1}{2} \int_{\tau_l}^{T_s + \tau_l} |\tilde{S}_k(t - \tau_l)|^2 \, dt \quad (7.7)$$

is the energy of the kth signal $s_k(t)$. Also, the constant K absorbs all the K_l's as well as the factor $\exp[\sum_{l=1}^{L_p}(1/2N_l) \int |\tilde{R}_l(t)|^2 \, dt]$, which is independent of k and thus has no bearing on the decision. Since the natural logarithm is a monotonic function of its argument, we can equivalently maximize (with respect to k)

$$\Lambda_k \triangleq \ln p\left(\{r_l(t)\}_{l=1}^{L_p} \big| s_k(t), \{\alpha_l\}_{l=1}^{L_p}, \{\theta_l\}_{l=1}^{L_p}, \{\tau_l\}_{l=1}^{L_p}\right)$$

$$= \sum_{l=1}^{L_p} \left[\text{Re}\left\{\frac{\alpha_l}{N_l} e^{-j\theta_l} y_{kl}(\tau_l)\right\} - \frac{\alpha_l^2 E_k}{N_l}\right] \quad (7.8)$$

where we have ignored the $\ln K$ term since it is independent of k.[6] The first bracketed term in the summation of (7.8) requires a complex weight ($\alpha_l e^{j\theta_l}$) to be applied to the lth cross-correlator output (scaled by the noise PSD N_l) and the second bracketed term is a bias dependent on the signal energy-to-noise ratio in the lth path. For constant envelope signal sets (i.e., $E_k = E$; $l = 1, 2, \ldots, M$), the bias can be omitted from the decision-making process.

A receiver that implements (7.8) as its decision statistic is illustrated in Fig. 7.1 and is generically referred to as a *RAKE receiver* [4,5] because of its structural

---

[6] For convenience, in what follows we shall use the notation Λ_k for all decision metrics associated with the kth signal regardless of any constants that will be ignored because they do not depend on k.

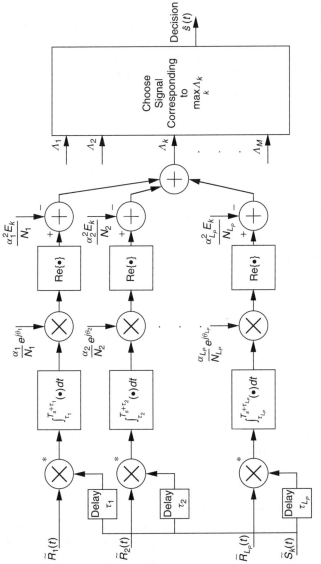

Figure 7.1. Complex form of optimum receiver for known amplitudes, phases, and delays: coherent detection. (The asterisk on the multiplier denotes complex-conjugate multiplication.)

similarity with the teeth on a garden rake.[7] Note that this receiver is, for the CSI conditions specified (i.e., perfect knowledge of all channel parameters), optimum *regardless of the statistics of these parameters*. We shall see shortly that as soon as we deviate from this ideal condition (i.e., one or more sets of parameters are unknown), the receiver structure will immediately depend on the channel parameter statistics.

We conclude this subsection by noting that if instead of the generalized fading channel model consisting of L_p independently received noisy replicas of the transmitted signal, we had assumed the random multipath channel model suggested by Turin [6], wherein the received signal would instead be of the form

$$r(t) = \sum_{l=1}^{L_p} \text{Re}\{\alpha_l \tilde{s}_k(t - \tau_l)e^{j\theta_l}\} + \text{Re}\{\tilde{n}(t)\}$$

$$= \sum_{l=1}^{L_p} \text{Re}\{\alpha_l \tilde{S}_k(t - \tau_l)e^{j(2\pi f_c t + \theta_l)}\} + \text{Re}\{\tilde{N}(t)e^{j2\pi f_c t}\}$$

$$= \text{Re}\{\tilde{R}(t)e^{j2\pi f_c t}\} \tag{7.9}$$

with $\tilde{N}(t)$ a complex AWGN processes with PSD $2N_0$ watts/Hz, the decision metric analogous to (7.8) would be

$$\Lambda_k \triangleq \sum_{l=1}^{L_p} \left[\text{Re}\left\{ \frac{\alpha_l}{N_0} e^{-j\theta_l} y_{kl}(\tau_l) \right\} - \frac{\alpha_l^2 E_k}{N_0} \right] \tag{7.10}$$

which is in agreement with Ref. 6. Since N_0 is now a constant independent of l, we can eliminate it from (7.10) in so far as the decision is concerned and rewrite the decision metric as

$$\Lambda_k \triangleq \sum_{l=1}^{L_p} [\text{Re}\{\alpha_l e^{-j\theta_l} y_{kl}(\tau_l)\} - \alpha_l^2 E_k] \tag{7.11}$$

For single-channel reception (i.e., $L_p = 1$), (7.8) or (7.11) simplifies to

$$\Lambda_k \triangleq \text{Re}\{\alpha e^{-j\theta} y_{kl}(\tau)\} - \alpha^2 E_k \tag{7.12}$$

which is identical to the decision metric for a purely AWGN channel except for the scaling of the first term by the known fading amplitude α and the second (bias) term by α^2. For the special case of constant-envelope signal sets, the second term becomes independent of k and can therefore be ignored, leaving as a decision metric $\Lambda_k = \alpha \,\text{Re}\{e^{-j\theta} y_{kl}(\tau)\}$. Since α now appears strictly as a

---

[7] Such a receiver is also considered to implement the *maximum-ratio combining (MRC)* form of diversity and is discussed further in Chapter 9 which deals with the performance of multichannel receivers.

multiplicative constant that is independent of k, it has no bearing on the decision and thus can also be eliminated from the decision metric. Hence, *for single-channel reception of constant envelope signal sets, the decision metric is identical to that for the pure AWGN channel, and knowledge of the fading amplitude does not aid in improving the performance.* It should be emphasized, however, that despite the lack of dependence of the optimum decision metric on knowledge of the channel fading amplitude, the error probability performance of this receiver does indeed depend on the fading amplitude statistics and will of course be worse for the fading channel than for the pure AWGN channel. On the other hand, for nonconstant envelope signal sets (e.g., M-QAM), the second term in (7.12) cannot be ignored and optimum performance requires perfect knowledge of the channel fading amplitude (typically provided by an AGC).

Finally, note that if in the generalized fading channel model all paths have equal noise PSD (i.e., $N_l = N_0$, $l = 1, 2, \ldots, L_p$), the decision metric of (7.8) reduces to that of (7.10).

7.2 THE CASE OF KNOWN PHASES AND DELAYS, UNKNOWN AMPLITUDES

When the amplitudes are unknown, the conditional probability of (7.5) must be averaged over their joint PDF to arrive at the decision metric. Assuming independent amplitudes with first-order PDFs $\{p_{\alpha_l}(\alpha_l)\}_{l=1}^{L_p}$, we obtain

$$p\big(\{r_l(t)\}_{l=1}^{L_p}\big|s_k(t), \{\theta_l\}_{l=1}^{L_p}, \{\tau_l\}_{l=1}^{L_p}\big)$$

$$= K \prod_{l=1}^{L_p} \int_0^\infty \exp\left[\frac{\alpha_l}{N_l} \mathrm{Re}\{e^{-j\theta_l} y_{kl}(\tau_l)\} - \frac{\alpha_l^2 E_k}{N_l}\right] p_{\alpha_l}(\alpha_l)\, d\alpha_l \quad (7.13)$$

We now consider the evaluation of (7.13) for Rayleigh and Nakagami-m fading.

7.2.1 Rayleigh Fading

For Rayleigh fading with channel PDFs,

$$p_{\alpha_l}(\alpha_l) = \frac{2\alpha_l}{\Omega_l} \exp\left(-\frac{\alpha_l^2}{\Omega_l}\right), \qquad \alpha_l \geq 0 \quad (7.14)$$

and $\Omega_l \triangleq E\{\alpha_l^2\}$, the integrals of (7.13) can be evaluated in closed form. In particular, using Eq. (3.462.5) of Ref. 7, we obtain

$$p\big(\{r_l(t)\}_{l=1}^{L_p}\big|s_k(t), \{\theta_l\}_{l=1}^{L_p}, \{\tau_l\}_{l=1}^{L_p}\big)$$

$$= K \prod_{l=1}^{L_p} (1+\overline{\gamma_{kl}})^{-1} \left\{1 + \sqrt{\pi} U_{kl} \exp\left(\frac{U_{kl}^2}{4}\right)\left[1 - Q\left(\frac{U_{kl}}{\sqrt{2}}\right)\right]\right\} \quad (7.15)$$

where $Q(x)$ is the Gaussian Q-function (see Chapter 4), $\overline{\gamma}_{kl} \triangleq \Omega_l E_k / N_l$ is the average SNR of the kth signal over the lth path, and

$$U_{kl} \triangleq \sqrt{\frac{E_k}{N_l} \frac{\overline{\gamma}_{kl}}{1 + \overline{\gamma}_{kl}}} \left[\frac{1}{E_k} \text{Re}\{e^{-j\theta_l} y_{kl}(\tau_l)\} \right] \qquad (7.16)$$

The combination of (7.15) and (7.16) agrees, after a number of corrections, with the results of Hancock and Lindsey [3, Eq. (28)] using a different notation.

The decision metric analogous to (7.8) is obtained by taking the natural logarithm of (7.15) and ignoring the $\ln K$ term, which results in

$$\Lambda_k = -\sum_{l=1}^{L_p} \ln(1 + \overline{\gamma}_{kl}) + \sum_{l=1}^{L_p} \ln \left\{ 1 + \sqrt{\pi} U_{kl} \exp\left(\frac{U_{kl}^2}{4}\right) \left[1 - Q\left(\frac{U_{kl}}{\sqrt{2}}\right) \right] \right\} \qquad (7.17)$$

The first summation in (7.17) is a bias, and the second summation is the decision variable that depends on the observation. For large average SNR (i.e., $\overline{\gamma}_{kl} \gg 1$), the decision metric above simplifies to (ignoring the $\ln \sqrt{\pi}$ term)

$$\Lambda_k = -\sum_{l=1}^{L_p} \ln \overline{\gamma}_{kl} + \sum_{l=1}^{L_p} \left(\ln U_{kl} + \frac{1}{4} U_{kl}^2 \right) \qquad (7.18)$$

A receiver that implements the decision rule based on the high SNR decision metric above is illustrated in Fig. 7.2.

7.2.2 Nakagami-m Fading

For Nakagami-m fading with channel PDFs,

$$p_{\alpha_l}(\alpha_l) = \frac{2}{\Gamma(m_l)} \left(\frac{m_l}{\Omega_l}\right)^{m_l} \alpha_l^{2m_l - 1} \exp\left(-\frac{m_l \alpha_l^2}{\Omega_l}\right), \qquad \alpha_l \geq 0 \qquad (7.19)$$

the integrals of (7.13) can be evaluated in closed form using Eq. (3.462.1), of Ref. 7 with the result

$$p(\{r_l(t)\}_{l=1}^{L_p} | s_k(t), \{\theta_l\}_{l=1}^{L_p}, \{\tau_l\}_{l=1}^{L_p})$$

$$= K \prod_{l=1}^{L_p} \left(\frac{\Gamma(2m_l)}{2^{m_l - 1} \Gamma(m_l)}\right) \left(\frac{m_l}{m_l + \overline{\gamma}_{kl}}\right)^{m_l} \exp\left(\frac{V_{kl}^2}{8}\right) D_{-2m_l}\left(-\frac{V_{kl}}{\sqrt{2}}\right) \qquad (7.20)$$

where

$$V_{kl} \triangleq \sqrt{\frac{E_k}{N_l} \left(\frac{\overline{\gamma}_{kl}}{m_l + \overline{\gamma}_{kl}}\right)} \left[\frac{1}{E_k} \text{Re}\{e^{-j\theta_l} y_{kl}(\tau_l)\} \right] \qquad (7.21)$$

and $D_p(x)$ is the parabolic cylinder function [7, Eq. (3.462.1) and Sec. 9.24].

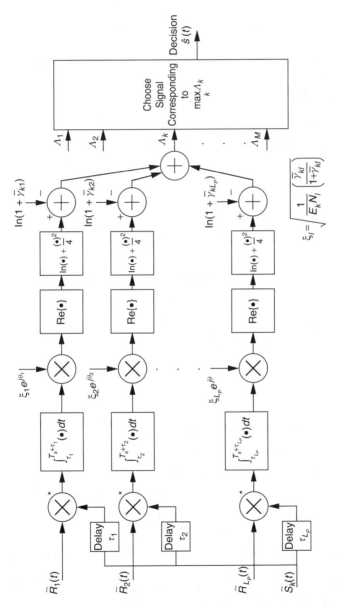

Figure 7.2. Complex form of optimum receiver for known phases and delays, unknown amplitudes: Rayleigh fading, high average SNR ($\overline{\gamma_{kl}} \gg 1$).

7.3 CASE OF KNOWN AMPLITUDES AND DELAYS, UNKNOWN PHASES

When the phases are unknown, the conditional probability of (7.5) must be averaged over their joint PDF to arrive at the decision metric. Assuming independent phases with PDFs specified over the interval $(0, 2\pi)$, we obtain

$$p\big(\{r_l(t)\}_{l=1}^{L_p} \big| s_k(t), \{\alpha_l\}_{l=1}^{L_p}, \{\tau_l\}_{l=1}^{L_p}\big)$$

$$= K \prod_{l=1}^{L_p} \int_0^{2\pi} \exp\left[\frac{\alpha_l}{N_l} \operatorname{Re}\{e^{-j\theta_l} y_{kl}(\tau_l)\} - \frac{\alpha_l^2 E_k}{N_l}\right] p_{\theta_l}(\theta_l) \, d\theta_l \quad (7.22)$$

For uniformly distributed phases as is typical of Rayleigh and Nakagami-m fading, (7.22) becomes

$$p\big(\{r_l(t)\}_{l=1}^{L_p} \big| s_k(t), \{\alpha_l\}_{l=1}^{L_p}, \{\tau_l\}_{l=1}^{L_p}\big)$$

$$= K \prod_{l=1}^{L_p} \exp\left(-\frac{\alpha_l^2 E_k}{N_l}\right) \frac{1}{2\pi} \int_0^{2\pi} \exp\left[\frac{\alpha_l}{N_l} \operatorname{Re}\{e^{-j\theta_l} y_{kl}(\tau_l)\}\right] d\theta_l$$

$$= K \prod_{l=1}^{L_p} \exp\left(-\frac{\alpha_l^2 E_k}{N_l}\right) \frac{1}{2\pi} \int_0^{2\pi} \exp\left\{\frac{\alpha_l}{N_l} |y_{kl}(\tau_l)| \cos[\theta_l - \arg(y_{kl}(\tau_l))]\right\} d\theta_l$$

$$= K \prod_{l=1}^{L_p} \exp\left(-\frac{\alpha_l^2 E_k}{N_l}\right) I_0\left(\frac{\alpha_l}{N_l} |y_{kl}(\tau_l)|\right) \quad (7.23)$$

Taking the natural logarithm of (7.23) and ignoring the $\ln K$ term, we obtain the decision metric

$$\Lambda_k = \sum_{l=1}^{L_p} \ln I_0\left(\frac{\alpha_l}{N_l} |y_{kl}(\tau_l)|\right) - \sum_{l=1}^{L_p} \frac{\alpha_l^2 E_k}{N_l} \quad (7.24)$$

which for constant envelope signal sets simplifies to (ignoring the bias term)

$$\Lambda_k = \sum_{l=1}^{L_p} \ln I_0\left(\frac{\alpha_l}{N_l} |y_{kl}(\tau_l)|\right) \quad (7.25)$$

An implementation of a receiver that bases its decisions on the metric of (7.24) is illustrated in Fig. 7.3.

For large arguments, the function $\ln I_0(x)$ is approximated by a scaled version of $|x|$, and thus for high SNR, the decision metric is similarly approximated by

$$\Lambda_k = \sum_{l=1}^{L_p} \frac{\alpha_l}{N_l} |y_{kl}(\tau_l)| \quad (7.26)$$

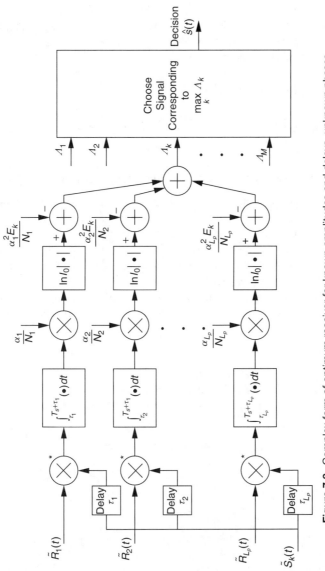

Figure 7.3. Complex form of optimum receiver for known amplitudes and delays, unknown phases.

7.4 CASE OF KNOWN DELAYS AND UNKNOWN AMPLITUDES AND PHASES

When only the delays are known, then the conditional probability of (7.5) must be averaged over both the unknown amplitudes and phases to arrive at the decision. Assuming, as was done in Section 7.3, the case of independent, identically distributed (i.i.d.) uniformly distributed phases, the conditional probability needed to compute the decision statistic is obtained by averaging (7.23) over the PDFs of the independent amplitudes, resulting in

$$p(\{r_l(t)\}_{l=1}^{L_p} | s_k(t), \{\tau_l\}_{l=1}^{L_p})$$
$$= K \prod_{l=1}^{L_p} \int_0^\infty \exp\left(-\frac{\alpha_l^2 E_k}{N_l}\right) I_0\left(\frac{\alpha_l}{N_l}|y_{kl}(\tau_l)|\right) p_{\alpha_l}(\alpha_l)\, d\alpha_l \quad (7.27)$$

7.4.1 One-Symbol Observation: Noncoherent Detection

In this subsection we consider the case where the observation interval of the received signal is one symbol in duration. Receivers that implement their decision rules based on statistics formed from one-symbol duration-correlations are referred to as *noncoherent receivers*. This is in direct contrast to the cases that will be considered next, wherein the observation of the received signal extends over two or more symbols, resulting in *differentially coherent receivers*. This distinction in terminology regarding the method of detection (i.e., noncoherent versus differentially coherent) employed by the receiver and its relation to the observation interval is discussed by Simon et al. [8, App. 7A] for AWGN channels.

7.4.1.1 Rayleigh Fading.
For the Rayleigh fading PDF of (7.14), the conditional probability of (7.27) can be evaluted in closed form. In particular, using Eq. (6.633.4) of Ref. 7, we obtain after some manipulation

$$p(\{r_l(t)\}_{l=1}^{L_p} | s_k(t), \{\tau_l\}_{l=1}^{L_p}) = K \prod_{l=1}^{L_p} (1+\overline{\gamma}_{kl})^{-1} \exp\left[\frac{(U'_{kl})^2}{4}\right] \quad (7.28)$$

where analogous to (7.16) for the coherent case,

$$U'_{kl} \triangleq \sqrt{\frac{E_k}{N_l}\frac{\overline{\gamma}_{kl}}{1+\overline{\gamma}_{kl}}}\left[\frac{1}{E_k}|y_{kl}(\tau_l)|\right] \quad (7.29)$$

Once again taking the natural logarithm of the likelihood of (7.28) and ignoring the $\ln K$ term, we obtain the decision metric

$$\Lambda_k = -\sum_{l=1}^{L_p} \ln(1+\overline{\gamma}_{kl}) + \sum_{l=1}^{L_p} \frac{E_k}{4N_l}\left(\frac{\overline{\gamma}_{kl}}{1+\overline{\gamma}_{kl}}\right)\left[\frac{1}{E_k}|y_{kl}(\tau_l)|\right]^2 \quad (7.30)$$

A receiver that implements a decision rule based on the metric of (7.30) is illustrated in Fig. 7.4.

For the special case of constant envelope signal sets, wherein the bias [first term of (7.30)] becomes independent of k and can be ignored, the decision metric becomes (ignoring the scaling by the energy E)

$$\Lambda_k = \sum_{l=1}^{L_p} \left(\frac{\overline{\gamma}_l}{1+\overline{\gamma}_l}\right) \frac{|y_{kl}(\tau_l)|^2}{N_l} \qquad (7.31)$$

where $\overline{\gamma}_l \triangleq \Omega_l E/N_l$. If, further, we assume that $N_l = N_0;\ l = 1, 2, \ldots, L_p$, (7.31) simplifies still further to (ignoring the scaling by N_0)

$$\Lambda_k = \sum_{l=1}^{L_p} \left(\frac{\overline{\gamma}_l}{1+\overline{\gamma}_l}\right) |y_{kl}(\tau_l)|^2 \qquad (7.32)$$

Finally, for a flat power delay profile (PDP), $\Omega_l = \Omega,\ l = 1, 2, \ldots, L_p$, then ignoring the scaling by $\overline{\gamma}/(1+\overline{\gamma})$, the decision metric is simply

$$\Lambda_k = \sum_{l=1}^{L_p} |y_{kl}(\tau_l)|^2 \qquad (7.33)$$

which is identical in *structure* to the optimum receiver for a pure AWGN multichannel; that is, each finger implements a complex cross-correlator matched to the delayed signal for that path followed by a square-law envelope detector with no postdetection weighting.

Methods for evaluating the average bit error probability (BEP) performance of multichannel receivers with square-law detection are discussed in Chapter 9. In general, the performance of the optimum receiver that implements the decision metric of (7.32) is difficult to evaluate using these methods because of the nonuniformity of the postdetection weights $\overline{\gamma}_l/(1+\overline{\gamma}_l)$. On the other hand, the performance of a receiver that implements the unweighted decision metric of (7.33), which for other than a uniform PDP would be suboptimum, is straightforward. In what follows we examine the BEP of the optimum receiver (for which results are obtained from computer simulation) and the BEP of the suboptimum receiver [for which results are obtained from the analysis of equal gain combining (EGC) diversity reception to be studied in Chapter 9][8] for the case of binary FSK and an exponential PDP described by $\overline{\gamma}_l = \overline{\gamma}_1 e^{-\delta(l-1)}$, $l = 1, 2, \ldots, L_p$.

---

[8] Simulation results were also obtained for the BEP of the suboptimum receiver as a means of verifying the simulation program and were shown to be in perfect agreement with the analytically obtained results.

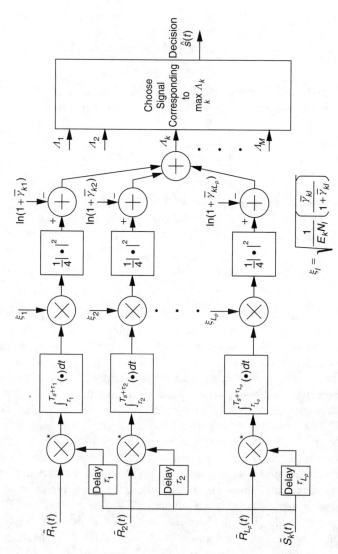

Figure 7.4. Complex form of optimum receiver for known delays, unknown amplitudes and phases: Rayleigh fading, one-symbol observation (noncoherent detection).

CASE OF KNOWN DELAYS AND UNKNOWN AMPLITUDES AND PHASES

TABLE 7.1 Average BEP Data for Optimum and Suboptimum Reception of Noncoherently Detected Binary FSK over Rayleigh Fading with an Exponential PDP[a]

L	$\bar{\gamma}_l$ (dB)								
	0	2	4	6	8	10	12	14	16

Optimum Case (Simulation Result) for Sample Size $= 10^8$ for $\delta = 0$

1	.33333	.27895	.22163	.16719	.12034	.08333	.05603	.03687	.023917
2	.25925	.19008	.12554	.07451	.03996	.01967	.00907	.00397	.001685
3	.20987	.13640	.07587	.03579	.01443	.00508	.00162	.00048	.000130
4	.17333	.10042	.04729	.01782	.00541	.00138	.00030	.00006	.000011

Optimum Case (Analysis Result) for $\delta = 0$

1	.33333	.27895	.22163	.16719	.12034	.08333	.05603	.03687	.023917
2	.25926	.19003	.12559	.07451	.03996	.01968	.00906	.00398	.001689
3	.20987	.13637	.07589	.03580	.01443	.00509	.00161	.00047	.000132
4	.17330	.10039	.04730	.01783	.00543	.00137	.00030	.00006	.000011

Optimum Case (Simulation Result) for Sample Size $= 10^8$ for $\delta = 0.1$

1	.33333	.27895	.22163	.16719	.12034	.08333	.05603	.03687	.023917
2	.26645	.19740	.13192	.07920	.04293	.02130	.00989	.00436	.001853
3	.22574	.15124	.08713	.04259	.01776	.00643	.00208	.00062	.000173
4	.19800	.12168	.06144	.02492	.00809	.00218	.00050	.00010	.000019

Suboptimum Case (Analysis) for $\delta = 0.1$

1	.33333	.27895	.22163	.16719	.12034	.08333	.05603	.03687	.023917
2	.26650	.19737	.13198	.07922	.04293	.02132	.00988	.00436	.001855
3	.22589	.15129	.08718	.04263	.01776	.00643	.00208	.00062	.000175
4	.19827	.12183	.06154	.02495	.00813	.00218	.00050	.00010	.000019

Optimum Case (Simulation Result) for Sample Size $= 10^8$ for $\delta = 0.5$

1	.33333	.27895	.22163	.16719	.12034	.08333	.05603	.03687	.023917
2	.29083	.22373	.15638	.09848	.05584	.02883	.01379	.00621	.002682
3	.27520	.20311	.13241	.07457	.03590	.01482	.00535	.00173	.000519
4	.26915	.19490	.12270	.06499	.02823	.01000	.00292	.00072	.000153

Suboptimum Case (Analysis) for $\delta = 0.5$

1	.33333	.27895	.22163	.16719	.12034	.08333	.05603	.03687	.023917
2	.29196	.22451	.15691	.09871	.05503	.02885	.01379	.00622	.002687
3	.27926	.20625	.13435	.07545	.03617	.01489	.00536	.00173	.000517
4	.27793	.20213	.12738	.06718	.02895	.01016	.00294	.00072	.000155

Optimum case (Simulation Result) for Sample Size $= 10^8$ for $\delta = 1$

1	.33333	.27895	.22163	.16719	.12034	.08333	.05603	.03687	.023917
2	.31137	.24835	.18213	.12149	.07322	.04001	.02011	.00941	.004179
3	.30802	.24323	.17511	.11273	.06385	.03149	.01354	.00512	.001731
4	.30755	.24244	.17400	.11127	.06219	.02981	.01216	.00421	.001234
5	.30739	.24230	.17381	.11106	.06190	.02957	.01190	.00402	.001129

(continued overleaf)

TABLE 7.1 (continued)

	$\bar{\gamma}_l$ (dB)							
L	0	2	4	6	8	10	12	14	16

	Suboptimum Case (Analysis) for $\delta = 1$							
1	.33333	.27895	.22163	.16719	.12034	.08333	.05603	.03687	.023917
2	.31594	.25204	.18467	.12277	.07375	.04022	.02015	.00943	.004190
3	.32195	.25612	.18497	.11872	.06666	.03252	.01382	.00517	.001738
4	.33243	.26714	.19457	.12518	.06958	.03290	.01314	.00443	.001277
5	.34279	.27886	.20600	.13440	.07551	.03583	.01417	.00464	.001260

	Optimum Case (Simulation Result) for Sample Size = 10^8 for $\delta = 2$							
1	.33333	.27895	.22163	.16719	.12034	.08333	.05603	.03687	.023917
2	.32884	.27202	.21125	.15300	.10242	.06318	.03588	.01878	.009146
3	.32885	.27181	.21098	.15249	.10190	.06248	.03500	.01783	.008290
4	.32878	.27184	.21097	.15253	.10191	.06246	.03500	.01784	.008256

	Suboptimum Case (Analysis) for $\delta = 2$							
1	.33333	.27895	.22163	.16719	.12034	.08333	.05603	.03687	.023917
2	.34317	.28588	.22268	.16062	.10665	.06507	.03653	.01897	.009197
3	.35916	.30381	.24044	.17584	.11776	.07180	.03969	.01986	.008999
4	.37163	.31859	.25610	.19048	.12976	.08040	.04509	.02281	.010394

[a]The simulation is accurate to 10^{-4}

Table 7.1 presents the numerical BEP data for the optimum and suboptimum receivers corresponding to values of δ equal to 0, 0.1, 0.5, 1.0, and 2.0. For each value of δ, the average SNR/bit of the first path, $\bar{\gamma}_1$, is allowed to vary over a range from 0 to 16 dB, and the number of paths, L_p, is varied from 1 to 4. For $\delta = 0$ (i.e., a uniform PDP), the simulation and analytical data are seen to agree exactly since in this case the suboptimum receiver corresponding to the decision metric of (7.33) is indeed optimum, as mentioned previously. For $\delta > 0$, the optimum receiver clearly outperforms (has a smaller BEP than) the suboptimum receiver, as it should. To illustrate the behavior of the optimum and suboptimum receivers as a function of the fading power decay factor, δ, and the number of paths, L_p, the simulation data in Table 7.1 are plotted in Figs. 7.5a–e and 7.6a–e, respectively. We observe from the curves in Fig. 7.5a–e that for fixed δ the performance of the optimum receiver always improves monotonically with increasing L_p over the entire range of $\bar{\gamma}_1$ considered. By contrast, the curves in Figs. 7.6a–e illustrate that for large δ, the performance of the suboptimum receiver can in fact degrade with increasing L_p as a result of the *noncoherent combining loss*, which is more prevalent at low SNR's. Comparing the various groups of curves within each set of figures also reveals that the improvement in BEP obtained by increasing L_p is larger when the fading power decay factor, δ, is smaller; that is, a uniform PDP stands to gain more from an increase in the number of combined paths than one with an exponentially decaying multipath and the same average SNR/bit of the first path.

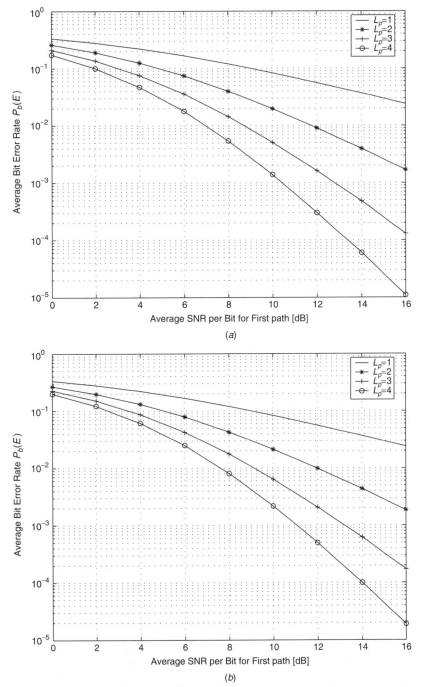

Figure 7.5. Average BEP performance for optimum reception of noncoherently detected binary FSK over Rayleigh fading with an exponential PDP: (a) $\delta = 0$; (b) $\delta = 0.1$; (c) $\delta = 0.5$; (d) $\delta = 1.0$; (e) $\delta = 2.0$. $m = 1, M = 2$.

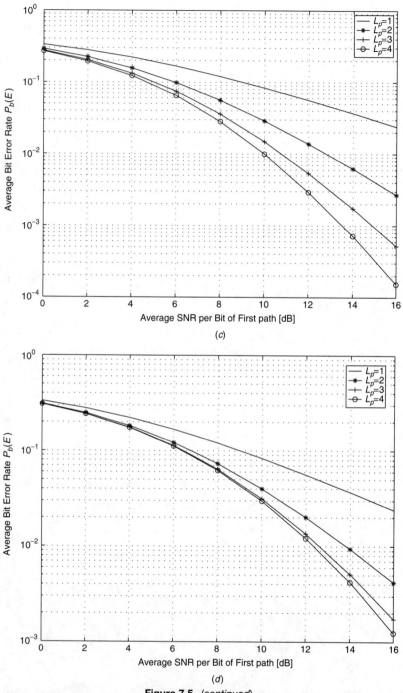

Figure 7.5. (continued)

CASE OF KNOWN DELAYS AND UNKNOWN AMPLITUDES AND PHASES   175

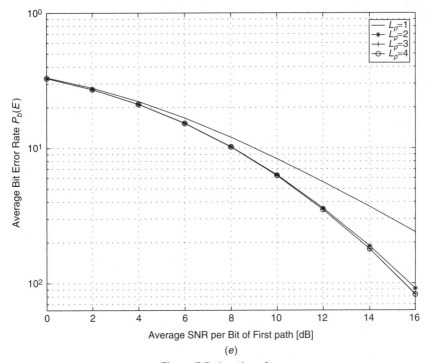

(e)

Figure 7.5. (continued)

To compare the behavior of the optimum and suboptimum receivers, Fig. 7.7a and b illustrate their performance for two different combinations of δ and L_p: namely, $\delta = 1, L_p = 5$ and $\delta = 2, L_p = 4$. Also illustrated in these figures are the corresponding results for $L_p = 1$, in which case the two receivers once again yield identical performance since the single scaling factor $\overline{\gamma}_1/(1 + \overline{\gamma}_1)$ in (7.32) is now inconsequential. We observe from these figures that the suboptimum receiver performs quite well with respect to its optimum counterpart but does in fact exhibit a noncoherent combining loss at sufficiently low SNR, as mentioned previously. As a further comparison of the behavior of the optimum and suboptimum BFSK receivers, Fig. 7.8 illustrates their performance with $L_p = 4$ and varying δ. Finally, Fig. 7.9 gives an analogous performance comparison for 4-ary FSK with $\delta = 1.0$ and varying L_p.

7.4.1.2 Nakagami-m Fading. For the Nakagami-m fading PDF of (7.19), the conditional probability of (7.27) can also be evaluated in closed form. In particular, using Eq. (6.631.1) of Ref. 7 we obtain [9]

$$p\big(\{r_l(t)\}_{l=1}^{L_p} \big| s_k(t), \{\tau_l\}_{l=1}^{L_p}\big) = K \prod_{l=1}^{L_p} \left(1 + \frac{\overline{\gamma}_{kl}}{m_l}\right)^{-m_l} {}_1F_1\left(m_l, 1; \frac{(V'_{kl})^2}{4}\right) \quad (7.34)$$

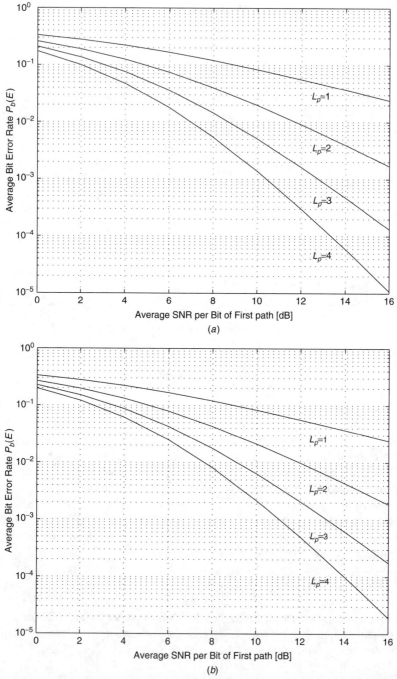

Figure 7.6. Average BEP performance for suboptimum reception of noncoherently detected binary FSK over Rayleigh fading with an exponential PDP: (a) $\delta = 0$; (b) $\delta = 0.1$; (c) $\delta = 0.5$; (d) $\delta = 1.0$; (e) $\delta = 2.0$; $m = 1$, $M = 2$.

CASE OF KNOWN DELAYS AND UNKNOWN AMPLITUDES AND PHASES 177

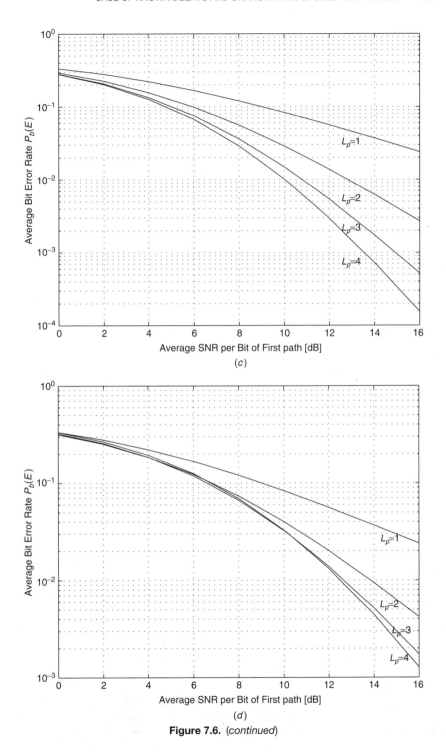

Figure 7.6. (*continued*)

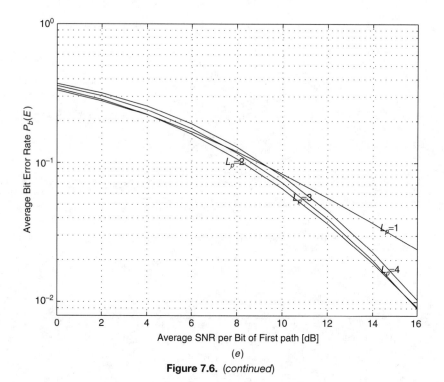

(e)

Figure 7.6. (continued)

where analogous to (7.21) for the coherent case

$$V'_{kl} \triangleq \sqrt{\frac{E_k}{N_l}\left(\frac{\overline{\gamma}_{kl}}{m_l + \overline{\gamma}_{kl}}\right)} \left[\frac{1}{E_k}|y_{kl}(\tau_l)|\right] \quad (7.35)$$

and $_1F_1(a, b; x)$ is Kummer's confluent hypergeometric function [7, Sec. 9.210], which has the property that for $x > 0$, $a > 0$, $_1F_1(a, 1; x)$ is a monotonically increasing function of x. Also, the larger a is, the greater the rate of increase. Finally, since $_1F_1(1, 1; x) = e^x$, then for $m_l = 1, l = 1, 2, \ldots, L_p$, the conditional probability of (7.34) reduces to (7.28), as it should.

The decision metric for this case is obtained by taking the natural logarithm of (7.34) with the result (ignoring the $\ln K$ term)

$$\Lambda_k = -\sum_{l=1}^{L_p} m_l \ln\left(1 + \frac{\overline{\gamma}_{kl}}{m_l}\right) + \sum_{l=1}^{L_p} \ln {}_1F_1\left(m_l, 1; \frac{(V'_{kl})^2}{4}\right) \quad (7.36)$$

Once again the first summation in (7.36) is a bias term, whereas the second summation has a typical term that is a nonlinearly processed sample (at time τ_l) of the cross-correlation modulus $y_{kl}(|\tau_l|)$. A receiver that implements a decision

CASE OF KNOWN DELAYS AND UNKNOWN AMPLITUDES AND PHASES

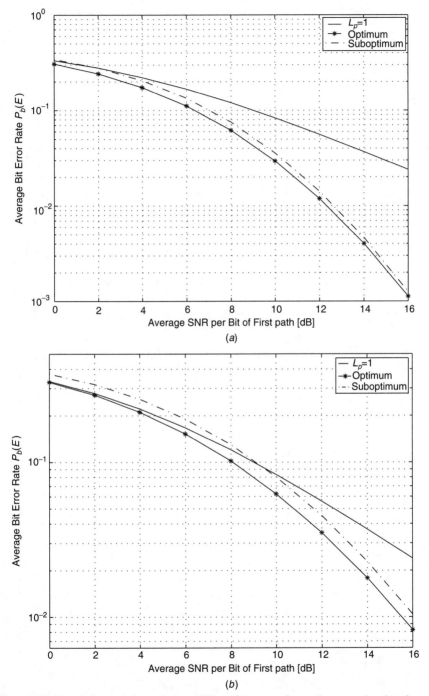

Figure 7.7. Comparison of the average BEP performance for optimum and suboptimum reception of noncoherently detected binary FSK over Rayleigh fading with an exponential PDP: (a) $\delta = 1.0$, $L_p = 5$; (b) $\delta = 2.0$, $L_p = 4$. $m = 1$, $M = 2$.

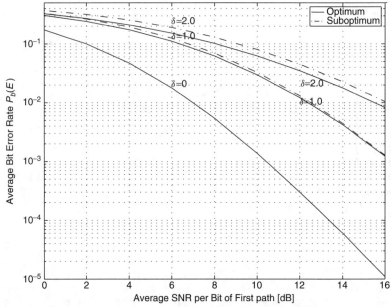

Figure 7.8. Comparison of the average BEP performance for optimum and suboptimum reception of noncoherently detected binary FSK over Rayleigh fading with an exponential PDP; $L_p = 4$, varying δ, $m = 1$, $M = 2$.

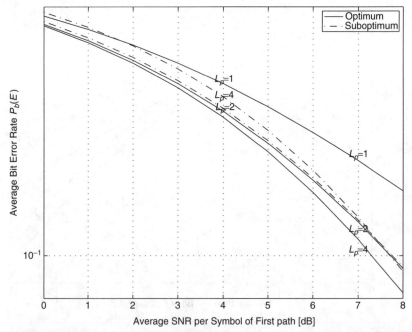

Figure 7.9. Comparison of the average BEP performance for optimum and suboptimum reception of noncoherently detected 4-ary FSK over Rayleigh fading with an exponential PDP; $\delta = 1.0$, varying L_p, $m = 1$, $M = 4$.

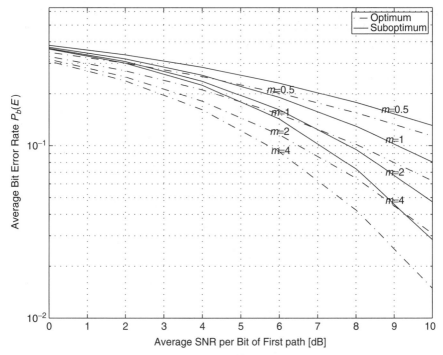

Figure 7.10. Comparison of the average BEP performance for optimum and suboptimum reception of noncoherently detected binary FSK over Nakagami-m fading with an exponential PDP; $\delta = 2.0$, $L_p = 4$, varying m.

rule based on (7.36) would be similar to Fig. 7.4, where, however, the square-law nonlinearity is replaced by the $\ln {}_1F_1(\cdot, \cdot; \cdot)$ nonlinearity and the bias is modified accordingly.

To compare the behavior of the optimum and suboptimum receivers, Fig. 7.10 illustrates their performance as a function of the m parameter for $\delta = 2$ and $L_p = 4$. Here we observe that the difference between the suboptimum and optimum performances increases with m (i.e., as the severity of the fading decreases).

7.4.2 Two-Symbol Observation: Conventional Differentially Coherent Detection

We assume here that in addition to the channel phases and amplitudes being unknown, the channel is sufficiently slowly varying that these parameters can be considered to be constant over a time interval that is at least two symbols in duration. Furthermore, we consider only constant envelope modulations, namely, M-PSK. For a purely AWGN channel, the optimum receiver has been shown [8, App. 7A] to implement differentially coherent detection which for M-PSK results in M-DPSK. What we seek here is the analogous optimum receiver when

182 OPTIMUM RECEIVERS FOR FADING CHANNELS

in addition to AWGN, fading with unknown amplitude is present on the received signal. The derivation of this optimum receiver to be presented here follows the development by Simon et al. [8, App. 7A].

We begin by rewriting (7.4) with integration limits corresponding to a $2T_s$-second observation, namely,

$$p\big(\{r_l(t)\}_{l=1}^{L_p} \big| s_k(t), \{\alpha_l\}_{l=1}^{L_p}, \{\theta_l\}_{l=1}^{L_p}, \{\tau_l\}_{l=1}^{L_p}\big)$$
$$= \prod_{l=1}^{L_p} K_l \exp\left[-\frac{1}{2N_l} \int_{\tau_l}^{2T_s+\tau_l} \left|\tilde{R}_l(t) - \alpha_l \tilde{S}_k(t-\tau_l)e^{j\theta_l}\right|^2 dt\right] \quad (7.37)$$

Defining the individual symbol energies of the kth signal as

$$E_{ki} = \frac{1}{2}\int_{iT_s}^{(i+1)T_s} |\tilde{S}_k(t)|^2\, dt = \frac{1}{2}\int_{iT_s+\tau_l}^{(i+1)T+\tau_l} |\tilde{S}_k(t-\tau_l)|^2\, dt, \quad i=0,1 \quad (7.38)$$

we obtain, analogous to (7.5),

$$p\big(\{r_l(t)\}_{l=1}^{L_p} \big| s_k(t), \{\alpha_l\}_{l=1}^{L_p}, \{\theta_l\}_{l=1}^{L_p}, \{\tau_l\}_{l=1}^{L_p}\big)$$
$$= K \prod_{l=1}^{L_p} \exp\left[\operatorname{Re}\left\{\frac{\alpha_l}{N_l} e^{-j\theta_l} y_{kl}(\tau_l)\right\} - \frac{\alpha_l^2(E_{k0}+E_{k1})}{N_l}\right]$$
$$= K \exp\left[\sum_{l=1}^{L_p} \operatorname{Re}\left\{\frac{\alpha_l}{N_l} e^{-j\theta_l} y_{kl}(\tau_l)\right\} - \sum_{l=1}^{L_p} \frac{\alpha_l^2(E_{k0}+E_{k1})}{N_l}\right] \quad (7.39)$$

where now

$$y_{kl}(\tau_l) \stackrel{\Delta}{=} \int_{\tau_l}^{2T_s+\tau_l} \tilde{R}_l(t) \tilde{S}_k^*(t-\tau_l)\, dt = \int_0^{2T_s} \tilde{R}_l(t+\tau_l) \tilde{S}_k^*(t)\, dt \quad (7.40)$$

Since we have assumed constant envelope M-PSK modulation, the kth complex baseband signal can be expressed as[9]

$$\tilde{S}_k(t) = \sqrt{\frac{E_s}{T_s}} e^{j\phi_k^{(i)}}, \quad iT_s \le t \le (i+1)T_s, \quad i=0,1 \quad (7.41)$$

where $E_{k0} = E_{k1} = E_s$ (the energy per symbol) and $\phi_k^{(i)}$ denotes the information phase transmitted in the ith symbol interval of the kth signal and ranges over the

---

[9] To avoid notational confusion with the channel fading phases, we use ϕ (as opposed to θ from Chapter 3) to denote the transmitted phases.

set $\beta_k = (2k-1)\pi/M$, $k = 1, 2, \ldots, M$. Substituting (7.41) into (7.40), we can rewrite (7.39) as

$$p\left(\{r_l(t)\}_{l=1}^{L_p} | s_k(t), \{\alpha_l\}_{l=1}^{L_p}, \{\theta_l\}_{l=1}^{L_p}, \{\tau_l\}_{l=1}^{L_p}\right)$$

$$= K \prod_{l=1}^{L_p} \exp\left[\mathrm{Re}\left\{\frac{\alpha_l}{N_l} e^{-j\theta_l} \left(y_{kl}^{(0)}(\tau_l) + y_{kl}^{(1)}(\tau_l)\right)\right\}\right]$$

$$= K \prod_{l=1}^{L_p} \exp\left\{\frac{\alpha_l}{N_l} \left|y_{kl}^{(0)}(\tau_l) + y_{kl}^{(1)}(\tau_l)\right| \cos\left[\theta_l - \arg\left(y_{kl}^{(0)}(\tau_l) + y_{kl}^{(1)}(\tau_l)\right)\right]\right\} \quad (7.42)$$

where we have absorbed the constant term $\exp(-2E_s \sum_{l=1}^{L_p} \alpha_l^2/N_l)$ in K and

$$y_{kl}^{(i)}(\tau_l) \triangleq \sqrt{\frac{E_s}{T_s}} \int_{iT_s+\tau_l}^{(i+1)T_s+\tau_l} \tilde{R}_l(t) e^{-j\phi_k^{(i)}} \, dt, \quad i = 0, 1 \quad (7.43)$$

As in Section 7.3, we first need to average (7.42) over the uniformly distributed statistics of the unknown channel phases. Proceeding as was done in (7.23), we arrive at the result

$$p\left(\{r_l(t)\}_{l=1}^{L_p} | s_k(t), \{\alpha_l\}_{l=1}^{L_p}, \{\tau_l\}_{l=1}^{L_p}\right)$$

$$= K \prod_{l=1}^{L_p} \exp\left(-\frac{\alpha_l^2 E_s}{N_l}\right) I_0\left(\frac{\alpha_l}{N_l} \left|y_{kl}^{(0)}(\tau_l) + y_{kl}^{(1)}(\tau_l)\right|\right) \quad (7.44)$$

Next, we must average over the statistics of the unknown amplitudes.

7.4.2.1 Rayleigh Fading. Following steps analogous to those taken in Section 7.4.1.1, we obtain

$$p\left(\{r_l(t)\}_{l=1}^{L_p} | s_k(t), \{\tau_l\}_{l=1}^{L_p}\right) = K \prod_{l=1}^{L_p} \exp\left[\frac{\Omega_l \left|y_{kl}^{(0)}(\tau_l) + y_{kl}^{(1)}(\tau_l)\right|^2}{4N_l}\right] \quad (7.45)$$

with the equivalent decision metric (ignoring the $\ln K$ term)

$$\Lambda_k = \sum_{l=1}^{L_p} \frac{\Omega_l}{4N_l} \left|y_{kl}^{(0)}(\tau_l) + y_{kl}^{(1)}(\tau_l)\right|^2$$

$$= \sum_{l=1}^{L_p} \frac{\bar{\gamma}_l}{4T_s} \left|\int_{\tau_l}^{T_s+\tau_l} \tilde{R}_l(t) e^{-j\phi_k^{(0)}} \, dt + \int_{T_s+\tau_l}^{2T_s+\tau_l} \tilde{R}_l(t) e^{-j\phi_k^{(1)}} \, dt\right|^2 \quad (7.46)$$

The decision rule based on the decision metric in (7.46) is to choose as the transmitted signal that pair of phases $\phi_k^{(0)} = \beta_{j_0}, \phi_k^{(1)} = \beta_{j_1}$ that results in the largest Λ_k. We note that adding an arbitrary phase, say β, to both $\phi_k^{(0)}$ and $\phi_k^{(1)}$ does not affect the decision metric, and thus in accordance with the decision rule above, the joint decision on $\phi_k^{(0)}$ and $\phi_k^{(1)}$ will be completely ambiguous. To resolve this phase ambiguity, we observe that although the decisions on $\phi_k^{(0)}$ and $\phi_k^{(1)}$ can each be ambiguous with an arbitrary phase β, the difference of these two decisions is not ambiguous at all. Thus, an appropriate solution is to encode the phase information as the difference between two successive transmitted phases (i.e., employ *differential phase encoding* at the transmitter). This is exactly the solution discussed in Section 3.5 for phase-ambiguity resolution on the pure AWGN channel (see also Simon et al. [8, App. 7A]). Mathematically speaking, we can set the arbitrary phase $\beta = -\phi_k^{(0)}$, in which case (7.46) becomes

$$\Lambda_k = \sum_{l=1}^{L_p} \frac{\overline{\gamma}_l}{4T_s} \left| \int_{\tau_l}^{T_s+\tau_l} \tilde{R}_l(t) e^{-j(\phi_k^{(0)}+\beta)} dt + \int_{T_s+\tau_l}^{2T_s+\tau_l} \tilde{R}_l(t) e^{-j(\phi_k^{(1)}+\beta)} dt \right|^2$$

$$= \sum_{l=1}^{L_p} \frac{\overline{\gamma}_l}{4T_s} \left| \int_{\tau_l}^{T_s+\tau_l} \tilde{R}_l(t) dt + \int_{T_s+\tau_l}^{2T_s+\tau_l} \tilde{R}_l(t) e^{-j(\phi_k^{(1)}-\phi_k^{(0)})} dt \right|^2$$

$$= \sum_{l=1}^{L_p} \frac{\overline{\gamma}_l}{4T_s} \left| \int_{\tau_l}^{T_s+\tau_l} \tilde{R}_l(t) dt + e^{-j\Delta\phi_k^{(1)}} \int_{T_s+\tau_l}^{2T_s+\tau_l} \tilde{R}_l(t) dt \right|^2 \quad (7.47)$$

where $\Delta\phi_k^{(i)} \triangleq \phi_k^{(i)} - \phi_k^{(i-1)}$ represents the information phase corresponding to the ith transmission interval, which ranges over the set of values $\beta_k = 2k\pi/M$, $k = 0, 1, \ldots, M - 1$. Expanding the squared magnitude in (7.47) and retaining only terms that depend on the information phase $\Delta\phi_k^{(1)}$, we obtain (ignoring other multiplicative constants)

$$\Lambda_k = \sum_{l=1}^{L_p} \overline{\gamma}_l \, \text{Re}\{\tilde{V}_{0l} \tilde{V}_{1l}^* e^{j\Delta\phi_k^{(1)}}\} \quad (7.48)$$

where

$$\tilde{V}_{il} \triangleq \int_{iT_s+\tau_l}^{(i+1)T_s+\tau_l} \tilde{R}_l(t) \, dt, \quad i = 0, 1 \quad (7.49)$$

A receiver that bases its decision rule on the decision metric of (7.48) is illustrated in Fig. 7.11. For a flat power delay profile and equal channel noise PSDs, the metric of (7.48) reduces to that corresponding to optimum reception in a pure AWGN environment (see Section 3.5).

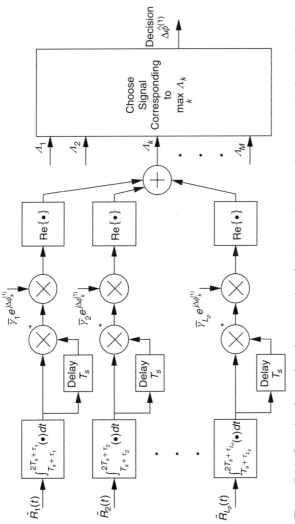

Figure 7.11. Complex form of optimum receiver for known phases, unknown amplitudes and delays: Rayleigh fading, two-symbol observation (conventional differentially coherent detection).

7.4.2.2 Nakagami-m Fading. By comparing the conditional probabilities of (7.23) and (7.44) corresponding, respectively, to noncoherent and differentially coherent detection, it is straightforward to show that for Nakagami-m fading, the decision metric becomes

$$\Lambda_k = \sum_{l=1}^{L_p} \ln {}_1F_1(m_l, 1; W_{kl}^2/4) \qquad (7.50)$$

where

$$W_{kl} \triangleq \sqrt{\frac{E_s}{N_l}\left(\frac{\overline{\gamma_l}}{m_l}\right)}\left[\frac{1}{E_s}|y_{kl}^{(0)}(\tau_l) + y_{kl}^{(1)}(\tau_l)|\right]$$

$$= \sqrt{\frac{E_s}{N_l}\left(\frac{\overline{\gamma_l}}{m_l}\right)}\left[\frac{1}{\sqrt{E_sT_s}}\left|\int_{\tau_l}^{T_s+\tau_l}\tilde{R}_l(t)e^{-j\phi_k^{(0)}}\,dt + \int_{T_s+\tau_l}^{2T_s+\tau_l}\tilde{R}_l(t)e^{-j\phi_k^{(1)}}\,dt\right|\right] \qquad (7.51)$$

As for the Rayleigh case, the decision metric of (7.50) in combination with (7.51) is ambiguous to an arbitrary phase shift β. With differential phase encoding employed at the transmitter, the unambiguous decision metric is still given by (7.50) with W_{kl} now defined as

$$W_{kl} \triangleq \sqrt{\frac{E_s}{N_l}\left(\frac{\overline{\gamma_l}}{m_l}\right)}\left(\frac{1}{\sqrt{E_sT_s}}|\tilde{V}_{0l} + e^{-j\Delta\phi_k^{(1)}}\tilde{V}_{1l}|\right)$$

$$= \sqrt{\frac{E_s}{N_l}\left(\frac{\overline{\gamma_l}}{m_l}\right)}\left[\frac{1}{E_sT_s}(|\tilde{V}_{0l}|^2 + |\tilde{V}_{1l}|^2 + 2\operatorname{Re}\{\tilde{V}_{0l}\tilde{V}_{1l}^*e^{j\Delta\phi_k^{(1)}}\})\right]^{1/2} \qquad (7.52)$$

Note that because of the nonlinear postdetection processing via the $\ln {}_1F_1(\cdot,\cdot;\cdot)$ function, the terms $|\tilde{V}_{0l}|^2$ and $|\tilde{V}_{1l}|^2$ cannot be ignored, nor can the other multiplicative factors in (7.52), despite the fact that they are all independent of k.

7.4.3 N_s-Symbol Observation: Multiple Symbol Differentially Coherent Detection

In Ref. 10, the authors considered differential detection of M-PSK over an AWGN channel based on an N_s-symbol ($N_s > 2$) observation of the received signal. The optimum receiver (see Fig. 3.18) was derived and shown to yield improved (monotonically with increasing N_s) performance relative to that attainable with the conventional (two-symbol observation) M-DPSK receiver. Our intent here is to generalize the results of Divsalar and Simon [10] (see also Simon et al. [8, Sec. 7.2]) to the fading multichannel with unknown amplitudes.

CASE OF KNOWN DELAYS AND UNKNOWN AMPLITUDES AND PHASES

Clearly, for $N_s > 2$, the decision metric and associated receiver derived here will reduce to those obtained in Section 7.4.2. Without going into great detail, it should be immediately obvious that for an N_s-symbol observation, the conditional probability of (7.44) generalizes to

$$p\left(\{r_l(t)\}_{l=1}^{L_p} \mid s_k(t), \{\alpha_l\}_{l=1}^{L_p}, \{\tau_l\}_{l=1}^{L_p}\right)$$

$$= K \prod_{l=1}^{L_p} \exp\left(-\frac{\alpha_l^2 E_s}{N_l}\right) I_0 \left(\frac{\alpha_l}{N_l} \left| \sum_{n=0}^{N_s-1} y_{kl}^{(n)}(\tau_l) \right| \right) \quad (7.53)$$

7.4.3.1 Rayleigh Fading. Averaging (7.53) over Rayleigh statistics for the unknown amplitudes results in the generalization of the decision metric in (7.46), namely,

$$\Lambda_k = \sum_{l=1}^{L_p} \frac{\overline{\gamma}_l}{4T_s} \sum_{n=0}^{N_s-1} \left| \int_{nT_s+\tau_l}^{(n+1)T_s+\tau_l} \tilde{R}_l(t) e^{-j\phi_k^{(n)}} dt \right|^2 \quad (7.54)$$

Using the same differential phase encoding rule as for the two-symbol observation case to resolve the phase ambiguity in (7.54), the unambiguous form of this decision metric becomes

$$\Lambda_k = \sum_{l=1}^{L_p} \frac{\overline{\gamma}_l}{4T_s} \sum_{n=0}^{N_s-1} \left| e^{-j \sum_{i=0}^{n} \Delta\phi_k^{(i)}} \int_{nT_s+\tau_l}^{(n+1)T_s+\tau_l} \tilde{R}_l(t) dt \right|^2 \quad (7.55)$$

where, by definition, $\Delta\phi_k^{(0)} = 0$. As before, expanding the squared magnitude and retaining only terms that depend on the information phases, we obtain (ignoring other multiplicative constants)

$$\Lambda_k = \sum_{l=1}^{L_p} \overline{\gamma}_l \operatorname{Re}\left\{ \sum_{\substack{i=0 \\ i<j}}^{N_s-1} \sum_{j=0}^{N_s-1} \tilde{V}_{il} \tilde{V}_{jl}^* \exp\left(j \sum_{n=i+1}^{j} \Delta\phi_k^{(n)}\right) \right\} \quad (7.56)$$

The decision rule based on the metric of (7.56) is to choose as the transmitted signal that sequence of phases $\Delta\phi_k^{(i)} = \beta_{j_i}$, $i = 1, \ldots, N_s - 1$, that results in the largest Λ_k. Once again note that for a flat power delay profile and equal noise PSDs, the decision metric of (7.56) becomes equal to that discussed by Divsalar and Simon [10] and Simon et al. [8, Sec. 7.2] for the fading-free AWGN channel.

7.4.3.2 Nakagami-m Fading. At this point it should be obvious to the reader how to extend the results of Section 7.4.2.2 to the N_s-symbol observation case. In particular, the decision metric of (7.50) applies, now with

$$W_{kl} = \sqrt{\frac{E_s}{N_l}\left(\frac{\overline{\gamma_l}}{m_l}\right)} \left[\frac{1}{\sqrt{E_s T_s}} \left| \sum_{n=0}^{|N_s-1|} \tilde{V}_{nl} \exp\left(-j \sum_{i=0}^{n} \Delta\phi_k^{(i)}\right) \right| \right]$$

$$= \sqrt{\frac{E_s}{N_l}\left(\frac{\overline{\gamma_l}}{m_l}\right)} \left[\frac{1}{E_s T_s} \left| \sum_{n=0}^{|N_s-1|} |\tilde{V}_{nl}|^2 \right.\right.$$

$$\left.\left. + 2\,\text{Re}\left\{ \sum_{\substack{i=0 \\ i<j}}^{N_s-1} \sum_{j=0}^{N_s-1} \tilde{V}_{il}\tilde{V}_{jl}^* \exp\left(j \sum_{n=i+1}^{j} \Delta\phi_k^{(n)}\right) \right\} \right| \right]^{1/2} \quad (7.57)$$

As for the two-symbol observation case, (7.57) cannot be simplified by ignoring the $|\tilde{V}_{nl}|^2$ terms or any of the multiplicative constants.

7.5 CASE OF UNKNOWN AMPLITUDES, PHASES, AND DELAYS

When all channel parameters are unknown, the conditional probability of (7.4) must be averaged over the statistics of the amplitudes, phases, and delays, all of which are assumed to be independent. Equivalently, since (7.23), (7.44), and (7.53) already represent the average over the unknown i.i.d. uniformly distributed phases, the desired likelihood functions can be obtained by averaging these equations over the statistics of the unknown amplitudes, for example, Rayleigh, Nakagami-m, and the unknown delays which, following Ref. 9, will be modeled over the interval (A, B) as i.i.d. uniformly distributed random variables, that is,

$$p_{\tau_l}(\tau_l) = \frac{1}{B-A}, \quad A \leq \tau_l \leq B, \quad l = 1, 2, \ldots, L_p \quad (7.58)$$

7.5.1 One-Symbol Observation: Noncoherent Detection

In Section 7.4.1 we derived likelihood functions and decision metrics for a one-symbol observation conditioned on the delays being known. Here we simply average these expressions over the i.i.d. uniform PDFs of (7.58) to arrive at the optimum noncoherent receiver for fading channels with all parameters unknown. In the most general case, the parameters that characterize the fading amplitude PDFs (e.g., Ω_l, m_l) might depend on the delay τ_l. However, to simplify matters, we shall assume, as in Ref. 9 that no such dependencies exist. This will enable us more easily to derive suboptimum receiver structures based on approximations to the optimum decision metrics.

7.5.1.1 Rayleigh Fading.
Starting with (7.28), then averaging over the PDFs in (7.58), results in the likelihood function

$$p(\{r_l(t)\}_{l=1}^{L_p}|s_k(t)) = K \prod_{l=1}^{L_p}(1+\overline{\gamma}_{kl})^{-1} \int_A^B \exp\left[\frac{1}{4E_kN_l}\left(\frac{\overline{\gamma}_{kl}}{1+\overline{\gamma}_{kl}}\right)|y_{kl}(\tau)|^2\right] d\tau \quad (7.59)$$

where we have absorbed the constant $(B-A)^{-L_p}$ in K. For the multipath fading model proposed by Turin [6] with constant envelope signals, equal noise PSDs, and a uniform power delay profile (as assumed in Ref. 9), (7.59) simplifies to

$$p(r(t)|s_k(t)) = K \left\{\int_A^B \exp\left[\frac{1}{4E_sN_0}\left(\frac{\overline{\gamma}}{1+\overline{\gamma}}\right)|y_k(\tau)|^2\right] d\tau\right\}^{L_p} \quad (7.60)$$

where, now analogous to (7.6),

$$y_k(\tau) \stackrel{\Delta}{=} \int_{\tau_l}^{T_s+\tau_l} \tilde{R}(t)\tilde{S}_k^*(t-\tau)\,dt \quad (7.61)$$

and the constant $(1+\overline{\gamma})^{-L_p}$ has been absorbed into K. Since the decision rule is based on choosing the largest (with respect to k) of the likelihood functions in (7.60), then since their integrands are always positive, it is sufficient to ignore the exponent L_p and, keeping the same notation for convenience, redefine the likelihoods as

$$p(r(t)|s_k(t)) = K \int_A^B \exp\left[\frac{1}{4E_sN_0}\left(\frac{\overline{\gamma}}{1+\overline{\gamma}}\right)|y_k(\tau)|^2\right] d\tau \quad (7.62)$$

To proceed further toward a simpler but suboptimum receiver, we must approximate the integral in (7.62). Following the approach taken in Ref. 9, the first step is to approximate the nonlinearity of the integrand (i.e., the exponential) by its behavior for small arguments, namely, $e^x \simeq 1+x$, which leads to a likelihood function [ignoring the constant term $K(B-A)$ and all other multiplicative constants]

$$p(r(t)|s_k(t)) = \int_A^B |y_k(\tau)|^2\,d\tau \quad (7.63)$$

To evaluate the performance of a receiver that uses a decision rule based on this likelihood function would require knowledge of the PDF of the integral in (7.63). Even when the cross-correlation function $y_k(\tau)$ is stationary and Gaussian, obtaining this PDF is not possible. To circumvent this problem, we proceed to the second step in the approximation, namely, to replace the integral by the discrete (Riemann) sum $\sum_{i=1}^{N}|y_k(\tau_i)|^2\Delta\tau$, where the τ_i's are equally spaced over the interval (A, B), with spacing $\Delta\tau$ chosen equal to the correlation time of the process $y_k(\tau)$.[10] When this is done, the suboptimal decision metric becomes (here

---

[10] Such a sample spacing results in a set of independent complex-valued Gaussian RVs for $\{y_k(\tau_i)\}_{i=1}^N$.

there is no need to take the natural logarithm of the likelihood function)

$$\Lambda_k = \sum_{i=1}^{N} |y_k(\tau_i)|^2 \tag{7.64}$$

Comparing (7.64) with (7.33), which corresponds to the case of known delays, we see that the two metrics are of identical form, the difference being in the sampling instants and number of samples taken of the cross-correlation function.

7.5.1.2 Nakagami-m Fading. Starting with (7.28), then averaging over the PDFs in (7.58) and assuming constant envelope signals, equal noise PSDs, and Turin's multipath fading model with i.i.d. fading channels, results in a likelihood function analogous to (7.62), namely,

$$p(r(t)|s_k(t)) = K \int_A^B {}_1F_1\left(m, 1; (4E_sN_0)^{-1}\left(\frac{\overline{\gamma}}{m+\overline{\gamma}}\right)|y_k(\tau)|^2\right) d\tau \tag{7.65}$$

Once again, to obtain a simple but suboptimum receiver, we follow the approach taken in Ref. 9 and approximate the nonlinearity of the integrand by its behavior for small arguments, namely, ${}_1F_1(m, 1; x) \simeq 1 + mx$, which, ignoring the constant term $K(B - A)$ and all other multiplicative constants, leads to a likelihood function identical to (7.63), and using the second step of approximation, a decision metric identical to (7.64). Hence, the suboptimal receiver for Nakagami-m fading would be identical to that for Rayleigh fading.

7.5.2 Two-Symbol Observation: Conventional Differentially Coherent Detection

By analogy with the results obtained in Section 7.4.2 and their relation to those in Section 7.4.1, we can immediately deduce from the foregoing, in particular (7.64), that for Rayleigh and Nakagami-m fading the suboptimum decision metric for conventional differential coherent detection with all fading parameters unknown becomes

$$\Lambda_k = \sum_{i=1}^{N} \left|y_k^{(0)}(\tau_i) + y_k^{(1)}(\tau_i)\right|^2 \tag{7.66}$$

Again comparing (7.66) with (7.46), which corresponds to the case of known delays, we see that the two metrics are of identical form, the difference being in the sampling instants and number of samples taken of the cross-correlation function. Hence, a receiver implementation based on the suboptimum decision metric of (7.66) for unknown delays and Rayleigh or Nakagami-m fading would be identical in *structure* to that based on the optimum decision of (7.46) for Rayleigh fading and known delays.

At this point, extension of the two-symbol observation results to multiple (more than two)-symbol observation differentially coherent detection should be obvious in light of the discussion in Section 7.4.3 and thus requires no further development here.

REFERENCES

1. G. L. Turin, "On optimal diversity reception," *IRE Trans. Inf. Theory*, vol. IT-7, July 1961, pp. 154–167.
2. G. L. Turin, "On optimal diversity reception, II," *IRE Trans. Commun. Syst.*, vol. CS-10, March 1962, pp. 22–31.
3. J. C. Hancock and W. C. Lindsey, "Optimum performance of self-adaptive systems operating through a Rayleigh-fading medium," *IEEE Trans. Commun. Syst.*, December 1963, pp. 443–453.
4. R. Price and P. E. Green, "A communication technique for multipath channels," *Proc. IEEE*, vol. 46, March 1958, pp. 555–570.
5. D. Brennan, "Linear diversity combining techniques," *Proc. IRE*, vol. 47, June 1959, pp. 1075–1102.
6. G. L. Turin, "Communication through noisy, random-multipath channels," *IRE Natl. Conv. Rec.*, pt. 4, 1956, pp. 154–166.
7. I. S. Gradshteyn and I. M. Ryzhik, *Table of Integrals, Series, and Products*, 5th ed. San Diego, CA: Academic Press, 1994.
8. M. K. Simon, S. M. Hinedi, and W. C. Lindsey, *Digital Communication Techniques: Signal Design and Detection*. Upper Saddle River, NJ: Prentice Hall, 1995.
9. U. Charash, "Reception through Nakagami fading multipath channels with random delays," *IEEE Trans. Commun.*, vol. COM-27, April 1979, pp. 657–670.
10. D. Divsalar and M. K. Simon, "Multiple symbol differentially coherent detection of MPSK," *IEEE Trans. Commun.*, vol. 38, March 1990, pp. 300–308.

8

PERFORMANCE OF SINGLE CHANNEL RECEIVERS

As alluded to in Chapters 4, 5, and 6, the alternative representations of the Gaussian and Marcum Q-functions and other related functions expressed in the *desired form* are the key mathematical tools for unifying evaluation of the average error probability performance of digital communication systems over the generalized fading channel. Before discussing the specific details of these performance evaluations later in this chapter and the ones that follow, we first present the appropriate expressions for evaluating the performance of these systems over the AWGN. We present these results in two forms: (1) the classical expression for average bit error probability (BEP) or symbol error probability (SEP) as originally reported by the contributing author(s) and the one most commonly understood and familiar to those working in the field, and (2) the expression based on the alternative representations of the above-mentioned functions given in Chapter 4. These expressions, together with the special integrals developed in Chapter 5, then form the basis for evaluating the performance of digital communication systems in a fading environment modeled as a single transmission channel. Extension of these results to multiple transmission channels and the accompanying multichannel (diversity) reception is discussed in Chapter 9.

8.1 PERFORMANCE OVER THE AWGN CHANNEL

The average BEP and SEP performances over the AWGN channel of the various modulation/detection schemes discussed in Chapter 3 are well documented in many recent textbooks on digital communications [1]–[7]. Since this section of the chapter is intended to serve as a prelude of what is yet to come, our intent here is merely to review these classical results without derivation and then put them in a form that will be particularly suitable for arriving at simple expressions for performance over the generalized fading channel. The reader who is interested in the details of the derivations is referred to the above-mentioned textbook references.

8.1.1 Ideal Coherent Detection

Following the hierarchy of Chapter 3, we shall first consider the error probability performance of digital communication systems that employ ideal coherent detection. As mentioned in that chapter, such idealized performance can never by obtained in practice; nevertheless, these results serve as a benchmark against which the performance of realistic communications systems can be compared.

8.1.1.1 Multiple Amplitude-Shift-Keying or Multiple Amplitude Modulation.
Referring to the signal model in Section 3.1.1 and the accompanying forms of the optimum receiver in Fig. 3.2, then in terms of the carrier amplitude A_c, the SEP for symmetrical M-AM is given by

$$P_s(E) = 2\left(\frac{M-1}{M}\right) Q\left(\sqrt{\frac{2A_c^2 T_s}{N_0}}\right) \tag{8.1}$$

where $Q(\cdot)$ is the Gaussian Q-function defined in (4.1). Since the average symbol energy E_s is related to A_c by

$$E_s = \frac{1}{M} \sum_{l=1}^{M} (2l - 1 - M)^2 A_c^2 T_s = A_c^2 T_s \frac{M^2 - 1}{3} \tag{8.2}$$

then in terms of E_s, the SEP becomes

$$P_s(E) = 2\left(\frac{M-1}{M}\right) Q\left(\sqrt{\frac{6E_s}{N_0(M^2 - 1)}}\right) \tag{8.3}$$

For binary AM ($M = 2$), (8.3) becomes the BEP

$$P_b(E) = Q\left(\sqrt{\frac{2E_b}{N_0}}\right) \tag{8.4}$$

In terms of the desired form of the Gaussian Q-function as given in (4.2), the error probabilities of (8.2) and (8.4) become, respectively,

$$P_s(E) = \frac{2}{\pi}\frac{M-1}{M} \int_0^{\pi/2} \exp\left[-\frac{E_s}{N_0(M^2-1)}\frac{3}{\sin^2\theta}\right] d\theta \tag{8.5}$$

and

$$P_b(E) = \frac{1}{\pi} \int_0^{\pi/2} \exp\left(-\frac{E_b}{N_0}\frac{1}{\sin^2\theta}\right) d\theta \tag{8.6}$$

To convert the M-ary symbol decisions to decisions on the information bits, one must employ a bit-to-symbol mapping at the transmitter and then invert this mapping at the output of the receiver of Fig. 3.2. For this purpose a Gray code mapping is appropriate, which has the property that in transitioning from one symbol to an adjacent symbol, only one out of the $\log_2 M$ bits changes. Such a mapping at the transmitter results in only a single bit error when an adjacent symbol error is committed at the receiver. Although an exact computation of average BEP is possible for any given M, it is most common to consider the case of large symbol SNR (E_s/N_0), for which the only significant symbol errors are those that occur in adjacent signal levels. For this case, the average BEP is approximated by [5, Chap. 4]

$$P_b(E) \simeq \frac{P_s(E)}{\log_2 M} \tag{8.7}$$

where $P_s(E)$ is determined from (8.3) and it is convenient to replace the symbol energy by $E_s = E_b \log_2 M$. Clearly, for the binary case, (8.7) is in agreement with (8.4) and is thus exact.

8.1.1.2 Quadrature Amplitude-Shift-Keying or Quadrature Amplitude Modulation.
Referring to the signal model in Section 3.1.2 and the accompanying forms of the optimum receiver in Fig. 3.3, the SEP for square QAM can be obtained immediately from the SEP of $\sqrt{M}$–AM by making the following observation. Since a QAM modulation is composed of the quadrature combination of two $\sqrt{M}$–AM modulations each with half the total power, and since a correct QAM decision is made only when a correct symbol decision is made independently on each of these modulations, the probability of correct symbol decision for QAM can be expressed as

$$\left. P_s(C) \right|_{\substack{M-\text{QAM} \\ E_s}} = \left[\left. P_s(C) \right|_{\substack{\sqrt{M}-\text{AM} \\ E_s/2}} \right]^2 \tag{8.8}$$

or, equivalently in terms of the SEP,

$$\left. P_s(E) \right|_{\substack{M-\text{QAM} \\ E_s}} = 1 - \left[1 - \left. P_s(E) \right|_{\substack{\sqrt{M}-\text{AM} \\ E_s/2}} \right]^2$$

$$= 2 \left. P_s(E) \right|_{\substack{\sqrt{M}-\text{AM} \\ E_s/2}} \left[1 - \tfrac{1}{2} \left. P_s(E) \right|_{\substack{\sqrt{M}-\text{AM} \\ E_s/2}} \right] \tag{8.9}$$

Substituting (8.3) into (8.9) gives the desired classical form of the SEP for QAM, namely [5, Chap. 10],

$$P_s(E) = 4 \left(\frac{\sqrt{M}-1}{\sqrt{M}} \right) Q \left(\sqrt{\frac{3E_s}{N_0(M-1)}} \right) \left[1 - \left(\frac{\sqrt{M}-1}{\sqrt{M}} \right) Q \left(\sqrt{\frac{3E_s}{N_0(M-1)}} \right) \right]$$

$$= 4 \left(\frac{\sqrt{M}-1}{\sqrt{M}} \right) Q \left(\sqrt{\frac{3E_s}{N_0(M-1)}} \right) - 4 \left(\frac{\sqrt{M}-1}{\sqrt{M}} \right)^2 Q^2 \left(\sqrt{\frac{3E_s}{N_0(M-1)}} \right) \tag{8.10}$$

For 4-QAM, (8.10) reduces to

$$P_s(E) = 2Q\left(\sqrt{\frac{E_s}{N_0}}\right) - Q^2\left(\sqrt{\frac{E_s}{N_0}}\right) \tag{8.11}$$

Using the desired forms of the Gaussian Q-function and its square as given by (4.2) and (4.9), respectively, in (8.10) and (8.11), we obtain

$$P_s(E) = \frac{4}{\pi}\left(\frac{\sqrt{M}-1}{\sqrt{M}}\right)\int_0^{\pi/2} \exp\left[-\frac{E_s}{N_0}\frac{3}{2(M-1)\sin^2\theta}\right]d\theta$$

$$-\frac{4}{\pi}\left(\frac{\sqrt{M}-1}{\sqrt{M}}\right)^2\int_0^{\pi/4} \exp\left[-\frac{E_s}{N_0}\frac{3}{2(M-1)\sin^2\theta}\right]d\theta \tag{8.12}$$

and

$$P_s(E) = \frac{2}{\pi}\int_0^{\pi/2} \exp\left(-\frac{E_S}{2N_0}\frac{1}{\sin^2\theta}\right)d\theta - \frac{1}{\pi}\int_0^{\pi/4} \exp\left(-\frac{E_S}{2N_0}\frac{1}{\sin^2\theta}\right)d\theta \tag{8.13}$$

Once again using a Gray code (now in two dimensions) to map the information bits into the QAM symbols, it is possible (but tedious) to obtain an exact closed-form result for the average bit error probability for arbitrary M.[1] One method for circumventing this difficulty is to use the approximate (valid for large-symbol SNR) relation between bit and symbol error probability of (8.7) together with (8.13) for the latter. However, as we shall see later in this chapter, to obtain the average BEP in the presence of fading wherein the instantaneous SNR can vary between zero and infinity, it is essential to have a BEP expression for AWGN that is valid for low as well as high SNR. Recently, using a signal space approach, Lu et al. [8] derived approximate expressions for the BEP of QAM and M-PSK (to be discussed next) in AWGN, which have the above-mentioned desirable properties, namely, they are quite accurate at both low and high SNR, and furthermore, are valid for all M. In particular, for QAM it is shown in Ref. 8 that

$$P_b(E) \simeq 4\left(\frac{\sqrt{M}-1}{\sqrt{M}}\right)\left(\frac{1}{\log_2 M}\right)\sum_{i=1}^{\sqrt{M}/2} Q\left((2i-1)\sqrt{\frac{3E_b\log_2 M}{N_0(M-1)}}\right) \tag{8.14}$$

We note that, for large E_b/N_0, the first term in the summation of (8.14) is dominant in which case this equation simplifies to

$$P_b(E) \simeq 4\left(\frac{\sqrt{M}-1}{\sqrt{M}}\right)\left(\frac{1}{\log_2 M}\right)Q\left(\sqrt{\frac{3E_b\log_2 M}{N_0(M-1)}}\right) \tag{8.15}$$

---

[1] Two examples of this exact BEP computation corresponding to $M = 16$ and $M = 64$ can be found in Eqs. (10.36a) and (10.36b), respectively, of Ref. 5.

Comparing (8.15) with (8.10) after ignoring the $Q^2(\cdot)$ term (valid for large E_b/N_0), we observe that this is exactly the result that would be obtained by applying the relation between the bit and symbol error probability as given in (8.7). Thus, we conclude that the remaining terms of the summation in (8.14) account for what is needed to make the expression accurate at low E_b/N_0.

8.1.1.3 M-ary Phase-Shift-Keying. Referring to the signal model in Section 3.1.3 and the accompanying optimum receiver in Fig. 3.4, then equating the carrier amplitude with the average symbol energy (the same for all symbols since M-PSK is a constant envelope modulation, i.e., $A_c = \sqrt{E_s/T_s}$), the classical form for the SEP of M-PSK is given by [5, Eq. (4.130)]

$$P_s(E) = 1 - \frac{2}{\pi} \int_0^\infty \exp\left[-\left(u - \sqrt{\frac{E_s}{N_0}}\right)^2\right] \left[\int_0^{u \tan(\pi/M)} \exp(-v^2)\,dv\right] du \tag{8.16}$$

which after some manipulation can be rewritten in terms of the Gaussian Q-function as

$$P_s(E) = Q\left(\sqrt{\frac{2E_s}{N_0}}\right)$$
$$+ \frac{2}{\sqrt{\pi}} \int_0^\infty \exp\left[-\left(u - \sqrt{\frac{E_s}{N_0}}\right)^2\right] Q\left(\sqrt{2}u \tan \frac{\pi}{M}\right) du \tag{8.17}$$

From the form in (8.17) we immediately see that for binary PSK ($M = 2$), the second term evaluates to zero [since $Q(\infty) = 0$] and hence the bit error probability is given by

$$P_b(E) = Q\left(\sqrt{\frac{2E_b}{N_0}}\right) \tag{8.18}$$

which agrees, as it should, with (8.4) corresponding to binary AM.

Another special case of (8.16) that yields a closed-form solution corresponds to QPSK ($M = 4$). Here the average SEP is given by

$$P_s(E) = 2Q\left(\sqrt{\frac{E_s}{N_0}}\right) - Q^2\left(\sqrt{\frac{E_s}{N_0}}\right) \tag{8.19}$$

which agrees with (8.11) and assuming a Gray code mapping of bits to symbols, the bit error probability is also given by (8.18). Thus, we see that the BEP of BPSK and QPSK are identical, the latter having the advantage of being half the bandwidth of the former.

If one now applies the desired form of the Gaussian Q-function to (8.17), then after considerable manipulation, the following result is obtained:

$$P_s(E) = \frac{1}{\pi} \int_0^{\pi/2} \exp\left(-\frac{E_s}{N_0}\frac{1}{\sin^2\theta}\right) d\theta + \frac{2}{\pi} \int_0^{\pi/2} \exp\left[-\frac{E_s}{N_0}\frac{\tan^2(\pi/M)}{\tan^2(\pi/M)+\sin^2\theta}\right] d\theta$$

$$- \frac{2}{\pi^2} \int_0^{\pi/2}\int_0^{\pi/2} \exp\left[-\frac{E_s}{N_0}\frac{\tan^2(\pi/M)+\sin^2\theta/\sin^2\phi}{\tan^2(\pi/M)+\sin^2\theta}\right] d\phi\, d\theta \quad (8.20)$$

Even though (8.20) has the desired form of finite integration limits that are independent of E_s/N_0 and integrands that are exponential in E_s/N_0, it does contain a term with a double integral, which leaves a bit to be desired. Fortunately, Pawula et al. [9] were able to simplify the symbol error probability analysis of M-PSK by considering the more general case of the distribution of the angle between two vectors when the variance of the noise components perturbing each vector are in general unequal. Although this analysis was directly applicable to differentially coherent detection of M-PSK, in the degenerate case where one of the two vectors is noise free, the M-DPSK problem becomes the coherent M-PSK problem and the symbol error probability is given by the single integral

$$P_s(E) = \frac{1}{\pi} \int_{-\pi/2}^{\pi/2-\pi/M} \exp\left(-\frac{E_s}{N_0}\frac{g_{\text{PSK}}}{\sin^2\theta}\right) d\theta, \qquad g_{\text{PSK}} = \sin^2\frac{\pi}{M} \quad (8.21)$$

Many years later Craig [10] arrived at a similar result as a special case of a generic method for evaluating the average error probability for arbitrary two-dimensional modulations transmitted over the AWGN channel. This method defined the origin of coordinates for each decision region by the associated *signal vector* as opposed to using a fixed coordinate system origin for all decision regions derived from the *received vector*. This shift in vector-space coordinate systems allowed the integrand of the two-dimensional integral describing the conditional (on the transmitted signal) probability of error to be independent of the transmitted signal. For the particular case of coherently detected M-PSK, Craig [10] obtained the average SEP as

$$P_s(E) = \frac{1}{\pi} \int_0^{(M-1)\pi/M} \exp\left(-\frac{E_s}{N_0}\frac{g_{\text{PSK}}}{\sin^2\theta}\right) d\theta \quad (8.22)$$

which is easily shown to be equivalent to (8.21). Note, however, that for $M = 2$ (BPSK), replacing θ by $-\theta$ in (8.20) and letting $E_s = E_b$ must yield (8.18). Equating the two results one immediately obtains the desired form of the Gaussian Q-function as in (4.2), which is also obtained from equating (8.18) with (8.22) under the same conditions.

To obtain a quick assessment of the error probability performance of a particular modulation/detection scheme and at the same time enable a simple comparison of its performance with that of other modulation/detection schemes, simple upper bounds on average SEP are quite useful, especially if they can be obtained in closed form. Furthermore, when further integrations (e.g., statistical averages) are necessary over the SNR variable, as is the case for fading channels, these bounds have even more significance in terms of coming up with simple answers. The form of $P_s(E)$ in (8.22) lends itself nicely to the above-mentioned purpose. It is straightforward to show that over the interval of integration in (8.22), the function $f(\theta) = 1/\sin^2\theta$ has a single minimum which occurs at $\theta = \pi/2$ and corresponds to $f(\pi/2) = 1$. Using this in the argument of the exponential in (8.22) establishes the inequality

$$\exp\left(-\frac{E_s}{N_0}\frac{g_{\text{PSK}}}{\sin^2\theta}\right) \leq \exp\left(-\frac{E_s}{N_0}g_{\text{PSK}}\right) \qquad (8.23)$$

which when used in the integrand of this same equation together with $g_{\text{PSK}} = \sin^2\pi/M$ leads to the simple (no integration) upper bound

$$P_s(E) \leq \frac{M-1}{M}\exp\left(-\frac{E_s}{N_0}\sin^2\frac{\pi}{M}\right) \qquad (8.24)$$

A well-known union bound for the SEP of coherent M-PSK is (e.g., Ref. 11, Prob. 5.2)

$$P_s(E) \leq 2Q\left(\sqrt{\frac{2E_s}{N_0}}\sin\frac{\pi}{M}\right) \qquad (8.25)$$

which applying the Chernoff bound to the Gaussian Q-function results in the union–Chernoff bound

$$P_s(E) \leq \exp\left(-\frac{E_s}{N_0}\sin^2\frac{\pi}{M}\right) \qquad (8.26)$$

Comparing (8.24) with (8.26), we observe that for any fixed M, the former is slightly tighter than the latter, the difference between the two becoming smaller as M increases. Futhermore, in the limit of large SNR, all three upper bounds [i.e., (8.24), (8.25) and (8.26)] become asymptotically tight with respect to the exact result as given by (8.22) (see Fig. 8.1).

A method for determining the exact BEP of M-PSK using a Gray code bit-to-symbol mapping[2] was first discovered by Lee [13] and is also discussed by

---

[2] Extension of these results to arbitrary bit-to-symbol mappings (e.g., natural binary and folded binary mappings) was considered by Irshid and Salous [12].

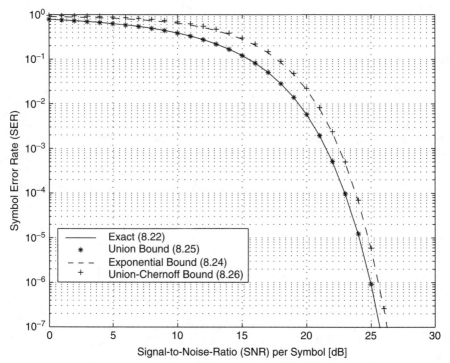

Figure 8.1. Comparison of three upper bounds on the symbol error probability of coherent M-PSK.

Simon et al. [5, pp. 211–212]. This method requires evaluating the probability that for a given transmitted phase, the received signal vector falls in each decision region (wedge of $2\pi/M$ radians centered around each of the signal points) within the circle. These probabilities, denoted by $P_k, k = 0, 1, 2, \ldots, M-1$, are given in the classical form [5, Eq. (4.198a)]

$$P_k = \frac{1}{\pi} \int_0^\infty \exp\left[-\left(u - \sqrt{\frac{E_s}{N_0}}\right)^2\right] \left[\int_{u\tan[(2k-1)\pi/M]}^{u\tan[(2k+1)\pi/M]} \exp(-v^2)\, dv\right] du,$$

$$k = 0, 1, 2, \ldots, M-1 \tag{8.27}$$

where the index k denotes the kth decision away from the one corresponding to the transmitted phase. [Note that (8.27) evaluated at $k = 0$ corresponds to the probability of the received signal vector falling in the correct decision region, i.e., the $2\pi/M$ wedge around the transmitted phase, which would then be the probability of a correct decision in agreement with 1 minus the symbol error probability of (8.16).]

In the same manner that (8.17) was derived from (8.16), the probabilities in (8.27) can be expressed in terms of the Gaussian Q-function. In particular,[3]

$$P_k = \frac{1}{\sqrt{\pi}} \int_0^\infty \exp\left[-\left(u - \sqrt{\frac{E_s}{N_0}}\right)^2\right] \left[Q\left(\sqrt{2}u \tan\left[(2k-1)\frac{\pi}{M}\right]\right)\right.$$
$$\left. - Q\left(\sqrt{2}u \tan\left[(2k+1)\frac{\pi}{M}\right]\right)\right] du \qquad (8.28)$$

The desired form of the Gaussian Q-function can be applied to (8.28). However, even though the resulting expression will be in the form of a single integral with finite $(0, \pi/2)$ limits, the integrand itself will still involve Gaussian Q-functions. Fortunately, the probabilities in (8.27) can be expressed as single integrals which are already in the desired form and do not involve Gaussian Q-functions. Analogous to (8.22), these probabilities can be obtained as [5, Eq. (4.198b)]

$$P_k = \frac{1}{2\pi} \int_0^{\pi(1-(2k-1)/M)} \exp\left\{-\frac{E_s}{N_0} \frac{\sin^2[(2k-1)\pi]/M}{\sin^2 \theta}\right\} d\theta$$
$$- \frac{1}{2\pi} \int_0^{\pi(1-(2k+1)/M)} \exp\left\{-\frac{E_s}{N_0} \frac{\sin^2[(2k+1)\pi]/M}{\sin^2 \theta}\right\} d\theta \qquad (8.29)$$

Having the set of probabilities $P_k, k = 0, 1, 2, \ldots, M-1$, in the desired form, we can now express the exact BEP of M-PSK in a similar desired form using the results of Lee [13]. In particular, the following results are obtained for $M = 4$, 8, and 16:

$$P_b(E) = \begin{cases} \frac{1}{2}(P_1 + 2P_2 + P_3), & M = 4 \\ \frac{1}{3}(P_1 + 2P_2 + P_3 + 2P_4 + 3P_5 + 2P_6 + P_7), & M = 8 \\ \frac{1}{2}\left(\sum_{k=1}^{8} P_k + \sum_{k=2}^{5} P_k + P_5 + 2P_6 + P_7\right), & M = 16 \end{cases} \qquad (8.30)$$

For $M = 4$ it is straightfoward to show that the result in (8.30) agrees with that in (8.18).

Although the approach of Lee [13] gives exact BEP results, it suffers from the fact that an explicit expression in terms of the P_k's of (8.29) must be obtained for each value of M. A simple solution around this difficulty is to again use the

---

[3] These probabilities were denoted by $S_k, k = 0, 1, \ldots, M-1$, and given in this form by Lee [13], who also observed them to have the symmetry property $S_m = S_{M-m}, m = (M/2) + 1, \ldots, M-1$.

approximate (valid for large symbol SNR) relation between bit and symbol error probability of (8.7) together now with (8.22) for the latter. However, as discussed for QAM, such an approximation is not useful for evaluating average BEP in the presence of fading since in this situation the instantaneous SNR can vary between zero and infinity. Thus, once again we turn to the results of Lu et al. [8], which give an approximate expression for the BEP of M-PSK in AWGN that is quite accurate at both low and high SNR and furthermore is valid for all M. In particular, for M-PSK, it is shown in Ref. 8 that

$$P_b(E) \simeq \frac{2}{\max(\log_2 M, 2)} \sum_{i=1}^{\max(M/4,1)} Q\left(\sqrt{\frac{2E_b \log_2 M}{N_0}} \sin \frac{(2i-1)\pi}{M}\right) \quad (8.31)$$

Here again for large E_b/N_0 and $M > 4$, the first term in the summation of (8.31) is dominant, in which case this equation simplifies to

$$P_b(E) \simeq \frac{2}{\log_2 M} Q\left(\sqrt{\frac{2E_b \log_2 M}{N_0}} \sin \frac{\pi}{M}\right) \quad (8.32)$$

which is precisely what would be obtained by applying the relation between the bit and symbol error probability as given in (8.7), using (8.25) for the latter. Thus, once again we conclude that the remaining terms of the summation in (8.31) account for what is needed to make the expression accurate at low E_b/N_0.

8.1.1.4 Differentially Encoded M-ary Phase-Shift-Keying and $\pi/4$-QPSK.
When differential phase encoding is applied to the transmitted M-PSK modulation but coherent detection is still used at the receiver, the evaluation of average SEP is a bit more complex than that considered in the preceding section. Since for differential phase encoding a correct decision on the information phase for the nth symbol interval will occur if both the nth and the $(n-1)$st received signal vectors fall k decision regions away from the correct one, $k = 0, 1, \ldots, M - 1$, then since these two adjacent receptions are independent, the probability of this occurring is

$$P_s(C) = \sum_{k=0}^{M-1} P_k^2 \quad (8.33)$$

independent of the particular value of the nth information phase. Thus, the average symbol error probability for coherently detected, differentially encoded M-PSK is

$$P_s(E) = 1 - P_s(C) = 1 - \sum_{k=0}^{M-1} P_k^2 \quad (8.34)$$

which can be expressed in terms of the average SEP for M-PSK without differential encoding [i.e., $P_s(E)|_{M-\text{PSK}}$ of (8.16)] as [5, Eq. (4.200)]

$$P_s(E) = 1 - \left[1 - P_s(E)|_{M-\text{PSK}}\right]^2 - \sum_{k=1}^{M-1} P_k^2$$

$$= 2P_s(E)|_{M-\text{PSK}} - \left[P_s(E)|_{M-\text{PSK}}\right]^2 - \sum_{k=1}^{M-1} P_k^2 \qquad (8.35)$$

Using (8.22) and (8.29) in (8.35), all terms involve only single integrals with finite integration limits that are independent of E_s/N_0 and integrands that are exponential in E_s/N_0. However, the fact that the second and third terms of (8.35) require that these integrals be squared still poses difficulties in terms of a simple extension of these results to the fading channel.

Two special cases of (8.35) are of interest. For coherent detection of differentially encoded BPSK, (8.35) together with (8.18) reduces to

$$P_b(E) = 2P_b(E)|_{\text{BPSK}} - 2\left[P_b(E)|_{\text{BPSK}}\right]^2$$

$$= 2Q\left(\sqrt{\frac{2E_b}{N_0}}\right) - 2Q^2\left(\sqrt{\frac{2E_b}{N_0}}\right) \qquad (8.36)$$

Since a desired form of the square of the Gaussian Q-function exists in (4.9), then (8.36) has the desired form

$$P_b(E) = \frac{2}{\pi} \int_0^{\pi/2} \exp\left(-\frac{E_b}{N_0}\frac{1}{\sin^2\theta}\right) d\theta + \frac{2}{\pi} \int_0^{\pi/4} \exp\left(-\frac{E_b}{N_0}\frac{1}{\sin^2\theta}\right) d\theta \qquad (8.37)$$

For differentially encoded QPSK, (8.35) simplifies to

$$P_s(E) = 4Q\left(\sqrt{\frac{E_s}{N_0}}\right) - 8Q^2\left(\sqrt{\frac{E_s}{N_0}}\right) + 8Q^3\left(\sqrt{\frac{E_s}{N_0}}\right) - 4Q^4\left(\sqrt{\frac{E_s}{N_0}}\right) \qquad (8.38)$$

Unfortunately, this special case cannot be put in the desired form due to the lack of such forms for the third and fourth powers of the Gaussian Q-function. Nevertheless, as we shall see shortly, it will still be possible obtain finite-limit single-integral expressions for the average error probability performance of differentially encoded QPSK in Rayleigh and Nakagami-m fading by making use of the alternative form of the Gaussian Q-function and the integrals developed in Section 5.4.3.

Finally, since as pointed out in Section 3.1.4.2, $\pi/4$-QPSK is a particular form of differentially encoded QPSK wherein the information phases are chosen to range over the set $\Delta\beta_i = \pi/4, 3\pi/4, 5\pi/4, 7\pi/4$ instead of the conventional $\Delta\beta_i = 0, \pi/2, \pi, 3\pi/2$, then since the receiver performance is independent of the

204 PERFORMANCE OF SINGLE CHANNEL RECEIVERS

choice of the information symbol set, coherently detected $\pi/4$-QPSK transmitted over a linear AWGN channel is also characterized by (8.38).

8.1.1.5 Offset QPSK or Staggered QPSK.
Referring to the signal model in Section 3.1.5 and the accompanying optimum receiver in Fig. 3.7, then noting the similarity of this receiver to the conventional QPSK receiver in Fig. 3.6, the classical form for the BEP of OQPSK is also given by (8.16). Stated another way, since in accordance with Fig. 3.7 independent decisions are made on the I and Q data bits, the time offset of these two channels has no effect on these decisions and hence on a linear AWGN channel with ideal coherent detection at the receiver, OQPSK has the same BEP performance as QPSK and also BPSK. The differences in performance between these three modulations comes about when the carrier demodulation is nonideal, as will be discussed shortly.

8.1.1.6 M-ary Frequency-Shift-Keying.
Consider first the case of orthogonal signaling using the M-FSK modulation described by the signal model in Section 3.1.6 and the receiver of Fig. 3.8. Assuming that the transmitted frequency in the nth symbol interval, f_n, is equal to $\xi_l = (2l - 1 - M)\Delta f/2$, the real parts of the integrate-and-dump (I&D) outputs, $\tilde{y}_{nk}, k = 1, 2, \ldots, M$, as given by (3.25) are independent, identically distributed (i.i.d.) Gaussian random variables with means as in (3.25) and variance $\sigma_n^2 = N_0 T_s/2$. The probability of a correct symbol decision is the probability that all $\text{Re}\{\tilde{y}_{nk}\}, k \neq l$, are less than $\text{Re}\{\tilde{y}_{nl}\}$. Thus, letting $A_c = \sqrt{E_s/T_s}$ and denoting $\text{Re}\{\tilde{y}_{nk}\}$ by z_{nk}, the probability of symbol error is given by [5, Eq. (4.92)]

$$P_s(E) = 1 - \int_{-\infty}^{\infty} \left[\int_{-\infty}^{z_{nl}} \frac{1}{\sqrt{2\pi\sigma_n^2}} \exp\left(-\frac{z_{nk}^2}{2\sigma_n^2}\right) dz_{nk} \right]^{M-1}$$

$$\times \frac{1}{\sqrt{2\pi\sigma_n^2}} \exp\left[-\frac{(z_{nl} - \sqrt{E_s T_s})^2}{2\sigma_n^2}\right] dz_{nl} \quad (8.39)$$

or in terms of the Gaussian Q-function,

$$P_s(E) = 1 - \int_{-\infty}^{\infty} \left[Q\left(-q - \sqrt{\frac{2E_s}{N_0}}\right) \right]^{M-1} \frac{1}{\sqrt{2\pi}} \exp\left(-\frac{q^2}{2}\right) dq \quad (8.40)$$

The corresponding bit error probability is given by [5, Eq. (4.96)]

$$P_b(E) = \frac{2^{k-1}}{2^k - 1} P_s(E), \quad k = \log_2 M \quad (8.41)$$

Unfortunately, for arbitrary M, (8.40) cannot be put in the desired form by using the form of the Gaussian Q-function in (4.2). The special case of binary orthogonal FSK ($M = 2$), however, does have a simple form, namely,

$$P_b(E) = Q\left(\sqrt{\frac{E_b}{N_0}}\right) \tag{8.42}$$

which can be put in the desired form,

$$P_b(E) = \frac{1}{\pi}\int_0^{\pi/2} \exp\left(-\frac{E_b}{2N_0}\frac{1}{\sin^2\theta}\right) d\theta \tag{8.43}$$

Another M-FSK case whose error probability performance can be put into the desired form corresponds to binary nonorthogonal FSK with cross-correlation given by (3.27). In particular, the BEP for such a modulation is given by[4]

$$P_b(E) = Q\left(\sqrt{\frac{E_b(1 - \sin 2\pi h/2\pi h)}{N_0}}\right)$$

$$= \frac{1}{\pi}\int_0^{\pi/2} \exp\left(-\frac{E_b}{2N_0}\frac{(1 - \sin 2\pi h/2\pi h)}{\sin^2\theta}\right) d\theta \tag{8.44}$$

where, as before, $h = \Delta f T_b$ is the frequency-modulation index. The minimum BEP is achieved when $h = 0.715$ (the value of h that maximizes the argument of the Gaussian Q-function), resulting in

$$P_b(E) = Q\left(\sqrt{\frac{E_b(1.217)}{N_0}}\right) = \frac{1}{\pi}\int_0^{\pi/2} \exp\left(-\frac{E_b}{2N_0}\frac{1.217}{\sin^2\theta}\right) d\theta \tag{8.45}$$

which is often approximated by

$$P_b(E) = Q\left(\sqrt{\frac{E_b(1 + 2/3\pi)}{N_0}}\right) = \frac{1}{\pi}\int_0^{\pi/2} \exp\left(-\frac{E_b}{2N_0}\frac{1 + 2/3\pi}{\sin^2\theta}\right) d\theta \tag{8.46}$$

8.1.1.7 Minimum-Shift-Keying. In Section 3.1.7 it was demonstrated that MSK was equivalent to pulse-shaped OQPSK, where the pulse shape was sinusoidal [see (3.33) and (3.34)]. Ignoring the implicit differential encoding at the transmitter (i.e., assuming that we are dealing with precoded MSK), the BEP of the receiver implemented as the one that's optimum for pulse-shaped OQPSK (e.g., Fig. 3.12) is independent of the shape of the pulse and is thus given by (8.18). In summary, the receivers for binary AM, BPSK, QPSK, OQPSK, and MSK all have identical BEP performance.

---

[4] This is a special case of the BEP for coherent detection of binary signals with arbitrary cross-correlation $-1 \leq \rho \leq 1$, which is given by $P_b(E) = Q(\sqrt{E_b(1-\rho)/N_0})$.

8.1.2 Nonideal Coherent Detection

We saw in the preceding section that many of the ideal coherent detection systems had identical error probability performances. In a practical system where the demodulation reference is nonideal (see Section 3.2), the performances of these systems whose receivers are designed on the basis of ideal coherent detection will differ from one another. In this section we present the results that enable one to assess these differences.

We begin with the simplest case of a BPSK system whose receiver has an imperfect carrier demodulation reference obtained from a Costas loop. The average BEP performance of such a BPSK system is given by [14][5]

$$P_b(E) = \int_{-\pi/2}^{\pi/2} P_b(E; \phi_c) p(\phi_c) \, d\phi_c \qquad (8.47)$$

where

$$P_b(E; \phi_c) = Q\left(\sqrt{\frac{2E_b}{N_0}} \cos \phi_c\right) \qquad (8.48)$$

is the conditional (on the loop phase error ϕ_c) BEP and for a Costas loop that tracks the doubled phase error process

$$p(\phi_c) = \frac{\exp(\rho_{eq} \cos 2\phi_c)}{\pi I_0(\rho_{eq})}, \qquad 0 \leq |\phi_c| \leq \frac{\pi}{2} \qquad (8.49)$$

is the phase error PDF in Tikhonov form [15]. Also, in (8.49),

$$\rho_{eq} = \frac{\rho_c S_L}{4} \qquad (8.50)$$

is the equivalent loop SNR with $\rho_c = (E_b/T_b)/N_0 B_L$ (B_L is the single-sided loop noise bandwidth) the loop SNR of a phase-locked loop (PLL) and

$$S_L = \frac{1}{1 + 1/(2E_b/N_0)} \qquad (8.51)$$

is called the *squaring loss* assuming ideal I&D arm filters for the Costas loop. Substituting (8.48) and (8.49) in (8.47) gives the classical result

$$P_b(E) = \int_{-\pi/2}^{\pi/2} Q\left(\sqrt{\frac{2E_b}{N_0}} \cos \phi_c\right) \frac{\exp(\rho_{eq} \cos 2\phi_c)}{\pi I_0(\rho_{eq})} \, d\phi_c \qquad (8.52)$$

which ordinarily is evaluated by numerical integration.

---

[5] This result assumes that the 180° phase ambiguity associated with the Costas loop is perfectly resolved. Methods for accomplishing this are beyond the scope of this discussion.

The evaluation of (8.52) can be simplified a bit by using the desired form of the Gaussian Q-function. In particular, using (4.2) in (8.52), we obtain the following development:

$$P_b(E) = \frac{1}{\pi^2 I_0(\rho_{eq})} \int_{-\pi/2}^{\pi/2} \int_0^{\pi/2} \exp\left(-\frac{E_b}{N_0 \sin^2 \theta} \cos^2 \phi_c\right)$$
$$\times \exp(\rho_{eq} \cos 2\phi_c) \, d\phi_c \, d\theta$$

$$= \frac{1}{\pi^2 I_0(\rho_{eq})} \int_{-\pi/2}^{\pi/2} \int_0^{\pi/2} \exp\left(-\frac{E_b}{2N_0 \sin^2 \theta}(1 + \cos 2\phi_c)\right)$$
$$\times \exp(\rho_{eq} \cos 2\phi_c) \, d\theta \, d\phi_c$$

$$= \frac{1}{\pi^2 I_0(\rho_{eq})} \int_0^{\pi/2} \exp\left(-\frac{E_b}{2N_0 \sin^2 \theta}\right)$$
$$\times \int_{-\pi/2}^{\pi/2} \exp\left\{\left(-\frac{E_b}{2N_0 \sin^2 \theta} + \rho_{eq}\right) \cos 2\phi_c\right\} d\phi_c \, d\theta$$

$$= \frac{1}{2\pi^2 I_0(\rho_{eq})} \int_0^{\pi/2} \exp\left(-\frac{E_b}{2N_0 \sin^2 \theta}\right)$$
$$\times \int_{-\pi}^{\pi} \exp\left\{\left(-\frac{E_b}{2N_0 \sin^2 \theta} + \rho_{eq}\right) \cos \Phi_c\right\} d\Phi_c \, d\theta \quad (8.53)$$

Finally, recognizing that the integral on Φ_c is in the form of a modified Bessel function of the first kind, we get the final desired result:

$$P_b(E) = \frac{1}{\pi} \int_0^{\pi/2} \exp\left(-\frac{E_b}{2N_0 \sin^2 \theta}\right) \frac{I_0(-(E_b/2N_0 \sin^2 \theta) + \rho_{eq})}{I_0(\rho_{eq})} d\theta \quad (8.54)$$

The form of (8.54) is interesting in that the Gaussian Q-function needed in the integrand of (8.52) has been replaced by a modified Bessel function with an argument related to both the equivalent loop SNR (ρ_{eq}) and the detection SNR (E_b/N_0).

For QPSK and an imperfect carrier demodulation reference obtained from a four-phase Costas loop with I&D arm filters, the appropriate expressions analogous to (8.47) through (8.52) are [14]

$$P_b(E) = \int_{-\pi/4}^{\pi/4} P_b(E; \phi_c) p(\phi_c) \, d\phi_c \quad (8.55)$$

where

$$P_b(E; \phi_c) = \frac{1}{2} Q\left(\sqrt{\frac{2E_b}{N_0}}(\cos \phi_c - \sin \phi_c)\right) + \frac{1}{2} Q\left(\sqrt{\frac{2E_b}{N_0}}(\cos \phi_c + \sin \phi_c)\right)$$
$$(8.56)$$

and

$$p(\phi_c) = \frac{2\exp(\rho_{eq}\cos 4\phi_c)}{\pi I_0(\rho_{eq})}, \qquad 0 \leq |\phi_c| \leq \frac{\pi}{4} \tag{8.57}$$

with

$$\rho_{eq} = \frac{\rho_c S_L}{16} \tag{8.58}$$

and

$$S_L = \frac{1}{1 + (9/4)/(E_b/N_0) + (3/2)/(E_b/N_0)^2 + (3/16)/(E_b/N_0)^3} \tag{8.59}$$

Unfortunately, substitution of (8.56) and (8.57) in (8.55) and using the desired form of the Gaussian Q-function does not provide for any further simplification, as before.

Consider now the additive Gaussian noise reference signal model of (3.39) as suggested by Fitz [16] to be characteristic of a large class of phase estimation techniques used to evaluate average error probability performance at moderate to high SNR. When used to demodulate the received signal in (3.38), the decision statistic for the nth symbol becomes equal to $\text{Re}\{\tilde{y}_{nk}\tilde{c}_r^*\}$, which is in the form of the real part of the product of two nonzero mean complex Gaussian random variables. The probability of error associated with such a generic decision statistic is discussed in Appendix 8A. When applied to BPSK modulation with $A_r = A_c(S_{1p} = S_{2p})$ and assuming that the signal and reference noises $\tilde{N}_{nk}$ and $\tilde{N}_r$ have equal power and are uncorrelated, then from (8A.5) together with (8A.7) and, in addition, $\theta_{1p} = \theta_{2p}$, the error probability becomes

$$P_b(E) = \tfrac{1}{2}[1 - Q_1(\sqrt{b},\sqrt{a}) + Q_1(\sqrt{a},\sqrt{b})] \tag{8.60}$$

where

$$a = \frac{E_b}{2N_0}(\sqrt{G}-1)^2, \qquad b = \frac{E_b}{2N_0}(\sqrt{G}+1)^2 \tag{8.61}$$

To tie the additive Gaussian noise reference and the Tikhonov-distributed phase error models together, we assume a phase reference generated by a PLL whose input has a signal power equal to that of the data-modulated (BPSK) signal. In this case, the SNR gain G of the former model is related to the loop bandwidth–bit time product $B_L T$ of the PLL by $G = 1/B_L T_b$ [16].[6] Using this equivalence, Fitz [16] shows that the error probability computed from (8.52) or any of its subsequent equivalent forms is virtually identical to that computed from the combination of (8.60) and (8.61).

---

[6] Equivalently, the loop SNR ρ_c, is related to the SNR gain G by $\rho_c \triangleq P/N_0 B_L = (1/B_L T_b) \times (PT_b/N_0) = GE_b/N_0$.

For QPSK modulation, the reference signal has twice the power of the signal in either the I or Q components [i.e., $A_r = \sqrt{2}A_c(S_{1p} = 2S_{2p})$]. Thus, detecting each of these components independently according to the decision variables $\text{Re}\{\tilde{y}_{nk}\tilde{c}_r^*\}$ and $\text{Im}\{\tilde{y}_{nk}\tilde{c}_r^*\}$, the two bit error probabilities will be equal, and hence the average bit error probability can again be obtained from (8A.5) together with (8A.7) using the same assumptions as above for the signal and reference noises. The result is given by (8.60), now with

$$a = \frac{E_b}{2N_0}(\sqrt{2G} - 1)^2, \quad b = \frac{E_b}{2N_0}(\sqrt{2G} + 1)^2 \quad (8.62)$$

Similar comparisons of average error probability computed from (8.55) through (8.57) and (8.60) together with (8.62) show excellent agreement [16].

Using a similar approach, the average BEP for offset QPSK and MSK can be computed as an arithmetic average of two terms in the form of (8.60). The reason for the two terms is that one them corresponds to decisions made on one (say, I) of the channels when during (in the middle of) the same detection interval there is a symbol transition on the other (say, Q) channel, while the other term corresponds to decisions made on one of the channels when during the same detection interval there is no symbol transition on the other channel. In particular,

$$P_b(E) = \tfrac{1}{4}[1 - Q_1(\sqrt{b_1}, \sqrt{a_1}) + Q_1(\sqrt{a_1}, \sqrt{b_1})]$$
$$+ \tfrac{1}{4}[1 - Q_1(\sqrt{b_2}, \sqrt{a_2}) + Q_1(\sqrt{a_2}, \sqrt{b_2})] \quad (8.63)$$

where the appropriate values of the parameters a and b are as follows:

$$a_1 = \frac{E_b}{2N_0}(\sqrt{2G} - 1)^2, \quad b_1 = \frac{E_b}{2N_0}(\sqrt{2G} + 1)^2$$
$$a_2 = \frac{E_b}{N_0}(G + 1 - \sqrt{2G}), \quad b_2 = \frac{E_b}{N_0}(G + 1 + \sqrt{2G}) \quad \text{(OQPSK)} \quad (8.64)$$

and

$$a_1 = \frac{E_b}{2N_0}(\sqrt{2G} - 1)^2, \quad b_1 = \frac{E_b}{2N_0}(\sqrt{2G} + 1)^2$$
$$a_2 = \frac{E_b}{N_0}\left(G + \frac{\pi^2 + 4}{2\pi^2} - \sqrt{2G}\right), \quad b_2 = \frac{E_b}{N_0}\left(G + \frac{\pi^2 + 4}{2\pi^2} + \sqrt{2G}\right) \quad \text{(MSK)} \quad (8.65)$$

8.1.3 Noncoherent Detection

In Section 3.3 the decision variables and the accompanying optimum receiver for noncoherent detection of an equal energy M-ary signaling set were presented [see (3.40) and Fig. 3.13]. It was concluded there that the most logical choice of modulation for this type of detection is M-FSK. Based on the matched filter outputs described by (3.41) and the assumption of orthogonal signals (corresponding to a minimum frequency spacing $\Delta f_{\min} = 1/T_s$, which is twice that for coherent detection), the SEP is given by

$$P_s(E) = \sum_{m=1}^{M-1}(-1)^{m+1}\binom{M-1}{m}\frac{1}{m+1}\exp\left[-\frac{m}{m+1}\left(\frac{E_s}{N_0}\right)\right] \tag{8.66}$$

and the corresponding BEP is obtained from (8.66) by the relation

$$P_b(E) = \frac{1}{2}\left(\frac{M}{M-1}\right)P_s(E) \tag{8.67}$$

For noncoherent detection of binary FSK, (8.66) reduces to

$$P_b(E) = \frac{1}{2}\exp\left(-\frac{E_b}{2N_0}\right) \tag{8.68}$$

The performance of nonorthogonal M-FSK is considerably more complicated to evaluate (see Simon et al. [5, Sec. 5.2.2]). For the binary nonorthogonal case, however, the result can be expressed in terms of the first-order Marcum Q-function as [17]

$$P_b(E) = Q_1(\sqrt{a}, \sqrt{b}) - \frac{1}{2}\exp\left(\frac{a+b}{2}\right)I_0(\sqrt{ab}) \tag{8.69}$$

which is equivalent to (8.60) and where

$$a = \frac{E_b}{2N_0}(1 - \sqrt{1-\rho^2}), \quad b = \frac{E_b}{2N_0}(1 + \sqrt{1-\rho^2}) \tag{8.70}$$

and ρ is the correlation coefficient of the two signals. For $\rho = 0$ (orthogonal signaling), the parameters a and b become $a = 0$ and $b = E_b/N_0$, and using the property of the Marcum Q-function in (4.22), we immediately obtain (8.68).

8.1.4 Partially Coherent Detection

8.1.4.1 Conventional Detection: One-Symbol Observation. In Section 3.4.1 the decision variables and the accompanying optimum receiver for partially coherent detection of an equal-energy M-ary signaling set were presented [see (3.43) and Fig. 3.14]. We observed there that for M-PSK modulation (including BPSK), the noncoherent term in the decision variables was independent of the information, and thus the decision is based entirely on the coherent term. Hence, the performance of Fig. 3.14 for partially coherent detection of

M-PSK would be equal to that of nonideal coherent detection of this same modulation, assuming a demodulation reference that produces a Tikhonov PDF for the phase error. For example, for BPSK, the performance would be given by

$$P_b(E) = \int_{-\pi}^{\pi} Q\left(\sqrt{\frac{2E_b}{N_0}} \cos \phi_c\right) \frac{\exp(\rho_c \cos \phi_c)}{2\pi I_0(\rho_c)} d\phi_c \qquad (8.71)$$

For orthogonal M-FSK modulation, both the noncoherent and coherent terms of the decision variables contribute to the decision. The resulting SEP is given by

$$P_s(E) = 1 - \int_{-\pi}^{\pi} \int_0^{\infty} y \exp\left(-\frac{c_1^2 + y^2}{2}\right) I_0(c_1 y)$$

$$\times [1 - Q_1(c_2, y)]^{M-1} \frac{\exp(\rho_c \cos \phi_c)}{2\pi I_0(\rho_c)} dy \, d\phi_c \qquad (8.72)$$

where

$$c_1^2 = \frac{2E_s}{N_0} + \frac{\rho_c^2}{2E_s/N_0} + 2\rho_c \cos \phi_c$$

$$c_2^2 = \frac{\rho_c^2}{2E_s/N_0} \qquad (8.73)$$

For the binary case, (8.72) can be expressed as

$$P_b(E) = \int_{-\pi}^{\pi} P(E; \phi_c) \frac{\exp(\rho_c \cos \phi_c)}{2\pi I_0(\rho_c)} d\phi_c \qquad (8.74)$$

where $P(E; \phi_c)$ is in the form of (8.69), now with

$$a = \frac{\rho_c^2}{4E_b/N_0}, \qquad b = \frac{4(E_b/N_0)^2 + \rho_c^2 + 4(E_b/N_0)\rho_c \cos \phi_c}{2E_b/N_0} \qquad (8.75)$$

Finally, for nonorthogonal BFSK, the BEP is once again given by (8.74) with $P(E; \phi_c)$ in the form of (8.69) and

$$a = \tfrac{1}{2}(\alpha_0^2 + \beta_0^2 + 2\alpha_0\beta_0 \cos \phi_c), \qquad b = \tfrac{1}{2}(\alpha_1^2 + \beta_1^2 + 2\alpha_1\beta_1 \cos \phi_c) \qquad (8.76)$$

with

$$\alpha_0 = \alpha_1 = \frac{\rho_c}{\sqrt{1-\rho^2}}\sqrt{\frac{1-\rho}{2E_b/N_0}}$$

$$\beta_0 = \sqrt{(1+\sqrt{1-\rho^2})(E_b/N_0)} \qquad (8.77)$$

$$\beta_1 = -\sqrt{(1-\sqrt{1-\rho^2})(E_b/N_0)}$$

8.1.4.2 Multiple-Symbol Detection.

Because of the memory introduced into the modulation by virtue of the fact that the carrier phase error ϕ_c is constant over many symbol intervals, the performance of conventional partially coherent detection schemes can be improved by increasing the observation interval beyond the duration of one symbol. This was pointed out in Section 3.4.2, and the optimum receiver for multiple-symbol partially coherent detection over the AWGN was shown in Fig. 3.15. It is of interest to specify the performance of that receiver in terms of the number of symbols, N_s, associated with the observation. Unlike the conventional case, the BEP for multiple-symbol detection cannot be obtained in closed form. However, based on block-by-block detection of N_s-symbol sequences, an upper union bound on the average BEP can be determined as follows.

For M-PSK, we first rewrite the decision variables of (3.46) in the form

$$z_{n\mathbf{k}} = \left| \sum_{i=0}^{N_s-1} \frac{1}{N_0} \tilde{y}_{n-i,k_i} + \frac{\rho_c}{2} \right|^2, \qquad k_i = 1, 2, \ldots, M \tag{8.78}$$

where the addition of the constant $(\rho_c/2)^2$ to $z_{n\mathbf{k}}$ in (3.46) has no bearing on the decision. Also since choosing the largest magnitude squared is equivalent to choosing the largest magnitude, we can consider instead the decision variables

$$z_{n\mathbf{k}} = \left| \sum_{i=0}^{N_s-1} \frac{1}{N_0} \tilde{y}_{n-i,k_i} + \frac{\rho_c}{2} \right|, \qquad k_i = 1, 2, \ldots, M \tag{8.79}$$

For any particular transmitted phase sequence, say $\boldsymbol{\beta} = (\beta_{k_0}, \beta_{k_1}, \ldots, \beta_{k_{N_s-1}})$, $z_{n\mathbf{k}}$ is a Rician random variable. Thus, the probability of choosing as the decision another phase sequence, say $\hat{\boldsymbol{\beta}} = (\hat{\beta}_{k_0}, \hat{\beta}_{k_1}, \ldots, \hat{\beta}_{k_{N_s-1}})$, which is equal to the probability that the corresponding decision variable, say $\hat{z}_{n\mathbf{k}}$, is greater than $z_{n\mathbf{k}}$, is statistically characterized by the probability of one Rican random variable exceeding another. Since the decision is made strictly between two sequences, the resulting probability is referred to as the *pairwise error probability*. Based on the characterization above, the pairwise error probability can be determined using the results pertinent to the noncoherent detection problem in Appendix 8A. In particular, it can be shown [5, Sec. 6.4.1] that this pairwise error probability (conditioned on the carrier phase error ϕ_c) is given by the generic form of (8A.5) with $A = 0$, namely,

$$\Pr\{\hat{z}_{n\mathbf{k}} > z_{n\mathbf{k}} | \phi_c\} = \tfrac{1}{2}[1 - Q_1(\sqrt{b}, \sqrt{a}) + Q_1(\sqrt{a}, \sqrt{b})] \tag{8.80}$$

with

$$\begin{Bmatrix} b \\ a \end{Bmatrix} = \frac{E_s}{2N_0} \left\{ N_s \left[1 + \frac{1}{N_s} \left(\frac{\rho_c}{E_s/N_0} \right) \cos \phi_c + \frac{N_s - |\delta| \cos \nu}{2N_s(N_s^2 - |\delta|^2)} \left(\frac{\rho_c}{E_s/N_0} \right)^2 \right] \right\}$$
$$\pm \frac{E_s}{N_0} \left[\sqrt{N_s^2 - |\delta|^2} + \frac{N_s \cos \phi_c - |\delta| \cos(\phi_c + \nu)}{\sqrt{N_s^2 - |\delta|^2}} \left(\frac{\rho_c}{E_s/N_0} \right) \right] \tag{8.81}$$

and

$$\delta \overset{\Delta}{=} \sum_{i=0}^{N_s-1} \exp[j(\beta_{k_i} - \hat{\beta}_{k_i})], \qquad \nu \overset{\Delta}{=} \arg \delta \qquad (8.82)$$

To determine the upper bound on average BEP from the pairwise error probability we first determine the number of bit errors that result from the erroneous sequence decision corresponding to a given pair of phase sequences and then average over all possible sequence pairs. Mathematically speaking, let **u** be the sequence of $b = N_s \log_2 M$ information bits that produces the transmitted phase sequence $\boldsymbol{\beta}$, and let $\hat{\mathbf{u}}$ be the sequence of b bits that results from the erroneously detected phase sequence $\hat{\boldsymbol{\beta}}$. Furthermore, let $w(\mathbf{u}, \hat{\mathbf{u}})$ be the Hamming distance between **u** and $\hat{\mathbf{u}}$ (i.e., the number of bit errors that result from the erroneous phase sequence decision). Then, the upper bound (conditioned on ϕ_c) on the average BEP is given by

$$P_b(E|\phi_c) \leq \frac{1}{b} \sum_{\boldsymbol{\beta} \neq \hat{\boldsymbol{\beta}}} w(\mathbf{u}, \hat{\mathbf{u}}), \Pr\{\hat{z}_{n\mathbf{k}} > z_{n\mathbf{k}}|\phi_c\}$$

$$= \frac{1}{N_s \log_2 M} \sum_{\boldsymbol{\beta} \neq \hat{\boldsymbol{\beta}}} w(\mathbf{u}, \hat{\mathbf{u}}) \Pr\{\hat{z}_{n\mathbf{k}} > z_{n\mathbf{k}}|\phi_c\} \qquad (8.83)$$

Finally, the upper bound on average BEP is obtained by averaging (8.83) over the Tikhonov PDF of (3.37).

8.1.5 Differentially Coherent Detection

8.1.5.1 M-ary Differential Phase-Shift-Keying. As discussed in Section 3.5 differentially coherent detection of M-ary PSK (M-DPSK) makes its phase decisions using a demodulation reference signal derived from the received signal in previous intervals. In the conventional case corresponding to a two-symbol observation, the previous matched filter output is used directly as the demodulation reference for the current matched filter output. Since, however, the assumption of a received carrier phase that is constant over a number of symbol intervals introduces memory into the modulation, then, as was true for the case of partially coherent detection, the performance can be improved by extending the observation beyond two symbol intervals.

Since a Tikhonov PDF with $\rho_c = 0$ corresponds to a uniform PDF, then in principle the results for differentially coherent detection should be obtainable from those for partially coherent detection with multiple (at least two)-symbol observation simply by setting $\rho_c = 0$. However, because of the presence of a coherent component in addition to the noncoherent component of the decision statistic for partially coherent detection, there was no formal requirement for assuming differential encoding at the transmitter. However, setting $\rho_c = 0$ in the decision statistic leaves only the noncoherent component, which without differential encoding is ambiguous insofar as making phase decisions (see the discussion in Section 3.5.1.1). Thus, for M-DPSK, it is a formal requirement that differential encoding be employed at the transmitter. In what follows we present the performance of classical (two-symbol observation) and multiple-symbol differential detection of M-PSK, keeping in mind that the results will be somewhat different than those obtained by simply setting $\rho_c = 0$ in the results of Section 8.1.4.2.

Conventional Detection: Two-Symbol Observation. The SEP of the optimum receiver (see Fig. 3.16) for conventional (two-symbol observation) differential detection of M-PSK over the AWGN in the desired form (a single integral with finite limits and an integrand that is Gaussian in the square root of SNR) was first determined by Pawula et al. [9]:

$$P_s(E) = \frac{\sin(\pi/M)}{2\pi} \int_{-\pi/2}^{\pi/2} \frac{\exp\{-(E_s/N_0)[1 - \cos(\pi/M)\cos\theta]\}}{1 - \cos(\pi/M)\cos\theta} d\theta$$

$$= \frac{\sqrt{g_{PSK}}}{2\pi} \int_{-\pi/2}^{\pi/2} \frac{\exp[-(E_s/N_0)(1 - \sqrt{1 - g_{PSK}}\cos\theta)]}{1 - \sqrt{1 - g_{PSK}}\cos\theta} d\theta \qquad (8.84)$$

where, as in (8.21), $g_{PSK} \triangleq \sin^2(\pi/M)$. For binary DPSK wherein $g_{PSK} = 1$, (8.84) simplifies to

$$P_b(E) = \frac{1}{2} \exp\left(-\frac{E_b}{N_0}\right) \qquad (8.85)$$

Assuming a Gray code bit-to-symbol mapping, the exact BEP of M-DPSK can be obtained using the method of Lee [13] combined with the results of Pawula et al. [9] (see also Simon et al. [5, App. 7B]). A summary of the results for $M = 4, 8, 16$, and 32 is given below:

$$P_b(E) = F\left(\frac{5\pi}{4}\right) - F\left(\frac{\pi}{4}\right), \qquad M = 4$$

$$P_b(E) = \frac{2}{3}\left[F\left(\frac{13\pi}{8}\right) - F\left(\frac{\pi}{8}\right)\right], \qquad M = 8$$

$$P_b(E) = \frac{1}{2}\left[F\left(\frac{13\pi}{16}\right) - F\left(\frac{9\pi}{16}\right) - F\left(\frac{3\pi}{16}\right) - F\left(\frac{\pi}{16}\right)\right], \qquad M = 16$$

$$P_b(E) = \frac{2}{5}\left[F\left(\frac{29\pi}{32}\right) - F\left(\frac{23\pi}{32}\right) - F\left(\frac{19\pi}{32}\right) - F\left(\frac{17\pi}{32}\right)\right.$$
$$\left. + F\left(\frac{13\pi}{32}\right) - F\left(\frac{9\pi}{32}\right) - F\left(\frac{3\pi}{32}\right) - F\left(\frac{\pi}{32}\right)\right], \qquad M = 16$$
$$(8.86)$$

where

$$F(\psi) = -\frac{\sin\psi}{4\pi} \int_{-\pi/2}^{\pi/2} \frac{\exp\{-[(E_b/N_0)\log_2 M](1 - \cos\psi\cos t)\}}{1 - \cos\psi\cos t} dt \qquad (8.87)$$

The bit error probability for the special case of $M = 4$ can also be written in the form of (8.60), where

$$a = (2 - \sqrt{2})\frac{E_b}{N_0}, \qquad b = (2 + \sqrt{2})\frac{E_b}{N_0} \tag{8.88}$$

Instead, using the alternative representation of the Marcum Q-function, the bit error probability becomes [see (8A.11)]

$$P_b(E) = \frac{1}{4\pi} \int_{-\pi}^{\pi} \left[\frac{1 - \zeta^2}{1 + 2\zeta \sin\theta + \zeta^2}\right] \exp\left\{-\left(1 + \frac{1}{\sqrt{2}}\right)\frac{E_b}{N_0}\right.$$

$$\left. \times [1 + 2\zeta \sin\theta + \zeta^2]\right\} d\theta, \qquad \zeta = \sqrt{\frac{2 - \sqrt{2}}{2 + \sqrt{2}}} \tag{8.89}$$

Finally, as was true for coherent detection of MPSK, for large symbol SNR, the BEP can be related to the SEP of (8.84) by the simple approximation of (8.7).

An alternative (simpler) form for the average SEP of M-DPSK has recently been found by Pawula [18] and is given by

$$P_s(E) = \frac{1}{\pi} \int_0^{(M-1)\pi/M} \exp\left(-\frac{E_s}{N_0}\frac{g_{\text{PSK}}}{1 + \sqrt{1 - g_{\text{PSK}}}\cos\theta}\right) d\theta \tag{8.90}$$

which, using simple trigonometric identities and the relation for g_{PSK} given previously, can be written as

$$P_s(E) = \frac{1}{2\pi} \int_0^{(M-1)\pi/M} \exp\left[-\frac{E_s}{N_0}\frac{\sin^2(\pi/M)}{\sin^2\theta + \sin^2(\theta + \pi/M)}\right] d\theta \tag{8.91}$$

For large M, the $\sin^2(\theta + \pi/M)$ term can be replaced by $\sin^2\theta$, which, further ignoring the factor of $\frac{1}{2}$ in front of the integral, results in the approximate relation [19]

$$P_s(E) \simeq \frac{1}{\pi} \int_0^{(M-1)\pi/M} \exp\left(-\frac{E_s}{2N_0}\frac{\sin^2(\pi/M)}{\sin^2\theta}\right) d\theta$$

$$= \frac{1}{\pi} \int_0^{(M-1)\pi/M} \exp\left(-\frac{E_s}{2N_0}\frac{g_{\text{PSK}}}{\sin^2\theta}\right) d\theta \tag{8.92}$$

Comparing (8.92) with (8.22), we immediately observe the well-known fact that for large M, M-PSK is 3 dB better than M-DPSK.

Another advantage of the form in (8.90), in contrast with that of (8.76), is that it lends itself nicely to obtaining a simple upper bound as was done for coherent M-PSK. In particular, the function $f(\theta) = 1/(1 + \sqrt{1 - g_{\text{PSK}}}\cos\theta)$ is monotonically increasing over the entire interval of the integration and thus can be lower bounded by its value at $\theta = 0$ resulting in $1/(1 + \sqrt{1 - g_{PSK}}) \leq f(\theta)$. Using this result in the integrand of (8.90) results in the simple (no integration) upper bound on average SEP:

$$\begin{aligned} P_s(E) &\leq \frac{M-1}{M} \exp\left(-\frac{E_s}{N_0} \frac{g_{PSK}}{1 + \sqrt{1 - g_{PSK}}}\right) \\ &= \frac{M-1}{M} \exp\left[-\frac{E_s}{N_0} \frac{\sin^2(\pi/M)}{1 + \sqrt{1 - \sin^2(\pi/M)}}\right] \\ &= \frac{M-1}{M} \exp\left[-\frac{E_s}{N_0}\left(1 - \cos\frac{\pi}{M}\right)\right] \\ &= \frac{M-1}{M} \exp\left(-\frac{2E_s}{N_0} \sin^2\frac{\pi}{2M}\right) \end{aligned} \quad (8.93)$$

Note the similarity of (8.93) with (8.24). Based on these bounds, one would conclude that for coherent M-PSK and M-DPSK to achieve the "same" SEP, the symbol SNRs should be related by

$$\left(\frac{E_s}{N_0}\right)_{M\text{-DPSK}} = \frac{\sin^2(\pi/M)}{2\sin^2(\pi/2M)} \left(\frac{E_s}{N_0}\right)_{M\text{-PSK}} \quad (8.94)$$

For $M = 2$, (8.93) gives the exact BEP performance of DPSK in agreement with (8.85). For $M > 2$, Pawula [20, Eq. (3)] had previously found an upper bound on this performance given by

$$P_s(E) \leq 2.06 \sqrt{\frac{1 + \cos(\pi/M)}{2\cos(\pi/M)}} Q\left(\sqrt{\frac{2E_s}{N_0}\left(1 - \cos\frac{\pi}{M}\right)}\right) \quad (8.95)$$

which applying the Chernoff bound to the Gaussian Q-function results in

$$\begin{aligned} P_s(E) &\leq 1.03 \sqrt{\frac{1 + \cos(\pi/M)}{2\cos(\pi/M)}} \exp\left[-\frac{E_s}{N_0}\left(1 - \cos\frac{\pi}{M}\right)\right] \\ &= 1.03 \sqrt{\frac{1 + \cos(\pi/M)}{2\cos(\pi/M)}} \exp\left(-\frac{2E_s}{N_0}\sin^2\frac{\pi}{2M}\right) \end{aligned} \quad (8.96)$$

Figure 8.2 illustrates a comparison of the exact evaluation of $P_s(E)$ from (8.84) or (8.91) with upper bounds obtained from (8.93), (8.95), and (8.96). As can be observed, the two exponential bounds [i.e., (8.93) and (8.96)] are reasonably tight at high SNR, whereas the Q-function bound of (8.95) is virtually a perfect match to the exact result over the entire range of SNR's illustrated.

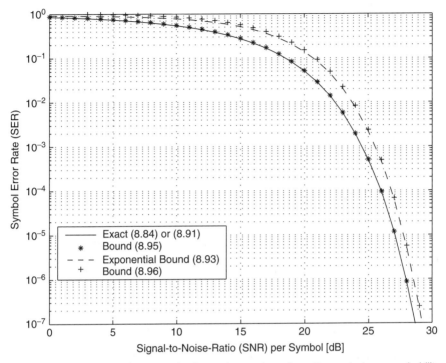

Figure 8.2. Comparison of exact evaluation and upper bounds on the symbol error probability of coherent 16-DPSK.

Multiple-Symbol Detection. In Section 3.5.1.2 we discussed the notion of multiple-symbol differential detection of M-PSK and developed the associated decision variables and optimum receiver (see Fig. 3.18 for a three-symbol observation, i.e., $N_s = 3$). The error probability performance of this receiver was first reported by Divsalar and Simon [21] and later included by Simon et al. [5, Sec. 7.2]. Since for differential detection a block of N_s symbols (phases) is observed in making a decision on $N_s - 1$ information symbols, then following the procedure developed for partially coherent detection, an upper bound on average BEP can be obtained analogous to (8.74), namely,

$$P_b(E) \leq \frac{1}{(N_s - 1)\log_2 M} \sum_{\beta \neq \hat{\beta}} w(\mathbf{u}, \hat{\mathbf{u}}) \Pr\{\hat{z}_{n\mathbf{k}} > z_{n\mathbf{k}}\} \quad (8.97)$$

where now $(N_s - 1)\log_2 M$ represents the number of bits corresponding to the information symbol sequence, β and $\hat{\beta}$ now refer to the correct and incorrect sequences associated with the information (prior to differential encoding) phases, and $\Pr\{\hat{z}_{n\mathbf{k}} > z_{n\mathbf{k}}\}$ is determined from the decision variables in (3.53) in the form of (8.80), now with

$$\left\{ \begin{matrix} b \\ a \end{matrix} \right\} = \frac{E_b \log_2 M}{2N_0} \left(N_s \pm \sqrt{N_s^2 - |\delta|^2} \right) \tag{8.98}$$

and δ now defined analogous (because of the differential encoding) to (8.82) by

$$\delta \triangleq \sum_{i=0}^{N_s-1} \exp \left[j \sum_{m=0}^{N_s-i-2} (\beta_{k_{i-m}} - \hat{\beta}_{k_{i-m}}) \right] \tag{8.99}$$

8.1.5.2 $\pi/4$-Differential QPSK. As discussed in Section 3.5.2, the only conceptual difference between $\pi/4$-DQPSK and conventional DQPSK is that the set of phases $\{\Delta\beta_k\}$ used to represent the information phases $\{\Delta\theta_n\}$ is $\Delta\beta_k = (2k-1)\pi/4, k = 1, 2, 3, 4$, for the former and $\Delta\beta_k = k\pi/4, k = 0, 1, 2, 3$, for the latter. Since the performance of the M-DPSK receiver of Fig. 3.16 is independent of the choice of the information symbol set, we can conclude immediately that $\pi/4$-DQPSK has an identical behavior to DQPSK on the ideal linear AWGN channel and hence is characterized by (8.84) and (8.86) with $M = 4$.

8.1.6 Generic Results for Binary Signaling

Although specific results for the BEP of binary signals transmitted over the AWGN have been given in previous sections, an interesting unification of some of these results into a single BEP expression is possible as discussed in Ref. 22. In particular, Wojnar [22] cites a result privately communicated to him by Lindner (see footnote 2 of Ref. 22), which states that the BEP of coherent, differentially coherent, and noncoherent detection of binary signals transmitted over the AWGN is given by the generic expression [see also (4.44)]

$$P_b(E) = \frac{\Gamma(b, a(E_b/N_0))}{2\Gamma(b)} = \frac{1}{2} Q_b \left(0, \sqrt{2a \frac{E_b}{N_0}} \right) \tag{8.100}$$

where $\Gamma(\bullet, \bullet)$ is the complementary incomplete gamma function [23, Eq. (8.350.2)], which for convenience is provided here as

$$\Gamma(\alpha, x) \triangleq \int_x^\infty e^{-t} t^{\alpha-1} \, dt \tag{8.101}$$

The parameters a and b depend on the particular form of modulation and detection and are presented in Table 8.1. We have also indicated in this table the specific equations to which (8.100) reduces in each instance. Although the result in (8.100) does not provide any new results relative to those indicated in Table 8.1, it does offer a nice unification of five different BEP expressions into a single one that can easily be programmed using standard mathematical software packages such

TABLE 8.1 Parameters *a* and *b* for Various Modulation/Detection Combinations

a \ *b*	$\frac{1}{2}$	1
$\frac{1}{2}$	Orthogonal coherent BFSK [Eq. (8.42)]	Orthogonal noncoherent BFSK [Eq. (8.68)]
1	Antipodal coherent BPSK [Eq. (8.18)]	Antipodal differentially coherent BPSK (DPSK) [Eq. (8.85)]
$0 \leq g \leq 1$	Correlated coherent binary signaling [Chapter 8, footnote 4]	—

as Mathematica. Furthermore, when evaluating the average BEP performance of these very same binary communication systems over the generalized fading channel, the form in (8.100) will also be helpful in unifying these results. This is discussed in Section 8.2 making use of the special integrals given in Section 5.3.

8.2 PERFORMANCE OVER FADING CHANNELS

In this section, we apply the special integrals evaluated in Chapter 5 to the AWGN error probability results presented in Section 8.1 to determine the performance of these same communication systems over generalized fading channels. Wherever possible, we shall again make use of the desired forms rather than the classical representations of the mathematical functions introduced in Chapter 4. By comparison with the level of detail presented in Section 8.1, the treatment here will be quite brief since indeed the entire machinery that allows determining the desired results has by this time been developed completely. Thus, for the most part we shall merely present the final results except for the few situations where further development is warranted.

When fading is present, the received carrier amplitude, A_c, is attenuated by the fading amplitude, α, which is a random variable (RV) with mean-square value $\overline{\alpha^2} = \Omega$ and probability density function (PDF) dependent on the nature of the fading channel. Equivalently, the received instantaneous signal power is attenuated by α^2, and thus it is appropriate to define the instantaneous SNR per bit by $\gamma \triangleq \alpha^2 E_b/N_0$ and the average SNR per bit by $\overline{\gamma} \triangleq \overline{\alpha^2} E_b/N_0 = \Omega E_b/N_0$. As such, conditioned on the fading, the BEP of any of the modulations considered in Section 8.1 is obtained by replacing E_b/N_0 by γ in the expression for AWGN performance. Denoting this conditional BEP by $P_b(E; \gamma)$, the average BEP in the presence of fading is obtained from

$$P_b(E) = \int_0^\infty P_b(E; \gamma) p_\gamma(\gamma) d\gamma \qquad (8.102)$$

where $p_\gamma(\gamma)$ is the PDF of the instantaneous SNR. On the other hand, if one is interested in the average SEP, the same relation as (8.102) applies using, instead, the conditional SEP in the integrand, which is obtained from the AWGN result with E_s/N_0 replaced by $\gamma \log_2 M$. Our goal in the remainder of this chapter is to evaluate (8.102) for the various modulation/detection schemes considered in Section 8.1 and the various fading channel models characterized in previous chapters. Because of the multitude of different signal–channel combinations, however, we shall only give explicit results for one or two of the fading channel models and then indicate how to obtain the rest of the results.

8.2.1 Ideal Coherent Detection

In this section we evaluate the average BEP of the various modulations considered in Section 8.1.1 when transmitted over the generalized fading channel and detected with an ideal phase coherent reference signal. The results will be obtained by applying the integrals presented in Section 5.1.1 to the appropriate expressions for BEP over the AWGN with the above-mentioned replacement of E_b/N_0 by γ.

8.2.1.1 Multiple Amplitude-Shift-Keying or Multiple Amplitude Modulation.
For M-AM the SEP over the AWGN channel is given by (8.3). To obtain the average SEP of M-AM over a Rayleigh fading channel, one first obtains the conditional SEP by replacing E_s/N_0 with $\gamma \log_2 M$ in (8.3) and then evaluates (8.102) for the Rayleigh PDF of (5.4). This type of evaluation was carried out in Chapter 5, in particular, comparing (8.102) with (5.1) and making use of (5.6), we obtain

$$P_s(E) = \left(\frac{M-1}{M}\right)\left(1 - \sqrt{\frac{3\overline{\gamma}_s}{M^2 - 1 + 3\overline{\gamma}_s}}\right) \qquad (8.103)$$

where $\overline{\gamma}_s \triangleq \overline{\gamma} \log_2 M$ denotes the average SNR per symbol. For the binary case, (8.103) becomes

$$P_b(E) = \frac{1}{2}\left(1 - \sqrt{\frac{\overline{\gamma}}{1+\overline{\gamma}}}\right) \qquad (8.104)$$

To obtain the remainder of the results for average SEP, one finds the particular integral in Section 5.1 corresponding to (5.1) for the fading channel of interest, multiplies it by $2(M-1)/M$ and substitutes $(6\log_2 M)/(M^2 - 1)$ for a^2. For example, for Nakagami-m fading, the appropriate integrals to use are (5.18a) and (5.18b). Thus, the average SEP of M-AM over a Nakagami-m fading channel is given by

$$P_s(E) = \left(\frac{M}{M-1}\right)\left[1 - \mu \sum_{k=0}^{m-1} \binom{2k}{k}\left(\frac{1-\mu^2}{4}\right)^k\right],$$

$$\mu \triangleq \sqrt{\frac{3\overline{\gamma}_s}{m(M^2-1) + 3\overline{\gamma}_s}}, \quad m \text{ integer} \qquad (8.105a)$$

which clearly reduces to (8.103) for $m = 1$ and

$$P_s(E) = \left(\frac{M-1}{M}\right) \frac{1}{\sqrt{\pi}} \frac{\sqrt{3\bar{\gamma}_s/m(M^2-1)}}{[(m(M^2-1)+3\bar{\gamma}_s)/m(M^2-1)]^{m+1/2}} \frac{\Gamma(m+\frac{1}{2})}{\Gamma(m+1)}$$
$$\times {}_2F_1\left(1, m+\frac{1}{2}; m+1; \frac{m(M^2-1)}{m(M^2-1)+3\bar{\gamma}_s}\right), \quad m \text{ noninteger}$$
(8.105b)

It is tempting to try evaluating the average BEP over the fading channel by using the asymptotic (large SNR) relation between the AWGN BEP and SEP as given in (8.7) to determine the conditional BEP needed in (8.102). Unfortunately, this procedure is inappropriate since, as mentioned earlier in the chapter, on the fading channel the symbol SNR of the AWGN SEP gets replaced by $\log_2 M$ times the instantaneous SNR per bit, γ, which is a RV varying between zero and infinity. Rather, one needs to compute the exact relation between AWGN BEP and SEP, substitute $\gamma \log_2 M$ for E_s/N_0, and then average over the PDF of γ. As mentioned in Section 8.1.1.1, this relation (i.e., the conditional BEP on γ) can be computed for any given M and a Gray code bit-to-symbol mapping.

8.2.1.2 Quadrature Amplitude-Shift-Keying or Quadrature Amplitude Modulation.
For QAM, the SEP over the AWGN channel is given by (8.10). To obtain the average SEP of M-AM over a Rayleigh fading channel, one proceeds as for the M-AM case by first obtaining the conditional SEP [i.e., replacing E_s/N_0 with $\gamma \log_2 M$ in (8.10)] and then evaluating an integral such as (8.102) for the Rayleigh PDF of (5.4). This type of evaluation involves two integrals that were developed in Chapter 5. In particular, comparing the two terms (8.102) with (5.1) and (5.28) and making use of (5.6) and (5.29), we obtain

$$P_s(E) = 2\left(\frac{\sqrt{M}-1}{\sqrt{M}}\right)\left(1 - \sqrt{\frac{1.5\bar{\gamma}_s}{M-1+1.5\bar{\gamma}_s}}\right)$$
$$- \left(\frac{\sqrt{M}-1}{\sqrt{M}}\right)^2 \left[1 - \sqrt{\frac{1.5\bar{\gamma}_s}{M-1+1.5\bar{\gamma}_s}}\left(\frac{4}{\pi}\tan^{-1}\sqrt{\frac{M-1+1.5\bar{\gamma}_s}{1.5\bar{\gamma}_s}}\right)\right]$$
(8.106)

which for 4-QAM reduces to

$$P_s(E) = \left(1 - \sqrt{\frac{\bar{\gamma}}{1+\bar{\gamma}}}\right) - \frac{1}{4}\left[1 - \sqrt{\frac{\bar{\gamma}}{1+\bar{\gamma}}}\left(\frac{4}{\pi}\tan^{-1}\sqrt{\frac{1+\bar{\gamma}}{\bar{\gamma}}}\right)\right] \quad (8.107)$$

To obtain the remainder of the results for average SEP, one finds the particular integrals in Section 5.1 corresponding to (5.1) and (5.28) for the fading channel of interest, multiplies the first by $4(\sqrt{M}-1)/\sqrt{M}$, the second by $4[(\sqrt{M}-1)/\sqrt{M}]^2$, and substitutes $(3\log_2 M)/(M-1)$ for a^2. For example, for Nakagami-m fading with m integer, the appropriate integrals to use are (5.18a) and (5.30). Thus, the average SEP of QAM over a Nakagami-m fading channel is given by

$$P_s(E) = 2\left(\frac{\sqrt{M}}{\sqrt{M}-1}\right)\left[1 - \mu\sum_{k=0}^{m-1}\binom{2k}{k}\left(\frac{1-\mu^2}{4}\right)^k\right]$$

$$-\left(\frac{\sqrt{M}}{\sqrt{M}-1}\right)^2\left(1 - \frac{4}{\pi}\mu\left\{\left(\frac{\pi}{2} - \tan^{-1}\mu\right)\sum_{k=0}^{m-1}\binom{2k}{k}\frac{1}{[4(1+c)]^k}\right.\right.$$

$$\left.\left. - \sin(\tan^{-1}\mu)\sum_{k=1}^{m-1}\sum_{i=1}^{k}\frac{T_{ik}}{(1+c)^k}[\cos(\tan^{-1}\mu)]^{2(k-i)+1}\right\}\right) \quad (8.108)$$

where

$$c = \frac{1.5\bar{\gamma}_s}{m(M-1)}, \qquad \mu \triangleq \sqrt{\frac{c}{1+c}} \quad (8.109)$$

and T_{ik} is defined in (5.32). Figure 8.3 is an illustration of the average SEP of 16-QAM as computed from (8.108) with m as a parameter.

To compute the average BEP performance, again one should not use the approximate asymptotic form of (8.7) but rather, determine either the exact relation between the AWGN BEP and SEP or the exact AWGN BEP directly (see

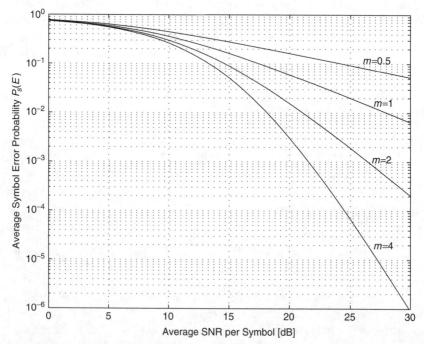

Figure 8.3. Average SEP of 16-QAM over a Nakagami-m channel versus the average SNR per symbol.

footnote 1 of this chapter), substituting $\gamma \log_2 M$ for E_s/N_0, and then average over the PDF of γ. Instead, one can use the approximate BEP expression obtained by Lu et al. [8] for the AWGN as in (8.14), which is accurate for a wide range of SNR's, again making the substitution $\gamma \log_2 M$ for E_s/N_0 followed by averaging over the PDF of γ. Using the alternative form of the Gaussian Q-function of (4.2), it is straightforward to show that the result of this evaluation is given by

$$P_b(E) \simeq 4 \left(\frac{\sqrt{M}-1}{\sqrt{M}} \right) \frac{1}{\log_2 M} \sum_{i=1}^{\sqrt{M}/2} \frac{1}{\pi} \int_0^{\pi/2} M_\gamma \left(-\frac{(2i-1)^2}{2\sin^2\theta} \frac{3E_b \log_2 M}{N_0(M-1)} \right) d\theta$$
(8.110)

where $M_\gamma(s)$ is again the MGF of the instantaneous fading power γ. For example, for a Rayleigh fading channel, we obtain, analogous to (8.106),

$$P_b(E) \simeq 2 \left(\frac{\sqrt{M}-1}{\sqrt{M}} \right) \frac{1}{\log_2 M} \sum_{i=1}^{\sqrt{M}/2} \left(1 - \sqrt{\frac{1.5(2i-1)^2 \bar{\gamma} \log_2 M}{M-1+1.5(2i-1)^2 \bar{\gamma} \log_2 M}} \right)$$
(8.111)

8.2.1.3 M-ary Phase-Shift-Keying.
For M-PSK, the classical form of the SEP over the AWGN channel is given by (8.17), and the desired form is given by (8.22). To obtain the average SEP of M-PSK over a Rayleigh fading channel, one first obtains the conditional SEP by replacing E_s/N_0 with $\gamma \log_2 M$ in (8.22) and then evaluates (8.102) for the Rayleigh PDF of (5.4). In particular, comparing (8.102) with (5.66) and making use of (5.68), we obtain

$$P_s(E) = \left(\frac{M-1}{M} \right) \left\{ 1 - \sqrt{\frac{g_{\text{PSK}} \bar{\gamma}_s}{1 + g_{\text{PSK}} \bar{\gamma}_s}} \frac{M}{(M-1)\pi} \right.$$
$$\left. \times \left[\frac{\pi}{2} + \tan^{-1} \left(\sqrt{\frac{g_{\text{PSK}} \bar{\gamma}_s}{1 + g_{\text{PSK}} \bar{\gamma}_s}} \cot \frac{\pi}{M} \right) \right] \right\}$$
(8.112)

where $g_{\text{PSK}} \triangleq \sin^2(\pi/M)$. For $M = 2$, (8.112) reduces to (8.104) since binary PSK and binary AM are identical.

For Rician fading, the average SEP is obtained from (5.67) together with (5.11), or equivalently, (5.13) [with the upper limit changed from $\pi/2$ to $(M-1)\pi/M$] with $a^2 = 2g_{\text{PSK}}$ and $\bar{\gamma}_s$ substituted for $\bar{\gamma}$, resulting in

$$P_s(E) = \frac{1}{\pi} \int_0^{(M-1)\pi/M} \frac{(1+K)\sin^2\theta}{(1+K)\sin^2\theta + g_{\text{PSK}} \bar{\gamma}_s}$$
$$\times \exp\left(-\frac{K g_{\text{PSK}} \bar{\gamma}_s}{(1+K)\sin^2\theta + g_{\text{PSK}} \bar{\gamma}_s} \right) d\theta$$
(8.113)

An equivalent result was reported by Sun and Reed [24, Eq. (11)].[7]

---

[7] It should be noted that an error occurs in Eqs. (10), (11), and (12) Ref. 24 in that the upper limit of their integrals should be $\pi/2 - \pi/M$ rather than $\pi/2$.

For Nakagami-m fading with m integer, the average SEP is obtained from (5.69) with the same substitutions for a^2 and $\bar{\gamma}$, resulting in the closed-form solution

$$P_s(E) = \frac{M-1}{M} - \frac{1}{\pi}\sqrt{\frac{(g_{PSK}\bar{\gamma}_s)/m}{1+(g_{PSK}\bar{\gamma}_s)/m}}$$

$$\times \left\{ \left(\frac{\pi}{2} + \tan^{-1}\alpha\right) \sum_{k=0}^{m-1} \binom{2k}{k} \frac{1}{[4(1+(g_{PSK}\bar{\gamma}_s)/m)]^k} \right.$$

$$\left. + \sin(\tan^{-1}\alpha) \sum_{k=1}^{m-1} \sum_{i=1}^{k} \frac{T_{ik}}{[1+(g_{PSK}\bar{\gamma}_s)/m]^k} [\cos(\tan^{-1}\alpha)]^{2(k-i)+1} \right\}$$

(8.114)

where, from (5.70),

$$\alpha \triangleq \sqrt{\frac{(g_{PSK}\bar{\gamma}_s)/m}{1+(g_{PSK}\bar{\gamma}_s)/m}} \cot\frac{\pi}{M} \quad (8.115)$$

and again T_{ik} is defined in (5.32). Figure 8.4 is an illustration of the average SEP as computed from (8.114) with m as a parameter.

Exact results for average BEP of 4-PSK, 8-PSK, and 16-PSK over Rayleigh fading channels can be obtained by averaging (8.30) over the fading PDF in (5.4). In particular, using a generalization of (5A.15) when the upper limit of the integral is $\pi[1 - (2k \pm 1)/M]$, we obtain

$$\bar{P}_k \triangleq \int_0^\infty P_k p_\gamma(\gamma)\, d\gamma = K_+ - K_-, \quad k = 0, 1, 2, \ldots, M-1 \quad (8.116)$$

where

$$K_\pm = \frac{1}{2}\left(\frac{2k \pm 1}{M}\right)\left[1 - \sqrt{\frac{g_{PSK}\bar{\gamma}_s}{1+g_{PSK}\bar{\gamma}_s}} \frac{M}{(2k \pm 1)\pi} \right.$$

$$\left. \times \tan^{-1}\left(\sqrt{\frac{1+g_{PSK}\bar{\gamma}_s}{g_{PSK}\bar{\gamma}_s}} \tan\frac{(2k \pm 1)\pi}{M}\right)\right] \quad (8.117)$$

Using $\bar{P}_k$ of (8.116) for P_k in (8.30) gives the desired results for $M = 4, 8$, and 16.

Similarly for Nakagami-m fading, $\bar{P}_k$ can be computed from (5A.22) as

$$\bar{P}_k = I_m\left(\frac{(2k+1)\pi}{M}, \frac{(2k-1)\pi}{M}; \bar{\gamma}_s\right), \quad k = 0, 1, 2, \ldots, M-1 \quad (8.118)$$

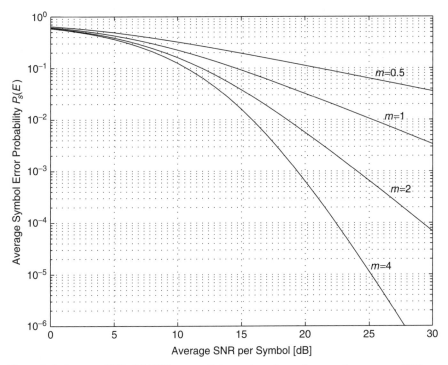

Figure 8.4. Average SEP of 8-PSK over a Nakagami-*m* channel versus the average SNR per symbol.

which again should be used in place of P_k in (8.30) to obtain average BEP.

For other values of M, one can again use the approximate AWGN result of Lu et al. [8] as given in (8.31), substituting $\gamma \log_2 M$ for E_s/N_0, followed by averaging over the PDF of γ. Using the alternative form of the Gaussian Q-function of (4.2), the end result of this evaluation is

$$P_b(E) \simeq \frac{2}{\max(\log_2 M, 2)} \sum_{i=1}^{\max(M/4,1)} \frac{1}{\pi}$$
$$\times \int_0^{\pi/2} M_\gamma \left(-\frac{1}{\sin^2 \theta} \frac{E_b \log_2 M}{N_0} \sin^2 \frac{(2i-1)\pi}{M} \right) d\theta \quad (8.119)$$

Specific results for the variety of fading channels being considered are easily worked out using the results of Chapter 5 and are left as exercises for the reader.

8.2.1.4 Differentially Encoded M-ary Phase-Shift-Keying and π/4-QPSK.

Consider first the case of differentially encoded QPSK for which the classical form of the SEP over the AWGN channel is given by (8.38). As pointed out in Section 8.1.1.4, the first two terms of (8.38) can be put in the desired form, but such a form is not available for the third and fourth terms. Nevertheless, using the results from Section 5.4.3, for Rayleigh and Nakagami fading, we are able to evaluate these terms in the form of a single integral with finite limits and an integrand composed of elementary functions; thus, we can obtain a solution for the average SEP in a similar form for these channels.

For the simpler Rayleigh case, making use of (5.6), (5.80), (5.82), and (5.84) with $a = 1$ and $\bar{\gamma}$ replaced by $\bar{\gamma}_s$, we obtain

$$P_s(E) = 4I_1 - 8I_2 + 8I_3 - 4I_4 \tag{8.120}$$

where

$$I_1 = \frac{1}{2}\left(1 - \sqrt{\frac{\bar{\gamma}_s/2}{1+\bar{\gamma}_s/2}}\right),$$

$$I_2 = \frac{1}{4}\left[1 - \sqrt{\frac{\bar{\gamma}_s/2}{1+\bar{\gamma}_s/2}}\left(\frac{4}{\pi}\tan^{-1}\sqrt{\frac{1+\bar{\gamma}_s/2}{\bar{\gamma}_s/2}}\right)\right],$$

$$I_3 = \frac{1}{\pi\bar{\gamma}_s}\int_0^{\pi/4} c(\phi)\left(1 - \sqrt{\frac{c(\phi)}{1+c(\phi)}}\right)d\phi$$

$$I_4 = \frac{1}{2\pi\bar{\gamma}_s}\int_0^{\pi/4}\left[1 - \sqrt{\frac{c(\phi)}{1+c(\phi)}}\left(\frac{4}{\pi}\tan^{-1}\sqrt{\frac{1+c(\phi)}{c(\phi)}}\right)\right]d\phi,$$

$$c(\phi) \triangleq \frac{\bar{\gamma}_s}{2}\left(\frac{\sin^2\phi}{\sin^2\phi + \bar{\gamma}_s/2}\right) \tag{8.121}$$

For Nakagami-m fading with m integer, the average SEP can similarly be obtained from (8.120) using (5.18a), (5.86), (5.88), and (5.91), again with $a = 1$ and $\bar{\gamma}$ replaced by $\bar{\gamma}_s$. Specifically, the I_k's needed in (8.120) are now given by

$$I_1 = \frac{1}{2}\left[1 - \mu\left(\frac{\bar{\gamma}_s}{2m}\right)\sum_{k=0}^{m-1}\binom{2k}{k}\left(\frac{1-\mu^2(\bar{\gamma}_s/2m)}{4}\right)^k\right],$$

$$\mu\left(\frac{\bar{\gamma}_s}{2m}\right) \triangleq \sqrt{\frac{\bar{\gamma}_s/2}{m+\bar{\gamma}_s/2}}$$

$$I_2 = \frac{1}{4} - \frac{1}{\pi}\sqrt{\frac{\bar{\gamma}_s/2}{1+\bar{\gamma}_s/2}}\left\{\left(\frac{\pi}{2} - \tan^{-1}\sqrt{\frac{\bar{\gamma}_s/2}{1+\bar{\gamma}_s/2}}\right)\sum_{k=0}^{m-1}\binom{2k}{k}\frac{1}{[4(1+\bar{\gamma}_s/2)]^k}\right.$$

$$- \sin\left(\tan^{-1}\sqrt{\frac{\bar{\gamma}_s/2}{1+\bar{\gamma}_s/2}}\right)\sum_{k=1}^{m-1}\sum_{i=1}^{k}\frac{T_{ik}}{(1+\bar{\gamma}_s/2)^k}$$

$$\times\left.\left[\cos\left(\tan^{-1}\sqrt{\frac{\bar{\gamma}_s/2}{1+\bar{\gamma}_s/2}}\right)\right]^{2(k-i)+1}\right\}$$

$$I_3 = \frac{1}{\pi}\int_0^{\pi/4}\left(\frac{2}{\bar{\gamma}_s}c(\phi)\right)^m\left(\frac{1-\mu(c(\phi))}{2}\right)^m$$

$$\times\sum_{k=0}^{m-1}\binom{m-1+k}{k}\left(\frac{1+\mu(c(\phi))}{2}\right)^k d\phi$$

$$I_4 = \frac{1}{\pi}\int_0^{\pi/4}\left(\frac{2}{\bar{\gamma}_s}c(\phi)\right)^m\left[\frac{1}{4} - \frac{1}{\pi}\sqrt{\frac{c(\phi)}{1+c(\phi)}}\left\{\left(\frac{\pi}{2} - \tan^{-1}\sqrt{\frac{c(\phi)}{1+c(\phi)}}\right)\right.\right.$$

$$\times\sum_{k=0}^{m-1}\binom{2k}{k}\frac{1}{[4(1+c(\phi))]^k} - \sin\left(\tan^{-1}\sqrt{\frac{c(\phi)}{1+c(\phi)}}\right)\sum_{k=1}^{m-1}\sum_{i=1}^{k}\frac{T_{ik}}{[1+c(\phi)]^k}$$

$$\times\left.\left.\left[\cos\left(\tan^{-1}\sqrt{\frac{c(\phi)}{1+c(\phi)}}\right)\right]^{2(k-i)+1}\right\}\right]d\phi \tag{8.122}$$

and $c(\phi)$ is still as defined in (8.121).

For the more general case of differentially encoded M-PSK, we need to evaluate the average of (8.35) over the fading PDF. Here we can only obtain the result in the simple desired form for Rayleigh fading. The average of the first term of (8.35) is given by (8.112) multiplied by 2, that is,

$$\int_0^\infty 2P_s(E)|_{M\text{-PSK}}\, p_\gamma(\gamma)d\gamma = 2\left(\frac{M-1}{M}\right)\left\{1 - \sqrt{\frac{g_{\text{PSK}}\bar{\gamma}_s}{1+g_{\text{PSK}}\bar{\gamma}_s}}\frac{M}{(M-1)\pi}\right.$$

$$\left.\times\left[\frac{\pi}{2} + \tan^{-1}\left(\sqrt{\frac{g_{\text{PSK}}\bar{\gamma}_s}{1+g_{\text{PSK}}\bar{\gamma}_s}}\cot\frac{\pi}{M}\right)\right]\right\} \tag{8.123}$$

The corresponding average of the second term is obtained from (5.99) with $a^2 = 2g_{PSK} = 2\sin^2 \pi/M$ and $\bar{\gamma}$ replaced by $\bar{\gamma}_s$, that is,

$$\int_0^\infty (P_s(E)|_{M\text{-PSK}})^2 p_\gamma(\gamma) d\gamma = \left(\frac{1}{\pi}\right)^2 \left(\frac{1}{g_{PSK}\bar{\gamma}_s}\right)$$

$$\times \int_0^{(M-1)\pi/M} c(\phi) \left[\frac{(M-1)\pi}{M} - \sqrt{\frac{c(\phi)}{1+c(\phi)}}\right.$$

$$\left.\times \tan^{-1}\left(\sqrt{\frac{1+c(\phi)}{c(\phi)}} \tan \frac{(M-1)\pi}{M}\right)\right] d\phi \quad (8.124)$$

where now

$$c(\phi) \triangleq g_{PSK}\bar{\gamma} \left(\frac{\sin^2 \phi}{\sin^2 \phi + g_{PSK}\bar{\gamma}}\right) \quad (8.125)$$

For the average of the third term, we must first square P_k of (8.29) and then make use of (5.100) through (5.102) with $a_\pm^2 \triangleq 2\sin^2(2k \pm 1)\pi/M$ and $\theta_\pm \triangleq \pi[1 - (2k \pm 1)/M]$ for $k = 0, 1, 2, \ldots, M-1$. The result is

$$\int_0^\infty P_k^2 p_\gamma(\gamma) d\gamma = L_+ + L_- - 2L_{+-}, \quad k = 0, 1, 2, \ldots, M-1 \quad (8.126)$$

where

$$L_\pm = \left(\frac{1}{2\pi}\right)^2 \left(\frac{1}{(\sin^2(2k \pm 1)\pi/M)\bar{\gamma}_s}\right) \int_0^{\pi(1-(2k\pm 1)/M)} c_\pm(\phi)$$

$$\times \left(\pi\left(1 - \frac{2k \pm 1}{M}\right) - \sqrt{\frac{c_\pm(\phi)}{1+c_\pm(\phi)}}\right.$$

$$\left.\times \tan^{-1}\left\{\sqrt{\frac{1+c_\pm(\phi)}{c_\pm(\phi)}} \tan\left[\pi\left(1 - \frac{2k \pm 1}{M}\right)\right]\right\}\right) d\phi \quad (8.127)$$

and

$$L_{+-} = \left(\frac{1}{2\pi}\right)^2 \left(\frac{1}{(\sin^2(2k+1)\pi/M)\bar{\gamma}_s}\right) \int_0^{\pi(1-(2k+1)/M)} c_{+-}(\phi)$$

$$\times \left(\pi\left(1 - \frac{2k - 1}{M}\right) - \sqrt{\frac{c_{+-}(\phi)}{1+c_{+-}(\phi)}}\right.$$

$$\left.\times \tan^{-1}\left\{\sqrt{\frac{1+c_{+-}(\phi)}{c_{+-}(\phi)}} \tan\left[\pi\left(1 - \frac{2k - 1}{M}\right)\right]\right\}\right) d\phi \quad (8.128)$$

with

$$c_{\pm}(\phi) \stackrel{\Delta}{=} (\sin^2(2k \pm 1)\pi/M)\bar{\gamma}_s \left(\frac{\sin^2 \phi}{\sin^2 \phi + (\sin^2(2k \pm 1)\pi/M)\bar{\gamma}_s} \right)$$

$$c_{+-}(\phi) \stackrel{\Delta}{=} (\sin^2(2k + 1)\pi/M)\bar{\gamma}_s \left(\frac{\sin^2 \phi}{\sin^2 \phi + (\sin^2(2k - 1)\pi/M)\bar{\gamma}_s} \right) \quad (8.129)$$

Finally, since as pointed out in Section 8.1.1.4, the performance of coherently detected $\pi/4$-QPSK transmitted over a linear AWGN channel is identical to that of differentially encoded QPSK, the same conclusion can be made for the fading channel. Hence, the SEP performance of coherently detected $\pi/4$-QPSK over the Rayleigh and Nakagami-m fading channels is also given by (8.120), together with (8.121) or (8.122), respectively.

8.2.1.5 Offset QPSK or Staggered QPSK. In Section 8.1.1.5 it was concluded that because of the similarity between conventional and offset QPSK receivers and the fact that time offset of the I and Q channels has no effect on the decisions made on the I and Q data bits, the BEP performances of these two modulation techniques on a linear AWGN channel with ideal coherent detection are identical. Thus, without further ado, we conclude that the same is true on the fading channel, and hence the error probability performance results of Sections 8.1.2.3 and 8.1.2.4 apply.

8.2.1.6 M-ary Frequency-Shift-Keying. In Section 8.1.1.6 we observed that the expression [see (8.40)] for the average SEP of orthogonal M-FSK involves the $(M - 1)$st power of the Gaussian Q-function. Since for M arbitrary an alternative form [analogous to (4.2)] is not available for $Q^{M-1}(x)$, (8.40) cannot be put in the desired form to allow simple evaluation of the average SEP on the generalized fading channel.[8] Despite this consequence, however, it is nevertheless possible to obtain simple-to-evaluate, asymptotically tight upper bounds on the average error probability performance of 4-ary FSK on the Rayleigh and Nakagami-m fading channels, as we shall show shortly. For the special case of binary FSK ($M = 2$), we can use the desired form in (8.43) (for orthogonal signals) or (8.44) (for nonorthogonal signals) to allow simple exact evaluation of average BEP on the generalized fading channel. Before moving on to the more difficult 4-ary FSK case, we first quickly dispense with the results for binary FSK since these follow immediately from the integrals developed in Chapter 5 or equivalently from the results obtained previously for binary AM and

---

[8] At the time this book was about to go to press, the authors learned of new, as yet unpublished work by Dong and Beaulieu [25] that using an M-dimensional extension of Craig's approach [10] obtains exact closed-form results for BEP and SEP of 3- and 4-ary orthogonal signaling in slow Rayleigh fading. Also shown in Ref. 25 is the fact that the results obtained for $M = 4$ can be used as close approximations to the exact results for values of $M > 4$. Finally, the MGF-based approach described in this chapter can also be used to extend this work to the generalized fading channel.

BPSK, replacing $\bar{\gamma}$ by $\bar{\gamma}/2$ for orthogonal BFSK and by $(\bar{\gamma}/2)[1 - (\sin 2\pi h)/2\pi h]$ for nonorthogonal BFSK. For example, for Rayleigh fading the average BEP of orthogonal BFSK is given by

$$P_b(E) = \frac{1}{2}\left(1 - \sqrt{\frac{\bar{\gamma}/2}{1+\bar{\gamma}/2}}\right) \tag{8.130}$$

whereas for Nakagami-m fading the analogous results are

$$P_b(E) = 2\left[1 - \mu \sum_{k=0}^{m-1} \binom{2k}{k}\left(\frac{1-\mu^2}{4}\right)^k\right], \quad \mu \triangleq \sqrt{\frac{\bar{\gamma}/2}{m+\bar{\gamma}/2}}, \quad m \text{ integer} \tag{8.131a}$$

and

$$P_b(E) = \frac{1}{2\sqrt{\pi}} \frac{\sqrt{\bar{\gamma}/2m}}{(1+\bar{\gamma}/2m)^{m+1/2}} \frac{\Gamma\left(m+\frac{1}{2}\right)}{\Gamma(m+1)}$$
$$\times {}_2F_1\left(1, m+\frac{1}{2}; m+1; \frac{1}{1+\bar{\gamma}/2m}\right), \quad m \text{ noninteger} \tag{8.131b}$$

For M-ary orthogonal FSK, the average SEP on the AWGN can be obtained from (8.40) as

$$P_s(E) = 1 - \int_{-\infty}^{\infty} \left[Q\left(-q - \sqrt{\frac{2E_s}{N_0}}\right)\right]^{M-1} \frac{1}{\sqrt{2\pi}} \exp\left(-\frac{q^2}{2}\right) dq$$
$$= \int_{-\infty}^{\infty} \left\{1 - \left[1 - Q\left(q + \sqrt{\frac{2E_s}{N_0}}\right)\right]^{M-1}\right\} \frac{1}{\sqrt{2\pi}} \exp\left(-\frac{q^2}{2}\right) dq$$
$$= \frac{1}{\sqrt{\pi}} \int_{-\infty}^{\infty} \left\{1 - \left[1 - Q\left(\sqrt{2}\left(u + \sqrt{\frac{E_s}{N_0}}\right)\right)\right]^{M-1}\right\} \exp(-u^2) du \tag{8.132}$$

and the corresponding BEP is obtained from (8.132) using (8.41). The most straightforward way of numerically evaluating (8.132) (and therefore the BEP derived from it) is to apply Gauss–Hermite quadrature [26, Eq. (25.4.46)], resulting in

$$P_s(E) \simeq \frac{1}{\sqrt{\pi}} \sum_{n=1}^{N_p} w_n \left\{1 - \left[1 - Q\left(\sqrt{2}\left(x_n + \sqrt{\frac{E_s}{N_0}}\right)\right)\right]^{M-1}\right\} \tag{8.133}$$

where $\{x_n; n = 1, 2, \ldots, N_p\}$ are the zeros of the Hermite polynomial of order N_p and w_n are the associated weight factors [26, Table 25.10]. A value of $N_p = 20$ is typically sufficient for excellent accuracy.

When slow fading is present, the average symbol error probability is obtained from (8.132) or (8.133) by first replacing E_s/N_0 with $\gamma = \alpha^2 E_s/N_0$ and then averaging over the PDF of γ, that is,

$$P_s(E) = \frac{1}{\sqrt{2\pi}} \int_{-\infty}^{\infty} \{1 - [1 - Q(y)]^{M-1}\}$$
$$\times \int_0^{\infty} \exp\left(-\frac{(y - \sqrt{2\gamma})^2}{2}\right) p_\gamma(\gamma) \, d\gamma \, dy \quad (8.134)$$

or approximately

$$P_s(E) \simeq \frac{1}{\sqrt{\pi}} \sum_{n=1}^{N_p} w_n \left\{1 - \int_0^{\infty} [1 - Q(\sqrt{2}(x_n + \sqrt{\gamma}))]^{M-1} p_\gamma(\gamma) \, d\gamma\right\} \quad (8.135)$$

Numerical evaluation of (8.134) and the associated bit error probability using (8.33) for Rayleigh and Nakagami-m fading channels is computationally intensive. Equation (8.135) does yield numerical values; however, its evaluation is very time consuming, especially for large values of m. Thus, tight upper bounds on the result in (8.134) which are simple to use and evaluate numerically are highly desirable.

Using Jensen's inequality [27], Hughes [28] derived a simple bound on the AWGN performance in (8.132). In particular, it was shown that

$$P_s(E) \leq 1 - \left[1 - Q\left(\sqrt{\frac{E_s}{N_0}}\right)\right]^{M-1} \quad (8.136)$$

which is tighter than the more common union upper bound [5, Eq. (4.97)],

$$P_s(E) \leq (M - 1)Q\left(\sqrt{\frac{E_s}{N_0}}\right) \quad (8.137)$$

Evaluation of an upper bound on average error probability for the fading channel by averaging the right-hand side of (8.136) (with E_s/N_0 replaced by γ_s) over the PDF of γ_s and using the conventional form for the Gaussian probability integral as in (4.1) is still computationally intensive. Using the alternative forms of the Gaussian Q-function and its square as in (4.2) and (4.9), respectively, it is possible to simplify the evaluation of this upper bound on performance. The details are as follows.

We begin by applying a binomial expansion to the Hughes bound of (8.136), which when averaged over the fading PDF results in

$$P_s(E) \leq \sum_{k=1}^{M-1} (-1)^{k+1} \binom{M-1}{k} I_k \qquad (8.138)$$

where

$$I_k \triangleq \int_0^\infty Q^k(\sqrt{\gamma_s}) p_{\gamma_s}(\gamma_s)\, d\gamma_s, \qquad k = 1, 2, \ldots, M-1 \qquad (8.139)$$

[Note that the result based on the union upper bound would simply be the first term ($k=1$) of (8.138)]. Using (4.2) and (4.9) and assuming a Nakagami-m channel with instantaneous SNR PDF given by (5.14), the integral in (8.139) can be evaluated for $M = 4$ ($k = 1, 2, 3$) either in closed form or in the form of a single integral with finite limits and an integrand composed of elementary functions (i.e., exponentials and trigonometrics). The results appear in Section 5.4.3.2 and are summarized here as follows:

$$I_1 = [P(c)]^m \sum_{k=0}^{m-1} \binom{m-1+k}{k} [1 - P(c)]^k,$$

$$P(c) \triangleq \frac{1}{2}\left(1 - \sqrt{\frac{c}{1+c}}\right), \qquad c \triangleq \frac{\bar{\gamma}_s}{2m} \qquad (8.140a)$$

$$I_2 = \frac{1}{4} - \frac{1}{\pi}\sqrt{\frac{c}{1+c}}\left\{\left(\frac{\pi}{2} - \tan^{-1}\sqrt{\frac{c}{1+c}}\right)\sum_{k=0}^{m-1}\binom{2k}{k}\frac{1}{[4(1+c)]^k}\right.$$

$$\left. - \sin\left(\tan^{-1}\sqrt{\frac{c}{1+c}}\right)\sum_{k=1}^{m-1}\sum_{i=1}^{k}\frac{T_{ik}}{(1+c)^k}\left[\cos\left(\tan^{-1}\sqrt{\frac{c}{1+c}}\right)\right]^{2(k-i)+1}\right\},$$

$$T_{ik} \triangleq \frac{\binom{2k}{k}}{\binom{2(k-i)}{k-i} 4^i [2(k-i)+1]}, \qquad c \triangleq \frac{\bar{\gamma}_s}{2m} \qquad (8.140b)$$

and

$$I_3 = \frac{1}{\pi}\int_0^{\pi/4}\left(\frac{2}{\bar{\gamma}_s}c(\phi)\right)^m [P(c(\phi))]^m \sum_{k=0}^{m-1}\binom{m-1+k}{k}[1-P(c(\phi))]^k\, d\phi,$$

$$c(\phi) \triangleq \frac{\bar{\gamma}_s}{2}\left(\frac{\sin^2\phi}{\sin^2\phi + \bar{\gamma}_s/2}\right) \qquad (8.140c)$$

PERFORMANCE OVER FADING CHANNELS 233

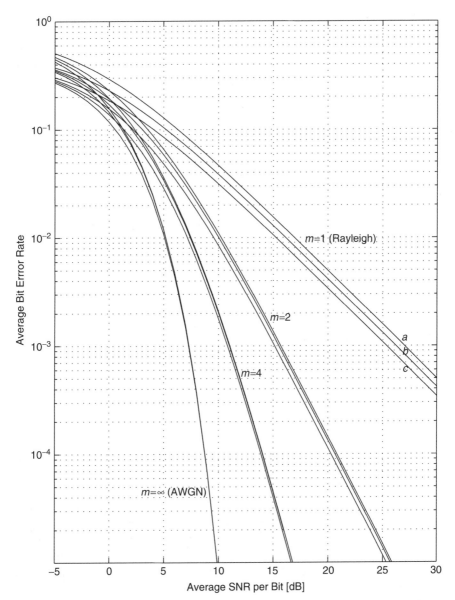

Figure 8.5. Average BEP of 4-ary orthogonal signals over a Nakagami-m channel versus the average SNR per bit: (a) union bound; (b) Hughes bound; (c) exact result.

Illustrated in Fig. 8.5 are curves for average bit error probability versus average bit SNR for 4-ary orthogonal signaling over the Nakagami-m fading channel, the special case of $m = 1$ corresponding to the Rayleigh channel. For each value of m, three curves are calculated. The first is the exact result obtained (with much computational power and time) by averaging (8.135) over the PDF in

(5.14). The second is the Hughes upper bound obtained from (8.138) together with (8.140a), (8.140b), and (8.140c). Finally, the third is the union upper bound obtained from the first term of (8.135) together with (8.140a). The curves labeled $m = \infty$ correspond to the nonfading (AWGN only) results. We observe, not surprisingly, that as m increases (the amount of fading decreases) the three results are asymptotically equal to each other. For Rayleigh fading (the smallest integer value of m) we see the most disparity between the three, with the Hughes bound falling approximately midway between the exact result and the union upper bound. More specifically, the "averaged" Hughes bound is 1 dB tighter than the union bound for high-average bit SNR values. As m increases, the difference between the Hughes bound and the exact results is at worst less than a few tenths of 1 dB over a wide range of average bit SNR's. Hence, for high values of m, we can conclude that it is accurate to use the former as a prediction of true system performance, with the advantage that the numerical results can be obtained instantaneously. Note also that for high values of m a slightly less accurate result can be obtained by using the union bound.

8.2.1.7 Minimum-Shift-Keying.
Following the same line of reasoning as discussed in Section 8.1.1.7 for the AWGN channel, we conclude here for the fading channel that the average BEP performance of the MSK receiver implemented as that which is optimum for half-sinusoidal pulse-shaped OQPSK is identical to that of AM, BPSK, QPSK, and conventional (rectangular pulse-shaped) OQPSK. As a result of this observation, no further discussion is necessary.

8.2.2 Nonideal Coherent Detection

To compute the average error probability performance of nonideal coherent receivers of BPSK, QPSK, OQPSK, and MSK modulations transmitted over a fading channel, we again follow the approach taken by Fitz [16] wherein the randomness of the demodulation reference signal is modeled as an additive Gaussian noise independent of the AWGN associated with the received signal. In the absence of fading, this model was introduced in Section 3.2, and the performance of the receiver based on this model was given in Section 8.1.2. When Rician fading is present, Fitz [16] proposes a suitable modification of the Gaussian noise reference signal model as follows.

Let $\eta_n = \eta_{In} + j\eta_{Qn}$ denote a complex Gaussian RV which represents the fading associated with the received signal in the nth symbol interval. In the most general case, when η_{In} and η_{Qn} are nonzero mean, $\alpha_n = |\eta_n|$ is a Rician RV, which is the case considered by Fitz. With reference to (3.38), the kth matched filter output in this symbol interval $\tilde{y}_{nk}$, $k = 1, 2, \ldots, M$, now becomes

$$\tilde{y}_{nk} = \tilde{s}_k \eta_n e^{j\theta_c} + \tilde{N}_{nk} = \overbrace{\tilde{s}_k(\bar{\eta}_{In} + j\bar{\eta}_{Qn})e^{j\theta_c}}^{\text{specular component}} + \overbrace{\tilde{s}_k(\xi_{In} + j\xi_{Qn})e^{j\theta_c}}^{\text{random component}} + \tilde{N}_{nk} \quad (8.141)$$

The reference signal is also assumed to be degraded by the channel fading. As such, the additive Gaussian noise model for this signal given in (3.39) is now modified to

$$\tilde{c}_r = \overbrace{A_r\sqrt{G_s}(\bar{\eta}_{In} + j\bar{\eta}_{Qn})e^{j\theta_c}}^{\text{specular component}} + \overbrace{A_r\sqrt{G_r}(\xi_{In} + j\xi_{Qn})e^{j\theta_c}}^{\text{random component}} + \tilde{N}_r \quad (8.142)$$

where G_s and G_r denote the SNR gains associated with its specular and random components, respectively[9] and $\xi_{in} \triangleq \eta_{in} - \bar{\eta}_{in}$, $i = I, Q$. In view of the complex Gaussian fading models above for the received signal and reference signal, the decision statistic for the nth symbol, namely, $\text{Re}\{\tilde{y}_{nk}\tilde{c}_r^*\}$, is, as was the case for the fading-free channel, in the form of the real part of the product of two nonzero mean complex Gaussian random variables; hence, the error probability analysis discussed in Appendix 8A is once again applicable. To apply Stein's analysis [29], we need to specify the first and second moments of $\tilde{y}_{nk}$ and $\tilde{c}_r$. These are computed as follows.

Assume that the real and imaginary components of the complex fading RV η_n have first and second moments

$$\bar{\eta}_I = m_I, \qquad \bar{\eta}_Q = m_Q, \qquad \text{var}(\eta_I) = \text{var}(\eta_Q) = \sigma^2 \quad (8.143)$$

Then the Rician factor K is given by

$$K = \frac{\text{specular power}}{\text{random power}} = \frac{(\bar{\eta}_I)^2 + (\bar{\eta}_Q)^2}{\text{var}(\eta_I) + \text{var}(\eta_Q)} = \frac{m_I^2 + m_Q^2}{2\sigma^2} \quad (8.144)$$

and the total power of η_n is given by

$$E\{|\eta_n|^2\} \triangleq \Omega = E\{\eta_I^2 + \eta_Q^2\} = 2\sigma^2 + m_I^2 + m_Q^2 = 2\sigma^2(1 + K) \quad (8.145)$$

For BPSK signaling, $\tilde{s}_k = A_c T_b a_n$ ($a_n = \pm 1$ represents the binary data) and $A_r = A_c = A$. Thus, from (8.141) and (8.142),

$$|\bar{\tilde{y}}_{nk}| = AT_b\sqrt{(\bar{\eta}_{In})^2 + (\bar{\eta}_{Qn})^2} = AT_b\sqrt{m_I^2 + m_Q^2} = \sqrt{\frac{K}{1+K}\Omega A^2 T_b^2}$$

$$\overline{|\tilde{y}_{nk} - \bar{\tilde{y}}_{nk}|^2} = (AT_b)^2 2\sigma^2 + \text{var}(\tilde{N}_{nk}) = \frac{1}{1+K}\Omega A^2 T_b^2 + N_0 T_b \quad (8.146a)$$

---

[9] Later we shall specifically consider (as does Fitz [16]) the slow-fading case, which implies that the fading changes slowly in comparison to the memory length of the phase estimator. This implies that $G_s = G_r = G$. For the moment, however, we shall allow the specular and random gains to maintain their individual identity.

and

$$|\bar{\tilde{c}}_r| = \sqrt{G_s}AT_b\sqrt{(\bar{\eta}_{In})^2 + (\bar{\eta}_{Qn})^2}$$

$$= \sqrt{G_s}AT_b\sqrt{m_I^2 + m_Q^2} = \sqrt{\frac{G_sK}{1+K}\Omega A^2 T_b^2}$$

$$\overline{|\tilde{c}_r - \bar{\tilde{c}}_r|^2} = G_r(AT_b)^2 2\sigma^2 + \mathrm{var}(\tilde{N}_{nk}) = \frac{G_r}{1+K}\Omega A^2 T_b^2 + N_0 T_b \quad (8.146b)$$

Letting $z_{1p} = \tilde{c}_r$, $z_{2p} = \tilde{y}_{nk}$, and $A^2 T_b = E_b$, then relating these moments to the parameters defined in (8A.3) and (8A.4), we get

$$S_{1p} = \frac{1}{2}|\bar{z}_{1p}|^2 = \frac{1}{2}|\bar{\tilde{c}}_r|^2 = \frac{1}{2}\frac{G_sK}{1+K}\Omega E_b T_b,$$

$$S_{2p} = \frac{1}{2}|\bar{z}_{2p}|^2 = \frac{1}{2}|\bar{\tilde{y}}_{nk}|^2 = \frac{1}{2}\frac{K}{1+K}\Omega E_b T_b$$

$$N_{1p} = \frac{1}{2}\overline{|z_{1p} - \bar{z}_{1p}|^2} = \frac{1}{2}\overline{|\tilde{c}_r - \bar{\tilde{c}}_r|^2} = \frac{N_0 T_b}{2}\left(\frac{G_r}{1+K}\frac{\Omega E_b}{N_0} + 1\right)$$

$$N_{2p} = \frac{1}{2}\overline{|z_{2p} - \bar{z}_{2p}|^2} = \frac{1}{2}\overline{|\tilde{y}_{nk} - \bar{\tilde{y}}_{nk}|^2} = \frac{N_0 T_b}{2}\left(\frac{1}{1+K}\frac{\Omega E_b}{N_0} + 1\right)$$

$$\rho_p = \rho_{cp} + j\rho_{sp} = \frac{1}{2\sqrt{N_{1p}N_{2p}}}\overline{(z_{1p} - \bar{z}_{1p})^*(z_{2p} - \bar{z}_{2p})}$$

$$= \frac{1}{2\sqrt{N_{1p}N_{2p}}}\overline{(\tilde{c}_r - \bar{\tilde{c}}_r)^*(\tilde{y}_{nk} - \bar{\tilde{y}}_{nk})}$$

$$= \frac{\frac{\sqrt{G_r}}{1+K}\frac{\Omega E_b}{N_0}}{\sqrt{\left(\frac{G_r}{1+K}\frac{\Omega E_b}{N_0}+1\right)\left(\frac{1}{1+K}\frac{\Omega E_b}{N_0}+1\right)}}, \quad \theta_{1p} = \theta_{2p}, \quad \phi = 0$$

(8.147)

Using these parameters in (8A.6a) gives the arguments of the Marcum Q-function in (8A.5) as [16, Eqs. (8a) and (8b)]

$$\begin{Bmatrix}a\\b\end{Bmatrix} = \frac{1}{2}\left(\frac{S_{1p}}{N_{1p}} + \frac{S_{2p}}{N_{2p}} \mp 2\sqrt{\frac{S_{1p}S_{2p}}{N_{1p}N_{2p}}}\right) = \frac{1}{2}\left(\sqrt{\frac{S_{1p}}{N_{1p}}} \mp \sqrt{\frac{S_{2p}}{N_{2p}}}\right)^2$$

$$= \frac{1}{2}\left(\sqrt{\frac{\frac{K}{1+K}G_s\bar{\gamma}}{1+\frac{1}{1+K}G_r\bar{\gamma}}} \mp \sqrt{\frac{\frac{K}{1+K}\bar{\gamma}}{1+\frac{1}{1+K}\bar{\gamma}}}\right)^2, \quad \bar{\gamma} \triangleq \frac{\Omega E_b}{N_0} \quad (8.148)$$

where once again we have elected to express the result in terms of the average fading SNR per bit $\bar{\gamma}$. Also since from (8.147) ρ_p is real, then from (8A.6a),

$$A = \frac{\rho_{cp}}{\sqrt{1-\rho_{sp}^2}} = \rho_p = \frac{\frac{\sqrt{G_r}}{1+K}\bar{\gamma}}{\sqrt{\left(\frac{G_r}{1+K}\bar{\gamma}+1\right)\left(\frac{1}{1+K}\bar{\gamma}+1\right)}} \qquad (8.149)$$

Finally, the average BEP for nonideal coherent detection of BPSK in a Rician fading environment is given by (8A.5), namely,

$$P_b(E) = \frac{1}{2}[1 - Q_1(\sqrt{b},\sqrt{a}) + Q_1(\sqrt{a},\sqrt{b})] - \frac{A}{2}\exp\left(-\frac{a+b}{2}\right)I_0(\sqrt{ab}) \qquad (8.150)$$

where a, b, and A are defined as above.

As mentioned earlier in footnote 9, we will be interested in the case of slow fading, which implies that $G_s = G_r = G$. Making this substitution in (8.148) and (8.149) gives the simplified results

$$\left\{\begin{matrix}a\\b\end{matrix}\right\} = \frac{1}{2}\left(\sqrt{\frac{\frac{K}{1+K}G\bar{\gamma}}{1+\frac{1}{1+K}G\bar{\gamma}}} \mp \sqrt{\frac{\frac{K}{1+K}\bar{\gamma}}{1+\frac{1}{1+K}\bar{\gamma}}}\right)^2 \qquad (8.151)$$

and

$$A = \frac{\frac{\sqrt{G}}{1+K}\bar{\gamma}}{\sqrt{\left(\frac{G}{1+K}\bar{\gamma}+1\right)\left(\frac{1}{1+K}\bar{\gamma}+1\right)}} \qquad (8.152)$$

which agrees with an unnumbered equation [between (8c) and (9)] in Ref. 16. Figure 8.6 is an illustration of average BEP as computed from (8.150) together with (8.151) and (8.152) for $K = 0$ and $K = 10$ and three nonideal coherence parameter values: $G = 3$, 10, and 20 dB. We observe from these numerical results that over a wide range of average SNR, the BEP is rather insensitive to the value of G, particularly for the higher value of K.

As a check on previous results, the no-fading case, which corresponds to $K \to \infty$, $\bar{\gamma} \to E_b/N_0$, would result in

$$\left\{\begin{matrix}a\\b\end{matrix}\right\} = \frac{1}{2}\left(\sqrt{G\frac{E_b}{N_0}} \mp \sqrt{\frac{E_b}{N_0}}\right)^2 = \frac{E_b}{2N_0}(\sqrt{G}\mp 1)^2, \qquad A = 0 \qquad (8.153)$$

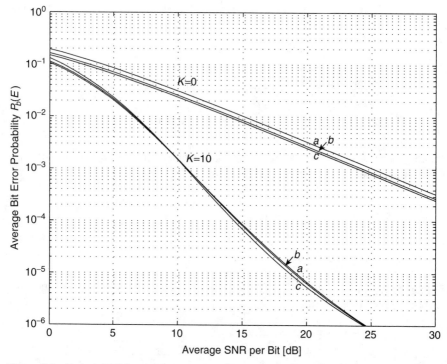

Figure 8.6. Average BEP for nonideal coherent detection of BPSK over a Rician channel versus the average SNR per bit: (a) $G = 3$ dB; (b) $G = 10$ dB; (c) $G = 20$ dB.

which agrees with (8.61). For Rayleigh fading ($K = 0$), the corresponding results are

$$\begin{Bmatrix} a \\ b \end{Bmatrix} = \begin{Bmatrix} 0 \\ 0 \end{Bmatrix}, \quad A = \frac{\sqrt{G\bar{\gamma}}}{\sqrt{(G\bar{\gamma}+1)(\bar{\gamma}+1)}} \tag{8.154}$$

Since $Q_1(0, 0) = 1$, from (8.150) we obtain

$$P_b(E) = \frac{1}{2}\left[1 - \frac{\sqrt{G\bar{\gamma}}}{\sqrt{(G\bar{\gamma}+1)(\bar{\gamma}+1)}}\right] \tag{8.155}$$

For a perfect phase reference (i.e., $G \to \infty$), (8.155) simplifies to

$$P_b(E) = \frac{1}{2}\left(1 - \sqrt{\frac{\bar{\gamma}}{1+\bar{\gamma}}}\right) \tag{8.156}$$

which is consistent with the result given in (8.104) for ideal coherent detection.

To extend the results above to other quadrature modulation schemes with I and Q carrier components that are independently modulated (e.g., QPSK, OQPSK,

MSK, and QAM), one merely recognizes that for such schemes the average BEP in the presence of fading can be expressed as

$$P_b(E) = \frac{1}{2}P_{bI}(E) + \frac{1}{2}P_{bQ}(E) \tag{8.157}$$

where $P_{bI}(E)$ and $P_{bQ}(E)$ are, respectively, the average BEPs for the I and Q data streams. Thus, Stein's analysis technique [29] of Appendix 8A can again be applied to evaluate $P_{bI}(E)$ and $P_{bQ}(E)$ separately and thereby arrive at a generalization to the fading channel case of the AWGN BEP results given by (8.62) through (8.65). For example, for QPSK results analogous to (8.151) and (8.152) are obtained by replacing G by $2G$, whereupon the former reduces to (8.62) when $K \to \infty$. The specific details for the remaining quadrature modulation schemes are left to the reader.

8.2.3 Noncoherent Detection

As alluded to previously, in a multipath environment it is often difficult in practice to achieve good carrier synchronization; in such instances it is necessary to employ a modulation for which noncoherent detection is possible. The most popular choice of such a modulation in fading channel applications is orthogonal M-FSK, whose error probability performance in AWGN was considered in Section 8.1.3. It is a simple matter now to extend these results to the fading channel. In particular, since each term of (8.66) is purely an exponential of the SNR, then applying the MGF-approach to this equation, we obtain the average SEP:

$$P_s(E) = \sum_{m=1}^{M-1}(-1)^{m+1}\binom{M-1}{m}\frac{1}{m+1}M_{\gamma_s}\left(-\frac{m}{m+1}\right) \tag{8.158}$$

where the moment generating function $M_{\gamma_s}(-s)$ is obtained from any of the results in Section 5.1 with $\bar{\gamma}$ replaced by the average symbol SNR $\bar{\gamma}_s$. Thus, for Rayleigh fading, using (5.5), we have

$$P_s(E) = \sum_{m=1}^{M-1}(-1)^{m+1}\binom{M-1}{m}\frac{1}{1+m(1+\bar{\gamma}_s)} \tag{8.159}$$

which for the special case of binary FSK simplifies to $P_b(E) = 1/(2+\bar{\gamma})$, in agreement with Proakis [6, Eq. (14-3-12)]. For Rician fading, using (5.11) gives

$$P_s(E) = \sum_{m=1}^{M-1}(-1)^{m+1}\binom{M-1}{m}\frac{1+K}{1+K+m(1+K+\bar{\gamma}_s)}$$

$$\times \exp\left(-\frac{Km\bar{\gamma}_s}{1+K+m(1+K+\bar{\gamma}_s)}\right) \tag{8.160}$$

which agrees with Sun and Reed [24, Eq. (8)], and reduces to (8.159) when $K = 0$. Finally for Nakagami-m fading, using (5.15), we obtain

$$P_s(E) = \sum_{l=1}^{M-1} (-1)^{l+1} \binom{M-1}{l} \frac{(l+1)^{m-1}}{[1+l(1+\overline{\gamma}_s/m)]^m} \quad (8.161)$$

where we have changed the summation index to avoid confusion with the Nakagami-m fading parameter. As expected, (8.161) reduces to (8.159) when $m = 1$. Before concluding this section, we note that the results for average BEP over a fading channel can be obtained, as was the case for the AWGN channel, by applying the relation between bit and symbol error probability given in (8.67) to the results above. We note furthermore that although we have specifically addressed M-FSK, the results above apply equally well to any M-ary orthogonal signaling scheme transmitted over a slow, flat fading channel and detected noncoherently at the receiver.

For nonorthogonal M-FSK, we observed in Section 8.1.3 that a simple analytical result for average BEP over the AWGN was possible for the binary case, namely, (8.69). To extend this result to the fading channel, we first rewrite it in the alternative form [see (9A.14) of Appendix 9A]

$$P_b(E) = \tfrac{1}{2}[1 - Q_1(\sqrt{b}, \sqrt{a}) + Q_1(\sqrt{a}, \sqrt{b})] \quad (8.162)$$

and then make use of the alternative representation of the Marcum Q-function in (4.16) and (4.19) to allow application of the MGF-based approach [see (8A.12)]. Using the definitions of a and b in (8.70), the result of this application produces

$$P_b(E) = \frac{1}{4\pi} \int_{-\pi}^{\pi} \frac{1-\zeta^2}{1+2\zeta \sin\theta + \zeta^2}$$
$$\times M_\gamma \left(-\frac{1}{4}(1+\sqrt{1-\rho^2})(1+2\zeta\sin\theta + \zeta^2) \right) d\theta,$$
$$\zeta \triangleq \sqrt{\frac{1-\sqrt{1-\rho^2}}{1+\sqrt{1-\rho^2}}} \quad (8.163)$$

where ρ is the correlation coefficient of the two signals. To obtain specific results for the various fading channels, one merely substitutes the appropriate MGF from Section 5.1 in (8.163), analogous to what was done previously for the orthogonal signaling case. The specific analytical results are left as an exercise for the reader. As an illustration of the numerical results that can be obtained from (8.163) after making the aforementioned substitutions, Figs. 8.7, 8.8, and 8.9 illustrate the average BEP performance for Nakagami-q (Hoyt), Nakagami-n (Rice), and Nakagami-m channels.

PERFORMANCE OVER FADING CHANNELS **241**

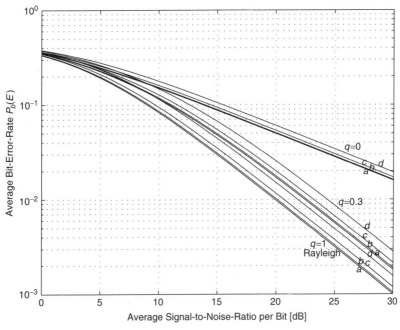

Figure 8.7. Average BEP of correlated BFSK over a Nakagami-q (Hoyt) channel: (a) $\rho = 0$; (b) $\rho = 0.2$; (c) $\rho = 0.4$; (d) $\rho = 0.6$.

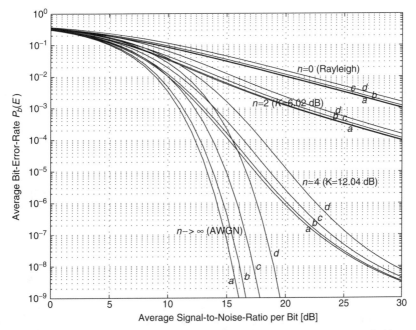

Figure 8.8. Average BEP of correlated BFSK over a Nakagami-n (Rice) channel: (a) $\rho = 0$; (b) $\rho = 0.2$; (c) $\rho = 0.4$; (d) $\rho = 0.6$.

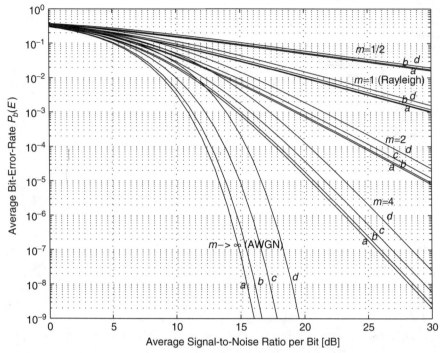

Figure 8.9. Average BEP of correlated BFSK over a Nakagami-m channel: (a) $\rho = 0$; (b) $\rho = 0.2$; (c) $\rho = 0.4$; (d) $\rho = 0.6$.

8.2.4 Partially Coherent Detection

In this section we apply the MGF-based approach to the AWGN results of Section 8.1.4 to predict the performance of partially coherent detection systems in the presence of fading. The steps to be followed parallel those of the previous sections and thus the presentation will be brief.

For BPSK with conventional (one-symbol observation) detection, the conditional (on a fixed phase error ϕ_c) BEP is in the form of a Gaussian Q-function as described by (8.71). Thus, first performing the averaging over the fading takes the form of (5.1), which is expressed in terms of the MGF of the fading as in (5.3). Finally, performing the averaging over the Tikhonov phase error PDF gives the desired result, namely,

$$P_b(E) = \int_{-\pi}^{\pi} \frac{1}{\pi} \int_0^{\pi/2} M_\gamma \left(-\frac{\cos^2 \phi_c}{\sin^2 \theta} \right) d\theta \frac{\exp(\rho_c \cos \phi_c)}{2\pi I_0(\rho_c)} d\phi_c \qquad (8.164)$$

For orthogonal and nonorthogonal BFSK, the results [see (8.74) together with (8.69)] are expressed in terms of the first-order Marcum Q-function. However, in these cases the ratio of the two arguments of this function [see (8.75) and (8.76)] are not independent of SNR, and thus the MGF-based approach is not useful here in allowing an easy evaluation of average BEP. Instead, one must resort to the brute force approach of replacing E_b/N_0 by γ in the a and b parameters and then performing the average over the PDF of γ as appropriate for the type of fading channel under consideration. A similar statement is made for the multiple-symbol detection case since again the ratio of the two arguments of the Marcum Q-function [see (8.81)] are not independent of SNR.

8.2.5 Differentially Coherent Detection

In this the final section of this chapter, we consider the characterization of the error probability performance of differentially detected M-ary phase-shift-keying when transmitted over a fading channel. This modulation/detection combination has received a lot of attention in the literature, particularly the $M = 4$ case (DQPSK), which has been adopted in the most recent North American and Japanese digital cellular system standards. For instance, Tjhung et al. [30] and Tanda [31] analyzed the average BEP of DQPSK over slow Rician and Nakagami-m fading channels, respectively. Later, Tellambura and Bhargava [32] presented an alternative unified BEP analysis of DQPSK over Rician and Nakagami-m fading channels. In keeping with the unifying theme of this book, our purpose in this section is to once again unify and add to the previous contributions by obtaining results for arbitrary values of M as well as for a broad class of fading channels. As in Section 8.1.5, we first focus on conventional (two-symbol observation) detection of M-PSK, for which, as noted there, the SEP is already in the desired form, namely, one that lends itself to immediate application of the MGF-based approach.

8.2.5.1 M-ary Differential Phase-Shift-Keying: Slow Fading

Conventional Detection: Two-Symbol Observation. With reference to (8.84), which gives the SEP of M-DPSK for the AWGN channel, we observe that the integrand is already an exponential function of the symbol SNR. Thus, unlike the cases where the integrand's dependence on SNR is through Gaussian and Marcum Q-functions, no alternative form is necessary here to allow averaging over the fading statistics of the channel. All that needs to be done is to replace E_s/N_0 by γ_s in the argument of the exponential and then average over the PDF of γ_s resulting in the MGF-based expression

$$P_s(E) = \frac{\sqrt{g_{\text{PSK}}}}{2\pi} \int_{-\pi/2}^{\pi/2} \frac{M_{\gamma_s}(-(1 - \sqrt{1 - g_{\text{PSK}}} \cos \theta))}{1 - \sqrt{1 - g_{\text{PSK}}} \cos \theta} \, d\theta,$$

$$g_{\text{PSK}} \stackrel{\Delta}{=} \sin^2(\pi/M) \qquad (8.165)$$

or the simpler form derived from (8.90),

$$P_s(E) = \frac{1}{\pi} \int_0^{(M-1)\pi/M} M_{\gamma_s}\left(-\frac{g_{\text{PSK}}}{1 + \sqrt{1 - g_{\text{PSK}}}\cos\theta}\right) d\theta \qquad (8.166)$$

The special case of binary DPSK wherein $g_{\text{PSK}} = 1$ simplifies to the closed-form result

$$P_b(E) = \tfrac{1}{2} M_\gamma(-1) \qquad (8.167)$$

Comparing (8.167) with the special case of (8.158) corresponding to $M = 2$, namely, $P_b(E) = \tfrac{1}{2} M_\gamma(-\tfrac{1}{2})$, then since, independent of the type of fading, the MGF of the fading SNR $M_\gamma(-s)$ is only a function of the product $s\bar{\gamma}$ [see, e.g., (5.5), (5.8), (5.11), and (5.15)], we conclude that *the BEP of noncoherent orthogonal FSK is 3 dB worse in average fading SNR than that of DPSK*. We remind the reader that this is the same conclusion reached when comparing these two modulation/detection schemes over the AWGN.

To obtain the average BEP corresponding to values of $M > 2$, we make use of the AWGN results in (8.86), which correspond to a Gray code bit-to-symbol mapping. Since each of the BEP results in (8.86) is expressed in terms of the function $F(\psi)$ defined in (8.87), which, analogous to (8.84), has an integrand with exponential dependence on symbol SNR, then clearly the average BEP over the fading channel can be obtained from (8.86) by replacing $F(\psi)$ with

$$\overline{F(\psi)} = -\frac{\sin\psi}{4\pi} \int_{-\pi/2}^{\pi/2} \frac{M_\gamma(-(\log_2 M)(1 - \cos\psi \cos t))}{1 - \cos\psi \cos t} \, dt \qquad (8.168)$$

or the simpler form [see Eq. (4.68)]

$$\overline{F(\psi)} = -\frac{1}{4\pi} \int_{-(\pi-\psi)}^{\pi-\psi} M_\gamma\left(-(\log_2 M)\frac{\sin^2\psi}{1 + \cos\psi \cos t}\right) dt \qquad (8.169)$$

The average BEP for the special case of DQPSK can, of course, be obtained from the first relation in (8.86) together with (8.168) or (8.169) with $M = 4$. In view of (8.89) for the AWGN channel, it can also be obtained in a form analogous to (8.163), namely,

$$P_b(E) = \frac{1}{4\pi} \int_{-\pi}^{\pi} \frac{1 - \zeta^2}{1 + 2\zeta \sin\theta + \zeta^2} M_\gamma\left(-\left(1 + \frac{1}{\sqrt{2}}\right)(1 + 2\zeta \sin\theta + \zeta^2)\right) d\theta,$$

$$\zeta \triangleq \sqrt{\frac{2 - \sqrt{2}}{2 + \sqrt{2}}} \qquad (8.170)$$

Using instead the alternative forms of the first-order Marcum Q-functions given in (4.20) and (4.21) corresponding to only positive values of the integration variable, an equivalent form to (8.170) can be obtained from Tellambura and Bhargava [32, Eq. (3)], namely,

$$P_b(E) = \frac{1}{2\pi} \int_0^\pi \frac{1}{\sqrt{2} - \cos\theta} M_\gamma(-(2 - \sqrt{2}\cos\theta))d\theta \quad (8.171)$$

Without further ado, we now give the specific results of the above, corresponding to Rayleigh, Rician, and Nakagami-m channels. These results, as well as those for the fading channels discussed previously, are taken from Ref. 33.

RAYLEIGH FADING. From (8.165) and (5.5), the average SEP of M-DPSK is given by

$$P_s(E) = \frac{\sin(\pi/M)}{2\pi} \int_{-\pi/2}^{\pi/2} \frac{1}{[1 - \cos(\pi/M)\cos\theta]\{1 + \bar{\gamma}_s[1 - \cos(\pi/M)\cos\theta]\}} d\theta \quad (8.172)$$

which is in agreement with Sun and Reed [24, Eq. (6)]. The corresponding binary DPSK result is

$$P_b(E) = \frac{1}{2(1 + \bar{\gamma})} \quad (8.173)$$

which agrees with Proakis [6, Eq. (14-3-10)]. For DQPSK, the average BEP is evaluated from (8.170) in closed form as

$$P_b(E) = \frac{1}{2}\left[1 - \frac{1}{\sqrt{(1 + 2\bar{\gamma})^2/(2\bar{\gamma}^2) - 1}}\right] \quad (8.174)$$

which agrees with an equivalent result obtained by Tjhung et al. [30, Eq. (18)] and Tanda [31, Eq. (13)], namely,

$$P_b(E) = \frac{1}{2\sqrt{1 + 4\bar{\gamma} + 2\bar{\gamma}^2}} \left[\frac{\sqrt{2\bar{\gamma}} + (\sqrt{2} - 1)(1 + 2\bar{\gamma} - \sqrt{1 + 4\bar{\gamma} + 2\bar{\gamma}^2})}{\sqrt{2\bar{\gamma}} - (\sqrt{2} - 1)(1 + 2\bar{\gamma} - \sqrt{1 + 4\bar{\gamma} + 2\bar{\gamma}^2})}\right] \quad (8.175)$$

or the one reported by Tellambura and Bhargava [32, Eq. (8)], namely,

$$P_b(E) = \frac{1}{2}\left(1 - \frac{\sqrt{2\bar{\gamma}}}{\sqrt{1 + 4\bar{\gamma} + 2\bar{\gamma}^2}}\right) \quad (8.176)$$

For other values of M, the average BEP is computed from (8.86) using

$$\overline{F(\psi)} = -\frac{\sin\psi}{4\pi} \int_{-\pi/2}^{\pi/2} \frac{1}{(1 - \cos\psi\cos t)[1 + \bar{\gamma}(\log_2 M)(1 - \cos\psi\cos t)]} dt \quad (8.177)$$

in place of $F(\psi)$.

RICIAN FADING. From (8.165) and (5.11), the average SEP of M-DPSK is given by

$$P_s(E) = \frac{\sin(\pi/M)}{2\pi} \int_{-\pi/2}^{\pi/2} \frac{1+K}{(1-\cos(\pi/M)\cos\theta)\{1+K+\overline{\gamma}_s[1-\cos(\pi/M)\cos\theta]\}}$$

$$\times \exp\left\{-\frac{K\overline{\gamma}_s[1-\cos(\pi/M)\cos\theta]}{1+K+\overline{\gamma}_s[1-\cos(\pi/M)\cos\theta]}\right\} d\theta \tag{8.178}$$

which is in agreement with Sun and Reed [24, Eq. (5)]. The corresponding binary DPSK result is

$$P_b(E) = \frac{1}{2}\left(\frac{1+K}{1+K+\overline{\gamma}}\right)\exp\left(-\frac{K\overline{\gamma}}{1+K+\overline{\gamma}}\right) \tag{8.179}$$

For DQPSK, the average BEP is most easily evaluated from (8.171), which produces

$$P_b(E) = \frac{e^{-K}}{2\pi} \int_0^\pi \frac{1+K}{(\sqrt{2}-\cos\theta)[1+K+\overline{\gamma}_s(2-\sqrt{2}\cos\theta)]}$$

$$\times \exp\left[-\frac{K(1+K)}{1+K+\overline{\gamma}_s(2-\sqrt{2}\cos\theta)}\right] d\theta \tag{8.180}$$

in agreement with Tellambura and Bhargava [32, Eq. (6)].

For other values of M, the average BEP is computed from (8.86) using

$$\overline{F(\psi)} = -\frac{\sin\psi}{4\pi}\int_{-\pi/2}^{\pi/2}\frac{1+K}{(1-\cos\psi\cos t)[1+K+\overline{\gamma}(\log_2 M)(1-\cos\psi\cos t)]}$$

$$\times \exp\left[-\frac{K\overline{\gamma}(\log_2 M)(1-\cos\psi\cos t)}{1+K+\overline{\gamma}(\log_2 M)(1-\cos\psi\cos t)}\right] dt \tag{8.181}$$

in place of $F(\psi)$.

NAKAGAMI-m FADING. From (8.165) and (5.15), the average SEP of M-DPSK is given by

$$P_s(E) = \frac{\sin(\pi/M)}{2\pi}\int_{-\pi/2}^{\pi/2}\frac{1}{[1-\cos(\pi/M)\cos\theta]\{1+(\overline{\gamma}_s/m)[1-\cos(\pi/M)\cos\theta]\}^m} d\theta \tag{8.182}$$

and is illustrated in Fig. 8.10 as a function of average symbol SNR and parameterized by m. The corresponding binary DPSK result is

$$P_b(E) = \frac{1}{2}\left(\frac{m}{m+\overline{\gamma}}\right)^m \tag{8.183}$$

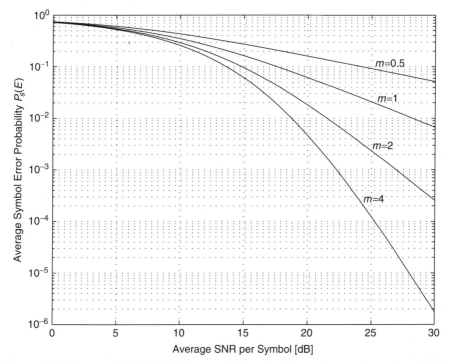

Figure 8.10. Average SEP of 8-DPSK over a Nakagami-*m* channel versus the average SNR per symbol.

which agrees with the expression attributed to Barrow [34] and later reported by Wojnar [22, Eq. (11)] and Crepeau [35, Eq. (B1)]. For DQPSK, the average BEP is evaluated from (8.171) as

$$P_b(E) = \frac{1}{2\pi} \left(\frac{m}{m+2\bar{\gamma}}\right)^m \int_0^\pi \frac{1}{(\sqrt{2} - \cos\theta)(1 - [\sqrt{2\bar{\gamma}}/(m+2\bar{\gamma})]\cos\theta)^m} d\theta \qquad (8.184)$$

which agrees with Tellambura and Bhargava [32, Eq. (7)].

Finally, the function necessary to compute average BEP for other values of M is given by

$$\overline{F(\psi)} = -\frac{\sin\psi}{4\pi} \int_{-\pi/2}^{\pi/2} \frac{1}{(1 - \cos\psi\cos t)} \left[\frac{m}{m + \bar{\gamma}(\log_2 M)(1 - \cos\psi\cos t)}\right]^m dt \qquad (8.185)$$

Multiple Symbol Detection. The upper bound on the BEP for the AWGN channel as given in (8.97) is easily extended to the fading channel case by recognizing that the form of the probability $\Pr\{\hat{z}_{n\mathbf{k}} > z_{n\mathbf{k}}\}$ as described by (8.80) is identical to (8.162), which characterizes noncoherent detection of orthogonal FSK. One can thus make use of the results in Section 8.2.3 to express each term in the sum of (8.97) in an MGF-based form analogous to (8.163), namely,

$$\Pr\{\hat{z}_{n\mathbf{k}} > z_{n\mathbf{k}}\} = \frac{1}{4\pi}\int_{-\pi}^{\pi}\frac{1-\zeta^2}{1+2\zeta\sin\theta+\zeta^2}M_\gamma\left(-\frac{\log_2 M}{4}(N_s+\sqrt{N_s-|\delta|^2})\right.$$

$$\left.\times(1+2\zeta\sin\theta+\zeta^2)\right)d\theta, \qquad \zeta \triangleq \sqrt{\frac{N_s-\sqrt{N_s-|\delta|^2}}{N_s+\sqrt{N_s-|\delta|^2}}}$$
(8.186)

with δ defined in (8.99). Substituting (8.186) into (8.97) gives the desired upper bound on BEP. It is left as an exercise for the reader to evaluate (8.186) for the various fading channels based on the same procedure as that stated at the end of Section 8.2.3.

8.2.5.2 M-ary Differential Phase-Shift-Keying: Fast Fading

Conventional Detection: Two-Symbol Observation. Until now in this chapter we have focused entirely on the performance of digital communication systems operating over slow-fading channels. For conventional differentially coherent detection of M-PSK, the assumption of slow fading is tantamount to assuming that the fading amplitude is constant over a duration of at least two symbol intervals. A suitable modification of this model for the case of fast fading is to assume that the fading amplitude is constant within the duration of a single symbol but varies from symbol to symbol. That is, the symbol intervals are each characterized by their own fading amplitude, which relative to one another satisfy a given discrete correlation function that is related to the nature of the fading channel (more about this later). Such a discrete fast-fading model is an approximation to the true channel behavior wherein the fading varies continuously with time. To understand fully the method used to evaluate the average error probability for this scenario, we must first review the system model discussed in Section 3.5, making the necessary modifications in notation to account for the presence of fast fading on the signal. We focus all our attention on the Rician channel (with results for the Rayleigh channel obtained as a special case) and develop only the binary DPSK case.

Consider a binary DPSK system transmitting information bits over an AWGN channel that is also perturbed by fast Rician fading. The normalized kth information bit at the input to the system is given by

$$x_k = e^{j\Delta\theta_k} \qquad (8.187)$$

where for binary transmission $\Delta\theta_k$ takes on values of 0 and π corresponding, respectively, to values of 1 and -1 for x_k. The input information bits are differentially encoded, resulting in the transmitted bit

$$v_k = \sqrt{2E_b}e^{j\theta_k} = \sqrt{2E_b}e^{j(\theta_{k-1}+\Delta\theta_k)} = v_{k-1}x_k \quad (8.188)$$

After passing through the fast-fading channel, the received information bit in the kth transmission interval is

$$w_k = G_k v_k + N_k \quad (8.189)$$

where G_k is the complex Gaussian fading amplitude associated with the kth received bit and N_k is a zero-mean complex Gaussian noise RV with correlation function $E\{N_k^* N_m\} = 2N_0\delta(k-m)$. Denoting the mean and variance of G_k by $\eta = E\{G_k\}$ and $\sigma^2 = \frac{1}{2}E\{|G_k - \eta|^2\}$ (both assumed to be independent of k), then for the assumed Rician channel, the magnitude of G_k, namely, $\alpha_k \triangleq |G_k|$, has PDF

$$p(\alpha_k) = \alpha_k \frac{2(1+K)}{\Omega} \exp\left[-K - \frac{\alpha_k^2(1+K)}{\Omega}\right] I_0\left(\frac{2\alpha_k}{\sqrt{\Omega}}\sqrt{K(1+K)}\right) \quad (8.190)$$

where $\Omega = E\{\alpha_k^2\} = 2\sigma^2(1+K)$. Furthermore, the adjacent complex fading amplitudes have correlation

$$\tfrac{1}{2}E\{(G_{k-1} - \eta)^*(G_k - \eta)\} = \rho\sigma^2, \quad 0 \leq \rho \leq 1 \quad (8.191)$$

where ρ is the fading correlation coefficient whose value depends on the fast-fading channel model that is assumed.

At the receiver the received signal w_k for the current bit interval is complex conjugate multiplied by the same signal, corresponding to the previous bit interval, and the real part of the resulting product forms the decision variable (which is multiplied by 2 for mathematical convenience)

$$z_k \triangleq 2\,\text{Re}\{w_k^* w_{k-1}\} = w_k^* w_{k-1} + w_k w_{k-1}^* \quad (8.192)$$

Comparison of z_k with a zero threshold results in the final decision on the transmitted bit x_k, namely, $\hat{x}_k = e^{j\Delta\hat{\theta}_k} = \text{sgn } z_k$, which is consistent in form with the decision rule given in (3.52).

We note that conditioned on the information bit x_k, the components w_{k-1} and w_k are complex Gaussian RVs (since both the fading amplitude and additive noise RVs are complex Gaussian). Thus, (8.192) represents a Hermitian quadratic form of complex variables. Although it is possible to use an MGF-based approach based on the conditional MGF of such a quadratic form first considered by Turin [36] and later reported by Schwartz et al. [37, App. B], in this particular case there is an easier way to proceed. Specifically, letting $D = z_k|_{x_k=1}$ denote the decision variable corresponding to transmission of a $+1$ information bit, then based on the decision rule above, the average BEP is given by

$$P_b(E) = \Pr\{D < 0\} \tag{8.193}$$

The solution to (8.193) is a special case of the problem considered in [6, App. B] which has also been reconsidered in alternative forms in Appendix 9A of this book. Specifically, letting $A = B = 0$, $C = 1$, $X_k = w_{k-1}$, $Y_k = w_k$ in (9A.2), the decision variable (8.192) is identical to that in (9A.1) when $L = 1$. Evaluating the various coefficients required in (9A.10) produces after much simplification the following results:

$$\eta = \frac{v_2}{v_1} = \frac{1 + K + \bar{\gamma}(1 + \rho)}{1 + K + \bar{\gamma}(1 - \rho)}, \quad a = 0, \quad b = \sqrt{\frac{2K\bar{\gamma}}{1 + K + \bar{\gamma}}} \tag{8.194}$$

where $\bar{\gamma} = \Omega E_b/N_0$ is, as before, the average fading SNR. Finally, substituting (8.194) in (9A.10) and recalling that $Q_1(0, b) = e^{-b^2/2}$, we obtain the desired average BEP

$$P_b(E) = \frac{1}{2}\left[\frac{1 + K + \bar{\gamma}(1 - \rho)}{1 + K + \bar{\gamma}}\right] \exp\left(-\frac{K\bar{\gamma}}{1 + K + \bar{\gamma}}\right) \tag{8.195}$$

The corresponding result for the Rayleigh ($K = 0$) channel is

$$P_b(E) = \frac{1}{2}\left[\frac{1 + \bar{\gamma}(1 - \rho)}{1 + \bar{\gamma}}\right] \tag{8.196}$$

As a check, the results presented earlier for slow fading can be obtained by letting $\rho = 1$ in (8.195) and (8.196), which results, respectively, in (8.179) and (8.173), as expected.

What is different about the fast-fading case in comparison with the slow-fading case is the limiting behavior of $P_b(E)$ as the average fading SNR approaches infinity. Letting $\bar{\gamma} \to \infty$ in (8.195) and (8.196) gives

$$\lim_{\bar{\gamma} \to \infty} P_b(E) = \frac{1 - \rho}{2} \exp(-K) \tag{8.197}$$

and

$$\lim_{\bar{\gamma} \to \infty} P_b(E) = \frac{1 - \rho}{2} \tag{8.198}$$

(i.e., *an irreducible bit error probability exists for any* $\rho \neq 1$). The amount of this irreducible error probability can be related (through the parameter ρ) to the ratio of the Doppler spread (fading bandwidth) of the channel to the data rate. The specific functional relationship between these parameters depends on the choice of the fading channel correlation model. Mason [39] has tabulated such relationships for various types of fast-fading processes of interest. These results are summarized in Table 2.1, where $f_d T_s = f_d T_b$ denotes the Doppler

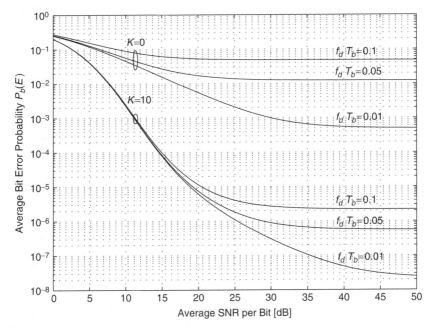

Figure 8.11. Average BEP of binary DPSK over a fast fading Rician channel versus the average SNR per bit: land mobile channel.

spread/data rate ratio, and in addition the variance of the fading process has, for convenience, been normalized to unity. For example, for the land mobile channel where $\rho = J_0(2\pi f_d T_b)$, Fig. 8.11 illustrates the average BEP as computed from (8.195) as a function of average bit SNR for Rician $K = 0$ (Rayleigh channel) and $K = 10$ with $f_d T_b$ as a parameter. As one would expect, as $f_d T_b$ diminishes, the irreducible error becomes smaller. Nevertheless, depending on the value of Rician factor, a Doppler spread of only 1% of the data rate can still cause a significant error floor.

8.2.5.3 $\pi/4$**-Differential QPSK.** From the conclusion drawn in Section 8.1.5.2 relative to the equivalence in behavior between DQPSK and $\pi/4$-DQPSK on the ideal linear AWGN channel, it is clear that the same statement can be made for the fading channel. Thus, without any additional detail, we conclude immediately that the error probability performance of $\pi/4$-DQPSK on the fading channel is characterized by the results of Section 8.2.5.1, namely, the generic BEP of (8.170) [or (8.171)] or the more specific results that followed these equations.

REFERENCES

1. P. Z. Peebles, Jr., *Digital Communication Systems*. Upper Saddle River, NJ: Prentice Hall, 1987.

2. B. Sklar, *Digital Communications: Fundamentals and Applications*. Upper Saddle River, NJ: Prentice Hall, 1988.
3. S. Haykin, *Digital Communications*. New York: Wiley, 1988.
4. R. E. Ziemer and W. H. Tranter, *Principles of Communications: Systems, Modulation, and Noise*, 3rd ed. Boston: Houghton Mifflin, 1990.
5. M. K. Simon, S. M. Hinedi, and W. C. Lindsey, *Digital Communication Techniques: Signal Design and Detection*. Upper Saddle River, NJ: Prentice Hall, 1995.
6. J. Proakis, *Digital Communications*, 3rd ed. New York, McGraw-Hill, 1995.
7. D. G. Messerschmitt and E. A. Lee, *Digital Communication*, 2nd ed. Norwell, MA: Kluwer Academic Publishers, 1994.
8. J. Lu, K. B. Letaief, J. C.-I. Chuang, and M. L. Liou, "M-PSK and M-QAM BER computation using signal-space concepts," *IEEE Trans. Commun.*, vol. 47, February 1999, pp. 181–184.
9. R. F. Pawula, S. O. Rice, and J. H. Roberts, "Distribution of the phase angle between two vectors perturbed by Gaussian noise," *IEEE Trans. Commun.*, vol. COM-30, August 1982, pp. 1828–1841.
10. J. W. Craig, "A new, simple and exact result for calculating the probability of error for two-dimensional signal constellations," *IEEE MILCOM'91 Conf. Rec.*, Boston, pp. 25.5.1–25.5.5.
11. G. L. Stüber, *Principles of Mobile Communication*. Norwell, MA: Kluwer Academic Publishers, 1996.
12. M. K. Irshid and I. S. Salous, "Bit error probability for coherent M-ary PSK systems" *IEEE Trans. Commun.*, vol. 39, March 1991, pp. 349–355.
13. P. J. Lee, "Computation of the bit error rate of coherent M-ary PSK with Gray code bit mapping," *IEEE Trans. Commun.*, vol. COM-34, May 1986, pp. 488–491.
14. W. C. Lindsey and M. K. Simon, *Telecommunication Systems Engineering*, Upper Saddle River, NJ: Prentice Hall, 1973.
15. V. I. Tikhonov, "The effect of noise on phase-locked oscillator operation," *Autom. Remote Control*, vol. 20, 1959, pp. 1160–1168. Translated from *Autom. Telemekh.*, Akademya Nauk, SSSR, vol. 20, September 1959.
16. M. P. Fitz, "Further results in the unified analysis of digital communication systems," *IEEE Trans. Commun.*, vol. 40, March 1992, pp. 521–532.
17. C. W. Helstrom, "The resolution of signals in white Gaussian noise," *IRE Proc.*, vol. 43, September 1955, pp. 1111–1118.
18. R. F. Pawula, "A new formula for MDPSK symbol error probability," *IEEE Commun. Lett.*, vol. 2, October 1998, pp. 271–272.
19. R. F. Pawula, "Generic error probabilities," *IEEE Trans. Commun.*, vol. 47, May 1999, pp. 697–702.
20. R. F. Pawula, "Asymptotics and error rate bounds for M-ary DPSK," *IEEE Trans. Commun.*, vol. COM-32, January 1984, pp. 93–94.
21. D. Divsalar and M. K. Simon, "Multiple-symbol differential detection of MPSK," *IEEE Trans. Commun.*, vol. 38, March 1990, pp. 300–308.
22. A. H. Wojnar, "Unknown bounds on performance in Nakagami channels," *IEEE Trans. Commun.*, vol. COM-34, January 1986, pp. 22–24.

23. I. S. Gradshteyn and I. M. Ryzhik, *Table of Integrals, Series, and Products*, 5th ed. San Diego CA: Academic Press, 1994.
24. J. S. Sun and I. S. Reed, "Performance of MDPSK, MPSK, and noncoherent MFSK in wireless Rician fading channels," *IEEE Trans. Commun.*, vol. 47, June 1999, pp. 813–816.
25. X. Dong and N. C. Beaulieu, "New analytical probability of error expressions for classes of orthogonal signals in Rayleigh fading," submitted to *IEEE Trans. Commun.*
26. M. Abramowitz and I. A. Stegun, *Handbook of Mathematical Functions with Formulas, Graphs, and Mathematical Tables*, 9th ed. New York: Dover Press, 1972.
27. A. Shiryayev, *Probability*. New York: Springer-Verlag, 1984, p. 190.
28. L. W. Hughes, "A simple upper bound on the error probability for orthogonal signals in white noise," *IEEE Trans. Commun.*, vol. 40, April 1992, p. 670.
29. S. Stein, "Unified analysis of certain coherent and noncoherent binary communication systems," *IEEE Trans. Inf. Theory*, vol. IT-10, January 1964, pp. 43–51.
30. T. T. Tjhung, C. Loo, and N. P. Secord, "BER performance of DQPK in slow Rician fading," *Electron. Lett.*, vol. 28, August 1992, pp. 1763–1765.
31. M. Tanda, "Bit error rate of DQPSK signals in slow Nakagami fading," *Electron. Lett.*, vol. 29, March 1993, pp. 431–432.
32. C. Tellambura and V. K. Bhargava, "Unified error analysis of DQPSK in fading channels," *Electron. Lett.*, vol. 30, December 1994, pp. 2110–2111.
33. M. K. Simon and M.-S. Alouini, "A unified approach to the probability of error for noncoherent and differentially coherent modulations over generalized fading channels," *IEEE Trans. Commun.*, vol. 46, December 1998, pp. 1625–1638.
34. B. B. Barrow, "Error probabilities for data transmission over fading radio paths," *Tech. Rep. TM-26*, SHAPE Air Defense Technical Center, 1962.
35. P. J. Crepeau, "Uncoded and coded performance of MFSK and DPSK in Nakagami fading channels," *IEEE Trans. Commun.*, vol. 40, March 1992, pp. 487–493.
36. G. L. Turin, "The characteristic function of Hermitian quadratic forms in complex normal variables," *Biometrika*, vol. 47, June 1960, pp. 199–201.
37. M. Schwartz, W. R. Bennett, and S. Stein, *Communication Systems and Techniques*. New York: McGraw-Hill, 1966.
38. R. F. Pawula, "Relations between the Rice I_e-function and Marcum Q-function with applications to error rate calculations," *Electron. Lett.*, vol. 31, September 28, 1995, pp. 1717–1719.
39. L. J. Mason, "Error probability evaluation of systems employing differential detection in a Rician fast fading environment and Gaussian noise," *IEEE Trans. Commun.*, vol. COM-35, January 1987, pp. 39–46.

APPENDIX 8A: STEIN'S UNIFIED ANALYSIS OF THE ERROR PROBABILITY PERFORMANCE OF CERTAIN COMMUNICATION SYSTEMS

The analysis of the error probability performance of differential and noncoherent detection as well as certain nonideal coherent detection systems on an AWGN channel is characterized by a decision statistic that is either in the form of the product of two complex Gaussian random variables or the difference of the squares of such variables. In what has now become a classic paper in the annals of communication theory literature, Stein [9] showed how, using a simple algebraic relation between the product and difference of square forms of the decision variable, the error probability of certain such binary systems could be analyzed by a unified approach. Our intent in this appendix is to summarize (without proof) the results found in Stein's original paper in a generic form that can easily be referenced in the main text, where it is applied to specific communication scenarios. This generic form will also be useful when extending Stein's results to M-ary communication systems [21] and fading channels [6] as well as certain nonideal coherent detection systems [16].

We start by considering two complex Gaussian variables, $z_{1p} = |z_{1p}|e^{j\Theta_{1p}}$ and $z_{2p} = |z_{2p}|e^{j\Theta_{2p}}$, that are in general correlated and whose sum and difference, $z_{1f} = (z_{1p} + z_{2p})/2$ and $z_{2f} = (z_{1p} - z_{2p})/2$, are also correlated complex Gaussian random variables.[1] A simple algebraic manipulation shows that

$$|z_{1f}|^2 - |z_{2f}|^2 = \left(\frac{z_{1p} + z_{2p}}{2}\right)\left(\frac{z_{1p} + z_{2p}}{2}\right)^* - \left(\frac{z_{1p} - z_{2p}}{2}\right)\left(\frac{z_{1p} - z_{2p}}{2}\right)^*$$

$$= 2\left(\frac{z_{1p}z_{2p}^* + z_{1p}^*z_{2p}}{4}\right) = \text{Re}\{z_{1p}^*z_{2p}\} \quad (8A.1)$$

Hence, a test of $|z_{1f}|^2 - |z_{2f}|^2$ or $\text{Re}\{z_{1p}^*z_{2p}\}$ against a zero threshold, which are typical of noncoherent FSK and differentially coherent PSK systems, respectively, would produce equivalent error probability performance expressions, that is,

$$P = \text{Pr}\{\text{Re}\{z_{1p}^*z_{2p}\} < 0\} \quad (8A.2a)$$

or

$$P = \text{Pr}\{|z_{1f}|^2 - |z_{2f}|^2 < 0\}$$
$$= \text{Pr}\{|z_{1f}|^2 < |z_{2f}|^2\}$$
$$= \text{Pr}\{|z_{1f}| < |z_{2f}|\} \quad (8A.2b)$$

To evaluate the error probability P, Stein used a succession of linear transformations to transform both the FSK and PSK models to a canonical problem that had a convenient solution. In particular, he showed that the solution to (8A.2a) or (8A.2b) could be expressed in terms of an equivalent noncoherent FSK problem based on two nonzero mean but *uncorrelated* complex Gaussian variables, t_1 and t_2 wherein the desired error probability could be stated

---

[1] The subscripts f and p refer, respectively, to FSK and PSK modulations, as will become clear shortly.

APPENDIX 8A: STEIN'S UNIFIED ANALYSIS OF THE ERROR PROBABILITY PERFORMANCE   255

as $P = \Pr\{|t_1|^2 < |t_2|^2\} = \Pr\{|t_1| < |t_2|\}$. By relating t_1 and t_2 to z_{1p}, z_{2p} and z_{1f}, z_{2f}, Stein arrived at the following generic results.

Define the first- and second-order moments of z_{1p}, z_{2p} by (using Stein's notation)

$$\bar{z}_{ip} \triangleq m_{ip} + j\mu_{ip} = |\bar{z}_{ip}|e^{\theta_{ip}}, \qquad i = 1, 2$$

$$S_{ip} \triangleq \tfrac{1}{2}|\bar{z}_{ip}|^2 = \tfrac{1}{2}(m_{ip}^2 + \mu_{ip}^2), \quad N_{ip} \triangleq \tfrac{1}{2}\overline{|z_{ip} - \bar{z}_{ip}|^2}, \qquad i = 1, 2$$

$$\rho_p\sqrt{N_{1p}N_{2p}} \triangleq \tfrac{1}{2}\overline{(z_{1p} - \bar{z}_{1p})^*(z_{2p} - \bar{z}_{2p})}, \qquad \rho_p = \rho_{cp} + j\rho_{sp}$$

$$\tfrac{1}{2}\overline{(z_{1p} - \bar{z}_{1p})(z_{2p} - \bar{z}_{2p})} = 0 \tag{8A.3}$$

and similarly for z_{1f} and z_{2f}. Finally, define the phase angle ϕ by

$$\phi = \arg(N_{1p} - N_{2p} - j2\rho_{sp}\sqrt{N_{1p}N_{2p}}) \tag{8A.4a}$$

or

$$\phi = \arg(\rho_{cf} + j\rho_{sf}) \tag{8A.4b}$$

for the problems characterized by (8A.2a) and (8A.2b), respectively. Then,

$$P = \tfrac{1}{2}[1 - Q_1(\sqrt{b}, \sqrt{a}) + Q_1(\sqrt{a}, \sqrt{b})] - \tfrac{A}{2}\exp\left(-\tfrac{a+b}{2}\right)I_0(\sqrt{ab}) \tag{8A.5}$$

where for the definition of P as in (8A.2a) we have

$$\left\{\begin{matrix}a\\b\end{matrix}\right\} = \frac{1}{2}\left[\frac{S_{1p} + S_{2p} + (S_{1p} - S_{2p})\cos\phi + 2\sqrt{S_{1p}S_{2p}}\sin(\theta_{1p} - \theta_{2p})\sin\phi}{N_{1p} + N_{2p} + \sqrt{(N_{1p} - N_{2p})^2 + 4\rho_{sp}^2 N_{1p}N_{2p}}}\right.$$

$$+ \frac{S_{1p} + S_{2p} - (S_{1p} - S_{2p})\cos\phi - 2\sqrt{S_{1p}S_{2p}}\sin(\theta_{1p} - \theta_{2p})\sin\phi}{N_{1p} + N_{2p} - \sqrt{(N_{1p} - N_{2p})^2 + 4\rho_{sp}^2 N_{1p}N_{2p}}}$$

$$\left.\mp \frac{2\sqrt{S_{1p}S_{2p}}\cos(\theta_{1p} - \theta_{2p})}{\sqrt{(1 - \rho_{sp}^2)N_{1p}N_{2p}}}\right] \tag{8A.6a}$$

$$A = \frac{\rho_{cp}}{\sqrt{1 - \rho_{sp}^2}}$$

and for the definition of P as in (8A.2b) we have

$$\left\{\begin{matrix}a\\b\end{matrix}\right\} = \frac{1}{2}\left[\frac{S_{1f} + S_{2f} + 2\sqrt{S_{1f}S_{2f}}\cos(\theta_{1f} - \theta_{2f} + \phi)}{N_{1f} + N_{2f} + 2\sqrt{N_{1f}N_{2f}|\rho_f|^2}}\right.$$
$$+ \frac{S_{1f} + S_{2f} - 2\sqrt{S_{1f}S_{2f}}\cos(\theta_{1f} - \theta_{2f} + \phi)}{N_{1f} + N_{2f} - 2\sqrt{N_{1f}N_{2f}|\rho_f|^2}}$$
$$\left.\mp \frac{2(S_{1f} - S_{2f})}{\sqrt{(N_{1f} + N_{2f})^2 - 4N_{1f}N_{2f}|\rho_f|^2}}\right] \quad (8A.6b)$$

$$A = \frac{N_{1f} - N_{2f}}{\sqrt{(N_{1f} + N_{2f})^2 - 4N_{1f}N_{2f}|\rho_f|^2}}$$

Several special cases of (8A.6a) and (8A.6b) are of interest. First, if z_{1p} and z_{2p} are uncorrelated (i.e., $|\rho_p| = 0$), $\phi = 0$ or π (depending, respectively, on whether $N_{1p} > N_{2p}$ or $N_{1p} < N_{2p}$. In either event, (8A.6a) simplifies to

$$\left\{\begin{matrix}a\\b\end{matrix}\right\} = \frac{1}{2}\left[\frac{S_{1p}}{N_{1p}} + \frac{S_{2p}}{N_{2p}} \mp 2\sqrt{\frac{S_{1p}}{N_{1p}}\frac{S_{2p}}{N_{2p}}}\cos(\theta_{1p} - \theta_{2p})\right], \quad A = 0$$
$$(8A.7)$$

A further special case of (8A.8) corresponds to $S_{1p} = S_{2p} \triangleq S_p$, $N_{1p} = N_{2p} \triangleq N_p$, $\theta_{1p} = \theta_{2p}$, in which case we obtain

$$\left\{\begin{matrix}a\\b\end{matrix}\right\} = \left\{\begin{matrix}0\\\frac{2S_p}{N_p}\end{matrix}\right\}, \quad A = 0 \quad (8A.8)$$

If for (8A.6b), z_{1f} and z_{2f} have equal noise power (i.e., $N_{1f} = N_{2f} \triangleq N_f$), then (8A.6b) simplifies to

$$\left\{\begin{matrix}a\\b\end{matrix}\right\} = \frac{1}{2N_f}\left[\frac{S_{1f} + S_{2f} - 2|\rho_f|\sqrt{S_{1f}S_{2f}}\cos(\theta_{1f} - \theta_{2f} + \phi)}{1 - |\rho_f|^2}\right.$$
$$\left.\mp \frac{S_{1f} - S_{2f}}{\sqrt{1 - |\rho_f|^2}}\right], \quad A = 0 \quad (8A.9)$$

If, in addition, z_{1f} and z_{2f} are uncorrelated, i.e., $|\rho_f| = 0$, then (8A.9) further simplifies to

$$\left\{\begin{matrix}a\\b\end{matrix}\right\} = \left\{\begin{matrix}\frac{S_{2f}}{N_f}\\\frac{S_{1f}}{N_f}\end{matrix}\right\}, \quad A = 0 \quad (8A.10)$$

APPENDIX 8A: STEIN'S UNIFIED ANALYSIS OF THE ERROR PROBABILITY PERFORMANCE

The generic result in (8A.5) can be simplified by using some of the alternative representations of classical functions given in Chapter 4. In particular, substituting (4.16), (4.19), and (4.65) in (8A.5) and combining terms, we arrive at the result

$$P = \frac{1}{4\pi} \int_{-\pi}^{\pi} \left[\frac{1 - A + 2\zeta A \sin\theta - \zeta^2(1 + A)}{1 + 2\zeta \sin\theta + \zeta^2} \right]$$

$$\times \exp\left[-\frac{b}{2}(1 + 2\zeta \sin\theta + \zeta^2) \right] d\theta, \qquad 0 \leq \zeta \stackrel{\Delta}{=} \sqrt{\frac{a}{b}} < 1 \qquad (8A.11)$$

which for $A = 0$ simplifies to

$$P = \frac{1}{4\pi} \int_{-\pi}^{\pi} \left[\frac{1 - \zeta^2}{1 + 2\zeta \sin\theta + \zeta^2} \right] \exp\left[-\frac{b}{2}(1 + 2\zeta \sin\theta + \zeta^2) \right] d\theta,$$

$$0 \leq \zeta \stackrel{\Delta}{=} \sqrt{\frac{a}{b}} < 1 \qquad (8A.12)$$

It should be noted that the specific form of the result in (8A.12) can be obtained from the work of Pawula [38], who cited certain relations between the Marcum Q-function and the Rice Ie-function, which is defined by

$$Ie(k, x) = \int_0^x \exp(-t) I_0(kt) \, dt \qquad (8A.13)$$

In particular, combining Eqs. (2a) and (2d) of Ref. 38 and making the substitutions $U = (b + a)/2$, $W = (b - a)/2$, and $V^2 = U^2 - W^2 = ab$ in these same equations, one arrives at the result

$$P = \frac{1}{2\pi} \int_0^{\pi} \left[\frac{1 - \zeta^2}{1 - 2\zeta \cos\theta + \zeta^2} \right] \exp\left[-\frac{b}{2}(1 - 2\zeta \cos\theta + \zeta^2) \right] d\theta,$$

$$0 \leq \zeta \stackrel{\Delta}{=} \sqrt{\frac{a}{b}} < 1 \qquad (8A.14)$$

which in view of the symmetry properties of the trigonometric functions over the intervals $(-\pi, 0)$ and $(0, \pi)$ can be shown to be identically equivalent to (8A.12).

Another (simpler) form of P can be obtained from the newer alternative form of the Marcum Q-function due to Pawula [19] and presented in (4.26) through (4.29). Using these relations in (8A.5) with $A = 0$, we obtain

$$P = \frac{1}{4\pi} \int_{-\pi}^{\pi} \exp\left\{ -\frac{b}{2}\left[\frac{(1 - \zeta^2)^2}{1 + 2\zeta \sin\theta + \zeta^2} \right] \right\} d\theta \qquad (8A.15)$$

or equivalently,

$$P = \frac{1}{2\pi} \int_0^\pi \exp\left\{-\frac{b}{2}\left[\frac{(1-\zeta^2)^2}{1 \pm 2\zeta\cos\theta + \zeta^2}\right]\right\} d\theta \qquad (8A.16)$$

The advantage of (8A.15) [or (8A.16)] is that simple upper and lower bounds on P are now readily obtainable by upper and lower bounding the exponential in the integrand by its maximum and minimum values, corresponding, respectively, to $\theta = -\pi/2$ and $\theta = \pi/2$, which immediately gives

$$\frac{1}{2}\exp\left[-\frac{b}{2}(1+\zeta)^2\right] \leq P \leq \frac{1}{2}\exp\left[-\frac{b}{2}(1-\zeta)^2\right] \qquad (8A.17)$$

A still tighter lower bound can be obtained from (8A.16) by the following sequence of steps:[2]

$$P = \frac{1}{2\pi}\int_0^\pi \exp\left\{-\frac{b}{2}\left[\frac{(1-\zeta^2)^2}{1+2\zeta\cos\theta+\zeta^2}\right]\right\} d\theta$$

$$= \frac{1}{2\pi}\int_0^\pi \exp\left\{-\frac{b}{2}\left[\frac{(1-\zeta^2)^2}{(1-\zeta)^2 + 4\zeta\cos^2(\theta/2)}\right]\right\} d\theta$$

$$= \frac{1}{2\pi}\int_0^\pi \exp\left\{-\frac{b}{2}\left[\frac{(1-\zeta^2)^2}{(1+\zeta)^2\cos^2(\theta/2) + (1-\zeta)^2\sin^2(\theta/2)}\right]\right\} d\theta$$

$$\geq \frac{1}{2\pi}\int_0^\pi \exp\left\{-\frac{b}{2}\left[\frac{(1-\zeta)^2}{\cos^2(\theta/2)}\right]\right\} d\theta$$

$$= \frac{1}{\pi}\int_0^{\pi/2}\exp\left\{-\frac{b}{2}\left[\frac{(1-\zeta)^2}{\sin^2(\theta/2)}\right]\right\} d\theta$$

$$= Q(\sqrt{b}(1-\zeta)) = Q(\sqrt{b} - \sqrt{a}) \qquad (8A.18)$$

where the last equality comes from the alternative form of the Gaussian Q-function in (4.2). What is particularly interesting about (8A.18) is that many authors have used this result as an asymptotic approximation to P (e.g., Turin [36, Eq. (A-3-4)]), where the additional constraints $b \gg 1, a \gg 1, b - a > 0$ were imposed. The result as presented in (8A.18) is stronger, in that it is a *strict lower bound* and as such does not require any asymptotic conditions on the parameters. Furthermore, since the upper bound of (8A.17) is, in fact, the Chernoff bound on the lower bound in (8A.18), we conclude that *the probability of error P is bounded between the Gaussian Q-function of the difference of the arguments and the Chernoff bound on this function.*

---

[2] This bound was derived and supplied to the authors by W. F. McGee of Ottawa, Canada.

＃ 9

PERFORMANCE OF MULTICHANNEL RECEIVERS

Many of the current and emerging wireless communication systems make use in one form or another of *diversity*: a classic and well-known concept [1–4] that has been used for the past half century to combat the effects of multipath fading. Indeed, diversity combining, in which two or more copies of the same information-bearing signal are combined skillfully to increase the overall signal-to-noise ratio (SNR), still offers one of the greatest potential for radio link performance improvement to many of the current and future wireless technologies. For example, to meet stringent quality of service requirements, spectrally efficient multilevel constellations need antenna (or space) diversity to reduce the fading-induced penalty on the SNR [5]. In addition, one of the most promising features of wideband CDMA systems is their ability to resolve additional multipaths [compared to "narrowband" (i.e., IS-95) CDMA systems], resulting in an increased diversity which can be exploited by RAKE reception. This particular application of diversity techniques is discussed in detail in Chapter 11. In this chapter we extend the MGF-based approach developed for the performance of single-channel receivers in Chapter 8 to the performance of diversity (i.e., multichannel) receivers. The coverage is broad in the sense that several combining techniques are presented and analyzed in terms of average combined SNR, outage probability, and average probability of error. A particular focus is put on how the performance of these techniques is affected by various channel fading characteristics, such as fading severity, power delay profile, and fading correlation. But first, to understand the concepts and terminology used, we summarize briefly the basic principles of diversity, then review the various types of combining techniques in the remainder of this introductory section.

9.1 DIVERSITY COMBINING

9.1.1 Diversity Concept

As mentioned above, diversity combining consists of receiving redundantly the same information-bearing signal over two or more fading channels, then combining these multiple replicas at the receiver to increase the overall received SNR. The intuition behind this concept is to exploit the low probability of concurrence of deep fades in all the diversity channels to lower the probability of error and of outage.

These multiple replicas can be obtained by extracting the signals via different radio paths:

- In space by using multiple receiver antennas (antenna or site diversity)
- In frequency by using multiple frequency channels which are separated by at least the coherence bandwidth of the channel (frequency hopping or multicarrier systems)
- In time by using multiple time slots which are separated by at least the coherence time of the channel (coded systems)
- Via multipath by resolving multipath components at different delays (direct-sequence spread-spectrum systems with RAKE reception)

9.1.2 Mathematical Modeling

The mathematical model considered in this chapter consists of a multilink channel where the transmitted signal is received over L independent slowly varying flat fading channels, as shown in Fig. 9.1. In the figure, l is the channel index, and $\{\alpha_l\}_{l=1}^{L}$, $\{\theta_l\}_{l=1}^{L}$, and $\{\tau_l\}_{l=1}^{L}$ are the random channel amplitudes, phases, and delays, respectively. We assume that the sets $\{\alpha_l\}_{l=1}^{L}$, $\{\theta_l\}_{l=1}^{L}$, and $\{\tau_l\}_{l=1}^{L}$ are mutually independent. The first channel is assumed to be the reference channel with delay $\tau_1 = 0$ and, without loss of generality, we assume that $\tau_1 < \tau_2 < \cdots < \tau_L$. We assume that the $\{\alpha_l\}_{l=1}^{L}$, $\{\theta_l\}_{l=1}^{L}$, and $\{\tau_l\}_{l=1}^{L}$ are all constant over at least a symbol interval.

When we talk about independent combined paths, we mean that the fading amplitudes $\{\alpha_l\}_{l=1}^{L}$ are assumed to be statistically independent random variables (RV's) where α_l has mean-square value $\overline{\alpha_l^2}$ denoted by Ω_l and a probability density function (PDF) described by any of the family of distributions [Rayleigh, Nakagami-n (Rice), or Nakagami-m] presented in Chapter 2. As we will see throughout the chapter, the multilink channel model used in our analyses is often sufficiently general to include the case where the different channels are not necessarily identically distributed or even distributed according to the same family of distributions. We call this type of multilink channel a *generalized multilink fading channel*.

After passing through the fading channel, each replica of the signal is perturbed by complex additive white Gaussian noise (AWGN) with a one-sided power

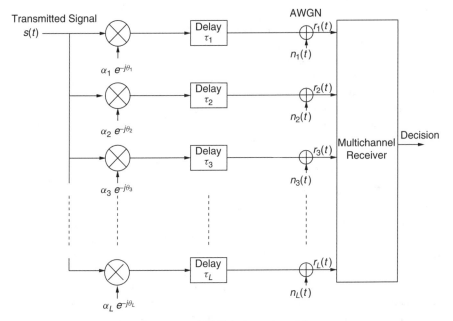

Figure 9.1. Multilink channel model.

spectral density denoted by $2N_l$ (W/Hz). The AWGN is assumed to be statistically independent from channel to channel and independent of the fading amplitudes $\{\alpha_l\}_{l=1}^{L}$. Hence, the instantaneous SNR per symbol of the lth channel is given by $\gamma_l = \alpha_l^2 E_s/N_l$, where E_s (J) is the energy per symbol and the SNR per symbol of the lth channel is given by $\overline{\gamma}_l = \Omega_l E_s/N_l$.

9.1.3 Brief Survey of Diversity Combining Techniques

Diversity techniques can first be classified according to the nature of the fading they are intended to mitigate. For instance, microdiversity schemes are designed to combat short-term multipath fading, whereas macrodiversity techniques mitigate the effect of long-term shadowing caused by obstructions such as buildings, trees, and hills. Diversity schemes can also be classified according to the type of combining employed at the receiver. At this point we should distinguish the classical pure combining schemes [1] from the more recently proposed hybrid techniques.

9.1.3.1 Pure Combining Techniques. There are four principal types of combining techniques, which depend essentially on the (1) complexity restrictions put on the communication system, and (2) amount of channel state information (CSI) available at the receiver.

Maximal Ratio Combining (MRC). As shown in Chapter 7, in the absence of interference, MRC is the optimal combining scheme (regardless of fading statistics) but comes at the expense of complexity since MRC requires knowledge of all channel fading parameters. Since knowledge of channel fading amplitudes is needed for MRC, this scheme can be used in conjunction with unequal-energy signals (e.g., M-QAM or any other amplitude or phase modulations). Furthermore, since knowledge of channel phases is also needed for MRC, this scheme is not practical for differentially coherent and noncoherent detection. Indeed, if channel phase estimates are obtained, the designer might as well go for coherent detection, thus achieving better performance.

Equal-Gain Combining (EGC). Although suboptimal, EGC with coherent detection is often an attractive solution since it does not require estimation of the fading amplitudes and hence results in reduced complexity relative to the optimum MRC scheme. However, EGC is often limited in practice to coherent modulations with equal-energy symbols (M-ary PSK signals). Indeed, for signals with unequal energy symbols such as M-QAM, estimation of the path amplitudes is needed anyway for automatic gain control (AGC) purposes, and thus for these modulations, MRC should be used to achieve better performance [3].

In many applications the phase of the received signal cannot be tracked accurately, and it is therefore not possible to perform coherent detection. In such scenarios, communication systems must rely on noncoherent detection techniques such as envelope or square-law detection of frequency-shift-keying (FSK) signals [6, Chap. 5] or on differentially coherent detection techniques such as differential phase-shift-keying (DPSK) [6, Chap. 7]. As explained above, MRC is not practical for such detection schemes, which are used, rather, in conjunction with postdetection EGC [3, Sec. 5.5.6; 7, Sec. 12.1].

Selection Combining (SC). The two former combining techniques (MRC and EGC) require all or some of the CSI (fading amplitude, phase, and delay) from all the received signals. In addition, a separate receiver chain is needed for each diversity branch, which adds to the overall receiver complexity. On the other hand, SC-type systems process only one of the diversity branches. Specifically, in its conventional form, the SC combiner chooses the branch with the highest SNR. In addition, since the output of the SC combiner is equal to the signal on only one of the branches, the coherent sum of the individual branch signals is not required. Therefore, the SC scheme can be used in conjunction with differentially coherent and noncoherent modulation techniques since it does not require knowledge of the signal phases on each branch as would be needed to implement MRC or EGC in a coherent system.

Switch and Stay Combining (SSC). For systems that use uninterrupted transmission, such as frequency-division multiple-access systems, SC in its conventional form may still be impractical since it requires simultaneous and continuous monitoring of all the diversity branches [3, p. 240]. Hence SC is often implemented in the form of switched or scanning diversity, in which rather than continually picking the best branch, the receiver selects a particular branch until its SNR drops below a predetermined threshold. When this happens the receiver switches to another branch. There are different variants of switched diversity [8], but in its simplest form the SSC receiver switches to, and stays with, the other branch, regardless of whether the SNR of that branch is above or below the predetermined threshold [9,10]. SSC diversity is obviously the least complex diversity scheme to implement and can be used in conjunction with coherent modulations as well as noncoherent and differentially coherent ones.

9.1.3.2 Hybrid Combining Techniques. Because of additional complexity constraints or because of the potential of a higher diversity gain with more sophisticated diversity schemes, newly proposed hybrid techniques have been receiving a great deal of attention in view of their promising offer to meet the specifications of emerging wideband communication systems. These schemes can be categorized into two groups: (1) generalized diversity schemes and (2) multidimensional diversity techniques.

Generalized Diversity Techniques. The complexity of MRC and EGC receivers depends on the number of diversity paths available, which can be quite high, especially for multipath diversity of wideband CDMA signals. In addition, MRC is sensitive to channel estimation errors, and these errors tend to be more important when the instantaneous SNR is low. On the other hand, SC uses only one path out of the L available multipaths and hence does not fully exploit the amount of diversity offered by the channel. Recently, a wave of papers have been published bridging the gap between these two extremes (MRC/EGC and SC) by proposing GSC, which adaptively combines (following the rules of MRC or EGC) the L_c strongest (highest SNR) paths among the L available ones. We denote such hybrid schemes as SC/MRC or SC/EGC-L_c/L. In the context of coherent wideband CDMA systems, these schemes offer less complex receivers than the conventional MRC RAKE receivers since they have a fixed number of fingers independent of the number of multipaths. More important, SC/MRC was shown to approach the performance of MRC, while SC/EGC was shown to outperform in certain cases conventional postdetection EGC since it is less sensitive to the "combining loss" of the very noisy (low-SNR) paths [11].

Multidimensional Diversity Techniques. Multidimensional diversity schemes involving the combination of two or more conventional means of realizing diversity (e.g., space and multipath) to provide better performance have recently received a great deal of attention. For example, in the context of wideband CDMA they are implemented in the form of two-dimensional RAKE receivers, consisting of an array of antennas, each followed by a conventional RAKE receiver. Furthermore, these schemes can take advantage of diversity from frequency and multipath, as is the case in multicarrier-RAKE CDMA systems [12] or from Doppler and multipath as proposed in Ref. 13. Composite microscopic plus macroscopic diversity can also be viewed as a two-dimensional diversity scheme. This type of diversity is used is systems originally proposed about a decade ago by Cox et al. [14] in conjunction with universal digital portable communications. These systems consist of several access ports (base stations) which continually track a mobile terminal. Each access port contains a multielement antenna array that employs microdiversity to reduce the effects of multipath fading. Macrodiversity is then performed at the output of the different access ports to mitigate the effects of shadowing. Two-dimensional diversity can be generalized to multidimensional diversity by simultaneous exploitation of, for example, space, frequency, and multipath diversity.

9.1.4 Complexity–Performance Trade-offs

Wireless system designers are in charge of developing sufficiently high performance systems that achieve a certain specified quality of service while meeting predetermined complexity constraints. The search for the appropriate system design typically involves trade-off studies among various modulation–coding–diversity scheme combinations. An informed decision/choice relies on a precise quantitative performance evaluation of these various combinations.

The objective of this chapter is to develop analytical methods and tools to assess accurately the performance of communication systems operating over wireless fading channels when various diversity techniques are employed to combat the effects of fading. An emphasis is put on the development of "generic" tools to address various performance measures (average combined or output SNR, outage probability, and average error rate), several modulation–diversity scheme combinations, and a variety of fading environments. In particular, analytical methods that are not limited to specific channel conditions are very important since the performance of diversity systems operating over such conditions is affected by various channel characteristics and parameters, such as:

- *Fading distribution on the various diversity branches and paths.* For example, for multipath diversity the statistics of the different paths may be characterized by different families of distributions.
- *Average fading power.* For example, in multipath diversity the average fading power is typically assumed to follow an exponentially decaying power delay profile with equispaced delays: $\overline{\gamma}_l = \overline{\gamma}_1 e^{-\delta(l-1)}$ ($l = 1, 2, \ldots, L$), where $\overline{\gamma}_1$ is the average SNR of the first (reference) propagation path and δ is the average fading power decay factor.
- *Severity of fading.* For example, fading in a macrocellular environment tends to follow a Rayleigh type of fading, whereas fading tends to be Rician or Nakagami-m in a microcellular type of environment.
- *Fading correlation.* For example, because of insufficient antenna spacing in small mobile units equipped with space antennas, diversity and, in this case, the maximum theoretical diversity gain cannot be achieved.

9.2 MAXIMAL-RATIO COMBINING

The performance of MRC over fading channels has long been of interest, as shown by the large number of papers published on this topic. With some special exceptions, most of the models for these systems typically assume either Rayleigh paths or independent identically distributed (i.i.d.) Nakagami or Rician paths. These idealizations are not always realistic since the average fading power [15,16] and the severity of fading [17–19] may vary from one path to another when, for example, multipath diversity is employed. In this section we consider a generalized multilink fading channel and derive expressions for the exact symbol error rate (SER) of linearly modulated signals over such channels [20,21]. The results of this section are applicable to systems that employ coherent demodulation and operate over independent paths.[1] As in Chapter 8 the approach to solving the problem takes advantage of the alternative integral representations [22,23] (see Chapter 4) of the probability of error of these signals over additive white Gaussian noise (AWGN) channels (i.e., the conditional SER), along with the Laplace transforms and/or Gauss–Hermite quadrature integrals of Chapter 5, to derive the SER expressions. Again these expressions involve a single finite-range integral whose integrand contains only elementary functions and which can therefore be easily evaluated. It should be noted that Tellambura et al. [24,25] and Dong et al. [26] also used these alternative representations to analyze the performance of several M-ary signals with MRC diversity reception. These works, which were done independently, have some of the same features as the MGF-based approach described in this section.

9.2.1 Receiver Structure

We consider the L-branch (finger) MRC receiver shown in Fig. 9.2. As mentioned earlier, this receiver is the optimal multichannel receiver regardless of the fading statistics on the various diversity branches since it results in a maximum-likelihood receiver (see Chapter 7). For equally likely transmitted symbols, the total conditional SNR per symbol, γ_t, at the output of the MRC combiner is given by [3, Eq. (5.98)]

$$\gamma_t = \sum_{l=1}^{L} \gamma_l \tag{9.1}$$

For coherent binary signals the conditional BER, $P_b(E|\{\gamma_l\}_{l=1}^{L})$, is given by

$$P_b(E|\{\gamma_l\}_{l=1}^{L}) = Q(\sqrt{2g\gamma_t}) \tag{9.2}$$

where $g = 1$ for coherent BPSK [6, Eq. (4.55)], $g = \frac{1}{2}$ for coherent orthogonal BFSK [6, Eq. (4.59)], $g = 0.715$ for coherent BFSK with minimum correlation [6, Eq. (4.63)], and $Q(\cdot)$ is the Gaussian Q-function. Our goal is to evaluate

---
[1] The independent assumption is relaxed for Nakagami-m fading channels in Section 9.6.

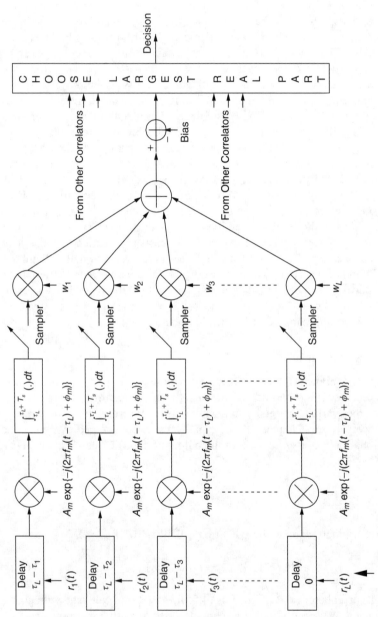

Figure 9.2. Coherent multichannel receiver structure. The weights w_l are such as $w_l = \alpha_l/N_l$ ($l = 1, 2, \ldots, L$) and the bias is set equal to $\sum_{l=1}^{L} (\alpha_l^2/N_l) E_m$ for MRC and $w_l = 1$ ($l = 1, 2, \ldots, L$) and the bias is set equal to 0 for coherent EGC. E_m is the energy of the mth symbol ($m = 1, 2, \ldots, M$).

the performance of the system in terms of users' average BER, and for this purpose the conditional BER (9.2) has to be statistically averaged over the random parameters $\{\gamma_l\}_{l=1}^{L}$. We now present two approaches to solving this problem: the classical PDF-based approach, then the MGF-based approach.

9.2.2 PDF-Based Approach

The classical approach relies on finding the PDF of γ_t, $p_{\gamma_t}(\gamma_t)$, then replacing the L-fold average by a single average over γ_t:

$$P_b(E) = \int_0^\infty Q(\sqrt{2g\gamma_t}) p_{\gamma_t}(\gamma_t) \, d\gamma_t \qquad (9.3)$$

This requires finding the distribution of γ_t in a simple form. If this is possible, it can lead to a closed-form expression for the average probability of error, as shown in the following example.

Example. Let us consider the MRC combining of L independent identically distributed (i.i.d.) Rayleigh fading paths. In this case the SNR per bit per path γ_l has an exponential PDF with average SNR per bit $\bar{\gamma}$:

$$p_{\gamma_l}(\gamma_l) = \frac{1}{\bar{\gamma}} e^{-\gamma_l/\bar{\gamma}} \qquad (9.4)$$

and the SNR per bit of the combined SNR γ_t has a chi-square PDF [7, Eq. (14-4-13)]

$$p_{\gamma_t}(\gamma_t) = \frac{1}{(L-1)!\,\bar{\gamma}^L} \gamma_t^{L-1} e^{-\gamma_t/\bar{\gamma}} \qquad (9.5)$$

Finally, the average probability of error can be found in closed form by successive integration by parts [7, Eq. (14-4-15)]:

$$P_b(E) = \left(\frac{1-\mu}{2}\right)^L \sum_{l=0}^{L-1} \binom{L-1+l}{l} \left(\frac{1+\mu}{2}\right)^l \qquad (9.6)$$

where

$$\mu = \sqrt{\frac{\bar{\gamma}}{1+\bar{\gamma}}} \qquad (9.7)$$

The PDF-based approach has some limitations. Indeed, finding the PDF of the combined SNR per bit γ_t in a simple form is typically feasible if the paths are i.i.d. However, finding the PDF of the combined SNR per bit γ_t is more difficult if the combined paths come from the same family of fading distribution (e.g., Rice) but have different parameters (e.g., different average fading powers (i.e., a nonuniform power delay profile) and/or different severity of fading parameters).[2] In addition, finding the PDF of the combined SNR per bit γ_t is intractable in a simple form if the paths have fading distributions coming from different families of distributions. We now show how the alternative representation of the Gaussian Q-function provides a simple and elegant MGF-based solution to many of these limitations.

9.2.3 MGF-Based Approach

9.2.3.1 Average Bit Error Rate of Binary Signals

Product Form Representation of the Conditional BER. Using the alternative representation of the Gaussian Q-function (4.2) in (9.2), the conditional BER (9.2), may be rewritten in a more desirable *product form* given by

$$P_b\left(E|\{\gamma_l\}_{l=1}^L\right) = \frac{1}{\pi} \int_0^{\pi/2} \exp\left(-\frac{g\gamma_t}{\sin^2\phi}\right) d\phi = \frac{1}{\pi} \int_0^{\pi/2} \prod_{l=1}^L \exp\left(-\frac{g\gamma_l}{\sin^2\phi}\right) d\phi \quad (9.8)$$

This form of the conditional BER is more desirable since we can first independently average over the individual statistical distributions of the γ_l's, and then perform the integral over ϕ, as described in more detail below.

Average BER with Multichannel Reception. To obtain the unconditional BER, $P_b(E)$, when multichannel reception is used, we must average the multichannel conditional BER, $P_b(E|\{\gamma_l\}_{l=1}^L)$, over the joint PDF of the instantaneous SNR sequence $\{\gamma_l\}_{l=1}^L$, namely, $p_{\gamma_1,\gamma_2,...,\gamma_L}(\gamma_1, \gamma_2, \ldots, \gamma_L)$. Since the RVs $\{\gamma_l\}_{l=1}^L$ are assumed to be statistically independent, $p_{\gamma_1,\gamma_2,...,\gamma_L}(\gamma_1, \gamma_2, \ldots, \gamma_L) = \prod_{l=1}^L p_{\gamma_l}(\gamma_l)$, and the averaging procedure results in

$$P_b(E) = \underbrace{\int_0^\infty \int_0^\infty \cdots \int_0^\infty}_{L-\text{fold}} P_b\left(\{\gamma_l\}_{l=1}^L\right) \prod_{l=1}^L p_{\gamma_l}(\gamma_l) d\gamma_1 d\gamma_2 \cdots d\gamma_L \quad (9.9)$$

Note that if the traditional integral representation of the Gaussian Q-function (4.1) were to be used in the $P_b(E|\{\gamma_l\}_{l=1}^L)$ term, (9.9) would result in an $(L+1)$-fold integral with infinite limits [one of these integrals comes from the classical definition of the Gaussian Q-function (4.1) in $P_b(E|\{\gamma_l\}_{l=1}^L)$], and a closed-form solution or an adequately efficient numerical integration method would not be available. Using the alternative product form representation of the conditional

---
[2] Note that a solution for this problem exists for Rayleigh [7, Eq. (14-5-26)] and Nakagami-m [27].

BER (9.8) in (9.9) yields

$$P_b(E) = \underbrace{\int_0^\infty \int_0^\infty \cdots \int_0^\infty}_{L\text{-fold}} \frac{1}{\pi}$$

$$\times \int_0^{\pi/2} \prod_{l=1}^L \exp\left(-\frac{g\gamma_l}{\sin^2\phi}\right) p_{\gamma_l}(\gamma_l) \, d\phi \, d\gamma_1 \, d\gamma_2 \cdots d\gamma_L \quad (9.10)$$

The integrand in (9.10) is absolutely integrable, and hence the order of integration can be interchanged. Thus, grouping terms of index l, we obtain

$$P_b(E) = \frac{1}{\pi} \int_0^{\pi/2} \prod_{l=1}^L M_{\gamma_l}\left(-\frac{g}{\sin^2\phi}\right) d\phi \quad (9.11)$$

where $M_{\gamma_l}(s) \triangleq \int_0^\infty p_{\gamma_l}(\gamma_l) e^{s\gamma_l} d\gamma_l$ is the MGF of the SNR per symbol γ_l associated with path l and is summarized in Table 9.1 (or equivalently, Table 2.2) for various channel models of interest. If the fading is identically distributed with the same fading parameter and the same average SNR per bit $\bar{\gamma}$ for all L channels, (9.11) reduces to

$$P_b(E) = \frac{1}{\pi} \int_0^{\pi/2} \left(M_\gamma\left(-\frac{g}{\sin^2\phi}\right)\right)^L d\phi \quad (9.12)$$

Hence, in all cases this approach reduces the $(L+1)$-fold integral with infinite limits of (9.9) (accounting for the infinite range integral coming from the traditional representation of the Gaussian Q-function) to a single finite-range integral (9.11) whose integrand contains only elementary functions such as exponentials and trigonometrics, and which can therefore easily be evaluated numerically. As a numerical example, Fig. 9.3 shows the average BER

TABLE 9.1 Moment Generating Function of the SNR per Symbol γ_l for Some Common Multipath Fading Channels

Type of Fading	Fading Parameter	$M_{\gamma_l}(s)$
Rayleigh		$(1 - s\bar{\gamma}_l)^{-1}$
Nakagami-q (Hoyt)	$0 \le q_l \le 1$	$\left(1 - 2s\bar{\gamma}_l + \frac{(2s\bar{\gamma}_l)^2 q_l^2}{(1+q_l^2)^2}\right)^{-1/2}$
Nakagami-n (Rice)	$0 \le n_l$	$\frac{(1+n_l^2)}{(1+n_l^2) - s\bar{\gamma}_l} \exp\left(\frac{n_l^2 s\bar{\gamma}_l}{(1+n_l^2) - s\bar{\gamma}_l}\right)$
Nakagami-m	$\frac{1}{2} \le m_l$	$\left(1 - \frac{s\bar{\gamma}_l}{m_l}\right)^{-m_l}$

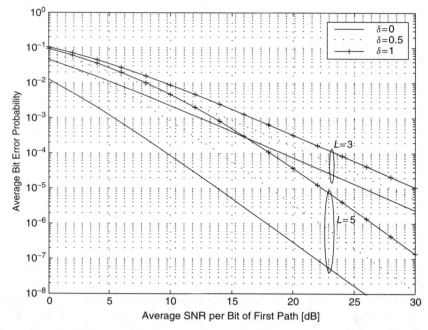

Figure 9.3. Average BER of BPSK with L-fold MRC diversity versus the SNR per bit of the first path for an L-path frequency-selective Nakagami-m ($m = 0.5$) fading channel with an exponentially decaying power delay profile. δ is the power decay factor.

performance of BPSK over a frequency-selective Nakagami-m ($m = 0.5$) fading channel with an exponentially decaying power delay profile when MRC RAKE reception is used.

It is interesting to mention at this point that the same final result (9.11) can be obtained without using the alternative representation of the Gaussian Q-function, but by starting with Eq. (17) of Ref. 28. Indeed, it has been pointed out to the authors by Mazo [29] that Eq. (17), which is expressed in terms of the characteristic function of γ_t (using our notations) can be rewritten in terms of the MGF of γ_t by changing the integration contour. The details of the procedure are described in an internal AT&T Bell Laboratories memorandum which was never submitted for publication [30]. Following that procedure and using the fact that the MGF of the sum of independent RV's is the product of the MGF's of the individual RVs [31, Sec. 7.4], [28, Eq. (17)] can be rewritten as (using our notations)

$$P_b(E) = \frac{1}{2\pi} \int_1^\infty \frac{\prod_{l=1}^L M_{\gamma_l}(-gy)}{y\sqrt{y-1}} dy \qquad (9.13)$$

which can be changed to the same single finite-range integral (9.11) by adopting the change of variables $y = 1/\sin^2 \phi$ [29].

9.2.3.2 Average Symbol Error Rate of M-PSK Signals

Product Form Representation of the Conditional SER. Similar to the binary case, using (8.22), the conditional SER for M-PSK, $P_s(E|\{\gamma_l\}_{l=1}^L)$ can be expressed as an integral of the desired product form,

$$P_s(E|\{\gamma_l\}_{l=1}^L) = \frac{1}{\pi}\int_0^{(M-1)\pi/M} \exp\left(-\frac{g_{\text{PSK}}\gamma_t}{\sin^2\phi}\right) d\phi$$

$$= \frac{1}{\pi}\int_0^{(M-1)\pi/M} \prod_{l=1}^L \exp\left(-\frac{g_{\text{PSK}}\gamma_l}{\sin^2\phi}\right) d\phi \quad (9.14)$$

where $g_{\text{PSK}} = \sin^2(\pi/M)$.

Average SER of M-PSK. Following the same steps as in (9.9) through (9.11), it can easily be shown that the average SER of M-PSK, $P_s(E)$, over generalized fading channels is given by

$$P_s(E) = \frac{1}{\pi}\int_0^{(M-1)\pi/M} \prod_{l=1}^L M_{\gamma_l}\left(-\frac{g_{\text{PSK}}}{\sin^2\phi}\right) d\phi \quad (9.15)$$

The result (9.15) generalizes the M-PSK average SER results of Proakis [32, Eq. (22)] and Chennakeshu and Anderson [33, Eq. (21)] for L independent identically distributed Rayleigh paths. It also gives an alternative approach for the performance evaluation of coherent M-PSK over frequency-selective channels characterized by a Rician dominant path with Rayleigh secondary paths [34,35].

Furthermore, by setting L to 1, the result (9.15) can be used to evaluate the average SER performance of M-PSK with single-channel reception, as shown in Chapter 8. This leads, for example, to the following results:

- *Rayleigh.* Substituting the MGF corresponding to Rayleigh fading in (9.15) (with $L = 1$), then using Eq. (2.562.1) of Ref. 36 yields the closed-form expression given by (8.112), which can also be found in Pauw and Schilling [37, Eq. (9)] and Ekanayake [38, Eq. (7)] and which agrees with the results obtained using various other methods [32, Eq. (22); 39, Eq. (36)].
- *Nakagami-n (Rice).* Substituting the MGF corresponding to Nakagami-n (Rice) fading in (9.15) leads to an expression for the SER of M-PSK which is easily shown to agree with Eq. (35) of Ref. 39.
- *Nakagami-m.* Substituting the MGF corresponding to Nakagami-m fading in (9.15) (with $L = 1$) gives the SER of M-PSK over a Nakagami-m channel as

$$P_s(E) = \frac{1}{\pi}\int_0^{(M-1)\pi/M} \left(1 + \frac{\overline{\gamma}\sin^2(\pi/M)}{m\sin^2\phi}\right)^{-m} d\phi \quad (9.16)$$

Note that (9.16) yields the same numerical values as Eq. (17) of Ref. 40 and Eq. (9) of Ref. 41 and it is much easier to compute for any arbitrary value of m.

9.2.3.3 Average Symbol Error Rate of M-AM Signals

Product-Form Representation of the Conditional SER. Recall that the conditional SER for M-AM, $P_s(E|\{\gamma_l\}_{l=1}^L)$, with signal points located symmetrically about the origin, is given by (8.3) as

$$P_s\left(E|\{\gamma_l\}_{l=1}^L\right) = \frac{2(M-1)}{M} Q(\sqrt{2g_{AM}\gamma_t}) \qquad (9.17)$$

where $g_{AM} = 3/(M^2 - 1)$. Using the alternative representation of the Gaussian Q-function (4.2) in (9.17), we obtain the conditional SER in the desired product form as

$$\begin{aligned}
P_s\left(E|\{\gamma_l\}_{l=1}^L\right) &= \frac{2(M-1)}{M\pi} \int_0^{\pi/2} \exp\left(-\frac{g_{AM}\gamma_t}{\sin^2\phi}\right) d\phi \\
&= \frac{2(M-1)}{M\pi} \int_0^{\pi/2} \prod_{l=1}^L \exp\left(-\frac{g_{AM}\gamma_l}{\sin^2\phi}\right) d\phi \qquad (9.18)
\end{aligned}$$

Average SER of M-AM. Following the same steps as in (9.9) through (9.11), it is straightforward to show that the average SER of M-AM over generalized fading channels is given by

$$P_s(E) = \frac{2(M-1)}{M\pi} \int_0^{\pi/2} \prod_{l=1}^L M_{\gamma_l}\left(-\frac{g_{AM}}{\sin^2\phi}\right) d\phi \qquad (9.19)$$

9.2.3.4 Average Symbol Error Rate of Square M-QAM Signals

Product Form Representation of the Conditional SER. Consider square M-QAM signals whose constellation size is given by $M = 2^k$ with k even. The conditional SER for square M-QAM is given by (8.10) as

$$\begin{aligned}
P_s\left(E|\{\gamma_l\}_{l=1}^L\right) &= 4\left(1 - \frac{1}{\sqrt{M}}\right) Q(\sqrt{2g_{QAM}\gamma_t}) \\
&\quad - 4\left(1 - \frac{1}{\sqrt{M}}\right)^2 Q^2(\sqrt{2g_{QAM}\gamma_t}) \qquad (9.20)
\end{aligned}$$

where $g_{QAM} = 3/2(M-1)$. Using the alternative representation of the Gaussian Q-function (4.2) as well as of its square (4.9), the conditional SER (9.20) may be rewritten in the more desirable product form given by

$$P_s\left(E|\{\gamma_l\}_{l=1}^L\right) = \frac{4}{\pi}\left(1-\frac{1}{\sqrt{M}}\right)\int_0^{\pi/2}\exp\left(-\frac{g_{QAM}\gamma_t}{\sin^2\phi}\right)d\phi$$
$$-\frac{4}{\pi}\left(1-\frac{1}{\sqrt{M}}\right)^2\int_0^{\pi/4}\exp\left(-\frac{g_{QAM}\gamma_t}{\sin^2\phi}\right)d\phi$$
$$=\frac{4}{\pi}\left(1-\frac{1}{\sqrt{M}}\right)\int_0^{\pi/2}\prod_{l=1}^L\exp\left(-\frac{g_{QAM}\gamma_l}{\sin^2\phi}\right)d\phi$$
$$-\frac{4}{\pi}\left(1-\frac{1}{\sqrt{M}}\right)^2\int_0^{\pi/4}\prod_{l=1}^L\exp\left(-\frac{g_{QAM}\gamma_l}{\sin^2\phi}\right)d\phi.$$

Average SER of M-QAM. Following the same steps as in (9.9) through (9.11) yields the average SER of M-QAM over generalized fading channels as

$$P_s(E) = \frac{4}{\pi}\left(1-\frac{1}{\sqrt{M}}\right)\int_0^{\pi/2}\prod_{l=1}^L M_{\gamma_l}\left(-\frac{g_{QAM}}{\sin^2\phi}\right)d\phi$$
$$-\frac{4}{\pi}\left(1-\frac{1}{\sqrt{M}}\right)^2\int_0^{\pi/4}\prod_{l=1}^L M_{\gamma_l}\left(-\frac{g_{QAM}}{\sin^2\phi}\right)d\phi \quad (9.21)$$

Of particular interest is the average SER performance of M-QAM with single-channel reception, which can be obtained by setting L to 1 in (9.21). For example, substituting the MGF corresponding to Rayleigh fading in (9.21) (with $L = 1$), then using again Eq. (2.562.1) of Ref. 36 yields a closed-form expression for the average SER of M-QAM over Rayleigh channels as given by (8.106), namely,

$$P_s(E) = 2\left(1-\frac{1}{\sqrt{M}}\right)\left(1-\sqrt{\frac{g_{QAM}\overline{\gamma}}{1+g_{QAM}\overline{\gamma}}}\right)$$
$$+\left(1-\frac{1}{\sqrt{M}}\right)^2\left[\frac{4}{\pi}\sqrt{\frac{g_{QAM}\overline{\gamma}}{1+g_{QAM}\overline{\gamma}}}\tan^{-1}\sqrt{\frac{1+g_{QAM}\overline{\gamma}}{g_{QAM}\overline{\gamma}}}-1\right] \quad (9.22)$$

Note that (9.22) matches the result obtained by Shayesteh and Aghamohammadi [39, Eq. (44)] for the particular case where $M = 16$. Furthermore, note that (9.22) can be alternatively obtained by averaging (9.20) over the Rayleigh PDF and by using a standard known integral involving the function $\text{erfc}^2(\cdot)$ [36, Eq. (8.258.2)]. In addition using (5A.4b) and (5A.21) in (9.21) we obtain the performance of M-QAM over L i.i.d. Rayleigh fading channels as

$$P_s(E) = 4\left(1 - \frac{1}{\sqrt{M}}\right)\left(\frac{1-\mu_c}{2}\right)^L \sum_{l=0}^{L-1} \binom{L-1+l}{l}\left(\frac{1+\mu_c}{2}\right)^l$$

$$- 4\left(1 - \frac{1}{\sqrt{M}}\right)^2 \left\{\frac{1}{4} - \frac{\mu_c}{\pi}\left[\left(\frac{\pi}{2} - \tan^{-1}\mu_c\right)\sum_{l=0}^{L-1} \frac{\binom{2l}{l}}{4(1+g_{QAM}\bar{\gamma})^l}\right.\right.$$

$$\left.\left. - \sin(\tan^{-1}\mu_c)\sum_{l=1}^{L-1}\sum_{i=1}^{l} \frac{T_{il}}{(1+g_{QAM}\bar{\gamma})^l}[\cos(\tan^{-1}\mu_c)]\right]\right\} \quad (9.23)$$

where

$$\mu_c = \sqrt{\frac{g_{QAM}\bar{\gamma}}{1+g_{QAM}\bar{\gamma}}} \quad (9.24)$$

and

$$T_{il} = \frac{\binom{2l}{l}}{\binom{2(l-i)}{l-i}\{4^i[2(l-i)+1]\}} \quad (9.25)$$

Note that (9.23) is equivalent to Eq. (15) of Ref. 42 and to Eq. (12) of Ref. 43, which involves a sum of Gauss hypergeometric functions.[3] Furthermore, using a partial fraction expansion on the integrand of (9.21), we obtain with the help of Eq. (2.562.1) of Ref. 36 the average SER of M-QAM over L Rayleigh fading channels with distinct average fading powers and with MRC reception as

$$P_s(E) = 2\left(1 - \frac{1}{\sqrt{M}}\right)\sum_{l=1}^{L} \rho_l\left(1 - \sqrt{\frac{g_{QAM}\bar{\gamma}_l}{1+g_{QAM}\bar{\gamma}_l}}\right) + \left(1 - \frac{1}{\sqrt{M}}\right)^2$$

$$\times \left[\frac{4}{\pi}\sum_{l=1}^{L} \rho_l\sqrt{\frac{g_{QAM}\bar{\gamma}_l}{1+g_{QAM}\bar{\gamma}_l}}\tan^{-1}\left(\sqrt{\frac{1+g_{QAM}\bar{\gamma}_l}{g_{QAM}\bar{\gamma}_l}}\right) - \sum_{l=1}^{L}\rho_l\right]$$

where

$$\rho_l = \left[\prod_{\substack{k=1\\k\neq l}}^{L}\left(1 - \frac{\bar{\gamma}_k}{\bar{\gamma}_l}\right)\right]^{-1} \quad (9.26)$$

which is equivalent to Eq. (10) of Ref. 42 and to Eq. (21) of Ref. 43.

---

[3] Equation (12) of Ref. 43 gives the same numerical result as the one given by (9.23) if a minor typo is corrected in Eq. (18) of Ref. 43 [the denominator should be $(2k+1)\sqrt{\pi}$ rather than $(2k-1)\sqrt{\pi}$].

Before concluding our discussion on the exact average SER evaluation of M-ary signals with MRC over independent fading paths, we should mention that the approach presented for M-PSK, M-AM, and M-QAM signals can be applied to any two-dimensional amplitude or phase linear modulation as shown in Ref. 26, since, based on Craig's approach [23], the conditional SER expressions of any of these constellations can be expressed as a summation of integrals in the desired exponential form (see Section 5.4.2).

9.2.4 Bounds and Asymptotic SER Expressions

In this section we are interested in determining simple closed-form bounds and asymptotic expressions (limit as the average SNR/symbol/channel approaches infinity) for the SER of M-ary signals with MRC reception.

9.2.4.1 Bounds. As discussed in Section 8.1.1.3, the integrand of the conditional SER of M-PSK as given by (9.14) has a single maximum that occurs at $\phi = \pi/2$. Thus, replacing the integrand by its maximum yields an upper bound for (9.15) given by

$$P_s(E) \leq \frac{M-1}{M} \prod_{l=1}^{L} M_{\gamma_l}(-g_{\text{PSK}}) \qquad (9.27)$$

Similarly, the average SER of M-AM is upper-bounded by

$$P_s(E) \leq \frac{M-1}{M} \prod_{l=1}^{L} M_{\gamma_l}(-g_{\text{AM}}) \qquad (9.28)$$

9.2.4.2 Asymptotic Results. Consider an M-ary communication system operating over an L-path slowly varying fading channel and assume that the L channels have independent Rician statistics that do not have to be identically distributed in that they can have different average symbol SNR's, $\overline{\gamma}_l, l = 1, 2, \ldots, L$, and different Rician factors $K_l, l = 1, 2, \ldots, L$. Let $P_s(E|\gamma_1, \gamma_2, \ldots, \gamma_L)$ denote the conditional (on the fading SNR's) SER of the system, and let $p_{\gamma_l}(\gamma_l), l = 1, 2, \ldots, L$, denote the PDF's of these SNR's, which for a Rician channel are given by

$$p_{\gamma_l}(\gamma_l) = \frac{1+K_l}{\overline{\gamma}_l} \exp\left(-K_l - \frac{1+K_l}{\overline{\gamma}_l}\gamma_l\right)$$

$$\times I_0\left(2\sqrt{\frac{\gamma_l K_l(1+K_l)}{\overline{\gamma}_l}}\right) \qquad (9.29)$$

Then, substituting (9.29) in (9.9), the exact average SER is given by

$$P_s(E) = \int_0^\infty \int_0^\infty \cdots \int_0^\infty \left(\prod_{l=1}^{L} p_{\gamma_l}(\gamma_l)\right) P_s(E|\gamma_1, \gamma_2, \ldots, \gamma_L) d\gamma_1 d\gamma_2 \cdots d\gamma_L$$

$$= \left[\prod_{l=1}^{L} \frac{1+K_l}{\bar{\gamma}_l}\right] \exp\left(-\sum_{l=1}^{L} K_l\right)$$

$$\times \int_0^\infty \int_0^\infty \cdots \int_0^\infty \prod_{l=1}^{L} \left[\exp\left(-\frac{1+K_l}{\bar{\gamma}_l}\gamma_l\right) I_0\left(2\sqrt{\frac{\gamma_l K_l(1+K_l)}{\bar{\gamma}_l}}\right)\right]$$

$$\times P_s(E|\gamma_1, \gamma_2, \ldots, \gamma_L) d\gamma_1 d\gamma_2 \cdots d\gamma_L \tag{9.30}$$

Introducing the shorthand notation

$$\Psi(L) \triangleq \prod_{l=1}^{L} \frac{\bar{\gamma}_l}{1+K_l}, \quad \Delta(L) \triangleq \sum_{l=1}^{L} K_l \tag{9.31}$$

(9.30) can be rewritten as

$$P_s(E) = \frac{\exp(-\Delta(L))}{\Psi(L)}$$

$$\times \int_0^\infty \int_0^\infty \cdots \int_0^\infty \prod_{l=1}^{L} \left[\exp\left(-\frac{1+K_l}{\bar{\gamma}_l}\gamma_l\right) I_0\left(2\sqrt{\frac{\gamma_l K_l(1+K_l)}{\bar{\gamma}_l}}\right)\right]$$

$$\times P_s(E|\gamma_1, \gamma_2, \ldots, \gamma_L) d\gamma_1 d\gamma_2 \cdots d\gamma_L \tag{9.32}$$

Abdel-Ghaffar and Pasupathy [44] have shown that for large $\bar{\gamma}_l, l = 1, 2, \ldots, L$, the average SER of (9.32) behaves as

$$P_s(E) \simeq \frac{\exp[-\Delta(L)]C(L,M)}{\Psi(L)} = \frac{\left[\prod_{l=1}^{L}(1+K_l)\right]\exp\left(-\sum_{l=1}^{L}K_l\right)C(L,M)}{\prod_{l=1}^{L}\bar{\gamma}_l} \tag{9.33}$$

or

$$P_s(E) \simeq \frac{\left[\prod_{l=1}^{L}(1+K_l)\right]\exp\left(-\sum_{l=1}^{L}K_l\right)C_b(L,M)}{\prod_{l=1}^{L}(\bar{\gamma}_b)_l},$$

$$(\bar{\gamma}_b)_l \triangleq \frac{\bar{\gamma}_l}{\log_2 M}, \quad C_b(L,M) \triangleq \frac{C(L,M)}{(\log_2 M)^L} \tag{9.34}$$

where $C(L,M)$ [or $C_b(L,M)$] is a term that depends on the modulation/detection/diversity scheme but is independent of $\bar{\gamma}_l, l = 1, 2, \ldots, L$. Thus, the entire dependence on the average SNR per channel is embedded in the product form of $\Psi(L)$ defined in (9.31) and for equal average SNRs would vary as $\bar{\gamma}^{-L}$ [or $(\bar{\gamma}_b)^{-L}$]. For Rayleigh channels, (9.34) simplifies to

$$P_s(E) \simeq \frac{C(L,M)}{\Psi(L)\big|_{K_l|_{l=1}^L=0}} = \frac{C(L,M)}{\prod_{l=1}^L \bar{\gamma}_l} \tag{9.35}$$

or

$$P_s(E) \simeq \frac{C_b(L,M)}{\prod_{l=1}^L (\bar{\gamma}_b)_l} \tag{9.36}$$

The evaluation of $C(L,M)$ for some special cases is of interest and is described by Theorems 2 and 3 in Abdel-Ghaffar and Pasupathy [44]. In particular, when the conditional SER $P_s(E|\gamma_1, \gamma_2, \ldots, \gamma_L)$ depends only on the sum of the γ_l's as is the case for coherent MRC, it is shown that $C(L,M)$ is computed from

$$C(L,M)|_{\text{MRC}} = \frac{1}{(L-1)!} \int_0^\infty \gamma_t^{L-1} P_s(E|\gamma_t) \, d\gamma_t \tag{9.37}$$

Thus, the asymptotic average SER for Rayleigh channels with MRC is

$$P_s(E) \simeq \frac{1}{(L-1)! \prod_{l=1}^L \bar{\gamma}_l} \int_0^\infty \gamma_t^{L-1} P_s(E|\gamma_t) \, d\gamma_t \tag{9.38}$$

which for equal average SNRs becomes

$$P_s(E) \simeq \frac{1}{(L-1)! \bar{\gamma}^L} \int_0^\infty \gamma_t^{L-1} P_s(E|\gamma_t) \, d\gamma_t \tag{9.39}$$

This is to be compared with the exact result obtained from (9.39) which can be written as

$$P_s(E) = \int_0^\infty p_{\gamma_t}(\gamma_t) P_s(E|\gamma_t) \, d\gamma_t \tag{9.40}$$

where for Rayleigh i.i.d. channels, $p_{\gamma_t}(\gamma_t)$ is given by (9.5) Substituting (9.5) into (9.40) gives

$$P_s(E) = \frac{1}{(L-1)! \bar{\gamma}^L} \int_0^\infty \gamma_t^{L-1} \exp\left(-\frac{\gamma_t}{\bar{\gamma}}\right) P_s(E|\gamma_t) \, d\gamma_t \tag{9.41}$$

Notice the similarity between (9.41) and (9.39), i.e., the exact result has an additional $\exp(-\gamma_t/\bar{\gamma})$ in its integrand.

$C(L,M)$ can be evaluated in closed form for a variety of different modulation/detection/diversity techniques of interest. In particular for M-PSK with MRC

Abdel-Ghaffar and Pasupathy [44] showed that

$$C(L, M) = \frac{1}{[2\sin(\pi/M)]^{2L}} \left[\binom{2L}{L} \frac{M-1}{M} - \sum_{l=1}^{L} \binom{2L}{L-l} (-1)^l \frac{\sin(2\pi l/M)}{\pi l} \right] \quad (9.42)$$

with the special case of $L = 1$ (i.e., no diversity) given by

$$C(1, M) = \frac{1}{2\sin^2(\pi/M)} \left(\frac{M-1}{M} + \frac{\sin(2\pi/M)}{2\pi} \right) \quad (9.43)$$

Substituting (9.43) in (9.36) and letting $\overline{\gamma}_1 = \overline{\gamma} = \overline{\gamma}_b \log_2 M$, we get

$$P_s(E) \simeq \frac{1}{2\overline{\gamma}_b (\log_2 M) \sin^2(\pi/M)} \left(\frac{M-1}{M} + \frac{\sin(2\pi/M)}{2\pi} \right) \quad (9.44)$$

which resembles Eq. (14-4-39) of Ref. 7 with the addition of the term $(\sin 2\pi/M)/2\pi$. For M-QAM with MRC, using the conditional SER expression (9.20) as well as the alternative representations of the Gaussian Q-function (4.2) and its square (4.9) in (9.41), we arrive at the following result for $C(L, M)$:

$$C(L, M) = \frac{1}{2^{2L}} \left(\frac{\sqrt{M}-1}{\sqrt{M}} \right)^2 \left(\frac{2(M-1)}{3} \right)^L \left[\binom{2L}{L} \left(\frac{2\sqrt{M}}{\sqrt{M}-1} - 1 \right) \right.$$

$$\left. - \frac{4}{\pi} \sum_{\substack{l=1 \\ l \text{ odd}}}^{L} \binom{2L}{L-l} \frac{(-1)^{(l+1)/2}}{l} \right] \quad (9.45)$$

For $M = 4$ (QPSK), this can be shown to check with the result in (9.42).

9.3 COHERENT EQUAL GAIN COMBINING

As mentioned in Section 9.2, MRC provides the maximum performance improvement relative to all other diversity combining techniques by maximizing the signal-to-noise ratio (SNR) at the combiner output. However, MRC has the highest complexity of all combining techniques since it requires knowledge of the fading amplitude in each signal branch. Alternative combining techniques such as EGC are often used in practice because of their reduced complexity relative to the optimum MRC scheme [3, Sec. 5.5]. Indeed, EGC weights each branch equally before combining, and therefore does not require estimation of the channel (path) fading amplitudes.

Our focus in this section is on evaluating the average SER for ideal coherent detection of M-PSK signals with EGC reception over Nakagami-m fading channels. Work related to this topic can be found in Refs. 25, 26, and 45 through 47.

More specifically, in Refs. 45 and 46 Abu-Dayya and Beaulieu employ an infinite series representation for the PDF of the sum of Nakagami-m and Rice random variables [48] to analyze the performance of binary modulations when used in conjunction with EGC. The same approach was adopted by Dong et al. [26] to extend the results to several two-dimensional constellations of interest. Another approach based on the Gil-Pelaez lemma [49] was recently proposed by Zhang [47] and lead to closed-form solutions for binary modulations with two or three branch EGC receivers over Rayleigh fading channels. This approach was extended to Nakagami-m fading channels in Ref. 50. In Ref. 25 Annamalai et al. use a frequency domain–based approach and Parseval's theorem to compute the average symbol error rate with EGC over Nakagami-m fading channels. In this section we use the alternative representation of the conditional SER to analyze the average SER of M-PSK signals with EGC reception over Nakagami-m fading channels. The approach leads to a final expression for the average SER in the form of a single finite-range integral and an integrand composed of tabulated functions [20,51].

9.3.1 Receiver Structure

The EGC receiver processes the L received replicas, weights them equally, then sums them to produce the decision statistic, as shown in Fig. 9.2. Note that estimation of the channel carrier phase is still required in this case since the weights applied to each branch in the combiner are complex quantities whose amplitudes are all set to 1 and whose phases are indeed equal to the negatives of these carrier-phase estimates. For equally likely transmitted symbols, it can be easily shown that the total conditional SNR per symbol, γ_{EGC}, at the output of the EGC combiner is given by [3, Eq. (5.108)]

$$\gamma_{\text{EGC}} = \frac{\left(\sum_{l=1}^{L} \alpha_l\right)^2 E_s}{\sum_{l=1}^{L} N_l} \tag{9.46}$$

where E_s (J) is the energy per symbol and N_l is the AWGN power spectral density on the lth path.

9.3.2 Average Output SNR

We consider the average combined SNR at the output of a coherent EGC receiver operating over a frequency-selective Nakagami-m fading channel with an exponentially decaying power delay profile ($\Omega_l = \Omega_1 e^{\delta(l-1)}, l = 1, 2, \ldots, l$). Assuming independent paths and the same AWGN power spectral density N_0, the average combined SNR can be written from (9.46) as

$$\bar{\gamma}_{\text{EGC}} = \frac{E_s}{LN_0} \left(\sum_{l=1}^{L} \Omega_l + \sum_{i=1}^{L} \sum_{\substack{j=1 \\ j \neq i}}^{L} E(\alpha_i) E(\alpha_j) \right) \tag{9.47}$$

For $\delta = 0$ [i.e., uniform power delay profile such as $\bar{\gamma}_l = \bar{\gamma}$ ($l = 1, 2, \ldots, L$)] it can easily be shown that (9.47) can be written as

$$\bar{\gamma}_{\text{EGC}} = \bar{\gamma}\left(1 + (L-1)\frac{[\Gamma(m+\frac{1}{2})]^2}{m[\Gamma(m)]^2}\right) \qquad (9.48)$$

which reduces for the Rayleigh case to

$$\bar{\gamma}_{\text{EGC}} = \bar{\gamma}\left(1 + (L-1)\frac{\pi}{4}\right) \qquad (9.49)$$

in agreement with Eq. (5.112) of Ref. 3. These two expressions should be contrasted with the average combined SNR at the output of an optimal MRC receiver, which for $\delta = 0$ is given, in view of (9.1), by [4, Eq. (6.70)]

$$\bar{\gamma}_{\text{MRC}} = L\bar{\gamma} \qquad (9.50)$$

regardless of the type of fading.

For $\delta > 0$ using geometric summations it can be shown that (9.47) reduces to

$$\bar{\gamma}_{\text{EGC}} = \frac{\bar{\gamma}_1}{L}\left[\frac{1 - e^{-L\delta}}{1 - e^{-\delta}} + \frac{2[\Gamma(m+\frac{1}{2})]^2}{m[\Gamma(m)]^2(1 - e^{-\delta/2})}\right.$$
$$\left. \times \left(e^{-\delta/2}\frac{1 - e^{-(L-1)\delta}}{1 - e^{-\delta}} - e^{-L\delta/2}\frac{1 - e^{-(L-1)\delta/2}}{1 - e^{-\delta/2}}\right)\right] \qquad (9.51)$$

which simplifies for the Rayleigh case ($m = 1$) to

$$\bar{\gamma}_{\text{EGC}} = \frac{\bar{\gamma}_1}{L}\left[\frac{1 - e^{-L\delta}}{1 - e^{-\delta}} + \frac{\pi}{2(1 - e^{-\delta/2})}\right.$$
$$\left. \times \left(e^{-\delta/2}\frac{1 - e^{-(L-1)\delta}}{1 - e^{-\delta}} - e^{-L\delta/2}\frac{1 - e^{-(L-1)\delta/2}}{1 - e^{-\delta/2}}\right)\right] \qquad (9.52)$$

Again these two last expressions should be contrasted with the average combined SNR at the output of an MRC receiver which can easily be shown, in view of (9.1), to be given by

$$\bar{\gamma}_{\text{MRC}} = \bar{\gamma}_1 \frac{1 - e^{-L\delta}}{1 - e^{-\delta}} \qquad (9.53)$$

regardless of the type of fading. These expressions are illustrated in Fig. 9.4, where the normalized output SNR $\bar{\gamma}_{\text{EGC}}/\bar{\gamma}_1$ of EGC is plotted as a function of

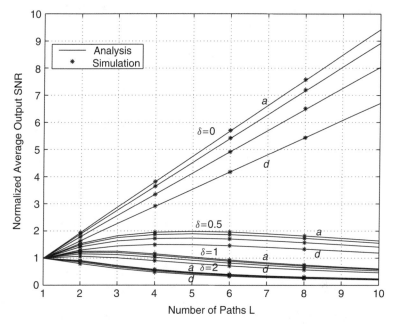

Figure 9.4. Normalized average output SNR $\bar{\gamma}_{\text{EGC}}/\bar{\gamma}_1$ of coherent EGC over Nakagami-m channels with an exponentially decaying power delay profile: (a) $m = 4$; (b) $m = 2$; (c) $m = 1$; (d) $m = 0.5$.

the number of paths for various values of the Nakagami-m parameter and the power decay factor δ. Note the combining loss when $\delta > 0$, which gets more accentuated as δ increases.

9.3.3 Exact Error Rate Analysis

9.3.3.1 Binary Signals. We begin our discussion by considering the performance of an EGC receiver when coherent binary BPSK or binary BFSK modulation is transmitted over a multilink channel with L paths. Conditioned on the fading amplitudes $\{\alpha_l\}_{l=1}^{L}$, the BER $P_b(E|\{\alpha_l\}_{l=1}^{L})$, of an EGC receiver is given by

$$P_b\left(E|\{\alpha_l\}_{l=1}^{L}\right) = Q(\sqrt{2g\gamma_{\text{EGC}}})$$

$$= Q\left(\sqrt{\frac{2gE_b}{\sum_{l=1}^{L} N_l}} \left(\sum_{l=1}^{L} \alpha_l\right)^2\right) \quad (9.54)$$

where as for the MRC case g is a modulation-dependent parameter such that $g = 1$ for BPSK, $g = \frac{1}{2}$ for orthogonal BFSK, and $g = 0.715$ for BFSK with minimum correlation [6].

The average BER $P_b(E)$ is obtained by averaging (9.54) over the joint PDF of the channel fading amplitudes $p_{\alpha_1,\alpha_2,\ldots,\alpha_L}(\alpha_1, \alpha_2, \ldots, \alpha_L)$, that is,

$$P_b(E) = \int_0^\infty \cdots \int_0^\infty Q\left(\sqrt{\frac{2gE_b}{\sum_{l=1}^L N_l}\left(\sum_{l=1}^L \alpha_l\right)^2}\right)$$

$$\times p_{\alpha_1,\alpha_2,\ldots,\alpha_L}(\alpha_1, \alpha_2, \ldots, \alpha_L)\, d\alpha_1\, d\alpha_2 \cdots d\alpha_L \quad (9.55)$$

The L-fold integral in (9.55) can be collapsed to a single integral, namely,

$$P_b(E) = \int_0^\infty Q\left(\sqrt{\frac{2gE_b}{\sum_{l=1}^L N_l}\alpha_t^2}\right) p_{\alpha_t}(\alpha_t)\, d\alpha_t \quad (9.56)$$

where $\alpha_t = \sum_{l=1}^L \alpha_l$ denotes the sum of the fading amplitudes after combining.

In general, there are two difficulties associated with analytically evaluating the average BER as expressed in (9.56). The first relates to the requirement of obtaining the PDF of the total fading RV α_t. When the fading amplitudes can be assumed independent (the case to be considered in this section), finding this PDF requires a convolution of the PDFs of the α_l's and can often be quite difficult to evaluate. The second difficulty has to do with the fact that the argument of the classical definition of the Gaussian Q-function in (4.1) appears in the lower limit of the integral, which is undesirable when trying to perform the average over α_t.

To circumvent these difficulties, we now propose a new method of solution based on the alternative representation of the Gaussian Q-function as given by (4.2), namely,

$$Q(x) = \frac{1}{\pi} \int_0^{\pi/2} \exp\left(-\frac{x^2}{2\sin^2\phi}\right) d\phi; \quad x \geq 0 \quad (9.57)$$

First, using (9.57) in (9.55) gives

$$P_b(E) = \int_0^\infty \cdots \int_0^\infty \frac{1}{\pi} \int_0^{\pi/2} \exp\left(-\frac{gE_b\left(\sum_{l=1}^L \alpha_l\right)^2}{\sum_{l=1}^L N_l \sin^2\phi}\right)$$

$$\times p_{\alpha_1}(\alpha_1) \cdots p_{\alpha_L}(\alpha_L)\, d\phi\, d\alpha_1\, d\alpha_2 \cdots d\alpha_L \quad (9.58)$$

Unfortunately, we cannot represent the exponential in (9.58) as a product of exponentials each involving only a single α_l because of the presence of the $\alpha_k \alpha_l$ cross-product terms. Hence, we cannot partition the L-fold integral into a product of one-dimensional integrals as is possible for MRC (see Ref. 21) and thus we must abandon this approach. Instead, we use the alternative representation of the Gaussian Q-function in (9.56), which gives after switching the order of integration

$$P_b(E) = \frac{1}{\pi} \int_0^{\pi/2} \int_0^\infty \exp\left(-\frac{A^2}{2\sin^2\phi}\alpha_t^2\right) p_{\alpha_t}(\alpha_t) \, d\alpha_t \, d\phi \quad (9.59)$$

where $A = \sqrt{2gE_b/\sum_{l=1}^L N_l}$. Although this maneuver cures the second difficulty by getting the total fading RV α_t out of the lower limit of the integral and into the integrand, it appears that we are still faced with the problem of determining the PDF of α_t. To get around this difficulty, we represent $p_{\alpha_t}(\alpha_t)$ in terms of its characteristic function $\Psi_{\alpha_t}(jv)$, which, because of the independence assumption on the fading channel amplitudes, becomes

$$p_{\alpha_t}(\alpha_t) = \frac{1}{2\pi} \int_{-\infty}^\infty \Psi_{\alpha_t}(jv) e^{-jv\alpha_t} \, dv$$

$$= \frac{1}{2\pi} \int_{-\infty}^\infty \left[\prod_{l=1}^L \Psi_{\alpha_l}(jv)\right] e^{-jv\alpha_t} \, dv \quad (9.60)$$

where $\Psi_{\alpha_l}(jv)$ is the characteristic function of the fading amplitude α_l corresponding to the lth path. Substituting (9.60) into (9.59) gives

$$P_b(E) = \frac{1}{2\pi^2} \int_0^{\pi/2} \int_{-\infty}^\infty \left[\prod_{l=1}^L \Psi_{\alpha_l}(jv)\right]$$

$$\times \underbrace{\int_0^\infty \exp\left(-\frac{A^2 \alpha_t^2}{2\sin^2\phi} - jv\alpha_t\right) d\alpha_t}_{J(v,\phi)} \, dv \, d\phi \quad (9.61)$$

The integral $J(v, \phi)$ can be obtained in terms of the complementary error function $\mathrm{erfc}(\cdot)$ as

$$J(v, \phi) = \sqrt{\frac{\pi}{2}\frac{\sin\phi}{A}} \exp\left(-\frac{\sin^2\phi}{2A^2}v^2\right)\left[1 + \mathrm{erfc}\left(j\frac{\sin\phi}{\sqrt{2}A}v\right)\right] \quad (9.62)$$

or alternatively, by separately evaluating its real and imaginary parts, namely [36, Eqs. (3.896.4) and (3.896.3)],

$$\int_0^\infty \exp\left(-\frac{A^2}{2\sin^2\phi}\alpha_t^2\right) \cos(v\alpha_t) \, d\alpha_t = \sqrt{\frac{\pi\sin^2\phi}{2A^2}} \exp\left(-\frac{\sin^2\phi}{2A^2}v^2\right)$$

$$\int_0^\infty \exp\left[-\frac{A^2}{2\sin^2\phi}\alpha_t^2\right] \sin(v\alpha_t) \, d\alpha_t = \frac{v\sin^2\phi}{A^2} \exp\left(-\frac{\sin^2\phi}{2A^2}v^2\right)$$

$$\times {}_1F_1\left(\frac{1}{2};\frac{3}{2};\frac{\sin^2\phi}{2A^2}v^2\right) \quad (9.63)$$

where $_1F_1(\cdot;\cdot;\cdot)$ is the Kummer confluent hypergeometric function [52, Eq. (13.1.2)]. Thus, letting

$$X(\phi) = \sqrt{\frac{\pi}{2}\frac{\sin\phi}{A}}$$

$$Y(v,\phi) = -\frac{v\sin^2\phi}{A^2}\,_1F_1\left(\frac{1}{2};\frac{3}{2};\frac{\sin^2\phi}{2A^2}v^2\right) \tag{9.64}$$

we can write the integral $J(v,\phi)$ in the form

$$J(v,\phi) = [X(\phi) + jY(v,\phi)]\exp\left(-\frac{\sin^2\phi}{2A^2}v^2\right)$$

$$= \sqrt{X^2(\phi) + Y^2(v,\phi)}\exp\left(j\tan^{-1}\frac{Y(v,\phi)}{X(\phi)}\right)$$

$$\times \exp\left(-\frac{\sin^2\phi}{2A^2}v^2\right) \tag{9.65}$$

In general, the characteristic function of a PDF will be a complex quantity, and hence the product of characteristic functions in (9.61) will also be complex. However, since the average BER is real, it is sufficient to consider only the real part of the right-hand side of (9.61), which yields

$$P_b(E) = \frac{1}{2\pi^2}\int_0^{\pi/2}\int_{-\infty}^{\infty}\mathrm{Re}\left\{\left[\prod_{l=1}^{L}\Psi_{\alpha_l}(jv)\right]J(v,\phi)\right\}dv\,d\phi \tag{9.66}$$

Expressing the characteristic function of each fading path PDF by

$$\Psi_{\alpha_l}(jv) = U_l(v) + jV_l(v)$$

$$= \sqrt{U_l^2(v) + V_l^2(v)}\exp\left(j\tan^{-1}\frac{V_l(v)}{U_l(v)}\right) \tag{9.67}$$

then substituting (9.65) and (9.67) into (9.66) gives

$$P_b(E) = \frac{1}{2\pi^2}\int_0^{\pi/2}\int_{-\infty}^{\infty}\mathcal{F}(v,\phi)\exp\left(-\frac{\sin^2\phi}{2A^2}v^2\right)dv\,d\phi \tag{9.68}$$

where

$$\mathcal{F}(v,\phi) = R(v,\phi)\cos\Theta(v,\phi)$$

$$R(v,\phi) = \sqrt{X^2(\phi) + Y^2(v,\phi)}\prod_{l=1}^{L}\sqrt{U_l^2(v) + V_l^2(v)}$$

$$\Theta(v,\phi) = \tan^{-1}\left(\frac{Y(v,\phi)}{X(\phi)}\right) + \sum_{l=1}^{L}\tan^{-1}\left(\frac{V_l(v)}{U_l(v)}\right)$$

$$+ \frac{\pi}{2}\left(L+1 - \text{sgn}(Y(v,\phi)) - \sum_{l=1}^{L}\text{sgn}(V_l(v))\right), \quad (9.69)$$

where sgn(·) denotes the sign function and the arctangent function is defined with respect to the standard principal value as available, for example, in the MATHEMATICA routine for that function.

The characteristic function corresponding to the Nakagami-m fading PDF can be evaluated with the help of Eq. (3.462.1) of Ref. 36 in terms of the parabolic cylinder function $D_{-v}(\cdot)$ [36, Secs. 9.24 and 9.25]

$$\Psi_{\alpha_l}(jv) = \frac{1}{2^{m_l-1}}\frac{\Gamma(2m_l)}{\Gamma(m_l)}D_{-2m_l}\left(-jv\sqrt{\frac{\Omega_l}{2m_l}}\right)\exp\left(-\frac{\Omega_l}{8m_l}v^2\right)$$

or alternatively, by separately evaluating its real and imaginary parts by using sine and cosine Fourier transforms found in Ref. 36 with the results [see Eq. (9.67)]

$$U_l(v) = A_l(v)\exp\left(-\frac{\Omega_l}{4m_l}v^2\right)$$
$$V_l(v) = B_l(v)\exp\left(-\frac{\Omega_l}{4m_l}v^2\right) \quad (9.70)$$

where

$$A_l(v) = {}_1F_1\left(\frac{1}{2} - m_l; \frac{1}{2}; \frac{v^2\Omega_l}{4m_l}\right)$$
$$B_l(v) = \frac{\Gamma(m_l + \frac{1}{2})}{\Gamma(m_l)}\sqrt{\frac{\Omega_l}{m_l}}\, v\, {}_1F_1\left(1 - m_l; \frac{3}{2}; \frac{v^2\Omega_l}{4m_l}\right) \quad (9.71)$$

Thus, the functions defined in (9.69) become

$$R(v,\phi) = \sqrt{X^2(\phi) + Y^2(v,\phi)}\prod_{l=1}^{L}\sqrt{A_l^2(v) + B_l^2(v)}$$

$$\times \exp\left(-\sum_{l=1}^{L}\frac{\Omega_l}{4m_l}v^2\right)$$

$$\Theta(v,\phi) = \tan^{-1}\left(\frac{Y(v,\phi)}{X(\phi)}\right) + \sum_{l=1}^{L}\tan^{-1}\left(\frac{B_l(v)}{A_l(v)}\right)$$

$$+ \frac{\pi}{2}\left(L+1 - \text{sgn}(Y(v,\phi)) - \sum_{l=1}^{L}\text{sgn}(B_l(v))\right), \quad (9.72)$$

with $X(\phi)$ and $Y(v, \phi)$ as defined in (9.64) and $A_l(v)$ and $B_l(v)$ as defined in (9.71). It is convenient in this case to absorb the exponential factor in $R(v, \theta)$ into the exponential factor in the integrand of (9.68). Hence, we can write the average BER of (9.68) as

$$P_b(E) = \frac{1}{2\pi^2} \int_0^{\pi/2} \int_{-\infty}^{\infty} \mathcal{F}_0(v, \phi) \exp\left[-\left(\frac{\sin^2\phi}{2A^2} + \sum_{l=1}^{L} \frac{\Omega_l}{4m_l}\right)v^2\right] dv\, d\phi \quad (9.73)$$

where $\mathcal{F}_0(v, \phi)$ is a normalized version of $\mathcal{F}(v, \phi)$ defined by

$$\mathcal{F}_0(v, \phi) = R_0(v, \phi) \cos \Theta(v, \phi) \quad (9.74)$$

with

$$R_0(v, \phi) = \sqrt{X^2(\phi) + Y^2(v, \phi)} \prod_{l=1}^{L} \sqrt{A_l^2(v) + B_l^2(v)} \quad (9.75)$$

and $\Theta(v, \phi)$ still as defined in (9.72). Finally, letting

$$\eta(\phi) = \frac{\sin^2\phi}{2A^2} + \sum_{l=1}^{L} \frac{\Omega_l}{4m_l} \quad (9.76)$$

and making the change of variables $x = \sqrt{\eta(\phi)}\, v$, the inner doubly infinite integral is of the form

$$\int_{-\infty}^{\infty} \mathcal{F}_0\left(\frac{x}{\sqrt{\eta(\phi)}}, \phi\right) e^{-x^2} dx \quad (9.77)$$

which can readily be evaluated by the Gauss–Hermite quadrature formula [52, Eq. (25.4.46)], yielding the desired final result in the form of a single finite-range integral on ϕ, namely,

$$P_b(E) = \frac{1}{2\pi^2} \int_0^{\pi/2} \frac{1}{\sqrt{\eta(\phi)}} \sum_{n=1}^{N_p} H_{x_n} \mathcal{F}_0\left(\frac{x_n}{\sqrt{\eta(\phi)}}, \phi\right) d\phi \quad (9.78)$$

where N_p is the order of the Hermite polynomial, $H_{N_p}(\cdot)$. Setting N_p to 20 is typically sufficient for excellent accuracy. In (9.78) x_n are the zeros of the N_pth-order Hermite polynomial, and H_{x_n} are the weight factors of the N_pth-order Hermite polynomial and are given by [52, Table 25.10]

$$H_{x_n} = \frac{2^{N_p-1} N_p! \sqrt{\pi}}{N_p^2 H_{N_p-1}^2(x_n)} \quad (9.79)$$

Both the zeros and the weights factors of the Hermite polynomial are tabulated in Table 25.10 of Ref. 52 for various polynomial orders N_p. Note that substituting (9.74), (9.75), and (9.76) in (9.78), it can be shown that if $N_l = N_0$ ($l = 1, 2, \ldots, L$), the average BER in (9.78) is solely a function of the various average SNR/bit/paths $\bar{\gamma}_l = \Omega_l E_b/N_0$. As a numerical example, Fig. 9.5 compares the

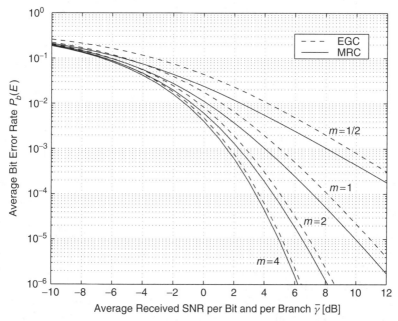

Figure 9.5. Average BER of B-PSK over Nakagami-m fading channels with MRC and coherent EGC. $L = 4$.

BER performance of BPSK with MRC and EGC over i.i.d. Nakagami-m fading channels. Note that EGC approaches the performance of MRC as m increases.

9.3.3.2 Extension to M-PSK Signals.
Recall that the SER for M-PSK over an AWGN is given by the integral expression (8.22), namely,

$$P_s(E|\alpha_t) = \frac{1}{\pi} \int_0^{(M-1)\pi/M} \exp\left(-\frac{g_{\text{PSK}} E_s}{N_0 \sin^2 \phi}\right) d\phi \qquad (9.80)$$

where $g_{\text{PSK}} = \sin^2(\pi/M)$ and E_s/N_0 is the received symbol SNR. For EGC reception in the presence of fading, the conditional SER is obtained from (9.78) by replacing E_s/N_0 by γ_{EGC}, which represents the instantaneous SNR per symbol after combining. Following the same steps as in Section 9.3.3.1, it is straightforward to show that the average SER is given by an equation analogous to (9.78), namely,

$$P_s(E) = \frac{1}{2\pi^2} \int_0^{(M-1)\pi/M} \frac{1}{\sqrt{\eta_{\text{PSK}}(\phi)}} \sum_{n=1}^{N_p} H_{x_n} \mathcal{F}_0\left(\frac{x_n}{\sqrt{\eta_{\text{PSK}}(\phi)}}, \phi\right) d\phi \qquad (9.81)$$

where

$$\eta_{\text{PSK}}(\phi) = \frac{\sin^2 \phi}{2A_{\text{PSK}}^2} + \sum_{l=1}^{L} \frac{\Omega_l}{4m_l}$$

$$A_{\text{PSK}} = \sqrt{\frac{2g_{\text{PSK}} E_s}{\sum_{l=1}^{L} N_l}}$$
(9.82)

and all other parameters and functions remain the same as for the binary signal case. It can also be shown that if $N_l = N_0$ ($l = 1, 2, \ldots, L$), then (9.81) is solely a function of the various average SNR/symbol/paths $\overline{\gamma}_l = \Omega_l E_s / N_0$.

9.3.4 Approximate Error Rate Analysis

As mentioned previously, one of the difficulties in evaluating (9.56) is the requirement of obtaining the PDF of the total fading RV α_t. Even use of the alternative representation of the Gaussian Q-function, which leads to (9.59), does not alleviate this problem. Although there is no known closed-form exact expression for the PDF of the sum of L i.i.d. Nakagami-m RVs, Nakagami [53] showed after rather complex calculations that such a sum can be approximated accurately by another Nakagami-m distribution [53, p. 22] with a parameter $_0m$ and average power $_0\Omega$ given by

$$_0m = mL$$

$$_0\Omega = L\Omega \left[1 + (L-1) \frac{\Gamma^2(m + \tfrac{1}{2})}{m \Gamma^2(m)} \right] \simeq L^2 \Omega \left(1 - \frac{1}{5m} \right) \quad (9.83)$$

Specifically, modeling the PDF $p_{\alpha_t}(\alpha_t)$ by a Nakagami-m PDF with parameters $_0m$ and $_0\Omega$, the average BER for binary signals as given by (9.56) can be evaluated in closed form as [54, App. A]

$$P_b(E) \simeq \frac{1}{2} \sqrt{\frac{g \overline{\gamma}_{\text{eq}}}{\pi (m + g \overline{\gamma}_{\text{eq}})}} \frac{\Gamma(Lm + \tfrac{1}{2})}{\Gamma(Lm + 1)} \left(\frac{m}{m + g \overline{\gamma}_{\text{eq}}} \right)^{Lm}$$

$$\times {}_2F_1 \left(1, Lm + \tfrac{1}{2}, Lm + 1; \frac{m}{m + g \overline{\gamma}_{\text{eq}}} \right), \quad (9.84)$$

where ${}_2F_1(\cdot; \cdot; \cdot; \cdot)$ is the Gauss hypergeometric function [52, Eq. (15.1.1)], $\overline{\gamma}$ is the SNR/symbol/path, and

$$\overline{\gamma}_{\text{eq}} = \frac{\overline{\gamma}}{L} \left[1 + (L-1) \frac{\Gamma^2(m + \tfrac{1}{2})}{m \Gamma^2(m)} \right] \simeq \overline{\gamma} \left(1 - \frac{1}{5m} \right) \quad (9.85)$$

For the special case where m is integer, it can be shown (using Ref. 54, App. A) that (9.84) simplifies to

$$P_b(E) \simeq \frac{1}{2}\left[1 - \sqrt{\frac{g\overline{\gamma}_{eq}}{m + g\overline{\gamma}_{eq}}} \sum_{l=0}^{Lm-1} \frac{\binom{2l}{l}}{[4(1 + g\overline{\gamma}_{eq}/m)]^l}\right] \qquad (9.86)$$

For M-PSK, the same approximate modeling of α_t can be used together with the conditional SER obtained from (9.80) to compute an approximate expression for the average SER. For an arbitrary noninteger m, no closed-form solution is available and the final average has to be computed numerically. However, when m is restricted to integer values, using the expression (5A.35), the average SER can be expressed in closed form as

$$P_s(E) \simeq \frac{M-1}{M} - \frac{T}{\pi}\sqrt{\frac{g_{PSK}\overline{\gamma}_{eq}}{m + g_{PSK}\overline{\gamma}_{eq}}} \sum_{l=0}^{Lm-1} \frac{\binom{2l}{l}}{[4(1 + g_{PSK}\overline{\gamma}_{eq}/m)]^l}$$

$$- \frac{2}{\pi}\sqrt{\frac{g_{PSK}\overline{\gamma}_{eq}}{m + g_{PSK}\overline{\gamma}_{eq}}} \sum_{l=0}^{Lm-1} \sum_{j=0}^{l-1} \binom{2l}{j} \frac{(-1)^{l+j}}{[4(1 + g_{PSK}\overline{\gamma}_{eq}/m)]^l}$$

$$\times \frac{\sin[2(l-j)T]}{2(l-j)} \qquad (9.87)$$

where

$$T = \frac{\pi}{2}\left\{1 + \frac{1}{2}\left(1 + \mathrm{sgn}\left[\left(1 + \frac{2g_{PSK}\overline{\gamma}_{eq}}{m}\right)\cos\left(\frac{2\pi}{M}\right) - 1\right]\right)\right.$$

$$\left. - \tan^{-1}\frac{2\sqrt{(g_{PSK}\overline{\gamma}_{eq}/m)(1 + g_{PSK}\overline{\gamma}_{eq}/m)}\sin(2\pi/M)}{(1 + 2g_{PSK}\overline{\gamma}_{eq}/m)\cos(2\pi/M) - 1}\right\} \qquad (9.88)$$

For $M = 2$ it can be shown that T as given in (9.88) becomes equal to $\pi/2$, and therefore (9.87) reduces to (9.86).

9.3.5 Asymptotic Error Rate Analysis

In this section we continue the discussion started in Section 9.2.4 and consider the asymptotic SER performance of M-PSK with coherent EGC reception. In this case, the conditional combined SNR depends on the γ_l's as given in (9.46), namely,

$$\gamma_t = \gamma_{\text{EGC}} = \frac{1}{L}\left(\sum_{l=1}^{L}\sqrt{\gamma_l}\right)^2 \qquad (9.89)$$

assuming that all the branches have the same AWGN power spectral density (i.e., $N_l = N_0, l = 1, 2, \ldots, L$). Then Abdel-Ghaffer and Pasupathy [44] showed that $C(L, M)$ for EGC is computed from

$$C(L,M)|_{\text{EGC}} = \frac{(2L)^L L!}{(2L)!} \frac{1}{(L-1)!} \int_0^\infty \gamma_t^{L-1} P_s(E|\gamma_t)\, d\gamma_t \qquad (9.90)$$

It is of interest to compare the asymptotic performance of coherent MRC and EGC receivers. Since the conditional SER of these two receivers would be equal when $\gamma_{\text{EGC}} = \gamma_{\text{MRC}}$, then comparing (9.90) with (9.37), we observe that

$$\frac{C(L,M)|_{\text{EGC}}}{C(L,M)|_{\text{MRC}}} = \frac{(2L)^L L!}{(2L)!} \simeq \frac{1}{\sqrt{2}}\left(\frac{e}{2}\right)^L, \qquad L \gg 1 \qquad (9.91)$$

where the latter approximation is obtained using Stirling's formula $x! \simeq \sqrt{2\pi}e^{-x}x^{x+1/2}, x \gg 1$. Thus, we conclude that the asymptotic SER degradation of EGC relative to MRC is given by (9.91). Hence, combining (9.91) and (9.42) immediately gives the desired expression for $C(L, M)$ for M-PSK with EGC reception, namely,

$$C(L,M) = \frac{(L/2)^L L!}{(2L)!} \frac{1}{[\sin(\pi/M)]^{2L}}$$

$$\times \left[\binom{2L}{L}\frac{M-1}{M} - \sum_{l=1}^{L}\binom{2L}{L-l}(-1)^l \frac{\sin(2\pi l/M)}{\pi l}\right] \qquad (9.92)$$

9.4 NONCOHERENT EQUAL-GAIN COMBINING

In this section we consider the performance of several differentially coherent and noncoherent modulations when used in conjunction with postdetection EGC [3, Sec. 5.5.6; 7, Sec. 12-1]. In Section 9.4.1 we present an approach based on the alternative representation of the generalized Marcum Q-function which applies to binary DPSK and FSK as well as DQPSK [55]. Then in Section 9.4.2 we present another approach, which applies to noncoherent orthogonal M-FSK [56].

9.4.1 DPSK, DQPSK, and BFSK: Exact and Bounds

There are a large number of papers dealing with the performance of noncoherent and differentially coherent communication and detection systems when used in conjunction with postdetection EGC over AWGN as well as fading channels. For example, Proakis [57] developed a generic expression for evaluating the BER for multichannel noncoherent and differentially coherent reception of binary signals over L independent AWGN channels. Further, Proakis [7, Sec. 14-4] provides closed-form expressions for the average BER of binary orthogonal square-law detected FSK and binary DPSK with multichannel reception over L i.i.d. Rayleigh fading channels. Lindsey [58] derived a general expression for the average BER of binary correlated FSK with multichannel communication over L independent Rician fading channels in which the strength of the scattered component is assumed to be constant for all the channels. Charash [59] analyzed the average BER performance of binary orthogonal FSK with multichannel reception over L i.i.d. Nakagami-m fading channels. More recently, Weng and Leung [60] derived a closed-form expression for the average BER of binary DPSK with multichannel reception over L i.i.d. Nakagami-m fading channels. Patenaude et al. [61] extended the results of Charash [59] and Weng and Leung [60] by providing a closed-form expression for the average BER performance of binary orthogonal square-law detected FSK and binary DPSK with multichannel reception over L independent but not necessarily identically distributed Nakagami-m fading channels. Their derivation is based on the characteristic function method and the resulting expression contains $(L-1)$th-order derivatives, which can be found for small L but which become more complicated to find as L increases.

In this section we present two unified approaches for the performance evaluation of such systems over generalized fading channels. The first approach, which is described in Section 9.4.1.2, exploits the alternative integral form of the Marcum Q-function as presented in Section 4.2 and the resulting alternative integral representation of the conditional BER as well as the Laplace transforms and/or Gauss–Hermite quadrature integration derived in Chapter 5 to independently average over the PDF of each channel that fades. In all cases, this approach leads to exact expressions of the average BER that involve a single finite-range integral whose integrand contains only elementary functions and which can therefore be easily computed numerically. The second approach, which is presented in Section 9.4.1.3, relies on the bounds on the generalized Marcum Q-function developed in Section 4.2 to derive tight closed-form bounds on the average BER of the systems under consideration.

9.4.1.1 Receiver Structures. We consider L branch (finger) postdetection EGC receivers, as shown in Figs. 9.6 and 9.7 for differentially coherent and noncoherent detection, respectively. Both receivers utilize M correlators to detect the maximum a priori transmitted symbol. Without loss of generality let us consider the mth symbol correlator. Each of the L received signals $r_l(t)$ is first delayed by $\tau_L - \tau_l$, then integrate and dump filtered followed by baud-rate sampling). These operations assume that the receiver is correctly time synchronized at every branch (i.e., perfect time delay $\{\tau_l\}_{l=1}^{L}$ estimates).

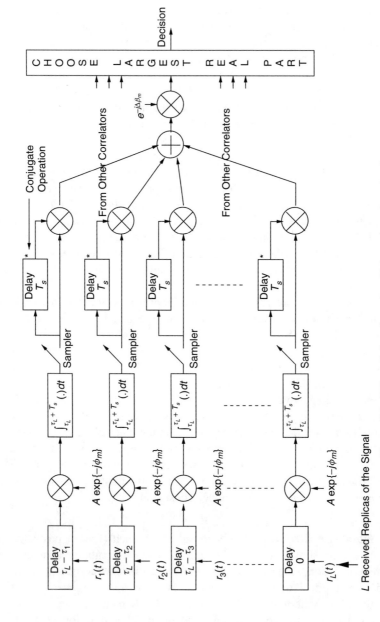

Figure 9.6. Differentially coherent multichannel receiver structure.

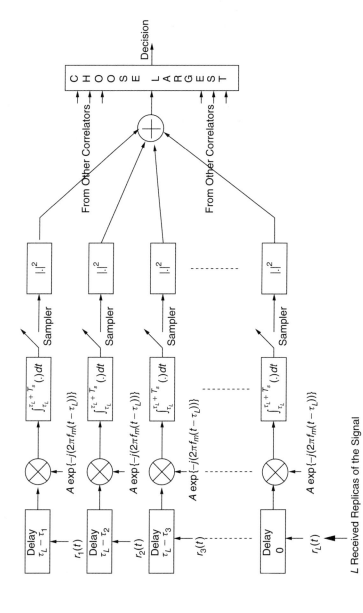

Figure 9.7. Noncoherent multichannel receiver structure.

For differentially coherent detection (see Fig. 9.6) the receiver takes, at every branch l, the difference of two adjacent transmitted phases to arrive at the decision $r_{m,l}$. For noncoherent detection (see Fig. 9.7) no attempt is made to estimate the phase, and the receiver yields the decision $r_{m,l}$ based on the squared envelope (i.e., square-law detection). Using EGC, the L decision outputs $\{r_{m,l}\}_{l=1}^{L}$ are summed to form the final decision variable r_m:

$$r_m = \sum_{l=1}^{L} r_{m,l}, \qquad m = 1, 2, \ldots, M \tag{9.93}$$

Last of all, the receiver selects the symbol corresponding to the maximum decision variable, as shown in Figs. 9.6 and 9.7.

For equally likely transmitted symbols, the total conditional SNR per bit, γ_t, at the output of the postdetection EGC combiner is given by Proakis [7, Sec. 12-1] as

$$\gamma_t = \sum_{l=1}^{L} \gamma_l \tag{9.94}$$

9.4.1.2 Exact Analysis of Average Bit Error Probability.
Many problems dealing with the BER performance of multichannel reception of differentially coherent and noncoherent detection of PSK and FSK signals in AWGN channels have a decision variable that is a quadratic form in complex-valued Gaussian random variables. Almost three decades ago, Proakis [57] developed a general expression for evaluating the BER when the decision variable is in that particular form. Indeed, the development and results originally obtained in Ref. 57 later appeared as Appendix B of Ref. 7 and have become a classic in the annals of communication system performance literature. The most general form of the BER expression [i.e., Ref. 7, Eq. (B-21)] obtained by Proakis was given in terms of the first-order Marcum Q-function and modified Bessel functions of the first kind. Although implied but not explicitly given in Refs. 7 and 57, this general form can be rewritten in terms of the generalized Marcum Q-function, $Q_l(\cdot, \cdot)$, as (see Appendix 9A)

$$
\begin{aligned}
P_b(L, \gamma_t; a, b, \eta) &= Q_1(a\sqrt{\gamma_t}, b\sqrt{\gamma_t}) - \left[1 - \frac{\sum_{l=0}^{L-1}\binom{2L-1}{l}\eta^l}{(1+\eta)^{2L-1}}\right]\exp\left[-\frac{(a^2+b^2)\gamma_t}{2}\right]I_0(ab\gamma_t) \\
&\quad + \frac{1}{(1+\eta)^{2L-1}}\left[\sum_{l=2}^{L}\binom{2L-1}{L-l}\eta^{L-l}[Q_l(a\sqrt{\gamma_t}, b\sqrt{\gamma_t}) - Q_1(a\sqrt{\gamma_t}, b\sqrt{\gamma_t})]\right] \\
&\quad - \frac{1}{(1+\eta)^{2L-1}}\left[\sum_{l=2}^{L}\binom{2L-1}{L-l}\eta^{L-1+l}[Q_l(b\sqrt{\gamma_t}, a\sqrt{\gamma_t}) - Q_1(b\sqrt{\gamma_t}, a\sqrt{\gamma_t})]\right]
\end{aligned}
\tag{9.95}
$$

where

$$\binom{2L-1}{L-l} = \frac{(2L-1)!}{(L-l)!(L+l-1)!}$$

denotes the binomial coefficient and all the modulation-dependent parameters are as defined previously. As a check for $L=1$ and $\eta=1$, the latter two summations in (9.95) do not contribute, and hence one immediately obtains the single-channel result (8.162), as expected. Note that although the form in (9.95) does not give the appearance of being much simpler than Eq. (B-21) of Ref. 7, we shall see shortly that it does have particular advantage for obtaining the average BER performance over generalized fading channels.

As in the single-channel reception case the parameters a and b in (9.95) are typically independent of SNR and furthermore, $b > a$. For instance, for noncoherent detection of equal energy, equiprobable, correlated binary signals, $\eta = 1$ and

$$a = \left(\frac{1 - \sqrt{1 - |\lambda|^2}}{2}\right)^{1/2}$$

$$b = \left(\frac{1 + \sqrt{1 - |\lambda|^2}}{2}\right)^{1/2} \quad (9.96)$$

where λ ($0 \leq |\lambda| \leq 1$) is the complex-valued cross-correlation coefficient between the two signals. The special case $\lambda = 0$ corresponds to orthogonal noncoherent BFSK for which $a = 0$ and $b = 1$. Furthermore, in the case of binary DPSK, $a = 0$, $b = \sqrt{2}$, and $\eta = 1$. Finally, $a = \sqrt{2 - \sqrt{2}}$, $b = \sqrt{2 + \sqrt{2}}$, and $\eta = 1$ correspond to DQPSK with Gray coding. At this point let us introduce again a modulation-dependent parameter $\zeta = a/b$ which is independent of SNR. With this in mind, we now show how the alternative integral representations of the generalized Marcum Q-function yields a desired product form representation of the conditional BER. In particular, it was shown in Section 4.2.2 that

$$Q_l(u,w) = \frac{1}{2\pi} \int_{-\pi}^{\pi} \frac{\zeta^{-(l-1)}(\cos[(l-1)(\phi + \pi/2)] - \zeta \cos[l(\phi + \pi/2)])}{1 + 2\zeta \sin\phi + \zeta^2}$$

$$\times \exp\left[-\frac{w^2}{2}(1 + 2\zeta \sin\phi + \zeta^2)\right] d\phi, \quad 0^+ \leq \zeta = \frac{u}{w} < 1$$
(9.97)

$$Q_l(u,w) = 1 - \frac{1}{2\pi} \int_{-\pi}^{\pi} \frac{\zeta^l(\cos[l(\phi + \pi/2)] - \zeta \cos[(l-1)(\phi + \pi/2)])}{1 + 2\zeta \sin\phi + \zeta^2}$$

$$\times \exp\left[-\frac{u^2}{2}(1 + 2\zeta \sin\phi + \zeta^2)\right] d\phi, \quad 0 \leq \zeta = \frac{w}{u} < 1$$

with the special case of $l = 1$ being given in (4.2.17) and (4.2.18). Now, using (4.13) and (9.97) in (9.95), it can be shown after tedious manipulations that *the* entire conditional BER expression (9.95) can be written as a single integral with an integrand that contains a single exponential factor in γ_t of the form

$$\exp\left[-\frac{b^2\gamma_t}{2}(1+2\zeta\sin\phi+\zeta^2)\right]$$

namely,

$$P_b(L,\gamma_t;a,b,\eta) = \frac{\eta^L}{2\pi(1+\eta)^{2L-1}} \int_{-\pi}^{\pi} \frac{f(L;\zeta,\eta;\phi)}{1+2\zeta\sin\phi+\zeta^2}$$
$$\times \exp\left[-\frac{b^2\gamma_t}{2}(1+2\zeta\sin\phi+\zeta^2)\right] d\phi, \qquad 0^+ \leq \zeta = \frac{a}{b} < 1$$
(9.98)

where

$$f(L;\zeta,\eta;\phi) = f_0(L;\zeta,\eta;\phi) + f_1(L;\zeta,\eta;\phi)$$

with[4]

$$f_0(L;\zeta,\eta;\phi) = \left[-\frac{(1+\eta)^{2L-1}}{\eta^L} + \sum_{l=1}^{L}\binom{2L-1}{L-l}(\eta^{-l}+\eta^{l-1})\right]\zeta(\zeta+\sin\phi)$$

$$f_1(L;\zeta,\eta;\phi) = \sum_{l=1}^{L}\binom{2L-1}{L-l}[(\eta^{-l}\zeta^{-l+1} - \eta^{l-1}\zeta^{l+1})\cos[(l-1)(\phi+\pi/2)]$$
$$- (\eta^{-l}\zeta^{-l+2} - \eta^{l-1}\zeta^{l})\cos[l(\phi+\pi/2)]] \qquad (9.99)$$

The form of the conditional BER in (9.98) has the advantage of being a single finite-range integral with limits independent of the conditional SNR and an integrand that can be written in a *product form*, such as

$$P_b(L,\gamma_t;a,b,\eta) = \frac{\eta^L}{2\pi(1+\eta)^{2L-1}} \int_{-\pi}^{\pi} \frac{f(L;\zeta,\eta;\phi)}{1+2\zeta\sin\phi+\zeta^2}$$
$$\times \prod_{l=1}^{L}\exp\left[-\frac{b^2\gamma_l}{2}(1+2\zeta\sin\phi+\zeta^2)\right] d\phi, \qquad 0^+ \leq \zeta = \frac{a}{b} < 1$$
(9.100)

Furthermore, the form of (9.100) is desirable since we can first independently average over the individual statistical distributions of the γ_l's, and then perform the integral over ϕ, as described in more detail below (Section 9.4.1.2). Before showing this, however, we first offer some simplifications of (9.95) and (9.98) for some special cases of interest.

---

[4] As the book was going to press, Mr. L.-F. Tsaur of Conexant Systems Inc, Newport Beach, CA pointed out to the authors that $f_0(L;\zeta,\eta;\phi)$ can be proven equal to zero for all values of L and η independent of ζ and ϕ.

Desired Product Form-Representation of the Conditional BER [Special Case ($\eta = 1$)]. For $\eta = 1$, and any $L \geq 1$, which corresponds to the case of multichannel detection of equal-energy correlated binary signals, the conditional BER expression (9.95) becomes

$$P_b(L, \gamma_t; a, b, 1) = Q_1(a\sqrt{\gamma_t}, b\sqrt{\gamma_t}) - \frac{1}{2}\exp\left[-\frac{(a^2+b^2)\gamma_t}{2}\right] I_0(ab\gamma_t)$$

$$+ \frac{1}{2^{2L-1}} \left[\sum_{l=1}^{L} \binom{2L-1}{L-l} \{[Q_l(a\sqrt{\gamma_t}, b\sqrt{\gamma_t}) - Q_l(b\sqrt{\gamma_t}, a\sqrt{\gamma_t})] \right.$$

$$\left. - [Q_1(a\sqrt{\gamma_t}, b\sqrt{\gamma_t}) - Q_1(b\sqrt{\gamma_t}, a\sqrt{\gamma_t})]\} \right]$$

$$= Q_1(a\sqrt{\gamma_t}, b\sqrt{\gamma_t}) - \frac{1}{2}\exp\left[-\frac{(a^2+b^2)\gamma_t}{2}\right] I_0(ab\gamma_t)$$

$$+ \frac{1}{2^{2L-1}} \sum_{l=1}^{L} \binom{2L-1}{L-l} [Q_l(a\sqrt{\gamma_t}, b\sqrt{\gamma_t}) - Q_l(b\sqrt{\gamma_t}, a\sqrt{\gamma_t})]$$

$$- \frac{1}{2}[Q_1(a\sqrt{\gamma_t}, b\sqrt{\gamma_t}) - Q_1(b\sqrt{\gamma_t}, a\sqrt{\gamma_t})], \quad (9.101)$$

where we have added back the $l = 1$ term in the sums of (9.95) since they have zero value anyway. However, comparing Eqs. (40) and (42) of Ref. 62 yields

$$Q_1(u, w) - \frac{1}{2}\exp\left(-\frac{u^2+w^2}{2}\right) I_0(uw) = \frac{1}{2}[1 - Q_1(w, u) + Q_1(u, w)]$$
(9.102)

Thus, combining (9.101) and (9.102) gives the simplified expression

$$P_b(L, \gamma_t; a, b, 1) = \frac{1}{2} + \frac{1}{2^{2L-1}} \sum_{l=1}^{L} \binom{2L-1}{L-l}$$

$$\times [Q_l(a\sqrt{\gamma_t}, b\sqrt{\gamma_t}) - Q_l(b\sqrt{\gamma_t}, a\sqrt{\gamma_t})] \quad (9.103)$$

which appears not to be given in Refs. 7 and 57. Setting $a = 0$ and $b = 1$ ($b = \sqrt{2}$) in (9.103), then using the relations [63, Eq. (9)]

$$Q_l(0, w) = e^{-w^2/2} \sum_{k=0}^{l-1} \frac{(w^2/2)^k}{k!}$$

$$Q_l(u, 0) = 1$$

along with the identity $\sum_{l=1}^{L} \binom{2L-1}{L-l} = 2^{2(L-1)}$, it can be shown that (9.103) reduces to the well-known expression for multichannel binary orthogonal FSK (binary DPSK) given by Proakis [7, Eq. (12-1-13)], namely,

$$P_b(L, \gamma_t; 0, \sqrt{2g}, 1) = \frac{1}{2^{2L-1}} e^{-g\gamma_t} \sum_{l=0}^{L-1} c_l (g\gamma_t)^l \quad (9.104)$$

where

$$c_l = \frac{1}{l!} \sum_{k=0}^{L-1-l} \binom{2L-1}{k}$$

$g = \frac{1}{2}$ for orthogonal binary FSK, and $g = 1$ for binary DPSK. Note that an alternative (equivalent) form to (9.104), involving the confluent hypergeometric function, $_1F_1(\cdot; \cdot; \cdot)$, and given by Charash [59, Eq. (32)] as

$$P_b(L, \gamma_t; 0, \sqrt{2g}, 1) = \frac{e^{-2g\gamma_t}}{2^L \Gamma(L)} \sum_{l=0}^{L-1} \frac{\Gamma(L+l)}{2^l \Gamma(l+1)} {}_1F_1(L+l; L; g\gamma_t) \quad (9.105)$$

has also been used in the literature for the BER of multichannel binary orthogonal FSK and binary DPSK [64,65].

The conditional BER expression (9.101) for the special case of $\eta = 1$ and any $L \geq 1$ can also be put in the desired product form. Indeed, it can be shown that in this particular case $f_0(L; \zeta, 1; \phi) = 0$, and hence (9.98) reduces to

$$P_b(L, \gamma_t; a, b, 1) = \frac{1}{2^{2L}\pi} \int_{-\pi}^{\pi} \frac{f_1(L; \zeta, 1; \phi)}{1 + 2\zeta \sin\phi + \zeta^2}$$

$$\times \exp\left[-\frac{b^2 \gamma_t}{2}(1 + 2\zeta \sin\phi + \zeta^2)\right] d\phi,$$

$$0^+ \leq \zeta = \frac{a}{b} < 1 \quad (9.106)$$

where the function $f_1(\cdot; \cdot, \cdot; \cdot)$ is now given by

$$f_1(L; \zeta, \eta; \phi) = \sum_{l=1}^{L} \binom{2L-1}{L-l} \{(\zeta^{-l+1} - \zeta^{l+1}) \cos[(l-1)(\phi + \pi/2)]$$

$$- (\zeta^{-l+2} - \zeta^l) \cos[l(\phi + \pi/2)]\} \quad (9.107)$$

Note that as $\zeta \to 0$, (9.106) assumes an indeterminate form, and thus an analytical expression for the limit is more easily obtained from (9.104) with g replaced by $b^2/2$. We point out further that the limit of (9.106) as $\zeta \to 0$ converges smoothly to the exact BER expression of (9.104). For example, numerical evaluation of (9.106) setting $\zeta = 10^{-3} (a = 10^{-3}, b = 1)$ gives an accuracy of five digits compared with numerical evaluation of (9.104) for the same system parameters. The representation (9.106) is therefore useful even in this specific case. This is particularly true for the performance of binary orthogonal FSK and binary DPSK, which cannot be obtained via the classical representation of (9.104) in the most general fading case but which can be solved using the desirable conditional BER expression (9.106), as we will show next.

Average BER. To obtain the unconditional BER, $\overline{P}_b(L, \{\overline{\gamma}_l\}_{l=1}^L, \{i_l\}_{l=1}^L; a, b, \eta)$, we must average the conditional BER, $P_b(L, \gamma_t; a, b, \eta)$, over the joint PDF of the instantaneous SNR sequence $\{\gamma_l\}_{l=1}^L$, namely, $p_{\gamma_1, \gamma_2, \ldots, \gamma_L}(\gamma_1, \gamma_2, \ldots, \gamma_L)$. Since the RV's $\{\gamma_l\}_{l=1}^L$ are assumed to be statistically independent, $p_{\gamma_1, \gamma_2, \ldots, \gamma_L}(\gamma_1, \gamma_2, \ldots, \gamma_L) = \prod_{l=1}^L p_{\gamma_l}(\gamma_l)$, and the averaging procedure results in

$$\overline{P}_b\left(L, \{\overline{\gamma}_l\}_{l=1}^L, \{i_l\}_{l=1}^L; a, b, \eta\right) = \underbrace{\int_0^\infty \int_0^\infty \cdots \int_0^\infty}_{L\text{-fold}} P_b(L, \gamma_t; a, b, \eta)$$

$$\times \left[\prod_{l=1}^L p_{\gamma_l}(\gamma_l)\right] d\gamma_1 \, d\gamma_2 \cdots d\gamma_L \quad (9.108)$$

If the classical representation of $P_b(L, \gamma_t; a, b, \eta)$, as given by Proakis [7, Eq. (B-21)] or equivalently, (9.95), were to be used, (9.108) would result in an $(L+1)$-fold integral with infinite limits [one of these integrals comes from the classical definition of the generalized Marcum Q-function in $P_b(L, \gamma_t; a, b, \eta)$] and an adequately efficient numerical integration method would not be available.

Using the desired product form representation of $P_b(L, \gamma_t; a, b, \eta)$, namely, (9.100) in (9.108), yields

$$\overline{P}_b\left(L, \{\overline{\gamma}_l\}_{l=1}^L, \{i_l\}_{l=1}^L; a, b, \eta\right)$$

$$= \frac{\eta^L}{2\pi(1+\eta)^{2L-1}} \underbrace{\int_0^\infty \int_0^\infty \cdots \int_0^\infty}_{L\text{-fold}} \int_{-\pi}^\pi \frac{f(L; \zeta, \eta; \phi)}{1 + 2\zeta \sin \phi + \zeta^2}$$

$$\times \left[\prod_{l=1}^L \exp\left[-\frac{b^2 \gamma_l}{2}(1 + 2\zeta \sin \phi + \zeta^2)\right]\right]$$

$$\times \left[\prod_{l=1}^L p_{\gamma_l}(\gamma_l)\right] d\phi \, d\gamma_1 \, d\gamma_2 \cdots d\gamma_L \quad (9.109)$$

The integrand in (9.109) is absolutely integrable and the order of integration can therefore be interchanged. Thus, grouping like terms, we have

$$\overline{P}_b\left(L, \{\overline{\gamma}_l\}_{l=1}^L, \{i_l\}_{l=1}^L; a, b, \eta\right)$$

$$= \frac{\eta^L}{2\pi(1+\eta)^{2L-1}} \int_{-\pi}^{\pi} \frac{f(L; \zeta, \eta; \phi)}{1 + 2\zeta \sin\phi + \zeta^2}$$

$$\times \left[\prod_{l=1}^{L} \int_0^\infty \exp\left[-\frac{b^2 \gamma_l}{2}(1 + 2\zeta \sin\phi + \zeta^2)\right] p_{\gamma_l}(\gamma_l) d\gamma_l\right] d\phi$$

$$= \frac{\eta^L}{2\pi(1+\eta)^{2L-1}} \int_{-\pi}^{\pi} \frac{f(L; \zeta, \eta; \phi)}{1 + 2\zeta \sin\phi + \zeta^2}$$

$$\times \prod_{l=1}^{L} M_{\gamma_l}\left(-\frac{a^2 + b^2 + 2ab \sin\phi}{2}\right) d\phi \qquad (9.110)$$

where $M_{\gamma_l}(s)$ is the MGF of the SNR per symbol associated with the lth path and is given in Table 9.1 for some channel models of interest. Note that if the fading is identically distributed with the same fading parameter i and the same average SNR per bit $\overline{\gamma}$ for all L channels, (9.110) reduces to

$$\overline{P}_b(L, \overline{\gamma}, i; a, b, \eta) = \frac{\eta^L}{2\pi(1+\eta)^{2L-1}} \int_{-\pi}^{\pi} \frac{f(L; \zeta, \eta; \phi)}{1 + 2\zeta \sin\phi + \zeta^2}$$

$$\times \left[M_{\gamma_l}\left(-\frac{a^2 + b^2 + 2ab \sin\phi}{2}\right)\right]^L d\phi \qquad (9.111)$$

Hence this approach reduces the $(L+1)$-fold integral with infinite limits of (9.108) to a single finite-range integral (9.110) whose integrand contains only elementary functions (i.e., no special functions) and which can therefore be easily evaluated numerically.

9.4.1.3 Bounds on Average Bit Error Probability.
While the results found above are exact, they still require numerical evaluation of an integral. Our goal in this section is to derive a closed-form upper bound on the BER performance. We start with the conditional BER expression given in (9.103). Recall that the arguments of the second generalized Marcum Q-function in (9.103) are in reverse order to those in the first one. Thus, if $a < b$ (the case of practical interest in communication problems), then using the tight upper bound (4.57a) for the first term and the tight lower bound of (4.64a) for the second would produce a tight upper bound on the conditional BER. The next step would then be to average the resulting conditional BER expression over the PDF's of the fading RV's. If we assume that the SNR's per bit per path $\gamma_l, l = 1, 2, \ldots, L$, are independent, unfortunately, the presence of the term in (4.64a) with the summation would prevent the desired partitioning of the integrand, thereby precluding averaging

on a channel-by-channel basis. To circumvent this difficulty, we simplify (with the understanding that it becomes negligibly looser, as shown in Figs. 4.2 through and 4.4) the lower bound of (4.64a) by ignoring this term (which is always positive), which is tantamount to using the lower bound on $Q_1(\alpha, \beta)$ as given by (4.31b) for $Q_m(\alpha, \beta)$. When this is done, the following upper bound on the conditional BER results:

$$P_b(E|\gamma_t) \leq \frac{1}{2^{2L-1}} \sum_{l=1}^{L} \binom{2L-1}{L-l} \left[\left(\frac{3}{2} + \frac{1}{\pi \zeta^{l-1}} \frac{1-\zeta^{l-1}}{1-\zeta} \right) \exp\left(-\frac{b^2 \gamma_t (1-\zeta)^2}{2} \right) \right.$$

$$\left. - \left(\frac{1}{2} + \frac{1}{\pi \zeta^{l-1}} \frac{1-\zeta^{l-1}}{1-\zeta} \right) \exp\left(-\frac{b^2 \gamma_t (1+\zeta)^2}{2} \right) \right] \quad (9.112)$$

where again ζ denotes $0^+ \leq \zeta \triangleq \alpha/\beta < 1$. Substituting $\gamma_t = \sum_{l=1}^{L} \gamma_l$ in (9.112) and averaging over the individual fading PDFs, $p_{\gamma_l}(\gamma_l)$, $l = 1, 2, \ldots, L$, gives the result for average BEP in the desired MGF product form, namely,

$$P_b(E) \leq \frac{1}{2^{2L-1}} \sum_{l=1}^{L} \binom{2L-1}{L-l} \left[\left(\frac{3}{2} + \frac{1}{\pi \zeta^{l-1}} \frac{1-\zeta^{l-1}}{1-\zeta} \right) \prod_{k=1}^{L} M_{\gamma_k}\left(\frac{-b^2(1-\zeta)^2}{2} \right) \right.$$

$$\left. - \left(\frac{1}{2} + \frac{1}{\pi \zeta^{l-1}} \frac{1-\zeta^{l-1}}{1-\zeta} \right) \prod_{k=1}^{L} M_{\gamma_k}\left(-\frac{b^2(1+\zeta)^2}{2} \right) \right] \quad (9.113)$$

where $M_{\gamma_l}(s)$ is again the MGF of the SNR per symbol associated with the lth path and is given in Table 9.1 for some channel models of interest.

As an example, consider the average BEP performance of DQPSK (for which $a = \sqrt{2 - \sqrt{2}}$ and $b = \sqrt{2 + \sqrt{2}}$) over a frequency-selective Nakagami-m channel with an arbitrary power delay profile. Since for a Nakagami-m channel with instantaneous SNR per bit per path PDF,

$$p_{\gamma_k}(\gamma_k) = \frac{m^m \gamma_k^{m-1}}{\bar{\gamma}_k^m \Gamma(m)} \exp\left(-\frac{m \gamma_k}{\bar{\gamma}_k} \right), \quad \gamma_k \geq 0 \quad (9.114)$$

the MGF is given by

$$M_{\gamma_k}(s) = \left(1 - \frac{s \bar{\gamma}_k}{m} \right)^{-m}, \quad s > 0 \quad (9.115)$$

then use of (9.115) in (9.113) immediately gives the upper bound on average BER in closed form. Figure 9.8 is a plot of the conditional BER $P_b(E|\gamma_t)$ versus SNR per bit per path (in dB) $\gamma = \gamma_t/L$ as determined from both the exact result of (9.103) and the purely exponential bound of (9.112) for values of $L = 2$ and 4. Similarly, Fig. 9.9 is a plot of average BER versus $\bar{\gamma}$ (assuming a uniform power delay profile for the Nakagami-m fading), as determined from both the exact

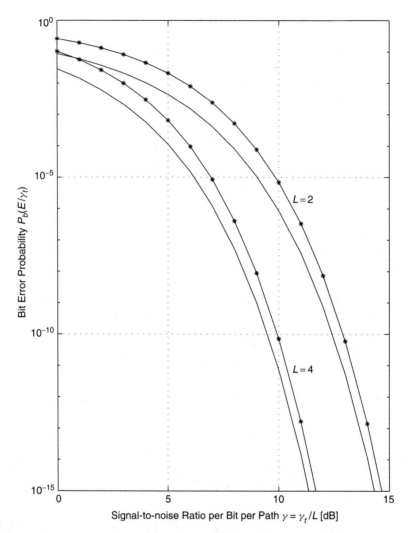

Figure 9.8. Bit error probability of D-QPSK with L-fold diversity postdetection EGC versus the SNR per bit per path for AWGN channels.

result in (9.110) combined with (9.115) and the upper bound of (9.113) combined with (9.115). The results are given for three values of m, 0.5, 1, and 2 with $m = 0.5$ corresponding to worst-case (single-sided Gaussian) fading and $m = 1$ corresponding to Rayleigh fading. As one can observe from both sets of plots, the upper bounds are quite tight over a wide range of values of the SNR. Finally, it should be pointed out that when evaluated numerically using MATHEMATICA software, the closed-form upper bounds of (9.112) and (9.113) offer a significant improvement in speed compared with the exact single finite-range integrals as given by (9.106) and (9.110), respectively.

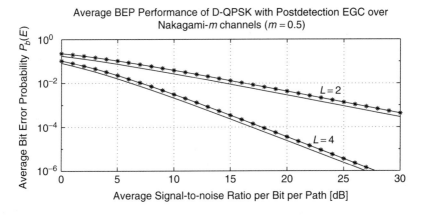

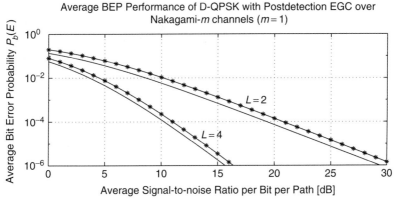

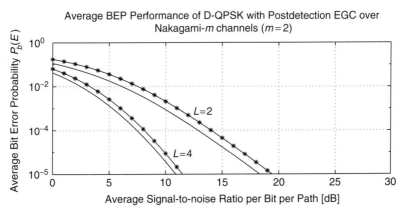

Figure 9.9. Average bit error probability of D-QPSK with L-fold diversity postdetection EGC versus the average SNR per bit per path for various Nakagami-m fading channels. The fading is assumed to be identically distributed over the L combined paths.

9.4.2 *M*-ary Orthogonal FSK

The BER of noncoherent M-ary orthogonal modulation [or equivalently, M-ary frequency-shift-keying (M-FSK)] operating over fading channels (both with and without diversity reception) has long been of interest. Hahn [66] and Lindsey [58] were the first to consider this problem for square-law combining (also called postdetection equal-gain combining) over Rayleigh and Rice fading channels, respectively. For the more general Nakagami-m channel [53], analogous results were obtained by Crepeau [67] for the case of no diversity, and more recently by Weng and Leung [68] for the case of square-law diversity combining.

In addition, during the last decade, the problem has been reexamined in the context of its application to the performance analysis of the reverse link of direct-sequence code-division multiple access (DS-CDMA) systems over frequency-selective fading channels and with RAKE reception [69–72]. By modeling the multiple access (MA) interference among users as an equivalent additive Gaussian noise process, as can be justified for many applications [73], these works evaluate performance as would be done for the traditional AWGN channel using a total noise variance equal to the sum of that due to the ever-present thermal noise and that due to the multiple access interference. However, for mathematical tractability, most of the channel models considered in all the studies cited typically assume a uniform power delay profile along the paths.

The unified approach presented in Section 9.4.1.2 for the analysis of noncoherent and differentially coherent communications over generalized fading channels does not apply for the case of noncoherent detection of M-ary ($M > 2$) orthogonal signals. In fact, even in the case of binary ($M = 2$) signaling, the average BER result can be obtained only as a limiting case of the generic result in (9.110) with $\zeta \to 0$, which cannot be expressed in closed form. However, since the MGF of the instantaneous combined SNR of a large variety of fading conditions is straightforward to evaluate, as mentioned earlier, it is still desirable to arrive at a generic expression for average BER in the form of an integral (preferably with finite limits) having an integrand that depends on this MGF. In this section we show that the original method used by Weng and Leung [68] to derive the BEP of noncoherent M-ary orthogonal modulation with square-law combining over i.i.d. Nakagami-m fading channels in fact enables this to occur. More specifically, in the next section we reformulate this method to apply to a generalized fading channel with an arbitrary MGF and thereby arrive at *exact* results for average BEP in the above-mentioned desired form. To avoid unnecessary repetition, we draw heavily on the results in Refs. 7 and 60, only going into detail where necessary to distinguish between the specific Nakagami-m and generalized fading cases. We then show that the desired generic result can be further simplified for the special cases of Nakagami-m fading with arbitrary power delay profile.

9.4.2.1 Exact Analysis of Average Bit Error Probability.
Consider a noncoherent M-ary orthogonal system operating over an L-path generalized fading channel and using an L-finger post detection EGC receiver with square-law

combining, as shown in Fig. 9.7. Denote the outputs of the M combiners by U_i ($i = 1, 2, \ldots, M$). Assume without loss in generality that the first combiner, i.e., (U_1) corresponds to the information-bearing signal, whereas the remaining $M-1$ combiner outputs contain noise only. Then the average BEP for such a system can be expressed as [60, Eq. (8)]

$$P_b(E) = \frac{2^{(\log_2 M)-1}}{2^{\log_2 M} - 1} P_s(E)$$

$$= \frac{2^{(\log_2 M)-1}}{2^{\log_2 M} - 1} \int_0^\infty \frac{1}{2\pi} \int_{-\infty}^\infty M_{U_1}(j\omega) e^{-j\omega u_1} g(u_1) \, d\omega \, du_1 \quad (9.116)$$

where $M_{U_1}(j\omega)$ is the MGF of the first combiner output U_1 and

$$g(u_1) = \sum_{i=1}^{M-1} \binom{M-1}{i} (-1)^{i+1} e^{-iu_1/2} \sum_{k=0}^{i(L-1)} \beta_{ki} \left(\frac{u_1}{2}\right)^k \quad (9.117)$$

In (9.117) the coefficients $\{\beta_{ki}\}$ are obtained via the expansion

$$\left(\sum_{k=0}^{L-1} \frac{x^k}{k!}\right)^i = \sum_{k=0}^{i(L-1)} \beta_{ki} x^k \quad (9.118)$$

and can be evaluated recursively by [66, Eq. (23) or equivalently 72, Eq. (32)]

$$\beta_{ki} = \sum_{n=k-L+1}^{k} \frac{\beta_{n(i-1)}}{(k-n)!} I_{[0,(i-1)(L-1)]}(i) \quad (9.119)$$

where $\beta_{00} = \beta_{0i} = 1$, $\beta_{k1} = 1/k!$, $\beta_{1i} = i$, and $I_{[a,b]}(i)$ is the indicator function defined by

$$I_{[a,b]}(i) = \begin{cases} 1, & a \leq i \leq b \\ 0, & \text{otherwise} \end{cases} \quad (9.120)$$

Since U_1 is of the form [60, Eq. (1)] (for more details, see Ref. 7, Sec. 14-4-3)

$$U_1 = \sum_{l=1}^{L} \left| \alpha_l \sqrt{\frac{2E_s}{N_0}} e^{-j\phi_l} + n_{l,1} \right|^2 \quad (9.121)$$

where α_l and ϕ_l are the fading amplitude and phase associated with the lth diversity path, E_s/N_0 is the symbol SNR, and $n_{l,1}$ ($l = 1, 2, \ldots, L$) are complex-valued i.i.d. zero-mean unit variance random variables, then conditioned on the fading, U_1 is a noncentral chi-square random variable with $2L$ degrees of freedom and MGF [6, Eq. (5A.8)]

$$M_{U_1}\left(s|\{\alpha\}_{l=1}^L\right) = \frac{1}{(1-2s)^L} \exp\left(\frac{2s}{1-2s} \frac{E_s}{N_0} \sum_{l=1}^L \alpha_l^2\right)$$

$$= \frac{1}{(1-2s)^L} \exp\left(\frac{2s}{1-2s} \sum_{l=1}^L \gamma_l\right) \quad (9.122)$$

where $\gamma_l = \alpha_l^2 E_s/N_0$ denotes the instantaneous symbol SNR of the lth diversity branch. Averaging (9.122) over the fading yields the unconditional MGF of U_1, namely,

$$M_{U_1}(s) = E\left\{M_{U_1}\left(s|\{\alpha\}_{l=1}^L\right)\right\}$$

$$= \frac{1}{(1-2s)^L} \int_0^\infty \int_0^\infty \cdots \int_0^\infty \exp\left(\frac{2s}{1-2s} \sum_{l=1}^L \gamma_l\right)$$

$$\times p_{\gamma_1,\gamma_2,\ldots,\gamma_L}(\gamma_1, \gamma_2, \ldots, \gamma_L) \, d\gamma_1 \, d\gamma_2 \cdots d\gamma_L$$

$$= \frac{1}{(1-2s)^L} \int_0^\infty \exp\left(\frac{2s}{1-2s}\gamma_t\right) p_{\gamma_t}(\gamma_t) \, d\gamma_t$$

$$= \frac{1}{(1-2s)^L} M_{\gamma_t}\left(\frac{2s}{1-2s}\right) \quad (9.123)$$

where $\gamma_t = \sum_{l=1}^L \gamma_l$ is, as before, the total instantaneous symbol SNR at the combiner output and $M_{\gamma_t}(s)$ is its MGF. Finally, substituting (9.123) in (9.116), reversing the order of integration and then integrating over u_1 [with the help of the integral in [Ref. 36, Eq. (3.351.3)] we obtain after some manipulation and a change of variables,

$$P_b(E) = \left(\frac{2^{(\log_2 M)-1}}{2^{\log_2 M}-1}\right) \frac{1}{2\pi} \sum_{i=1}^{M-1} \binom{M-1}{i} (-1)^{i+1} \sum_{k=0}^{i(L-1)} \beta_{ki}$$

$$\times \int_{-\infty}^\infty \frac{1}{(1-jv)^L} \frac{k!}{(i+jv)^{k+1}} M_{\gamma_t}\left(\frac{jv}{1-jv}\right) dv \quad (9.124)$$

To put (9.124) in the form of an integral with finite limits, we now make the change of variables $v = \tan\theta$. Furthermore, recognizing that $P_b(E)$ is real, it is only necessary to take the real part of the right-hand side of (9.124) since the imaginary part must equate to zero. Finally, after making this change of variables, performing some routine complex algebraic manipulations, and taking advantage of the fact that the resulting integrand is an even function of θ, we obtain the desired generic result, namely,

$$P_b(E) = \left(\frac{2^{(\log_2 M)-1}}{2^{\log_2 M}-1}\right) \frac{1}{\pi} \int_0^{\pi/2} \sum_{i=1}^{M-1} \binom{M-1}{i} (-1)^{i+1} \sum_{k=0}^{i(L-1)} \beta_{ki}$$
$$\times \left[\operatorname{Re}\left\{M_{\gamma_t}\left(\frac{j\tan\theta}{1-j\tan\theta}\right)\right\} - \operatorname{Im}\left\{M_{\gamma_t}\left(\frac{j\tan\theta}{1-j\tan\theta}\right)\right\} \sin\Phi_k\right]$$
$$\times \frac{k!(\cos\theta)^{L-2}}{(i^2+\tan^2\theta)^{(k+1)/2}} d\theta \qquad (9.125)$$

where $\operatorname{Re}\{\cdot\}$ and $\operatorname{Im}\{\cdot\}$ denote the real and imaginary parts, respectively, and

$$\Phi_k \triangleq L\theta - (k+1)\tan^{-1}\left(\frac{\tan\theta}{i}\right) \qquad (9.126)$$

For binary modulation, $M=2$ and $\beta_{k1}=1/k!$, and hence (9.125) simplifies further to

$$P_b(E) = \frac{1}{\pi} \int_0^{\pi/2} \sum_{k=0}^{L-1} \left[\operatorname{Re}\left\{M_{\gamma_t}\left(\frac{j\tan\theta}{1-j\tan\theta}\right)\right\} \cos[(L-k-1)\theta]\right.$$
$$\left. - \operatorname{Im}\left\{M_{\gamma_t}\left(\frac{j\tan\theta}{1-j\tan\theta}\right)\right\} \sin[(L-k-1)\theta]\right]$$
$$\times \frac{(\cos\theta)^{L-2}}{(1+\tan^2\theta)^{(k+1)/2}} d\theta \qquad (9.127)$$

which can be shown to agree numerically with (9.110), letting $\eta=1, b=1$, and then taking the limit as $\zeta \to 0$. Also recall that the conditional BER of binary DPSK [7, Eq. (12-1-13)] has the same functional dependence on the combined SNR γ_t, with the exception that γ_t is replaced by $2\gamma_t$ in the DPSK expression. However, since $M_{\gamma_t}(s)$ is only a function of the normalized versions of its argument, (i.e., $\{s\bar{\gamma}_l\}_{l=1}^L$), $M_{\gamma_t}(s) = E[e^{2\gamma_t s}] = E[e^{\gamma_t(2s)}]$, which from the foregoing is identical to $M_{\gamma_t}(s)$ with $\{\bar{\gamma}_l\}_{l=1}^L$ replaced by $\{2\bar{\gamma}_l\}_{l=1}^L$. Thus, we conclude that (9.127) also applies to the average BEP of DPSK with postdetection EGC when $\{\bar{\gamma}_l\}_{l=1}^L$ is replaced by $\{2\bar{\gamma}_l\}_{l=1}^L$.

Consider now the special case where the fading is independent from path to path but not necessarily identically distributed (i.e., each path is allowed to have arbitrary fading statistics). Because of the independence assumption, the MGF $M_{\gamma_t}(s)$ partitions into the product

$$M_{\gamma_t}(s) = \prod_{l=1}^L M_{\gamma_l}(s) \qquad (9.128)$$

where, as mentioned previously, $M_{\gamma_l}(s)$ is readily available for a variety of different channel models and a summary of these MGF results is included in

Table 9.1. In view of the above, the generic result for average BEP as given by (9.125) now becomes

$$P_b(E) = \left(\frac{2^{(\log_2 M)-1}}{2^{\log_2 M} - 1}\right) \frac{1}{\pi} \sum_{i=1}^{M-1} \binom{M-1}{i} (-1)^{i+1} \sum_{k=0}^{i(L-1)} \beta_{ki}$$

$$\times \int_0^{\pi/2} \left[\text{Re}\left\{\prod_{l=1}^{L} M_{\gamma_l}\left(\frac{j\tan\theta}{1-j\tan\theta}\right)\right\} \cos\Phi_k \right.$$

$$\left. - \text{Im}\left\{\prod_{l=1}^{L} M_{\gamma_l}\left(\frac{j\tan\theta}{1-j\tan\theta}\right)\right\} \sin\Phi_k \right] \frac{k!(\cos\theta)^{L-2}}{(i^2 + \tan^2\theta)^{(k+1)/2}} d\theta \quad (9.129)$$

As a further specialization of (9.129), consider the case where all paths have Nakagami-m fading with the same fading parameter m as was done in Ref. 68, now with, however, an arbitrary power delay profile. Since for this type of fading the MGF of γ_l is given by (see Table 9.1)

$$M_{\gamma_l}(s) = \left(1 - \frac{s\bar{\gamma}_l}{m}\right)^{-m} \quad (9.130)$$

then using this in (9.129) and simplifying the complex algebra, we arrive at the result

$$P_b(E) = \left(\frac{2^{(\log_2 M)-1}}{2^{\log_2 M} - 1}\right) \frac{1}{\pi} \int_0^{\pi/2} \sum_{i=1}^{M-1} \binom{M-1}{i} (-1)^{i+1}$$

$$\times \sum_{k=0}^{i(L-1)} \beta_{ki} \frac{k!A\cos(\Phi_k + m\Theta)}{(i^2 + \tan^2\theta)^{(k+1)/2}(\cos\theta)^{(2m-1)L+2}} d\theta \quad (9.131)$$

where

$$A \triangleq \prod_{l=1}^{L} \left[\left(1 + \left(1 + \frac{\bar{\gamma}_l}{m}\right)\tan^2\theta\right)^2 + \left(\frac{\bar{\gamma}_l}{m}\tan\theta\right)^2\right]^{-m/2},$$

$$\Theta \triangleq \sum_{l=1}^{L} \tan^{-1}\frac{(\bar{\gamma}_l/m)\tan\theta}{1 + (1 + \bar{\gamma}_l/m)\tan^2\theta} \quad (9.132)$$

For binary modulation, $M = 2$ and $\beta_{k1} = 1/k!$, and hence (9.131) simplifies further to

$$P_b(E) = \frac{1}{\pi}\int_0^{\pi/2} \sum_{k=0}^{L-1} \frac{A\cos[(L-k-1)\theta + m\Theta]}{(\cos\theta)^{(2m-1)L+1-k}} d\theta \quad (9.133)$$

which agrees numerically with the result obtained from substituting (9.130) into (9.110), letting $\eta = 1, b = 1$, and then taking the limit as $\zeta \to 0$. Also again because of the similarity of the conditional BER for binary orthogonal signaling and DPSK, (9.133) also applies for the average BEP of DPSK with $\overline{\gamma}_l$ replaced by $2\overline{\gamma}_l$. Hence for the i.i.d. case ($\overline{\gamma}_l = \overline{\gamma}, l = 1, 2, \ldots, L$), (9.133) with the substitution of $\overline{\gamma}_l$ by $2\overline{\gamma}_l$ is equivalent to the closed-form expression given by Weng and Leung [60, Eq. (11)].

9.4.2.2 Numerical Examples. Figure 9.10 is a plot of $P_b(E)$ for binary orthogonal FSK [obtained from (9.133)] versus the average SNR per bit of the first path $\overline{\gamma}_1$ for different values of m and an exponentially decaying power delay profile [$\overline{\gamma}_l = \overline{\gamma}_1 \exp(-\delta(l-1)), l = 1, 2, \ldots, L$, and δ is the average fading power decay factor]. In view of our previous remarks, in this figure as well

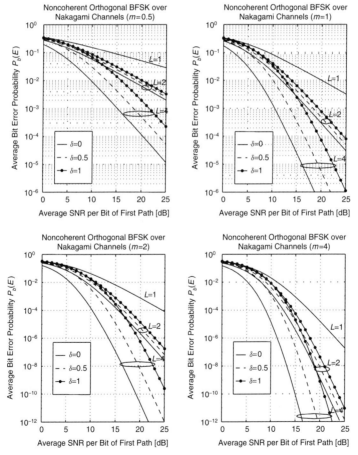

Figure 9.10. Average bit error probability $P_b(E)$ of binary FSK with square-law combining versus the average SNR per bit of the first path $\overline{\gamma}_1$ over Nakagami-m channels with an exponentially decaying power delay profile.

as in the remaining figures concerned with binary orthogonal FSK, the average BEP of DPSK may be found by shifting the curves 3 dB to the left. Analogous to Fig. 9.10, Figs. 9.11 and 9.12 are plots for the BEP $P_b(E)$ of 4-ary and 8-ary orthogonal FSK [obtained from (9.131) with $M = 4$ and $M = 8$] versus the average SNR per bit of the first path $\overline{\gamma}_1 / \log_2 M$. From these figures there are two observations worth noting. First, in comparison with a system operating over a uniform power delay profile, the BEP deterioration due to the exponentially decaying power delay profile can be quite important in all cases, and this degradation is more accentuated for a large number of combined paths, as one may expect. However, note that in these figures, once the first path average SNR $\overline{\gamma}_1$ is fixed (as it is typically done in this type of illustrations; see, e.g., Ref. 54), the total average SNR is smaller for channels with an exponentially

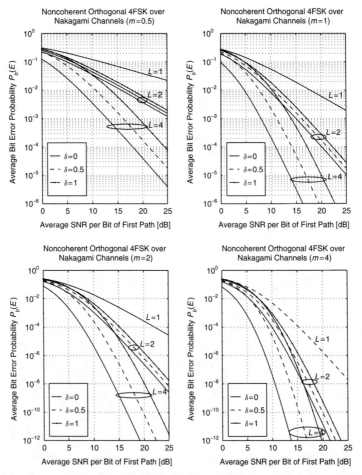

Figure 9.11. Average bit error probability $P_b(E)$ of 4-ary FSK with square-law combining versus the average SNR per bit of the first path $\overline{\gamma}_1/2$ over Nakagami-m channels with an exponentially decaying power delay profile.

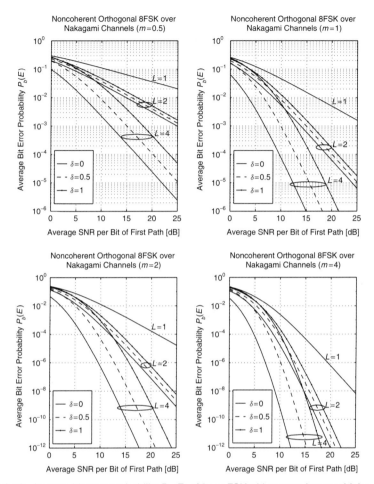

Figure 9.12. Average bit error probability $P_b(E)$ of 8-ary FSK with square-law combining versus the average SNR per bit of the first path $\bar{\gamma}_1/3$ over Nakagami-m channels with an exponentially decaying power delay profile.

decaying power delay profile ($\delta > 0$) than for channels with a uniform power delay profile ($\delta = 0$), which explains in part the relatively important performance degradation due to the exponentially decaying power delay profile. Second, the use of M-ary ($M > 2$) instead of binary signaling still improves the BEP performance, especially for channels subject to a low amount of fading (i.e., high fading parameter m).

9.5 OUTAGE PROBABILITY PERFORMANCE

As discussed in Chapter 1, in addition to the average error rate, outage probability, P_{out}, is another standard performance criterion of diversity systems operating over fading channels [2–4]. It is defined as the probability that the instantaneous error rate exceeds a specified value or equivalently, that the combined SNR γ_t falls below a certain specified threshold γ_{th}, that is,

$$P_{\text{out}} = \int_0^{\gamma_{\text{th}}} p_{\gamma_t}(\gamma_t)\,d\gamma_t \tag{9.134}$$

In other words, P_{out} is the cumulative distribution function (CDF) of γ_t, $P_{\gamma_t}(\gamma_t)$, evaluated at γ_{th}. Motivated by the fact that finding the PDF of γ_t in closed form is often restricted to some simple cases (such as i.i.d. diversity paths) while the MGF of γ_t, $M_{\gamma_t}(s)$, can be obtained in a simple form for a wide variety of fading conditions, as discussed in previous sections, in this section we develop an MGF-based approach for the outage probability evaluation of diversity systems over generalized fading channels [74]. This approach relies on a simple and accurate algorithm for numerical inversion of Laplace transforms of CDFs [75], which is summarized in Section 9B.1.

9.5.1 MRC and Noncoherent EGC

Consider a multilink channel where the transmitted signal is received over L independent slowly varying flat fading channels. Recall that for equally likely transmitted symbols, the total conditional SNR per symbol, γ_t, at the output of an MRC combiner or a postdetection (differentially coherent or noncoherent) EGC combiner is given by (9.1) or equivalently, (9.94). Since $p_{\gamma_t}(\gamma_t) = dP_{\gamma_t}(\gamma_t)/d\gamma_t$ and $P_{\gamma_t}(0) = 0$, we have [36, p. 1178]

$$\hat{P}_{\gamma_t}(s) = \frac{\hat{p}_{\gamma_t}(s)}{s} \tag{9.135}$$

where the "$\,\hat{}\,$" denotes the Laplace transform operator. Using the relation $\hat{p}_{\gamma_t}(s) = M_{\gamma_t}(-s)$ and applying the numerical technique described in Appendix 9B, we obtain the desired result:

$$P_{\text{out}} = P_{\gamma_t}(\gamma_{\text{th}}) = \frac{2^{-K} e^{A/2}}{\gamma_{\text{th}}} \sum_{k=0}^{K} \binom{K}{k} \sum_{n=0}^{N+k} \frac{(-1)^n}{\alpha_n}$$

$$\times \operatorname{Re}\left\{ \frac{M_{\gamma_t}\left(-\dfrac{A+2\pi jn}{2\gamma_{\text{th}}}\right)}{\dfrac{A+2\pi jn}{2\gamma_{\text{th}}}} \right\} + E(A, K, N) \tag{9.136}$$

where the overall error term $E(A, K, N)$ is approximately bounded by

$$|E(A, K, N)| \lesssim \frac{e^{-A}}{1 + e^{-A}}$$

$$+ \left| \frac{2^{-K} e^{A/2}}{\gamma_{\text{th}}} \sum_{k=0}^{K} (-1)^{N+1+k} \binom{K}{k} \text{Re} \left\{ \frac{M_{\gamma_t}\left(-\frac{A + 2\pi j(N+k+1)}{2\gamma_{\text{th}}}\right)}{\frac{A + 2\pi j(N+k+1)}{2\gamma_{\text{th}}}} \right\} \right|$$
(9.137)

Because of the independence assumption on the $\{\gamma_l\}_{l=1}^{L}$, the MGF of γ_t is the product of the MGF of the γ_l's [31, Sec. 7.4], which are available in Table 9.1. Hence (9.136) and (9.137) yield simple expressions for the outage probability and the corresponding numerical error, which can easily be computed using standard mathematical packages such as Mathematica.

9.5.2 Coherent EGC

Recall that for coherent EGC with equally likely transmitted symbols, the conditional combined SNR per symbol, γ_{EGC}, is given by (9.46), namely,

$$\gamma_{\text{EGC}} = \frac{\left(\sum_{l=1}^{L} \alpha_l\right)^2 E_s}{LN_0}$$
(9.138)

where we have now assumed that the AWGN noise power spectral density is the same for all the diversity paths. Since the outage probability P_{out} is defined by

$$P_{\text{out}} \triangleq \Pr\{0 \leq \gamma_{\text{EGC}} \leq \gamma_{\text{th}}\}$$
(9.139)

and since $\alpha_l \geq 0$, $l =, 1, 2, \ldots, L$, (9.139) can be rewritten as

$$P_{\text{out}} = \Pr\{0 \leq \alpha_t \leq \alpha_{\text{th}}\}$$
(9.140)

where $\alpha_t \triangleq \sum_{l=1}^{L} \alpha_l$ and $\alpha_{\text{th}} \triangleq \sqrt{L\gamma_{\text{th}}/(E_s/N_0)}$. Consequently, using the numerical technique described in Section 9B.1, for independent combined paths the outage probability of coherent EGC receivers can be computed using (9.136), and the corresponding numerical error can be estimated from (9.137), where in all these expressions γ_t, γ_l, and γ_{th} are replaced by α_t, α_l, and α_{th}, respectively. At this point let us restrict our multilink channel to L independent slowly varying Nakagami-m flat fading channels with fading parameter m_l ($0.5 \leq m_l$) and average fading power $\Omega_l = E[\alpha_l^2]$. The MGF of α_l can be expressed in this case with the help of Eq. (3.462.1) of Ref. 36 in terms of the parabolic cylinder function $D_{-v}(\cdot)$ [36, Secs. 9.24 and 9.25]

$$M_{\alpha_l}(s) = \frac{\Gamma(2m_l)}{2^{m_l-1}\Gamma(m_l)} \exp\left(\frac{\Omega_l s^2}{8m_l}\right) D_{-2m_l}\left(\frac{-s\sqrt{\Omega_l}}{\sqrt{2m_l}}\right) \quad (9.141)$$

or alternatively, in terms of the more common confluent hypergeometric functions $_1F_1[\cdot, \cdot; \cdot]$ by using Eq. (3.462.1) of Ref. 36, resulting in

$$M_{\alpha_l}(s) = \frac{\Gamma(2m_l)}{2^{2m_l-1}\Gamma(m_l)} \left[\frac{\sqrt{\pi}}{\Gamma(m_l + \frac{1}{2})} {}_1F_1\left(m_l, \frac{1}{2}; \frac{\Omega_l s^2}{4m_l}\right) \right.$$
$$\left. + \frac{\sqrt{\pi \Omega_l} s}{\Gamma(m_l)\sqrt{m_l}} {}_1F_1\left(m_l + \frac{1}{2}, \frac{3}{2}; \frac{\Omega_l s^2}{4m_l}\right) \right] \quad (9.142)$$

Note that similar to MRC and postdetection EGC, our numerical results confirm that the outage probability of coherent EGC systems is solely a function of the various average SNRs/bit/paths $\overline{\gamma}_l = \Omega_l E_s/N_0$, $l = 1, 2, \ldots, L$.

9.5.3 Numerical Examples

As examples, Figs. 9.13 and 9.14 show the impact of an exponentially decaying power delay profile ($\overline{\gamma}_l = e^{-\delta(l-1)}\overline{\gamma}_1$) on the outage probability performance of MRC (or postdetection EGC) RAKE receivers over frequency-selective Nakagami-m and Rician fading channels, respectively. These curves were generated with A set to equal $10 \ln 10 \simeq 23.026$ to guarantee a discretization error of less than 10^{-10}. In addition, the parameters N and K were chosen so that the resulting truncation error is negligible compared to the computed outage probability, as shown in Tables 9.2 and 9.3.[5] The numerical results show that the power delay profile induces a nonnegligible degradation in the outage probability and has therefore to be taken into account for accurate prediction of the outage probability performance of RAKE receivers. Furthermore, in both sets of curves it is clear that the effect of the power decay factor δ becomes more important as the number of combined paths increases, as expected.

Figure 9.15 shows the impact of an exponentially decaying power delay profile on the outage probability performance of coherent EGC RAKE receivers over Nakagami-m fading channels. For these sets of curves the parameter A was set equal to $8 \ln 10 \simeq 18.4$ to guarantee a discretization error of less than 10^{-8}, and the parameters N and K were again chosen so that the truncation error is negligible compared to the actual computed outage probabilities, as shown in Table 9.4. Finally, Fig. 9.16 compares the outage probability of MRC and coherent EGC receivers over Nakagami-m fading channels with an exponentially

---

[5] In Tables 9.2, through 9.4 the truncation error is set equal to zero if it is smaller than the precision limit of the computer.

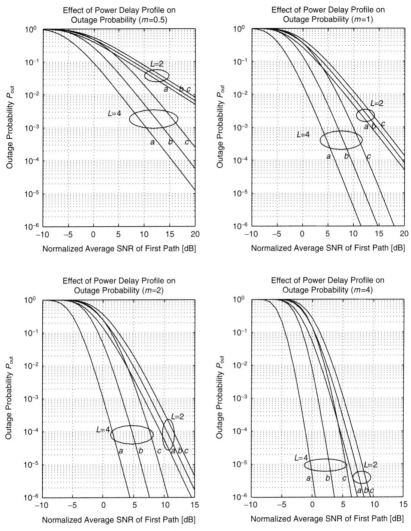

Figure 9.13. Outage probability with MRC or postdetection EGC RAKE reception ($L = 2$ and 4) versus normalized average SNR of the first path $\overline{\gamma}_1/\gamma_{th}$ over a Nakagami-m fading channel with an exponentially decaying power delay profile: (a) $\delta = 0$; (b) $\delta = 0.5$; (c) $\delta = 1$.

decaying power delay profile. Note that regardless of the power decay factor δ the outage probability of EGC approaches the performance of MRC for channels with low amounts of fading (high m). This behavior can be explained by the following arguments. For channels with high amounts of fading, coherent EGC combines very "noisy" branches and therefore suffers a serious penalty with respect to MRC. On the other hand, for channels with low amounts of fading, coherent EGC takes full advantage of the combining of all the "clean" paths and therefore approaches the performance of MRC.

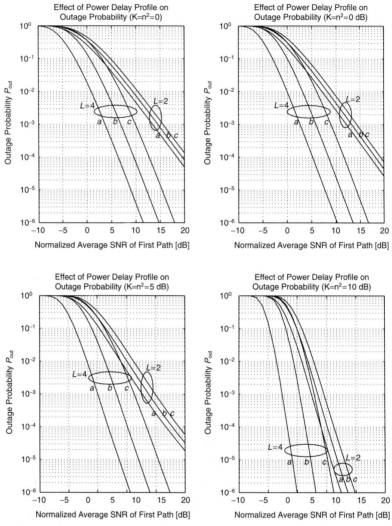

Figure 9.14. Outage probability with MRC or postdetection EGC reception ($L = 2$ and 4) versus normalized average SNR of the first path $\bar{\gamma}_1/\gamma_{th}$ over a Rician fading channel with an exponentially decaying power delay profile: (a) $\delta = 0$; (b) $\delta = 0.5$; (c) $\delta = 1$.

9.6 IMPACT OF FADING CORRELATION

In studying the performance of diversity systems, the usual assumption as made by us in all previous sections, is that the combined signals are independent of one another. However, as discussed in Refs. 76 and 77, there are a number of real-life scenarios in which this assumption is not valid, for example, insufficient antenna spacing in small mobile units equipped with space and polarization antenna diversity [78,79]. Furthermore, for multipath diversity over frequency-selective

TABLE 9.2 Truncation Error Estimates of MRC Outage Probability with 10^{-10} Discretization Error, $K = 11$ and $N = 20$ (Nakagami-m Fading) for Various Values of $\bar{\gamma}_1/\gamma_{th}$

(L, m, δ)	-10 dB	-5 dB	0 dB	5 dB	10 dB
(2,1,0)	4.6×10^{-10}	4.0×10^{-10}	5.0×10^{-11}	5.2×10^{-12}	5.3×10^{-13}
(2,1,1)	3.8×10^{-9}	7.0×10^{-10}	1.2×10^{-10}	1.4×10^{-11}	1.4×10^{-12}
(2,4,0)	3.9×10^{-8}	4.7×10^{-11}	5.2×10^{-15}	5.4×10^{-20}	2.8×10^{-23}
(2,4,1)	1.6×10^{-8}	1.0×10^{-9}	4.3×10^{-13}	1.1×10^{-17}	9.5×10^{-22}
(4,1,0)	6.0×10^{-10}	5.5×10^{-12}	9.8×10^{-14}	1.1×10^{-15}	1.1×10^{-17}
(4,1,1)	3.1×10^{-9}	1.2×10^{-9}	8.3×10^{-12}	2.7×10^{-13}	4.1×10^{-15}
(4,4,0)	4.5×10^{-9}	2.7×10^{15}	8.4×10^{-22}	0	3.0×10^{-36}
(4,4,1)	3.1×10^{-8}	1.3×10^{-9}	3.6×10^{-14}	6.7×10^{-21}	6.4×10^{-27}

TABLE 9.3 Truncation Error Estimates of MRC Outage Probability with 10^{-10} Discretization Error, $K = 15$ and $N = 30$ (Rician Fading) for Various Values of $\bar{\gamma}_1/\gamma_{th}$

$(L, n^2 \, dB, d)$	-10 dB	-5 dB	0 dB	5 dB	10 dB
(2,0,0)	1.0×10^{-14}	1.3×10^{-15}	1.9×10^{-16}	1.3×10^{-17}	5.6×10^{-18}
(2,0,1)	3.4×10^{-14}	5.6×10^{-15}	5.5×10^{-17}	0	1.7×10^{-18}
(2,10,0)	5.3×10^{-15}	3.8×10^{-9}	0	0	0
(2,10,1)	3.0×10^{-14}	2.9×10^{-9}	0	0	0
(4,0,0)	4.4×10^{-16}	1.6×10^{-16}	1.7×10^{-18}	5.4×10^{-20}	4.2×10^{-22}
(4,0,1)	1.0×10^{-14}	4.2×10^{-15}	1.1×10^{-16}	3.4×10^{-18}	5.4×10^{-20}
(4,10,0)	7.3×10^{-8}	1.7×10^{-13}	0	0	0
(4,10,1)	3.1×10^{-15}	7.2×10^{-9}	2.3×10^{-20}	0	0

channels, correlation coefficients up to 0.6 between adjacent and second adjacent paths in the channel impulse response of frequency-selective channels were observed by Turin et al. [82,83], and Bajwa [81]. These early observations were recently confirmed by the propagation campaign of Patenaude et al. [82,83], who, based on a thorough statistical analysis of several macrocellular, microcellular, and indoor wideband channel impulse responses, reported correlation coefficients sometimes higher than 0.8 with no significant reduction in the correlation even for large path delay differences. As a result, the maximum theoretical diversity gain promised by RAKE reception cannot be achieved, and hence any analysis must be revamped to account for the effect of correlation between the combined signals.

Along these lines, several correlation models have been proposed [53,76,77], and using these models several authors [40,61,72,76,77,84–89,91,92] have analyzed special cases of the performance of various systems, corresponding to specific detection, modulation, and diversity combining schemes. For instance, Pierce and Stein [76] considered the performance of binary coherent and noncoherent systems over correlated identically distributed Rayleigh fading channels. In particular, they obtained the average BER of coherent BPSK when used in conjunction with MRC, and of noncoherent BFSK when used with postdetection EGC. Miyagaki et al. [40] analyzed the outage probability and the average SER

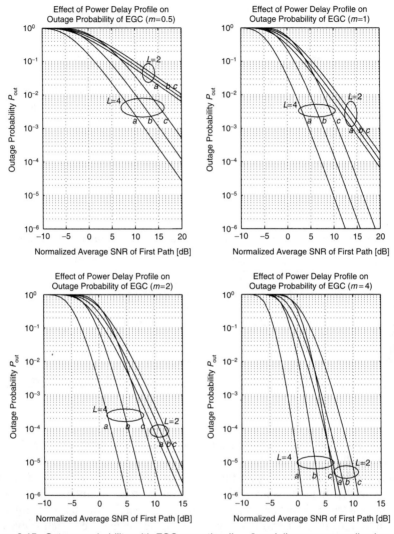

Figure 9.15. Outage probability with EGC reception ($L = 2$ and 4) versus normalized average SNR of the first path $\bar{\gamma}_1/\gamma_{th}$ over a Nakagami-m fading channel with an exponentially decaying power delay profile: (a) $\delta = 0$; (b) $\delta = 0.5$; (c) $\delta = 1$.

performance of M-PSK for various dual-branch diversity receivers over correlated identically distributed Nakagami-m fading channels [53]. Al-Hussaini and Al-Bassiouni [84] obtained a closed-form expression for the average BER of noncoherent BFSK with dual-branch MRC reception over correlated nonidentically distributed Nakagami-m fading channels. More recently, Aalo [77] analyzed the outage probability and the average BER of various coherent, differentially coherent, and noncoherent binary modulations with multichannel MRC reception over identically distributed Nakagami-m fading channels with two correlation

IMPACT OF FADING CORRELATION 319

TABLE 9.4 Truncation Error Estimates of EGC Outage Probability with 10^{-8} Discretization Error, $K = 11$ and $N = 15$ (Nakagami-m Fading) for Various Values of $\bar{\gamma}_1/\gamma_{th}$

(L,m,δ)	−10 dB	−5 dB	0 dB	5 dB	10 dB
(2,1,0)	8.3×10^{-10}	1.0×10^{-10}	6.0×10^{-12}	5.6×10^{-13}	3.6×10^{-16}
(2,1,1)	9.0×10^{-10}	2.4×10^{-10}	1.7×10^{-11}	1.5×10^{-12}	8.1×10^{-14}
(2,4,0)	1.2×10^{-9}	4.3×10^{-12}	6.2×10^{-17}	1.3×10^{-20}	3.3×10^{-24}
(2,4,1)	1.8×10^{-8}	5.9×10^{-10}	8.8×10^{-15}	4.3×10^{-19}	1.0×10^{-22}
(4,1,0)	3.1×10^{-11}	1.6×10^{-13}	1.7×10^{-15}	1.7×10^{-17}	1.7×10^{-19}
(4,1,1)	3.9×10^{-8}	6.6×10^{-10}	1.7×10^{-12}	6.5×10^{-15}	7.1×10^{-17}
(4,4,0)	5.9×10^{-7}	3.6×10^{-15}	0	0	0
(4,4,1)	5.1×10^{-7}	1.2×10^{-10}	0.2×10^{-14}	1.3×10^{-20}	0

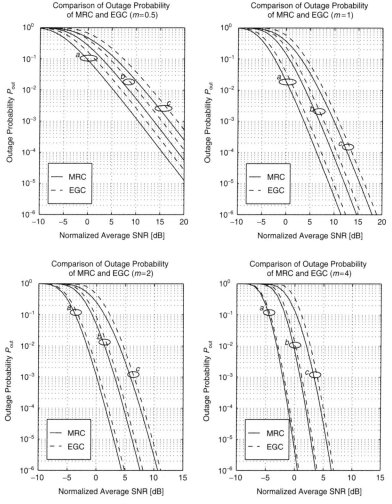

Figure 9.16. Comparison of the outage probability of MRC and EGC ($L = 4$) versus normalized average SNR of the first path $\bar{\gamma}_1/\gamma_{th}$ over a Nakagami-m fading channel with an exponentially decaying power delay profile: (a) $\delta = 0$; (b) $\delta = 0.5$; (c) $\delta = 1$.

models: the constant (equal) correlation model and the exponential correlation model. These results were extended to M-ary orthogonal signals in Ref. 72. In addition, Patenaude et al. [61] provided closed-form expressions for the average BER of orthogonal noncoherent BFSK with postdetection EGC reception over two correlated nonidentical and also D equicorrelated identically distributed Nakagami-m channels. Moreover, Zhang [85,86] and Lombardo et al. [87,88] presented analyses of binary signals over arbitrary correlated Nakagami-m fading channels with MRC and postdetection EGC. Finally, Cho and Lee [89], as well as Win et al. [90], considered the effect of fading correlation on the average SER performance of M-ary modulations with MRC reception.

In this section we use the MGF-based approach for the performance of diversity systems over fading channels combined with mathematical studies on the multivariate gamma distribution, to obtain general results for the exact average error rate and outage probability over Nakagami-m fading channels with arbitrary power delay and fading correlation profiles [93]. The results are applicable to M-ary coherent modulations when used in conjunction with MRC, as well as differentially coherent and noncoherent modulations when used in conjunction with postdetection EGC. More specifically, we consider four channel correlation models (models A, B, C, and D) of interest. For every model we give the PDF for the combined SNR per symbol, γ_t, as well as its MGF. In what follows, $p_i(\gamma_t)$ denotes the PDF of the combined SNR per symbol, with the index i identifying the model type and is hence equal to a, b, c, or d. In addition, the corresponding MGF of γ_t is denoted by $M_i(s)$ and is given by

$$M_i(s) = \int_0^\infty p_i(\gamma_t) e^{s\gamma_t} \, d\gamma_t \qquad (9.143)$$

Note that although model D is the most general model under consideration and as such incorporates models A, B, and C as special cases, direct derivations of the MGFs for these various models are interesting in their own right and will therefore be developed separately. As a double check, we show that the general MGF result obtained for model D will reduce to the MGF results obtained directly for models A, B, and C. Extensions of this MGF-based analysis to correlated Rician fading channels can be found in Refs. 91 and 92.

9.6.1 Model A: Two Correlated Branches with Nonidentical Fading

Model A was proposed by Nakagami [53, Sec. 6.4] and corresponds to the scenarios of dual-diversity reception over correlated Nakagami-m channels which are not necessarily identically distributed. This may therefore apply to small terminals equipped with space or polarization diversity where antenna spacing is insufficient to provide independent fading among signal paths.

9.6.1.1 PDF. In this case the PDF of the combined signal envelope, $p_a(r_t)$, is given by [53, Eq. (142)]

$$p_a(r_t) = \frac{2r_t\sqrt{\pi}}{\Gamma(m)[\sigma_1\sigma_2(1-\rho)]^m} \left[\frac{r_t^2}{2\beta}\right]^{m-1/2}$$
$$I_{m-1/2}(\beta r_t^2)e^{-\alpha r_t^2}, \qquad r_t \geq 0 \qquad (9.144)$$

where $I_v(\cdot)$ denotes the vth-order modified Bessel function [52, Sec. 10.2],

$$\rho = \frac{\operatorname{cov}(r_1^2, r_2^2)}{\sqrt{\operatorname{var}(r_1^2)\operatorname{var}(r_2^2)}}, \qquad 0 \leq \rho < 1 \qquad (9.145)$$

is the envelope correlation coefficient between the two signals[6] and the parameters σ_d ($d = 1, 2$), α, and β are defined as follows:

$$\sigma_d = \frac{\Omega_d}{m}, \qquad d = 1, 2, \qquad (9.146)$$

$$\alpha = \frac{\sigma_1 + \sigma_2}{2\sigma_1\sigma_2(1-\rho)} \qquad (9.147)$$

$$\beta^2 = \frac{(\sigma_1 - \sigma_2)^2 + 4\sigma_1\sigma_2\rho}{4\sigma_1^2\sigma_2^2(1-\rho)^2} \qquad (9.148)$$

where Ω_d, $d = 1, 2$, is the average fading power of the dth channel. By using a standard transformation of random variables, it can be shown that the PDF of the combined SNR per symbol, $p_a(\gamma_t)$, is given by

$$p_a(\gamma_t) = \frac{\sqrt{\pi}}{\Gamma(m)} \left[\frac{m^2}{\overline{\gamma}_1\overline{\gamma}_2(1-\rho)}\right]^m \left(\frac{\gamma_t}{2\beta'}\right)^{m-1/2} I_{m-1/2}(\beta'\gamma_t)e^{-\alpha'\gamma_t}, \qquad \gamma_t \geq 0 \qquad (9.149)$$

where the parameters α' and β' are normalized versions of the parameters α and β, and are given by

$$\alpha' \triangleq \frac{\alpha}{E_s/N_0} = \frac{m(\overline{\gamma}_1 + \overline{\gamma}_2)}{2\overline{\gamma}_1\overline{\gamma}_2(1-\rho)} \qquad (9.150)$$

$$\beta' \triangleq \frac{\beta}{E_s/N_0} = \frac{m((\overline{\gamma}_1 + \overline{\gamma}_2)^2 - 4\overline{\gamma}_1\overline{\gamma}_2(1-\rho))^{1/2}}{2\overline{\gamma}_1\overline{\gamma}_2(1-\rho)} \qquad (9.151)$$

---

[6] We use the envelope correlation coefficient throughout this chapter as a measure of the degree of correlation between the fading signals since, as pointed out in Refs. 76 and 77, experimental data on the correlation between fading signals are typically given in terms of this figure because of its relative ease of measurement.

For the case of identical channels ($\bar{\gamma}_1 = \bar{\gamma}_2 = \bar{\gamma}$), (9.149) reduces to

$$p_a(\gamma_t) = \frac{\sqrt{\pi}}{\Gamma(m)\sqrt{1-\rho}} \left(\frac{m}{\bar{\gamma}}\right)^{m+1/2} \left(\frac{\gamma_t}{2\sqrt{\rho}}\right)^{m-1/2}$$
$$\times \exp\left[-\frac{m\gamma_t}{(1-\rho)\bar{\gamma}}\right] I_{m-1/2}\left(\frac{m\sqrt{\rho}\gamma_t}{\bar{\gamma}(1-\rho)}\right), \quad \gamma_t \geq 0 \quad (9.152)$$

For the Rayleigh fading case ($m = 1$) using the identity [52, Eq. (10.2.13)]

$$\sqrt{\frac{\pi}{2z}} I_{1/2}(z) = \frac{\sinh z}{z}$$

where $\sinh(\cdot)$ denotes the hyperbolic sine function [52, Sec. 4.5], it can be shown that (9.149) reduces to

$$p_a(\gamma_t) = \frac{\exp\left[-\frac{\bar{\gamma}_1 + \bar{\gamma}_2 - \sqrt{(\bar{\gamma}_1 + \bar{\gamma}_2)^2 - 4\bar{\gamma}_1\bar{\gamma}_2(1-\rho)}}{2\bar{\gamma}_1\bar{\gamma}_2(1-\rho)}\gamma_t\right] - \exp\left[-\frac{\bar{\gamma}_1 + \bar{\gamma}_2 + \sqrt{(\bar{\gamma}_1 + \bar{\gamma}_2)^2 - 4\bar{\gamma}_1\bar{\gamma}_2(1-\rho)}}{2\bar{\gamma}_1\bar{\gamma}_2(1-\rho)}\gamma_t\right]}{\sqrt{(\bar{\gamma}_1 + \bar{\gamma}_2)^2 - 4\bar{\gamma}_1\bar{\gamma}_2(1-\rho)}}, \quad \gamma_t \geq 0$$
(9.154)

which itself reduces to the well-known expression for the case of identical Rayleigh channels originally derived in Ref. [94], which can also be found in Eq. (40) of Ref. 76:

$$p_a(\gamma_t) = \frac{1}{2\sqrt{\rho}\,\bar{\gamma}} \left\{ \exp\left[-\frac{\gamma_t}{(1+\sqrt{\rho})\bar{\gamma}}\right] - \exp\left[-\frac{\gamma_t}{(1-\sqrt{\rho})\bar{\gamma}}\right] \right\}, \quad \gamma_t \geq 0$$
(9.155)

9.6.1.2 MGF. Using the Laplace transform [36, Eq. (110)], it can be shown after some manipulations that the MGF of $p_a(\gamma_t)$ is given by

$$M_a(s; \bar{\gamma}_1, \bar{\gamma}_2; m; \rho) \triangleq M_a(s) = \left[1 - \frac{(\bar{\gamma}_1 + \bar{\gamma}_2)}{m}s + \frac{(1-\rho)\bar{\gamma}_1\bar{\gamma}_2}{m^2}s^2\right]^{-m}, \quad s \geq 0$$
(9.156)

It should be noted that for model A, using the MGF-based approach together with (9.156) leads to an *exact* expression for average probability of error [equivalent to Eq. (54) of Ref. 40], which has the advantage of being expressed in terms of a single finite-range integral with a much simpler integrand than that in Eq. (54) of Ref. 40, and hence easier to compute for any arbitrary value of the fading parameter m.

9.6.2 Model B: *D* Identically Distributed Branches with Constant Correlation

Model B was proposed by Aalo [77, Sec. II-A] for identically distributed Nakagami-m channels (i.e., all channels are assumed to have the same average SNR per symbol $\bar{\gamma}$ and the same fading parameter m). This model assumes that the envelope correlation coefficient ρ is the same between all the channel pairs $(d, d' = 1, 2, \ldots, D)$, that is,

$$\rho = \rho_{dd'} = \frac{\operatorname{cov}(r_d^2, r_{d'}^2)}{\sqrt{\operatorname{var}(r_d^2)\operatorname{var}(r_{d'}^2)}}, \qquad d \neq d', \quad 0 \leq \rho < 1 \tag{9.157}$$

and may therefore correspond to the scenario of multichannel reception from closely spaced diversity antennas or three antennas placed on an equilateral triangle.

9.6.2.1 PDF.
Based on the work of Gurland [95], Aalo showed that the PDF of γ_t is given in this case by [77, Eq. (18)][7]

$$p_b(\gamma_t) = \frac{\left(\frac{m\gamma_t}{\bar{\gamma}}\right)^{Dm-1} \exp\left(-\frac{m\gamma_t}{(1-\sqrt{\rho})\bar{\gamma}}\right) \times {}_1F_1\left(m, Dm; \frac{Dm\sqrt{\rho}\gamma_t}{(1-\sqrt{\rho})(1-\sqrt{\rho}+D\sqrt{\rho})\bar{\gamma}}\right)}{(\bar{\gamma}/m)(1-\sqrt{\rho})^{m(D-1)}(1-\sqrt{\rho}+D\sqrt{\rho})^m \Gamma(Dm)}, \qquad \gamma_t \geq 0 \tag{9.158}$$

where ${}_1F_1(\cdot, \cdot; \cdot)$ is the confluent hypergeometric function [52 p. 503]. For $D = 2$, using Eq. (13.6.3) of Ref. 52, namely,

$${}_1F_1(m, 2m; 2z) = \Gamma(m + 1/2)\left(\frac{z}{2}\right)^{-m+1/2} I_{m-1/2}(z) e^z \tag{9.159}$$

as well as the identities (6.1.12) and (6.1.8) of Ref. 52, yielding

$$\Gamma\left(m + \frac{1}{2}\right) = \frac{1 \cdot 3 \cdot 5 \cdot 7 \cdots (2m-1)}{2^m}\sqrt{\pi} \tag{9.160}$$

it can be shown after some manipulations that (9.158) reduces to (9.152) of model A, as expected.

---

[7] It should be noted at this point that in [Eq. (18) of Ref. 77] [or equivalently, Eq. (41) of Ref. 72] the symbol ρ is used to denote the correlation coefficient of the underlying Gaussian processes that produce the fading on the channels. This correlation coefficient is equal to the square root of the power correlation coefficient. Based on the work of Lawson and Uhlenbeck [96, p. 62], it is shown in by Pierce and Stein [76, App. V] that for all practical purposes, the power correlation coefficient, can be assumed to be equal to the envelope correlation coefficient which is denoted by ρ throughout this section, so as to follow what seems to be the more conventional usage of this symbol.

9.6.2.2 MGF.
Substituting (9.158) in (9.143), then using the Laplace transform [36, Eq. (4)],

$$\int_0^\infty x^{b-1} {}_1F_1(a; c; kx) e^{-sx} dx = \frac{\Gamma(b)}{s^b} {}_2F_1\left(a, b; c; \frac{k}{s}\right), \quad |s| > |k|, \quad b > 0, \quad s > 0, \quad s > k \quad (9.161)$$

together with the identity [52, Eq. (15.1.8); 77, Eq. (A-5)],

$$_2F_1(a, b; b; z) = (1 - z)^{-a} \quad (9.162)$$

it can be shown that

$$M_b(s; \overline{\gamma}; m; \rho; D) \triangleq M_b(s) = \left(1 - \frac{\overline{\gamma}(1 - \sqrt{\rho} + D\sqrt{\rho})}{m} s\right)^{-m}$$
$$\times \left(1 - \frac{\overline{\gamma}(1 - \sqrt{\rho})}{m} s\right)^{-m(D-1)}, \quad s \geq 0 \quad (9.163)$$

For $D = 2$, as a check, it can easily be shown that (9.163) agrees with (9.156) for $\overline{\gamma}_1 = \overline{\gamma}_2$.

9.6.3 Model C: *D* Identically Distributed Branches with Exponential Correlation

Model C was also proposed by Aalo [Sec. II-B] for identically distributed Nakagami-*m* channels (i.e., all channels are assumed to have the same average SNR per symbol $\overline{\gamma}$ and the same fading parameter *m*). This model assumes an exponential envelope correlation coefficient $\rho_{dd'}$ between any pair of channels ($d, d' = 1, 2, \ldots, D$) as given by

$$\rho_{dd'} = \frac{\text{cov}(r_d^2, r_{d'}^2)}{\sqrt{\text{var}(r_d^2) \text{var}(r_{d'}^2)}} = \rho^{|d-d'|}, \quad 0 \leq \rho \leq 1, \quad (9.164)$$

and may therefore correspond to the scenario of multichannel reception from equispaced diversity antennas in which the correlation between the pairs of combined signals decays as the spacing between the antennas increases.

9.6.3.1 PDF.
Based on the work of Kotz and Adams [97], Aalo showed that the PDF of γ_t can be very well approximated by a gamma distribution given by

[77, Eq. (19)][8]

$$p_c(\gamma_t) = \frac{\gamma_t^{(mD^2/r_\rho)-1} \exp(-mD\gamma_t/r_\rho\overline{\gamma})}{\Gamma(mD^2/r_\rho)(r_\rho\overline{\gamma}/mD)^{mD^2/r_\rho}}, \qquad \gamma_t \geq 0 \qquad (9.165)$$

where[9]

$$r_\rho = D + \frac{2\sqrt{\rho}}{1-\sqrt{\rho}}\left(D - \frac{1-\rho^{D/2}}{1-\sqrt{\rho}}\right) \qquad (9.166)$$

9.6.3.2 MGF. Substituting (9.165) in (9.143), then using the Laplace transform [36, Eq. (3.381.4)], it can be shown that

$$M_c(s;\overline{\gamma};m;\rho;D) \triangleq M_c(s) = \left(1 - \frac{r_\rho\overline{\gamma}}{mD}s\right)^{-mD^2/r_\rho}, \qquad s \geq 0 \qquad (9.167)$$

9.6.4 Model D: D Nonidentically Distributed Branches with Arbitrary Correlation

In this section, we treat a very general model in which the combined branches may not be identically distributed and also may have an arbitrary correlation. More specifically, this model assumes that the branches have an arbitrary average SNR per symbol $\overline{\gamma}_d$ and the same fading parameter m. The envelope correlation coefficient between any channel pair $(d, d' = 1, 2, \ldots, D)$ is denoted by $\rho_{dd'}$. The generality of this model may correspond to the impulse response of a frequency-selective channel with correlated paths and a nonuniform power delay profile.

9.6.4.1 MGF. The PDF of the combined SNR corresponding to this model was not previously found in a simple form. However, the joint distribution of the $\{\gamma_d\}_{d=1}^D$ can be deduced from the work of Krishnamoorthy and Parthasarathy [98] and can be expressed in terms of the generalized Laguerre polynomials. Their derivation is based on a relatively simple form for the joint MGF of a D-variate gamma distribution. Based on that derivation and defining $\beta_d = m\gamma_d/\overline{\gamma}_d$ as the normalized SNR per symbol per branch, we can express the MGF corresponding to this model as

$$M_d(s;\{\overline{\gamma}_d\}_{d=1}^D;m;[\rho_{dd'}];D) \triangleq M_d(s) = E_{\gamma_1,\gamma_2,\ldots,\gamma_D}\left[\exp\left(s\sum_{d=1}^D \gamma_d\right)\right]$$

$$= E_{\beta_1,\beta_2,\ldots,\beta_D}\left[\exp\left(s\sum_{d=1}^D \frac{\overline{\gamma}_d}{m}\beta_d\right)\right] \qquad (9.168)$$

---
[8] Based on the work of Kotz and Adams [97], Aalo [77] points out that the approximation (9.165) is valid for high values of D but is still accurate for values of D as small as 5.
[9] We remind the reader that contrary to its usage in Refs. 72 and 77, in this chapter the coefficient ρ denotes the envelope correlation coefficient.

Using the result Eq. (2.3) of Ref. 98 we can rewrite (9.168) as

$$M_d(s) = \prod_{d=1}^{D}\left(1 - \frac{s\bar{\gamma}_d}{m}\right)^{-m}$$

$$\times \left|\begin{bmatrix} 1 & \sqrt{\rho_{12}}\left(1-\frac{m}{s\bar{\gamma}_2}\right)^{-1} & \cdots & \sqrt{\rho_{1D}}\left(1-\frac{m}{s\bar{\gamma}_D}\right)^{-1} \\ \sqrt{\rho_{12}}\left(1-\frac{m}{s\bar{\gamma}_1}\right)^{-1} & 1 & \cdots & \sqrt{\rho_{2D}}\left(1-\frac{m}{s\bar{\gamma}_D}\right)^{-1} \\ \cdot & \cdot & \cdot & \cdot \\ \cdot & \cdot & \cdot & \cdot \\ \sqrt{\rho_{1D}}\left(1-\frac{m}{s\bar{\gamma}_1}\right)^{-1} & \sqrt{\rho_{2D}}\left(1-\frac{m}{s\bar{\gamma}_2}\right)^{-1} & \cdots & 1 \end{bmatrix}\right|_{D\times D}^{-m}$$

(9.169)

where $|[M]|_{D\times D}$ denotes the determinant of the $D \times D$ matrix M. For the case of identical channels ($\bar{\gamma}_d = \bar{\gamma}$), (9.169) reduces after some manipulation to

$$M_d(s) = \left(-\frac{s\bar{\gamma}}{m}\right)^{-mD} \left|\begin{bmatrix} 1-\frac{m}{s\bar{\gamma}} & \sqrt{\rho_{12}} & \cdots & \sqrt{\rho_{1D}} \\ \sqrt{\rho_{12}} & 1-\frac{m}{s\bar{\gamma}} & \cdots & \sqrt{\rho_{2D}} \\ \cdot & \cdot & \cdot & \cdot \\ \sqrt{\rho_{1D}} & \sqrt{\rho_{2D}} & \cdots & 1-\frac{m}{s\bar{\gamma}} \end{bmatrix}\right|_{D\times D}^{-m}$$

(9.170)

9.6.4.2 Special Cases of Interest

Dual Correlation Model (Model A). For $D = 2$, as a check, it is straightforward to show that (9.169) reduces to (9.156) of model A.

Intraclass Correlation or Constant Correlation Model (Model B). A correlation matrix M is called a Dth-order intraclass correlation matrix if it has the following structure [99, p. 14]:

$$M = \begin{bmatrix} a & b & \cdot & \cdot & \cdot & b \\ b & a & b & \cdot & \cdot & b \\ b & b & a & b & \cdot & b \\ \cdot & \cdot & \cdot & \cdot & \cdot & \cdot \\ b & \cdot & \cdot & \cdot & b & a \end{bmatrix}_{D\times D}$$

(9.171)

with $b \geq -a/(D-1)$. Eigenvalues of this type of matrices can be found in closed form and consequently their determinant can be written as [99, p. 21]

$$\det M = (a-b)^{D-1}[a + b(D-1)] \qquad (9.172)$$

Applying this property for $a = 1 - m/s\bar{\gamma}$ and $b = \sqrt{\rho}$, we get the MGF as

$$M_d(s) = M_b(s) = \left(1 - \frac{\bar{\gamma}(1 - \sqrt{\rho} + D\sqrt{\rho})}{m}s\right)^{-m}$$
$$\times \left(1 - \frac{\bar{\gamma}(1 - \sqrt{\rho})}{m}s\right)^{-m(D-1)} \quad (9.173)$$

which is in agreement with (9.163), as expected.

Exponential Correlation Model (Model C). For the exponential correlation model treated by Aalo (model C), $\bar{\gamma}_d = \bar{\gamma}$ for $d = 1, 2, \ldots, D$ and $\rho_{dd'} = \rho^{|d-d'|}$. Substituting this in (9.169) and using the algebraic technique presented in Appendix III of Ref. 76, it can easily be shown that the MGF is in this case given by

$$M_d(s) = \left(-\frac{s\bar{\gamma}}{m}\right)^{-mD} \prod_{d=1}^{D} \left(\frac{1-\rho}{1+\rho+2\sqrt{\rho}\cos\theta_d}\right)^{-m} \quad (9.174)$$

where θ_d ($d = 1, 2, 3, \ldots, D$) are the D solutions of the transcendental equation given by

$$\tan(D\theta_d) = \frac{-\sin\theta_d}{[(1+\rho)/(1-\rho)]\cos\theta_d + [2\sqrt{\rho}/(1-\rho)]} \quad (9.175)$$

Contrary to (9.167), which is accurate only for large values of D, (9.174) represents the exact MGF of the combined SNR at the output of an exponentially correlated Nakagami-m fading channel for an arbitrary value of D.

Tridiagonal Correlation Model. A correlation matrix M is called a Dth-order tridiagonal correlation matrix if it has the following structure [99, p. 16]:

$$M = \begin{bmatrix} a & b & 0 & \cdot & \cdot & 0 \\ b & a & b & 0 & \cdot & 0 \\ 0 & b & a & b & 0 & 0 \\ \cdot & \cdot & \cdot & \cdot & \cdot & \cdot \\ 0 & \cdot & \cdot & 0 & b & a \end{bmatrix}_{D\times D} \quad (9.176)$$

In this case the determinant can be shown to be given by [99, p. 21]

$$\det M = \prod_{d=1}^{D}\left[a + 2b\cos\left(\frac{d\pi}{D+1}\right)\right] \quad (9.177)$$

Applying this property for $a = 1 - m/s\bar{\gamma}$ and $b = \sqrt{\rho}$, and substituting the resulting determinant in (9.170), we get

$$M_d(s) = \prod_{d=1}^{D} \left\{ 1 - \frac{s\bar{\gamma}}{m} \left[1 + 2\sqrt{\rho} \cos\left(\frac{d\pi}{D+1}\right) \right] \right\}^{-m} \quad (9.178)$$

with [76, App. III]

$$\rho \leq \frac{1}{4\cos^2[\pi/(D+1)]} \quad (9.179)$$

to ensure that the matrix M as given by (9.176) is nonsingular and nonnegative. This model is useful for the accurate performance evaluation of space diversity with a "nearly" perfect antenna array in which the signal received at any antenna

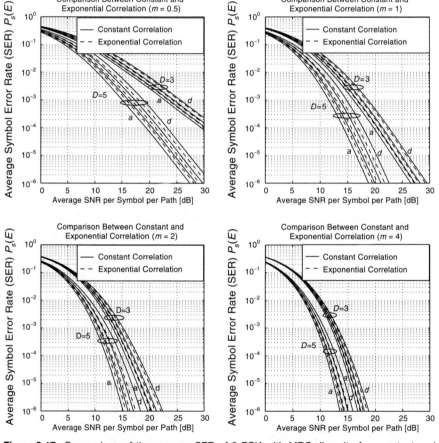

Figure 9.17. Comparison of the average SER of 8-PSK with MRC diversity for constant and exponential fading correlation profiles and various values of the correlation coefficient: (a) $\rho = 0$; (b) $\rho = 0.2$; (c) $\rho = 0.4$; (d) $\rho = 0.6$. $\bar{\gamma}_d = \bar{\gamma}$ for $d = 1, 2, \ldots, D$.

is weakly correlated with that received at any adjacent antenna ($\rho \ll 1$), but beyond the adjacent antenna the correlation is zero.

9.6.5 Numerical Examples

With the MGFs for the variety of correlation models in hand, we are now in a position to get the average probability of error and outage probability for MRC and postdetection EGC. For example, the average SER results of 8-PSK with third-order ($D = 3$) and fifth-order ($D = 5$) MRC diversity under constant and exponential correlation profile are shown in Fig. 9.17. On the other hand, Fig. 9.18 presents the same comparison (constant versus exponential correlation)

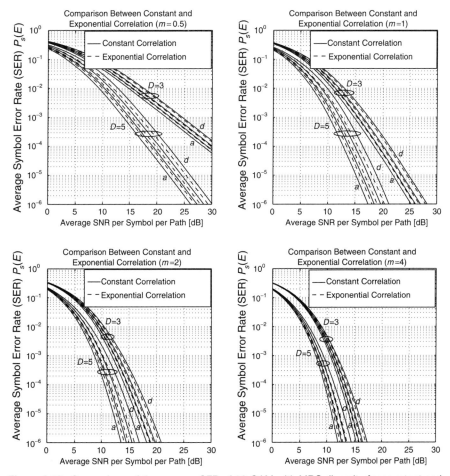

Figure 9.18. Comparison of the average SER of 16-QAM with MRC diversity for constant and exponential fading correlation profiles and various values of the correlation coefficient: (a) $\rho = 0$; (b) $\rho = 0.2$; (c) $\rho = 0.4$; (d) $\rho = 0.6$. $\overline{\gamma}_d = \overline{\gamma}$ for $d = 1, 2, \ldots, D$.

but for the the average SER of 16-QAM. For both of these figures constant correlation suffers a minor performance degradation compared to exponential correlation but the performance difference is more noticeable for a larger number of diversity paths and higher correlation between these paths.

Figure 9.19 shows the average BER of BPSK with MRC RAKE reception over an exponentially decaying PDP $[\overline{\gamma}_d = \overline{\gamma}_1 \exp(-\delta(d-1)), d = 1, 2, \ldots, D]$ and an exponential correlation profile across the multipaths. The corresponding outage probability curves obtained by using the numerical technique presented in Section 9.5 are presented in Fig. 9.20. Clearly, both the average SNR unbalance and the fading correlation induce a nonnegligible degradation in performance compared to a diversity system with i.i.d. fading across the combined paths. Furthermore, for the chosen parameters the effect of power delay profile is more

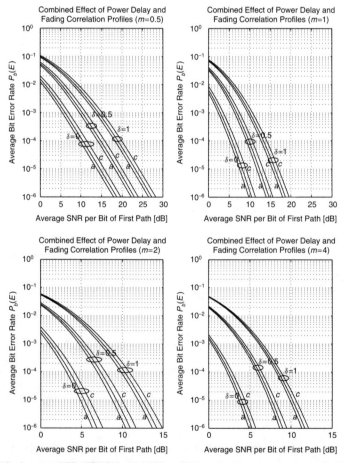

Figure 9.19. Average BER of BPSK with MRC reception ($D = 5$) over an exponentially decaying power delay profile and an exponential correlation profile across the multipaths for various values of the correlation coefficient: (a) $\rho = 0$; (b) $\rho = 0.2$; (c) $\rho = 0.4$.

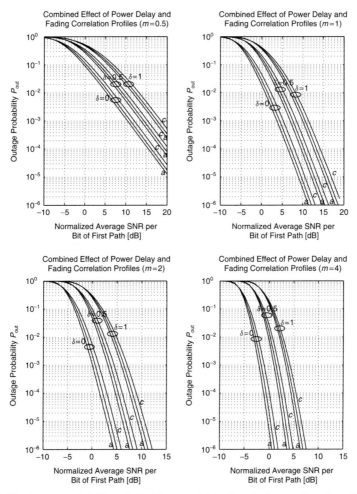

Figure 9.20. Outage probability with MRC or postdetection EGC reception ($D = 4$) over an exponentially decaying PDP and an exponential correlation profile across the multipaths for various values of the correlation coefficient: (a) $\rho = 0$; (b) $\rho = 0.2$; (c) $\rho = 0.4$.

important than the impact of correlation. However, note that in Figs. 9.19 and 9.20, once the first path average SNR $\overline{\gamma}_1$ is fixed (as is typically done in this type of illustrations; see, e.g., Ref. 54), the total average SNR is smaller for channels with an exponentially decaying power delay profile ($\delta > 0$) than for channels with a uniform power delay profile ($\delta = 0$), which explains in part the relatively important performance degradation due to the exponentially decaying power delay profile. Figure 9.21 is a plot of $P_b(E)$ for binary orthogonal FSK with dual diversity over correlated unbalanced ($\overline{\gamma}_2 = e^{-\delta}\overline{\gamma}_1$) Nakagami-$m$ fading channels. On the other hand, Fig. 9.21 shows the average BEP performance of binary orthogonal FSK with square-law combining over a multilink channel

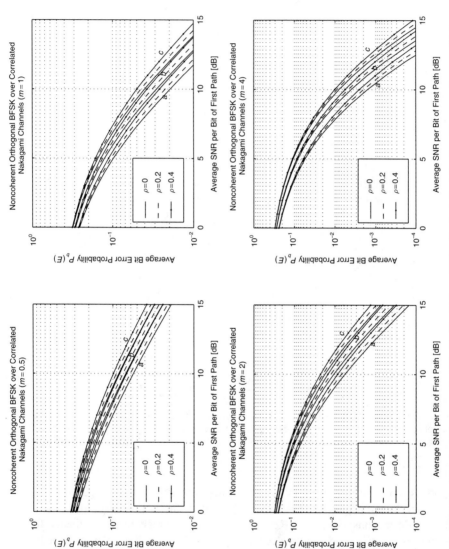

Figure 9.21. Average bit error probability $P_b(E)$ of binary FSK with dual diversity ($L = 2$) square-law combining versus the average SNR per bit of the first path $\bar{\gamma}_1$ over unbalanced correlated Nakagami-m channels with an exponentially decaying power delay profile: (a) $\delta = 0$; (b) $\delta = 0.5$; (c) $\delta = 1$.

with $L = 5$, an exponentially decaying power delay profile, and an exponential correlation profile (i.e., $\rho_{ij} = \rho^{|i-j|}$, $1 \leq i < j \leq L$). In both figures, for the parameters of interest, the BEP degradation induced by the power delay profile is higher than the degradation due to the fading correlation profile, where the degradation here is with respect to a system operating over a uniform power delay profile with independent multipaths. Furthermore, comparing Figs. 9.21 and 9.22, we conclude that this deterioration is more noticeable as the number of combined paths increases.

9.7 SELECTION COMBINING

Of the three types of linear diversity combining MRC, EGC, and SC normally employed in receivers of digital signals transmitted over multipath fading channels, SC is the least complicated of the three, since it only processes one of the diversity branches. Specifically, the combiner chooses the branch with the highest signal-to-noise ratio (or equivalently, with the strongest signal assuming equal noise power among the branches) [100, Sec. 10-4]. To obtain significant diversity gain, independent fading in the channels should be achieved. However, as mentioned previously, this is not always realized in practice because, for example, of insufficient antenna spacing in small terminals equipped with space antenna diversity and as a result, the maximum theoretical diversity gain cannot be achieved. In addition, the diversity branches in a practical system may have unequal average SNRs due to different noise figures or feedline lengths [101,102].

Hence, it is important to assess the effect of correlation and average SNR unbalance on the outage probability and average error probability of an SC diversity receiver, in particular, a dual-diversity (two-branch) SC receiver, which is the specific case to be considered in this section. Some special cases of the performance of various modulation schemes with dual SC over independent and correlated Rayleigh and Nakagami-m slow-fading channels have been reported in the literature [103–107]. For instance, Blanco [103] studied the performance of noncoherent BFSK with dual SC over i.i.d. Nakagami-m fading channels. In Ref. 108 Al-Hussaini and Al-Bassiouni analyzed the effect of fading correlation and average SNR unbalance on the BER performance of BFSK with dual SC over Nakagami-m fading channels. Adachi et al. [104] analyzed the performance of DQPSK over correlated unequal average power Rayleigh fading channels. Okui [105] studied the probability of co-channel interference for selection diversity reception in the Nakagami-m fading channel, whereas Wan and Chen [106] presented simulation results for DQPSK with dual SC over correlated Rayleigh fading channels. Fedele et al. [109] analyzed the performance of M-ary DPSK with dual SC over independent and correlated Nakagami-m fading channels. Finally, Ugweje and Aalo [107] analyzed the average BER performance of DPSK and BPSK with dual SC over correlated Nakagami-m fading channels. In this section we use the unifying MGF-based approach

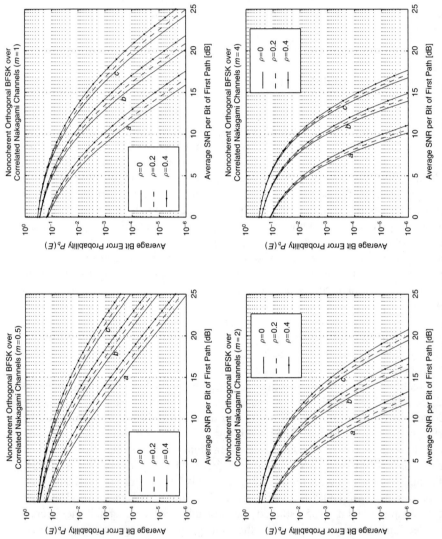

Figure 9.22. Average bit error probability $P_b(E)$ of binary FSK with square-law combining ($L = 5$) versus the average SNR per bit of the first path $\overline{\gamma}_1$ over correlated Nakagami-m channels with an exponential fading correlation profile and an exponentially decaying power delay profile: (a) $\delta = 0$; (b) $\delta = 0.5$; (c) $\delta = 1$.

SELECTION COMBINING   335

combined with the results presented in Chapter 6 to assess the performance of dual SC over independent and correlated slow Rayleigh and Nakagami-*m* fading channels [110,111].

9.7.1 MGF of Output SNR

Recall that the PDF of the output SNR, γ_{SC}, of a dual SC over correlated Nakagami-*m* fading channels is given by (6.21), namely,

$$p_{\gamma_{SC}}(\gamma) = \frac{(m/\overline{\gamma}_1)^m \gamma^{m-1}}{\Gamma(m)} \exp\left(-\frac{m\gamma}{\overline{\gamma}_1}\right) [1 - Q_m(A_1\sqrt{2\rho\gamma}, A_2\sqrt{2\gamma})]$$
$$+ \frac{(m/\overline{\gamma}_2)^m \gamma^{m-1}}{\Gamma(m)} \exp\left(-\frac{m\gamma}{\overline{\gamma}_2}\right) [1 - Q_m(A_2\sqrt{2\rho\gamma}, A_1\sqrt{2\gamma})] \quad (9.180)$$

where A_l is given by $A_l = \sqrt{m/\overline{\gamma}_l(1-\rho)}$. Hence, the MGF of the output SNR can be written as

$$M_{\gamma_{SC}}(s) = I_1(s; \overline{\gamma}_1, \overline{\gamma}_2, \rho, m) + I_2(s; \overline{\gamma}_1, \overline{\gamma}_2, \rho, m) \quad (9.181)$$

where

$$I_1(s; \overline{\gamma}_1, \overline{\gamma}_2, \rho, m) = \frac{(m/\overline{\gamma}_1)^m}{\Gamma(m)} \int_0^\infty \gamma^{m-1} \exp\left[-\left(\frac{m}{\overline{\gamma}_1} - s\right)\gamma\right]$$
$$\times [1 - Q_m(A_1\sqrt{2\rho\gamma}, A_2\sqrt{2\gamma})] \, d\gamma \quad (9.182)$$

$$I_2(s; \overline{\gamma}_1, \overline{\gamma}_2, \rho, m) = \frac{(m/\overline{\gamma}_2)^m}{\Gamma(m)} \int_0^\infty \gamma^{m-1} \exp\left[-\left(\frac{m}{\overline{\gamma}_2} - s\right)\gamma\right]$$
$$\times [1 - Q_m(A_2\sqrt{2\rho\gamma}, A_1\sqrt{2\gamma})] \, d\gamma \quad (9.183)$$

Using the identity from Ref. 105 [Eq. (6)], a straightforward change of integration variables in (9.182) and (9.183) allows $I_1(s; \overline{\gamma}_1, \overline{\gamma}_2, \rho, m)$ and $I_2(s; \overline{\gamma}_1, \overline{\gamma}_2, \rho, m)$ to be expressed in closed form as

$$I_1(s; \overline{\gamma}_1, \overline{\gamma}_2, \rho, m) = \frac{2^{3m}\Gamma(2m)m^{2m}X_1^{-2m}[W_1(1+W_1)]^{-m}}{\Gamma(m)\Gamma(m+1)(\overline{\gamma}_1\overline{\gamma}_2)^m(1-\rho)^m}$$
$$\times {}_2F_1\left[1-m, m; 1+m; \frac{1}{2}\left(1-\frac{1}{W_1}\right)\right] \quad (9.184)$$

$$I_2(s; \overline{\gamma}_1, \overline{\gamma}_2, \rho, m) = \frac{2^{3m}\Gamma(2m)m^{2m}X_2^{-2m}[W_2(1+W_2)]^{-m}}{\Gamma(m)\Gamma(m+1)(\overline{\gamma}_1\overline{\gamma}_2)^m(1-\rho)^m}$$
$$\times {}_2F_1\left[1-m, m; 1+m; \frac{1}{2}\left(1-\frac{1}{W_2}\right)\right] \quad (9.185)$$

where $_2F_1(\cdot, \cdot; \cdot; \cdot)$ is the Gauss hypergeometric function [52, Chap. 15], and X_i, Y_i, and W_i are given by

$$X_1 = 2(a-s)$$
$$X_2 = -2(a+s)$$
$$Y_1 = Y_2 = 2[(b-s)^2 - c^2]^{1/2}$$
$$W_1 = \frac{Y_1}{X_1} = \left[\left(\frac{b-s}{a-s}\right)^2 - \left(\frac{c}{a-s}\right)^2\right]^{1/2} \quad (9.186)$$
$$W_2 = \frac{Y_2}{X_2} = -\left[\left(\frac{b-s}{a+s}\right)^2 - \left(\frac{c}{a+s}\right)^2\right]^{1/2}$$

with a, b, and c given by

$$a = \frac{m(\overline{\gamma}_2 - \overline{\gamma}_1)}{\overline{\gamma}_1 \overline{\gamma}_2 (1-\rho)}, \quad b = \frac{m(\overline{\gamma}_1 + \overline{\gamma}_2)}{\overline{\gamma}_1 \overline{\gamma}_2 (1-\rho)}, \quad c = \frac{2m\sqrt{\rho}}{\sqrt{\overline{\gamma}_1 \overline{\gamma}_2}(1-\rho)} \quad (9.187)$$

9.7.2 Average Output SNR

9.7.2.1 General Case. Using the well-known result that the first moment of γ_{SC} is equal to its statistical average [112, Eq. (5-67)],

$$\overline{\gamma}_{SC} = \left.\frac{dM_{\gamma_{SC}}(s)}{ds}\right|_{s=0} \quad (9.188)$$

we obtain after substituting (9.181) in (9.188) and using the differentiation formula given by Eq. (15.2.1) of Ref. 52 we obtain after much manipulation the final desired closed-form result as

$$\overline{\gamma}_{SC} = K(K_1 + K_2) \quad (9.189)$$

where

$$K = \frac{2^m m^{2m+1} \Gamma(2m)}{\Gamma(m)\Gamma(m+1)(\overline{\gamma}_1 \overline{\gamma}_2)^m (1-\rho)^m}$$

$$K_1 = \frac{(b^2 - c^2 - ab)(b^2 - c^2 + a\sqrt{b^2 - c^2})^{-m}}{(b^2 - c^2)(\sqrt{b^2 - c^2} - a)}$$

$$\times \left[\left(\frac{1}{2} + \frac{a}{2\sqrt{b^2 - c^2}}\right)^{m-1} - {}_2F_1\left(1-m, m, 1+m, \frac{1}{2} - \frac{a}{2\sqrt{b^2 - c^2}}\right)\right]$$

$$+ \frac{(b^2 - c^2 + ab)(b^2 - c^2 - a\sqrt{b^2 - c^2})^{-m}}{(b^2 - c^2)(\sqrt{b^2 - c^2} + a)} \left[\left(\frac{1}{2} - \frac{a}{2\sqrt{b^2 - c^2}} \right)^{m-1} \right.$$

$$\left. - {}_2F_1\left(1 - m, m, 1 + m, \frac{1}{2} + \frac{a}{2\sqrt{b^2 - c^2}}\right) \right] \quad (9.190)$$

$$K_2 = (b^2 - c^2 + a\sqrt{b^2 - c^2})^{-m-1}\left(2b + \frac{b^2 - c^2 + ab}{\sqrt{b^2 - c^2}}\right)$$

$$\times {}_2F_1\left(1 - m, m; 1 + m; \frac{1}{2} - \frac{a}{2\sqrt{b^2 - c^2}}\right)$$

$$+ (b^2 - c^2 - a\sqrt{b^2 - c^2})^{-m-1}\left(2b + \frac{b^2 - c^2 - ab}{\sqrt{b^2 - c^2}}\right)$$

$$\times {}_2F_1\left(1 - m, m; 1 + m; \frac{1}{2} + \frac{a}{2\sqrt{b^2 - c^2}}\right)$$

Using the integral representation of the Gauss hypergeometric function as given by Eq. (15.1.25) of Ref. 52, it can easily be shown that ${}_2F_1(0, 1, 2, x) = 1$ for all x. Hence for the Rayleigh fading ($m = 1$) case, (9.189) reduces to the much simpler formula

$$\overline{\gamma}_{SC} = \frac{1}{2}\left(\overline{\gamma}_1 + \overline{\gamma}_2 + \frac{\overline{\gamma}_1^2 + \overline{\gamma}_2^2 - 2\rho\overline{\gamma}_1\overline{\gamma}_2}{\sqrt{(\overline{\gamma}_1 + \overline{\gamma}_2)^2 - 4\rho\overline{\gamma}_1\overline{\gamma}_2}}\right) \quad (9.191)$$

9.7.2.2 Special Cases. For the equal average SNR ($\overline{\gamma}_1 = \overline{\gamma}_2 = \overline{\gamma}$) but correlated fading case ($\rho \neq 0$), we have from (9.187) $a = 0$, $b = 2m/[\overline{\gamma}(1 - \rho)]$, and $c = 2m\sqrt{\rho}/[\overline{\gamma}(1 - \rho)]$. This, combined with the identity, Eq. (15.1.26), of Ref. 52 relating the Gauss hypergeometric function to the gamma function as well as the duplication formula of Ref. 52 [Eq. (6.1.18)] of the gamma function, leads to the average SNR in the greatly simplified form given by

$$\overline{\gamma}_{SC} = \overline{\gamma}\left[1 + \frac{\Gamma(2m)\sqrt{1 - \rho}}{2^{2m-1}\Gamma(m)\Gamma(m + 1)}\right] \quad (9.192)$$

which further reduces for the special Rayleigh fading case ($m = 1$) to

$$\overline{\gamma}_{SC} = \overline{\gamma}\left(1 + \frac{\sqrt{1 - \rho}}{2}\right) \quad (9.193)$$

For the unequal average SNR ($\overline{\gamma}_1 \neq \overline{\gamma}_2$) with uncorrelated fading ($\rho = 0$) case, we have from (9.187) $a = m(\overline{\gamma}_2 - \overline{\gamma}_1)/\overline{\gamma}_1\overline{\gamma}_2$, $b = m(\overline{\gamma}_1 + \overline{\gamma}_2)/\overline{\gamma}_1\overline{\gamma}_2$, and $c = 0$. Substituting this in (9.190) and using the linear transformation formula of Ref. 52 [Eq. (15.3.3)], it can be shown after some algebraic manipulation that

(9.189) reduces to

$$\overline{\gamma}_{SC} = \overline{\gamma}_1 + \overline{\gamma}_2 - \frac{(\overline{\gamma}_1\overline{\gamma}_2)^{m+1}\Gamma(2m+1)}{(m+1)\Gamma(m)\Gamma(m+1)(\overline{\gamma}_1+\overline{\gamma}_2)^{2m+1}}$$
$$\times \left[{}_2F_1\left(1, 2m+1; m+2; \frac{\overline{\gamma}_1}{\overline{\gamma}_1+\overline{\gamma}_2}\right) \right.$$
$$\left. + {}_2F_1\left(1, 2m+1; m+2; \frac{\overline{\gamma}_2}{\overline{\gamma}_1+\overline{\gamma}_2}\right) \right] \quad (9.194)$$

Using the series definition of the Gauss hypergeometric function as given in Eq. (15.1.1) of Ref. 52 it can easily be shown that ${}_2F_1(1, 3, 3, x) = 1/(1-x)$. Hence for the Rayleigh fading case ($m = 1$), (9.194) reduces to

$$\overline{\gamma}_{SC} = \overline{\gamma}_1 + \overline{\gamma}_2 - \frac{\overline{\gamma}_1\overline{\gamma}_2}{\overline{\gamma}_1+\overline{\gamma}_2} \quad (9.195)$$

Finally, for the equal average SNR ($\overline{\gamma}_1 = \overline{\gamma}_2 = \overline{\gamma}$) with uncorrelated fading ($\rho = 0$) case, we have in (9.187), $a = 0$, $b = 2m/\overline{\gamma}$, and $c = 0$. This combined again with Eqs. (15.1.26) and (6.1.18) of Ref. 52 leads to

$$\overline{\gamma}_{SC} = \overline{\gamma}\left[1 + \frac{\Gamma(2m)}{2^{2m-1}\Gamma(m)\Gamma(m+1)}\right] \quad (9.196)$$

which reduces for the Rayleigh case ($m = 1$) to $\overline{\gamma}_{SC} = 1.5\overline{\gamma}$, in agreement with Eq. (6.62) of Ref. 4 or equivalently, Eq. (5.86) of Ref. 3.

9.7.2.3 Numerical Examples.
As an example, Fig. 9.23 plots the first branch normalized average SNR, $\overline{\gamma}_{SC}/\overline{\gamma}_1$, versus the correlation coefficient ρ for an equal average dual SC receiver [(a) $\overline{\gamma}_1 = \overline{\gamma}_2$] as well as for an unbalanced dual SC receiver [(b) $\overline{\gamma}_1 = 2\overline{\gamma}_2$, (c) $\overline{\gamma}_1 = 5\overline{\gamma}_2$, and (d) $\overline{\gamma}_1 = 10\overline{\gamma}_2$]. The average SNR degrades quite rapidly as the correlation coefficient ρ increases, especially for the equal average SNR case and for low values of m. In particular, when $\rho = 0$ (uncorrelated fading), the average SNR is maximum, and in the limit of fully fading correlation ($\rho = 1$), the average output SNR approaches the average SNR of a single branch (i.e., without diversity), as expected. In addition, for a fixed fading correlation, the average SNR decreases as the severity of fading decreases (i.e., m decreases), which may seem surprising at first glance. However, as m increases the distribution becomes more skewed, which reduces the effective area of integration and explains this dependence of the average SNR on the fading parameter m.

9.7.3 Outage Probability

The outage probability, P_{out}, is defined as the probability that the SC output SNR $\gamma = \max(\gamma_1, \gamma_2)$ falls below a given threshold, say γ_{th}. Since this probability is

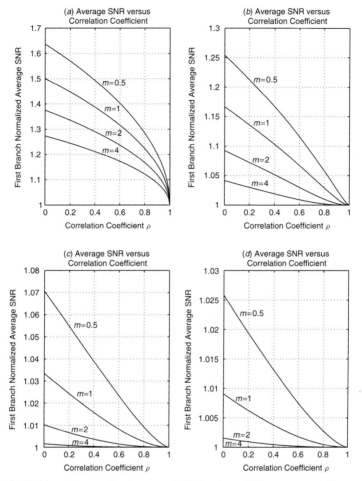

Figure 9.23. First branch normalized average SNR ($\bar{\gamma}_{sc}/\bar{\gamma}_1$) of SC versus correlation coefficient (ρ) for correlated Nakagami-m fading channels with (a) $\bar{\gamma}_1 = \bar{\gamma}_2$, (b) $\bar{\gamma}_1 = 2\bar{\gamma}_2$, (c) $\bar{\gamma}_1 = 5\bar{\gamma}_2$, and (d) $\bar{\gamma}_1 = 10\bar{\gamma}_2$.

simply the probability that neither γ_1 nor γ_2 exceeds the threshold, γ_{th}, then by inspection the outage probability is obtained by replacing γ with γ_{th} in the CDF expression given in (6.16), yielding for Rayleigh fading:

- Case 1: Identical channels ($\bar{\gamma}_1 = \bar{\gamma}_2 = \bar{\gamma}$)

$$P_{out} = 1 - 2\exp\left(-\frac{\gamma_{th}}{\bar{\gamma}}\right) + \frac{1-\rho}{2\pi}\int_{-\pi}^{\pi}\frac{1}{1+\rho+2\sqrt{\rho}\sin\theta}$$
$$\times \exp\left(-\frac{2\gamma_{th}}{\bar{\gamma}}\frac{1+\sqrt{\rho}\sin\theta}{1-\rho}\right)d\theta \quad (9.197)$$

Note that this result is equivalent to Eq. (10-10-7) of Ref. 100 which is expressed in terms of the Marcum Q-function. Note that even for this simpler case of identical channels, Tan and Beaulieu's result [113, Eq. (4)] (or equivalently, the result found in Ref. 114) does not simplify considerably since it is still an infinite series of squares of integrals. Furthermore, in the limiting case of uncorrelated branches (i.e., $\rho = 0$), (9.197) reduces to $P_{\text{out}} = [1 - \exp(-\gamma_{\text{th}}/\overline{\gamma})]^2$, as expected.

- Case 2: Nonidentical channels ($\overline{\gamma}_1 \neq \overline{\gamma}_2$)

$$P_{\text{out}} = 1 - G(H(\gamma_{\text{th}}, \overline{\gamma}_1), H(\gamma_{\text{th}}, \overline{\gamma}_2)|\rho)$$
$$+ \frac{1}{2\pi} \int_{-\pi}^{\pi} \exp\left(-\gamma_{\text{th}} \frac{\overline{\gamma}_1 + \overline{\gamma}_2 + 2\sqrt{\rho\overline{\gamma}_1\overline{\gamma}_2}\sin\theta}{\overline{\gamma}_1\overline{\gamma}_2(1-\rho)}\right)$$
$$\times \frac{(1-\rho^2)\overline{\gamma}_1\overline{\gamma}_2 + \sqrt{\rho}(1-\rho)\sqrt{\overline{\gamma}_1\overline{\gamma}_2}(\overline{\gamma}_1 + \overline{\gamma}_2)\sin\theta}{(\rho\overline{\gamma}_2 + 2\sqrt{\rho\overline{\gamma}_1\overline{\gamma}_2}\sin\theta + \overline{\gamma}_1)(\overline{\gamma}_2 + 2\sqrt{\rho\overline{\gamma}_1\overline{\gamma}_2}\sin\theta + \rho\overline{\gamma}_1)} d\theta$$
(9.198)

where $G(H(\gamma_{\text{th}}, \overline{\gamma}_1)$ and $H(\gamma_{\text{th}}, \overline{\gamma}_2)|\rho)$ are as given in (6.15). Note that (9.198) is equivalent to Eq. (10-10-3) of Ref. 100, which is expressed in terms of the Marcum Q-function. Furthermore, in the limiting case of uncorrelated branches (i.e., $\rho = 0$), (9.198) together with (6.15) reduces to $P_{\text{out}} = [1 - \exp(-\gamma_{\text{th}}/\overline{\gamma}_1)][1 - \exp(-\gamma_{\text{th}}/\overline{\gamma}_2)]$, as expected.

The outage probability expressions for Nakagami-m fading are obtained immediately from the CDF expressions (6.26) and (6.30) by replacing γ with γ_{th}. Since no further simplifications are possible and in the interest of brevity, we shall not write down the specific results for the two cases of identical and nonidentical channels considered previously for Rayleigh fading.

9.7.3.1 Numerical Example.
Figure 9.24 compares the outage probability of dual-branch MRC and SC for various values of the fading parameter m, correlation coefficient ρ, and average SNR unbalance. In this figure the SC outage probability results are based on (9.197), whereas the MRC outage probability results are obtained by substituting (9.156) in (9.136).

9.7.4 Average Probability of Error

9.7.4.1 BDPSK and Noncoherent BFSK.
Recall that the conditional BER for BDPSK and noncoherent BFSK is given by [7, Eqs. (5-2-69) and (5-4-47)]

$$P_b(E|\gamma) = \tfrac{1}{2}\exp(-g\gamma) \tag{9.199}$$

where g is again a modulation constant (i.e., $g = 1$ for BDPSK and $g = \tfrac{1}{2}$ for orthogonal BFSK). Averaging (9.199) over the PDF of the SC output (6.18), we obtain the following expression for the average BER:

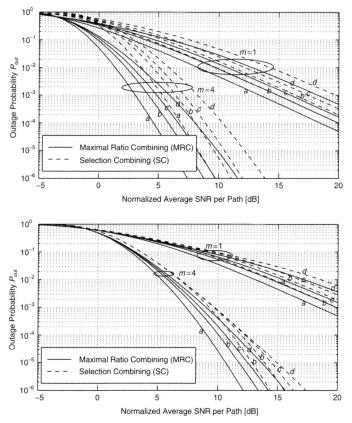

Figure 9.24. Comparison of outage probability with MRC and SC versus average SNR of the first branch for various values of the correlation coefficient: (a) $\rho = 0$; (b) $\rho = 0.5$; (c) $\rho = 0.7$; (d) $\rho = 0.9$. (Top) equal average branch SNRs ($\bar{\gamma}_1 = \bar{\gamma}_2$); (bottom) unequal average branch SNRs ($\bar{\gamma}_1 = 10\bar{\gamma}_2$).

$$P_b(E) = \frac{1}{2}\left[G((1+g\bar{\gamma}_1)^{-1}, (1+g\bar{\gamma}_2)^{-1}|\rho) - \frac{1}{2\pi}\int_{-\pi}^{\pi}\frac{h_1(\theta|\rho)h_2(\theta|\rho)}{g+h_1(\theta|\rho)}d\theta\right] \tag{9.200}$$

The integral term in (9.200) can be evaluated in closed form by first expanding the integrand into a partial fraction expansion, then making use of a well-known definite integral. In particular, identifying $h_1(\theta|\rho)$ and $h_2(\theta|\rho)$ from (6.17), it is straightforward to show that

$$\frac{h_1(\theta|\rho)h_2(\theta|\rho)}{g+h_1(\theta|\rho)} = \frac{\bar{\gamma}_1 + \bar{\gamma}_2 + 2\sqrt{\rho\bar{\gamma}_1\bar{\gamma}_2}\sin\theta}{g(1-\rho)\bar{\gamma}_1\bar{\gamma}_2 + \bar{\gamma}_1 + \bar{\gamma}_2 + 2\sqrt{\rho\bar{\gamma}_1\bar{\gamma}_2}\sin\theta}$$
$$\times \frac{(1-\rho^2)\bar{\gamma}_1\bar{\gamma}_2 + \sqrt{\rho}(1-\rho)\sqrt{\bar{\gamma}_1\bar{\gamma}_2}(\bar{\gamma}_1+\bar{\gamma}_2)\sin\theta}{(\rho\bar{\gamma}_2 + 2\sqrt{\rho\bar{\gamma}_1\bar{\gamma}_2}\sin\theta + \bar{\gamma}_1)(\bar{\gamma}_2 + 2\sqrt{\rho\bar{\gamma}_1\bar{\gamma}_2}\sin\theta + \rho\bar{\gamma}_1)} \tag{9.201}$$

which is in the form

$$\frac{h_1(\theta|\rho)h_2(\theta|\rho)}{g+h_1(\theta|\rho)} = \frac{\overline{\gamma}_1+\overline{\gamma}_2}{\overline{\gamma}_1+\overline{\gamma}_2+g\overline{\gamma}_1\overline{\gamma}_2}, \qquad \rho=0 \tag{9.202}$$

$$\frac{h_1(\theta|\rho)h_2(\theta|\rho)}{g+h_1(\theta|\rho)} = \frac{A+B\sin\theta+C\sin^2\theta}{(a_1+b\sin\theta)(a_2+b\sin\theta)(a_3+b\sin\theta)}$$

$$= \frac{c_1}{a_1+b\sin\theta}+\frac{c_2}{a_2+b\sin\theta}+\frac{c_3}{a_3+b\sin\theta}, \qquad \rho\neq 0 \tag{9.203}$$

with

$$a_1 = g(1-\rho)\overline{\gamma}_1\overline{\gamma}_2+\overline{\gamma}_1+\overline{\gamma}_2$$
$$a_2 = \rho\overline{\gamma}_2+\overline{\gamma}_1$$
$$a_3 = \rho\overline{\gamma}_1+\overline{\gamma}_2$$
$$b = 2\sqrt{\rho\overline{\gamma}_1\overline{\gamma}_2} \tag{9.204}$$
$$A = (1-\rho^2)\overline{\gamma}_1\overline{\gamma}_2(\overline{\gamma}_1+\overline{\gamma}_2)$$
$$B = 2\sqrt{\rho}(1-\rho^2)(\overline{\gamma}_1\overline{\gamma}_2)^{3/2}+\sqrt{\rho}(1-\rho)\sqrt{\overline{\gamma}_1\overline{\gamma}_2}(\overline{\gamma}_1+\overline{\gamma}_2)^2$$
$$C = 2\rho(1-\rho)\overline{\gamma}_1\overline{\gamma}_2(\overline{\gamma}_1+\overline{\gamma}_2)$$

The coefficients of the partial fraction expansion are readily determined as

$$c_1 = \frac{b^2A-a_1bB+a_1^2C}{(a_1-a_2)(a_1-a_3)b^2}$$

$$c_2 = -\frac{b^2A-a_2bB+a_2^2C}{(a_1-a_2)(a_2-a_3)b^2} \tag{9.205}$$

$$c_3 = \frac{b^2A-a_3bB+a_3^2C}{(a_1-a_3)(a_2-a_3)b^2}$$

Finally, substituting (9.202) and (9.203) into (9.200) and making use of the definite integral [36, Eq. (3.661.4)],

$$\frac{1}{2\pi}\int_{-\pi}^{\pi}\frac{1}{a+b\sin\theta}d\theta = \frac{1}{\sqrt{a^2-b^2}}, \qquad a\geq b \tag{9.206}$$

we get the desired closed-form result[10]

$$P_b(E) = \frac{1}{2}\sum_{i=1}^{3}\beta_i\frac{1}{1+g\overline{\gamma}_i}, \quad \left(\overline{\gamma}_3 \triangleq \frac{\overline{\gamma}_1\overline{\gamma}_2}{\overline{\gamma}_1+\overline{\gamma}_2}, \beta_1=\beta_2=1, \beta_3=-1\right), \qquad \rho=0 \tag{9.207}$$

---

[10] Note from (9.204) that it can easily be shown that $a_i \geq b$ for $i=1,2,3$. Hence, (9.206) applies.

$$P_b(E) = \frac{1}{2}\left[G((1+g\bar{\gamma}_1)^{-1},(1+g\bar{\gamma}_2)^{-1}|\rho) - \sum_{i=1}^{3}\frac{c_i}{\sqrt{a_i^2 - b^2}}\right], \quad \rho \neq 0 \tag{9.208}$$

Note that for the special case of $\bar{\gamma}_1 = \bar{\gamma}_2$ and $g = 1$ (BDPSK), (9.207) is in agreement with Eq. (5.88) of Ref. 3.

9.7.4.2 Coherent BPSK and BFSK.
Recall that based on an alternative representation of the Gaussian Q-function as given in (4.2), the conditional BER of BPSK and BFSK can be written in the integral form

$$P_b(E|\gamma) = \frac{1}{\pi}\int_0^{\pi/2} \exp\left(-\frac{g\gamma}{\sin^2\theta}\right) d\theta \tag{9.209}$$

where $g = 1$ for BPSK, $g = \frac{1}{2}$ for orthogonal BFSK, and $g = 0.715$ for BFSK with minimum correlation. Recognizing the analogy between (9.209) and (9.199) in terms of its functional dependence on γ, we can immediately write the average BER as

$$P_b(E) = \frac{1}{\pi}\int_0^{\pi/2}\left[\sum_{i=1}^{3}\beta_i\frac{1}{1+g(\theta)\bar{\gamma}_i}\right]d\theta$$

$$= \frac{1}{2}\left(1 - \sum_{i=1}^{3}\beta_i\sqrt{\frac{g\bar{\gamma}_i}{1+g\bar{\gamma}_i}}\right), \quad \rho = 0 \tag{9.210}$$

$$P_b(E) = \frac{1}{\pi}\int_0^{\pi/2}\left[G((1+g(\theta)\bar{\gamma}_1)^{-1},(1+g(\theta)\bar{\gamma}_2)^{-1}|\rho)\right.$$

$$\left. - \sum_{i=1}^{3}\frac{c_i(\theta)}{\sqrt{a_i^2(\theta) - b^2}}\right]d\theta, \quad \rho \neq 0 \tag{9.211}$$

where now $g(\theta) = g/\sin^2\theta$, $a_2(\theta) = a_2$, $a_3(\theta) = a_3$ and $a_1(\theta)$, $c_1(\theta)$, $c_2(\theta)$, and $c_3(\theta)$ are obtained by substituting $g(\theta)$ for g in (9.204) and (9.205), respectively, and the same substitution is made in $G((1+g\bar{\gamma}_1)^{-1},(1+g\bar{\gamma}_2)^{-1}|\rho)$.

For identical channels ($\bar{\gamma}_1 = \bar{\gamma}_2 = \bar{\gamma}$) using the Adachi et al. approximate expression of the PDF [104, Eq. (42)],

$$p_\gamma(\gamma) \simeq \frac{2}{(1+\rho)\bar{\gamma}}(e^{-\gamma/\bar{\gamma}} - e^{-2\gamma/(1-\rho)\bar{\gamma}}) \tag{9.212}$$

it can be shown that the average BER of BPSK can be written in closed form as

$$P_b(E) = \frac{1}{1+\rho}\left[\frac{1+\rho}{2} - \sqrt{\frac{g\bar{\gamma}}{1+g\bar{\gamma}}} + \frac{1-\rho}{2}\sqrt{\frac{(1-\rho)g\bar{\gamma}}{2+(1-\rho)g\bar{\gamma}}}\right] \tag{9.213}$$

Since it would be useful to know the relative accuracy improvements of the exact expressions (9.210) and (9.211) over the approximation (9.213), we plot all of them in Fig. 9.25. Note that the approximation tightly upper-bounds the exact BER expression and the bound gets tighter as the average SNR increases and as the correlation coefficient decreases. Extension of the average probability of error performance to Nakagami-m fading is omitted here but can be found in Ref. 110.

9.7.4.3 Numerical Example.

We present in this section various numerical examples to illustrate the effect of (1) the severity of fading, (2) branch correlation, and (3) branch average SNR unbalance on the performance of the system. Figures 9.26 and 9.27 plot the average BER of BDPSK versus the average SNR of the first branch for various values of the fading parameter m and correlation coefficient ρ for equal average branch SNR's and unequal average branch SNR's, respectively. Figures 9.28 and 9.29 plot the average BER of BPSK versus the average SNR of the first branch for various values of the fading parameter m and correlation coefficient ρ for equal average branch SNR's and unequal average branch SNR's, respectively. Note in all figures that the diversity gain decreases with the increase of the correlation coefficient, as expected. Note also that the effect of branch correlation is more important for channels with a lower amount of fading (higher m parameter).

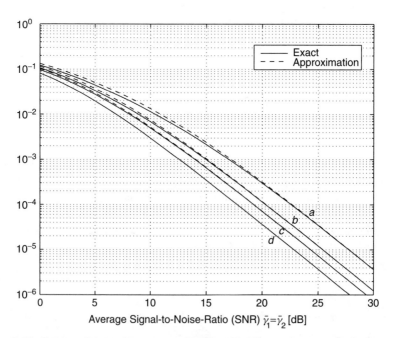

Figure 9.25. Exact and approximate average BER of BPSK versus average SNR of the first branch for equal average branch SNRs ($\bar{\gamma}_1 = \bar{\gamma}_2$) and for various values of the correlation coefficient: (a) $\rho = 0.9$; (b) $\rho = 0.7$; (c) $\rho = 0.5$; (d) $\rho = 0$.

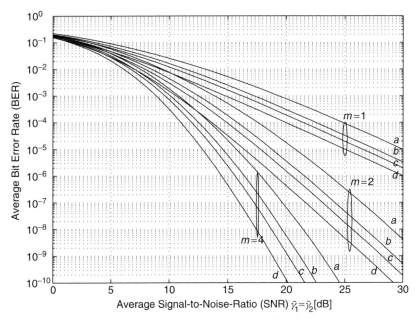

Figure 9.26. Average BER of BDPSK versus average SNR of the first branch for equal average branch SNRs ($\bar{\gamma}_1 = \bar{\gamma}_2$) and for various values of the correlation coefficient: (a) $\rho = 0.9$; (b) $\rho = 0.7$; (c) $\rho = 0.5$; (d) $\rho = 0$.

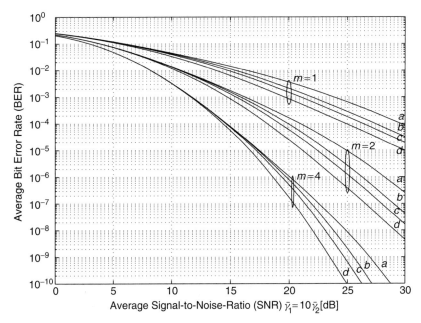

Figure 9.27. Average BER of BDPSK versus average SNR of the first branch for unequal average branch SNRs ($\bar{\gamma}_1 = 10\bar{\gamma}_2$) and for various values of the correlation coefficient: (a) $\rho = 0.9$; (b) $\rho = 0.7$; (c) $\rho = 0.5$; (d) $\rho = 0$.

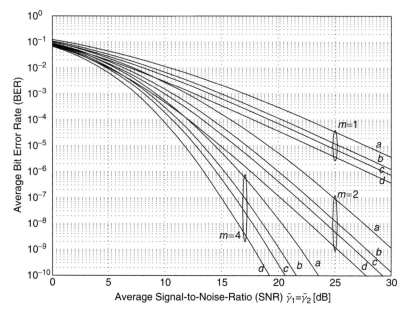

Figure 9.28. Average BER of BPSK versus average SNR of the first branch for equal average branch SNRs ($\overline{\gamma}_1 = \overline{\gamma}_2$) and for various values of the correlation coefficient: (a) $\rho = 0.9$; (b) $\rho = 0.7$; (c) $\rho = 0.5$; (d) $\rho = 0$.

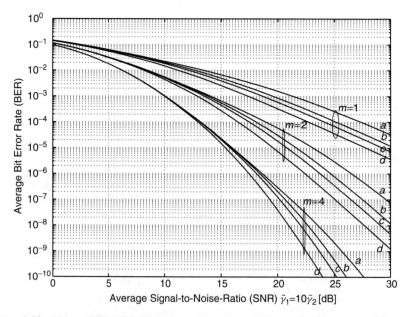

Figure 9.29. Average BER of BPSK versus average SNR of the first branch for unequal average branch SNRs ($\overline{\gamma}_1 = 10\overline{\gamma}_2$) and for various values of the correlation coefficient: (a) $\rho = 0.9$; (b) $\rho = 0.7$; (c) $\rho = 0.5$; (d) $\rho = 0$.

Finally, comparing the equal average branch SNR figures (Figs. 9.26 and 9.28) with the unequal average branch SNR figures (Figs. 9.27 and 9.29), observe that: (1) unbalance in the average branch SNR's always leads to lower overall system performance, and (2) the effect of correlation is more important for equal average branch SNR's.

Figure 9.30 compares the average BER performance of BPSK with dual-branch MRC and SC diversity for equal and unequal average SNR. In this figure, the SC average BER results are based on (9.210) and (9.211), whereas the MRC average BER results are based on the MGF given by (9.156). On the other hand, in both of these figures note that MRC outperforms SC, as expected, and that the diversity gain of MRC compared to SC is more important for channels with a low amount of fading (high m), regardless of the correlation between the two branches.

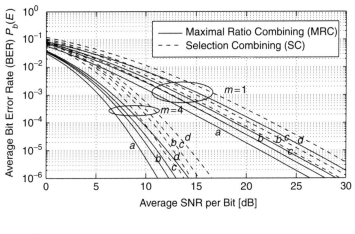

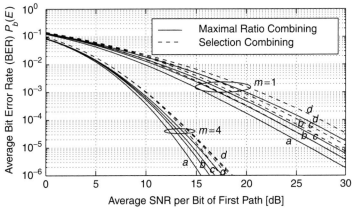

Figure 9.30. Comparison of average BER of BPSK with MRC and SC for various values of the correlation coefficient: (a) $\rho = 0$; (b) $\rho = 0.5$; (c) $\rho = 0.7$; (d) $\rho = 0.9$. (Top) equal average branch SNRs ($\bar{\gamma}_1 = \bar{\gamma}_2$); (bottom) unequal average branch SNRs ($\bar{\gamma}_1 = 10\bar{\gamma}_2$).

9.8 SWITCHED DIVERSITY

In this section we focus on the performance evaluation and optimization of *switch and stay combining (SSC)* systems over a wide variety of fading conditions in conjunction with several communication types of practical interest [115]. In these dual-branch diversity systems the receiver switches to, and stays with, the other branch regardless of whether the SNR of that branch is above or below the predetermined threshold (see Fig. 9.31).

The setting of the predetermined threshold is an additional important system design issue for SSC. For instance, if this threshold level is chosen too high, the switching unit is almost continually switching between the two antennas, which results not only in a poor diversity gain but also in an undesirable increase in the rate of the switching transients on the transmitted data stream. On the other hand, if this threshold level is chosen too low, the switching unit is almost locked to one of the diversity branches, even when the SNR level is quite low, and again there is little diversity gain achieved. Hence, another goal of this section is to determine the optimal switching threshold as a function of channel characteristic, performance measure, and modulation type. Work related to this topic can be found in Refs. 8 through 10, 103, and 116 through 120.

9.8.1 Performance of SSC over Independent Identically Distributed Branches

Let γ_{SSC} denote the SNR per symbol of the SSC combiner output, and let γ_T denote the predetermined switching threshold. Following the mode of operation of SSC as described earlier, we derive in this section the CDF, $P_{\gamma_{SSC}}(\gamma)$, PDF, $p_{\gamma_{SSC}}(\gamma)$, and MGF, $M_{\gamma_{SSC}}(\gamma)$, of γ_{SSC} assuming i.i.d. branches.

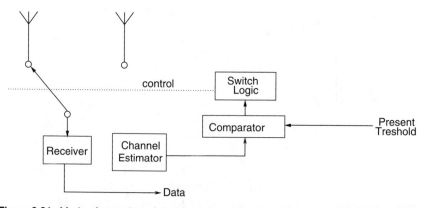

Figure 9.31. Mode of operation of dual-branch switched and stay combining (SSC) diversity.

9.8.1.1 SSC Output Statistics

CDF. The CDF of γ_{SSC} can be written as [9,10]

$$P_{\gamma_{SSC}}(\gamma) = \begin{cases} \Pr\{(\gamma_1 \leq \gamma_T) \text{ and } (\gamma_2 \leq \gamma)\}, & \gamma < \gamma_T \\ \Pr\{(\gamma_T \leq \gamma_1 \leq \gamma) \text{ or } (\gamma_1 \leq \gamma_T \\ \text{ and } \gamma_2 \leq \gamma)\}, & \gamma \geq \gamma_T \end{cases} \quad (9.214)$$

which can be expressed in terms of the CDF of the individual branches, $P_\gamma(\gamma)$, as

$$P_{\gamma_{SSC}}(\gamma) = \begin{cases} P_\gamma(\gamma_T)P_\gamma(\gamma), & \gamma < \gamma_T \\ P_\gamma(\gamma) - P_\gamma(\gamma_T) + P_\gamma(\gamma)P_\gamma(\gamma_T), & \gamma \geq \gamma_T \end{cases} \quad (9.215)$$

Using the one-branch CDFs given in Table 9.5, we can write the CDF of γ_{SSC} over Rayleigh channels as

$$P_{\gamma_{SSC}}(\gamma) = \begin{cases} 1 - (e^{-\gamma_T/\bar{\gamma}} + e^{-\gamma/\bar{\gamma}}) + e^{-(\gamma_T+\gamma)/\bar{\gamma}}, & \gamma < \gamma_T \\ 1 - 2e^{-\gamma/\bar{\gamma}} + e^{-(\gamma_T+\gamma)/\bar{\gamma}}, & \gamma \geq \gamma_T \end{cases} \quad (9.216)$$

over Nakagami-m as

$$P_{\gamma_{SSC}}(\gamma) = \begin{cases} \left(1 - \dfrac{\Gamma(m, (m/\bar{\gamma})\gamma_T)}{\Gamma(m)}\right)\left(1 - \dfrac{\Gamma(m, (m/\bar{\gamma})\gamma)}{\Gamma(m)}\right), & \gamma < \gamma_T \\ \dfrac{\Gamma(m, (m/\bar{\gamma})\gamma_T) - \Gamma(m, (m/\bar{\gamma})\gamma)}{\Gamma(m)} \\ + \left(1 - \dfrac{\Gamma(m, (m/\bar{\gamma})\gamma_T)}{\Gamma(m)}\right)\left(1 - \dfrac{\Gamma(m, (m/\bar{\gamma})\gamma)}{\Gamma(m)}\right), & \gamma \geq \gamma_T \end{cases}$$
$$(9.217)$$

TABLE 9.5 Statistics of the SNR per Symbol γ for the Three Multipath Fading Models Under Consideration

Model	PDF, $p_\gamma(\gamma)$	CDF, $P_\gamma(\gamma)$	MGF $M_\gamma(s)$
Rayleigh	$\dfrac{1}{\bar{\gamma}}e^{-\gamma/\bar{\gamma}}$	$1 - e^{-\gamma/\bar{\gamma}}$	$(1-s\bar{\gamma})^{-1}$
Nakagami-n	$\dfrac{(1+n^2)e^{-n^2}}{\bar{\gamma}}e^{[(1+n^2)/\bar{\gamma}]\gamma}$ $\times I_0\left(2n\sqrt{\dfrac{1+n^2}{\bar{\gamma}}\gamma}\right)$	$1 - Q_1\left(n\sqrt{2}, \sqrt{\dfrac{2(1+n^2)}{\bar{\gamma}}\gamma}\right)$	$\dfrac{1+n^2}{1+n^2-s\bar{\gamma}}$ $\times \exp\left(\dfrac{s\bar{\gamma}n^2}{1+n^2-s\bar{\gamma}}\right)$
Nakagami-m	$\dfrac{(m/\bar{\gamma})^m \gamma^{m-1}}{\Gamma(m)}e^{-m\gamma/\bar{\gamma}}$	$1 - \dfrac{\Gamma(m, (m/\bar{\gamma})\gamma)}{\Gamma(m)}$	$\left(1 - \dfrac{s\bar{\gamma}}{m}\right)^{-m}$

and over Nakagami-n (Rice) as

$$P_{\gamma_{\text{ssc}}}(\gamma) = \begin{cases} \left(1 - Q_1\left(\sqrt{2n^2}, \sqrt{\dfrac{2(1+n^2)\gamma}{\overline{\gamma}}}\right)\right) \\ \quad \times \left(1 - Q_1\left(\sqrt{2n^2}, \sqrt{\dfrac{2(1+n^2)\gamma_T}{\overline{\gamma}}}\right)\right) & \gamma < \gamma_T \\[2pt] Q_1\left(\sqrt{2n^2}, \sqrt{\dfrac{2(1+n^2)\gamma_T}{\overline{\gamma}}}\right) - Q_1\left(\sqrt{2n^2}, \sqrt{\dfrac{2(1+n^2)\gamma}{\overline{\gamma}}}\right) \\[2pt] \quad + \left(1 - Q_1\left(\sqrt{2n^2}, \sqrt{\dfrac{2(1+n^2)\gamma}{\overline{\gamma}}}\right)\right) \\ \quad \times \left(1 - Q_1\left(\sqrt{2n^2}, \sqrt{\dfrac{2(1+n^2)\gamma_T}{\overline{\gamma}}}\right)\right) & \gamma \geq \gamma_T \end{cases}$$
(9.218)

PDF. Differentiating $P_{\gamma_{\text{ssc}}}(\gamma)$ with respect to γ we get the PDF of the SSC output in terms of the CDF, $P_\gamma(\gamma)$, and the PDF, $p_\gamma(\gamma)$, of the individual branches as

$$p_{\gamma_{\text{ssc}}}(\gamma) = \frac{dP_{\gamma_{\text{ssc}}}(\gamma)}{d\gamma} = \begin{cases} P_\gamma(\gamma_T) p_\gamma(\gamma), & \gamma < \gamma_T \\ [1 + P_\gamma(\gamma_T)] p_\gamma(\gamma), & \gamma \geq \gamma_T \end{cases} \quad (9.219)$$

which can be written for Rayleigh fading as

$$p_{\gamma_{\text{ssc}}}(\gamma) = \begin{cases} \dfrac{1}{\overline{\gamma}}(1 - e^{-\gamma_T/\overline{\gamma}}) e^{-\gamma/\overline{\gamma}}, & \gamma < \gamma_T \\[4pt] \dfrac{1}{\overline{\gamma}}(2 - e^{\gamma_T/\overline{\gamma}}) e^{-\gamma/\overline{\gamma}}, & \gamma \geq \gamma_T \end{cases} \quad (9.220)$$

for Nakagami-m fading as

$$p_{\gamma_{\text{ssc}}}(\gamma) = \begin{cases} \left(1 - \dfrac{\Gamma(m, (m/\overline{\gamma})\gamma_T)}{\Gamma(m)}\right) \dfrac{(m/\overline{\gamma})^m \gamma^{m-1}}{\Gamma(m)} e^{-(m/\overline{\gamma})\gamma}, & \gamma < \gamma_T \\[6pt] \left(2 - \dfrac{\Gamma(m, (m/\overline{\gamma})\gamma_T)}{\Gamma(m)}\right) \dfrac{(m/\overline{\gamma})^m \gamma^{m-1}}{\Gamma(m)} e^{-(m/\overline{\gamma})\gamma}, & \gamma \geq \gamma_T \end{cases}$$
(9.221)

and for Nakagami-n (Rice) fading as

$$p_{\gamma_{SSC}}(\gamma) = \begin{cases} \dfrac{1+n^2}{\bar{\gamma}} \exp\left(-n^2 - \dfrac{1+n^2}{\bar{\gamma}}\gamma\right) I_0\left(2n\sqrt{\dfrac{1+n^2}{\bar{\gamma}}\gamma}\right) \\ \quad \times \left(1 - Q_1\left(n\sqrt{2}, \sqrt{\dfrac{2(1+n^2)}{\bar{\gamma}}\gamma_T}\right)\right), & \gamma < \gamma_T \\[2ex] \dfrac{1+n^2}{\bar{\gamma}} \exp\left(-n^2 - \dfrac{1+n^2}{\bar{\gamma}}\gamma\right) I_0\left(2n\sqrt{\dfrac{1+n^2}{\bar{\gamma}}\gamma}\right) \\ \quad \times \left(2 - Q_1\left(n\sqrt{2}, \sqrt{\dfrac{2(1+n^2)}{\bar{\gamma}}\gamma_T}\right)\right), & \gamma \geq \gamma_T \end{cases} \quad (9.222)$$

MGF. The MGF of γ_{SSC} can be expressed in terms of the individual branch MGFs as

$$M_{\gamma_{SSC}}(s) = P_\gamma(\gamma_T) M_\gamma(s) + \int_{\gamma_T}^{\infty} p_\gamma(\gamma) e^{s\gamma} \, d\gamma \qquad (9.223)$$

For Rayleigh fading, (9.223) simplifies to

$$M_{\gamma_{SSC}}(s) = (1 - s\bar{\gamma})^{-1} (1 - e^{-\gamma_T/\bar{\gamma}} + e^{-(1-s\bar{\gamma})(\gamma_T/\bar{\gamma})}) \qquad (9.224)$$

For Nakagami-m fading using Eq. (3.381.3) of Ref. 36, (9.223) can be expressed in terms of the complementary incomplete gamma function as

$$M_{\gamma_{SSC}}(s) = \left(1 - \dfrac{s\bar{\gamma}}{m}\right)^{-m} \\ \times \left[1 + \dfrac{\Gamma(m, (1 - (\bar{\gamma}/m)s)(m\gamma_T)/\bar{\gamma})) - \Gamma(m, (m/\bar{\gamma})\gamma_T)}{\Gamma(m)}\right] \qquad (9.225)$$

For Nakagami-n (Rice) fading (9.223) can be expressed in terms of the first-order Marcum Q-function as

$$M_{\gamma_{SSC}}(s) = \left(1 - \dfrac{s\bar{\gamma}}{1+n^2}\right)^{-1} \exp\left(\dfrac{s\bar{\gamma}n^2}{1+n^2 - s\bar{\gamma}}\right) \\ \times \left[1 - Q_1\left(n\sqrt{2}, \sqrt{\dfrac{2(1+n^2)\gamma_T}{\bar{\gamma}}}\right) \right. \\ \left. + Q_1\left(n\sqrt{\dfrac{2(1+n^2)}{1+n^2 - s\bar{\gamma}}}, \sqrt{2\left(\dfrac{1+n^2}{\bar{\gamma}} - s\right)\gamma_T}\right)\right] \qquad (9.226)$$

9.8.1.2 Average Output SNR

Analysis. The average SNR at the SSC output can be obtained by averaging γ over $p_{\gamma_{\text{SSC}}}(\gamma)$ as given by (9.219), yielding

$$\overline{\gamma}_{\text{SSC}} = P_\gamma(\gamma_T) \int_0^\infty \gamma p_\gamma(\gamma) \, d\gamma + \int_{\gamma_T}^\infty \gamma p_\gamma(\gamma) \, d\gamma$$

$$= P_\gamma(\gamma_T)\overline{\gamma} + \int_{\gamma_T}^\infty \gamma p_\gamma(\gamma) \, d\gamma \qquad (9.227)$$

Differentiating (9.227) with respect to γ_T and setting the result to zero, it can easily be shown that $\overline{\gamma}_{\text{SSC}}$ is maximized when the switching threshold is set to $\gamma_T^* = \overline{\gamma}$. For Rayleigh fading, substituting the one-branch CDF and PDFs given in Table 9.5 in (9.227), we get

$$\overline{\gamma}_{\text{SSC}} = \overline{\gamma}\left(1 + \frac{\gamma_T}{\overline{\gamma}} e^{-\gamma_T/\overline{\gamma}}\right) \qquad (9.228)$$

which reduces for the optimal threshold case to

$$\overline{\gamma}_{\text{SSC}}^* = \overline{\gamma}(1 + e^{-1}) = 1.368\overline{\gamma} \qquad (9.229)$$

Similarly, for the Nakagami-m fading case, the average output SNR can be found as

$$\overline{\gamma}_{\text{SSC}} = \overline{\gamma}\left(1 + \frac{[(m/\overline{\gamma})\gamma_T]^m e^{-(m/\overline{\gamma})\gamma_T}}{\Gamma(m+1)}\right) \qquad (9.230)$$

with the simplification to

$$\overline{\gamma}_{\text{SSC}}^* = \overline{\gamma}\left(1 + \frac{m^{m-1} e^{-m}}{\Gamma(m)}\right) \qquad (9.231)$$

when using the optimum switching threshold. For Nakagami-n (Rice) fading using the identity [36, Sec. 8.486]

$$I_0(z) = I_2(z) + \frac{2}{z} I_1(z) \qquad (9.232)$$

the average output SNR can be expressed in closed form in terms of generalized Marcum Q-functions as

$$\bar{\gamma}_{\text{SSC}} = \bar{\gamma}\left[1 - Q_1\left(n\sqrt{2}, \sqrt{\frac{2(n^2+1)}{\bar{\gamma}}}\gamma_T\right)\right]$$
$$+ \frac{\bar{\gamma}}{n^2+1}\left[Q_2\left(n\sqrt{2}, \sqrt{\frac{2(n^2+1)}{\bar{\gamma}}}\gamma_T\right)\right.$$
$$\left.+ n^2 Q_3\left(n\sqrt{2}, \sqrt{\frac{2(n^2+1)}{\bar{\gamma}}}\gamma_T\right)\right] \quad (9.233)$$

which reduces to

$$\bar{\gamma}_{\text{SSC}}^* = \bar{\gamma}\left[1 - Q_1\left(n\sqrt{2}, \sqrt{2(n^2+1)}\right)\right]$$
$$+ \frac{\bar{\gamma}}{n^2+1}\left[Q_2\left(n\sqrt{2}, \sqrt{2(n^2+1)}\right) + n^2 Q_3\left(n\sqrt{2}, \sqrt{2(n^2+1)}\right)\right] \quad (9.234)$$

when the optimum switching threshold is used.

Comparison with MRC and SC. For comparison purposes recall that the average SNR at the output of a dual-branch MRC diversity system is given by (9.50)

$$\bar{\gamma}_{\text{MRC}} = 2\bar{\gamma} \quad (9.235)$$

regardless of the fading model. On the other hand, the CDF of a dual-branch SC output is given by [3, Sec. 5.5.2]

$$P_{\gamma_{\text{SC}}}(\gamma) = [P_\gamma(\gamma)]^2 \quad (9.236)$$

which when differentiated with respect to γ gives the PDF

$$p_{\gamma_{\text{SC}}}(\gamma) = 2p_\gamma(\gamma)P_\gamma(\gamma) \quad (9.237)$$

Hence the average output SNR of a dual-branch SC is given by

$$\bar{\gamma}_{\text{SC}} = 2\int_0^\infty \gamma p_\gamma(\gamma)P_\gamma(\gamma)\,d\gamma \quad (9.238)$$

Using the one-branch PDFs and CDFs given in Table 9.5 for Rayleigh fading (9.238) reduces to [3, Eq. (5.86)]

$$\bar{\gamma}_{\text{SC}} = 1.5\bar{\gamma} \quad (9.239)$$

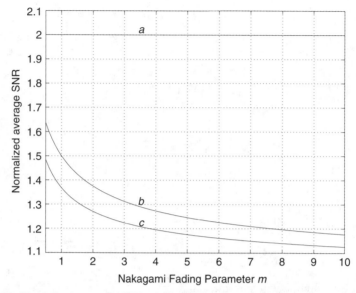

Figure 9.32. Normalized average SNR of (a) MRC ($\overline{\gamma}_{MRC}/\overline{\gamma}$), (b) SC ($\overline{\gamma}_{SC}/\overline{\gamma}$), and (c) SSC ($\overline{\gamma}_{SSC}/\overline{\gamma}$) versus the Nakagami-$m$ fading parameter m.

for Rayleigh fading channels. Similarly, the average output SNR of SC can be obtained in closed form for Nakagami-m fading by using the one-branch PDFs and CDFs given in Table 9.5 as well as Eq. (6.455) of Ref. 36 and Eq. (15.1.25) of Ref. 52, yielding

$$\overline{\gamma}_{SC} = \overline{\gamma}\left(1 + \frac{\Gamma(2m+1)}{2^{2m}(m\,\Gamma(m))^2}\right) \qquad (9.240)$$

As an example, Fig. 9.32 plots the normalized average output SNR, $\overline{\gamma}_{MRC}/\overline{\gamma}$, $\overline{\gamma}_{SC}/\overline{\gamma}$, and $\overline{\gamma}_{SSC}/\overline{\gamma}$ versus the Nakagami-m fading parameter for MRC, SC, and SSC with optimum threshold, respectively.

9.8.1.3 Outage Probability. As before, the outage probability, P_{out}, is defined as the probability that the combiner output SNR falls below a given threshold, γ_{th}, and is therefore obtained by replacing γ with γ_{th} in the CDF expressions given previously, that is,

$$P_{out}^{SSC} = \Pr\{\gamma_{SSC} < \gamma_{th}\} = P_{\gamma_{SSC}}(\gamma_{th}) \qquad (9.241)$$

Similarly, the outage probability of dual-branch SC systems can easily be deduced from (9.236) as

$$P_{out}^{SC} = [P_\gamma(\gamma_{th})]^2 \qquad (9.242)$$

Note that if we substitute γ_T for γ_{th} in (9.241), then using (9.215), (9.241) reduces to (9.242). Since SC can be viewed as an optimal implementation of any switched diversity system, we can conclude that the optimal switching threshold in the minimum outage probability sense is given by $\gamma_T^* = \gamma_{th}$. Hence, using (9.216), (9.217), and (9.218), we can write the outage probability of SC and SSC systems with optimal switching thresholds as

$$P_{out}^{SSC} = P_{out}^{SC} = (1 - e^{\gamma_{th}/\bar{\gamma}})^2 \tag{9.243}$$

for Rayleigh fading,

$$P_{out}^{SSC} = P_{out}^{SC} = \left(\frac{\Gamma(m) - \Gamma(m, (m/\bar{\gamma})\gamma_{th})}{\Gamma(m)}\right)^2 \tag{9.244}$$

for Nakagami-m fading, and

$$P_{out}^{SSC} = P_{out}^{SC} = \left[1 - Q_1\left(n\sqrt{2}, \sqrt{\frac{2(1+n^2)}{\bar{\gamma}}\gamma_{th}}\right)\right]^2 \tag{9.245}$$

for Nakagami-n (Rice) fading.

For comparison purposes we also derive the outage probability of dual-branch MRC receivers based on known results for the central and noncentral chi-square distributions [6, App. 5A]. For Rayleigh fading the outage probability is given by

$$P_{out}^{MRC} = 1 - \left(1 + \frac{\gamma_{th}}{\bar{\gamma}}\right) e^{-\gamma_{th}/\bar{\gamma}} \tag{9.246}$$

while for Nakagami-m it is given by

$$P_{out}^{MRC} = \frac{\Gamma(2m) - \Gamma(2m, (m/\bar{\gamma})\gamma_{th})}{\Gamma(m)} \tag{9.247}$$

and finally, for Nakagami-n (Rice) fading it can be expressed as

$$P_{out}^{MRC} = 1 - Q_2\left(n\sqrt{2}, \sqrt{\frac{2(1+n^2)}{\bar{\gamma}}\gamma_{th}}\right) \tag{9.248}$$

Figures 9.33 and 9.34 compare the outage probability of dual-branch MRC and SC/SSC with optimal switching thresholds versus the normalized threshold SNR $\gamma_{th}/\bar{\gamma}$ for Nakagami-m and Nakagami-n (Rice) fading channels, respectively. The

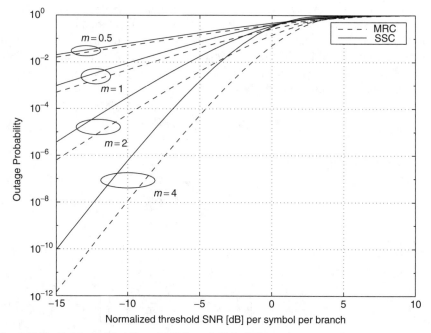

Figure 9.33. Outage probability of MRC and SSC (SC) versus normalized threshold SNR $(\gamma_{th}/\bar{\gamma})$ for Nakagami-m fading channel.

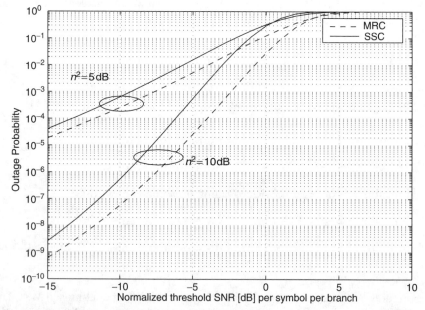

Figure 9.34. Outage probability of MRC and SSC (SC) versus normalized threshold SNR $(\gamma_{th}/\bar{\gamma})$ for Nakagami-n (Rice) fading channel.

difference of diversity gain between MRC and SSC is about 2 dB in Rayleigh fading and it increases as the fading environment improves, that is, as m or n (or equivalently, the Rice factor) increases.

9.8.1.4 Average Probability of Error

Analysis. In previous sections an MGF-based approach was taken for evaluation of the average error rate over fading channels, which although specifically explored for MRC, EGC, and SC is also applicable for SSC. Indeed, it was shown that the key to evaluating the average error rate of systems operating over fading channels is expressing the MGF of the combiner output in a form that is both simple and suitable for single integration. Since we have already derived the MGF of the SSC output SNR in Section 9.8.1.1, the evaluation of average error rate over the fading channel can be accomplished as before. As an example, and in view of the alternative conditional SER expressions presented in Chapter 8, the average SER of M-PSK signals is given by

$$P_s(E) = \frac{1}{\pi} \int_0^{(M-1)\pi/M} M_{\gamma_{\text{SSC}}}\left(-\frac{g_{\text{PSK}}}{\sin^2 \phi}\right) d\phi \qquad (9.249)$$

where $g_{\text{PSK}} = \sin^2(\pi/M)$.

For the particular case of BPSK over Rayleigh fading, the average BER can in fact be found in closed form in terms of the Gaussian Q-function, as we now show. Indeed, the average BER of BPSK using SSC can be written as

$$P_b(E) = \int_0^\infty Q(\sqrt{2\gamma}) p_{\gamma_{\text{SSC}}}(\gamma) d\gamma$$

$$= P_\gamma(\gamma_T) \int_0^\infty \frac{1}{\bar{\gamma}} e^{-\gamma/\bar{\gamma}} Q(\sqrt{2\gamma}) d\gamma + \int_{\gamma_T}^\infty \frac{1}{\bar{\gamma}} e^{-\gamma/\bar{\gamma}} Q(\sqrt{2\gamma}) d\phi \qquad (9.250)$$

which can be found in closed form with the help of (5A.4) and Eq. (24) of Ref. 121 as

$$P_b(E) = \left(1 - \frac{1}{2} e^{-\gamma_T/\bar{\gamma}}\right)\left(1 - \sqrt{\frac{\bar{\gamma}}{1+\bar{\gamma}}}\right)$$

$$+ \frac{\gamma_T}{2}\left[1 - 2e^{-1}Q(\sqrt{2\gamma_T}) - \sqrt{\frac{\gamma_T}{1+\gamma_T}}(1 - 2Q(\sqrt{2(1+\gamma_T)}))\right] \qquad (9.251)$$

However, the form given by (9.249) has the advantage of leading to a generic expression for the optimum switching threshold in a minimum average error rate sense, as shown next.

Optimum Threshold. Let us first focus on the binary case ($M = 2$). Using (9.223) in (9.249), we obtain

$$P_b(E) = \frac{1}{\pi} \int_0^{\pi/2} P_\gamma(\gamma_T) M_\gamma \left(-\frac{1}{\sin^2 \phi}\right) d\phi$$
$$+ \frac{1}{\pi} \int_0^{\pi/2} \left[\int_{\gamma_T}^\infty p_\gamma(\gamma) e^{-\gamma/\sin^2 \phi} d\gamma\right] d\phi \quad (9.252)$$

For $\gamma_T = 0$, $P_\gamma(\gamma_T) = 0$ and hence the first term of (9.252) vanishes, resulting in

$$P_b(E) = \frac{1}{\pi} \int_0^{\pi/2} M_\gamma \left(-\frac{1}{\sin^2 \phi}\right) d\phi \quad (9.253)$$

which is the BER performance of a single-branch (no diversity) receiver. On the other hand, as γ_T tends to infinity, the second term of (9.252) vanishes, and since $P_\gamma(\gamma_T) = 1$ in the first term, (9.252) reduces again to the average BER performance of a single-branch (no diversity) receiver. Since the average BER is a continuous function of γ_T, there exists an optimal value of γ_T for which the average BER is minimal. This optimal value γ_T^* is a solution of the equation

$$\left.\frac{dP_b(E)}{d\gamma_T}\right|_{\gamma_T=\gamma_T^*} = 0 \quad (9.254)$$

Substituting (9.252) in (9.254), we get

$$\frac{1}{\pi} \int_0^{\pi/2} p_\gamma(\gamma_T^*) M_\gamma \left(-\frac{1}{\sin^2 \phi}\right) d\phi - \frac{1}{\pi} \int_0^{\pi/2} p_\gamma(\gamma_T^*) e^{-\gamma_T^*/\sin^2 \phi} d\phi = 0 \quad (9.255)$$

which after simplification reduces to

$$\frac{1}{\pi} \int_0^{\pi/2} M_\gamma \left(-\frac{1}{\sin^2 \phi}\right) d\phi - Q(\sqrt{2\gamma_T^*}) = 0 \quad (9.256)$$

where we have used the alternative representation of the Gaussian Q-function in (4.2). Solving for γ_T^* in (9.256) leads to

$$\gamma_T^* = \frac{1}{2}\left[Q^{-1}\left(\frac{1}{\pi} \int_0^{\pi/2} M_\gamma \left(-\frac{1}{\sin^2 \phi}\right) d\phi\right)\right]^2 \quad (9.257)$$

where $Q^{-1}(\cdot)$ denotes the inverse Gaussian Q-function. Substituting the single-branch MGF's given in Table 9.5 for Rayleigh and Nakagami-m fading in

(9.257), then using the trigonometric integrals derived in Appendix 5A, we get the optimum threshold for BPSK over Rayleigh fading as

$$\gamma_T^* = \frac{1}{2}\left[Q^{-1}\left(\frac{1}{2}\left(1 - \sqrt{\frac{\overline{\gamma}}{1+\overline{\gamma}}}\right)\right)\right]^2 \qquad (9.258)$$

and for Nakagami-m fading as

$$\gamma_T^* = \frac{1}{2}\left[Q^{-1}\left(\frac{\sqrt{\overline{\gamma}/\pi m}}{2(1+\overline{\gamma}/m)^{m+1/2}}\frac{\Gamma(m+\frac{1}{2})}{\Gamma(m+1)}\,{}_2F_1\left(1, m+\frac{1}{2}; m+1; \frac{1}{1+\overline{\gamma}/m}\right)\right)\right]^2 \qquad (9.259)$$

For Nakagami-n (Rice) fading, it can be shown using the integrals given by Eqs. (6) and (7) of Ref. 122 that the optimum threshold is given by

$$\gamma_T^* = \frac{1}{2}\left\{Q^{-1}\left[Q_1(a,b) - \frac{1}{2}\left(1 + \sqrt{\frac{p}{1+p}}\right)e^{-(a^2+b^2)/2}I_0(ab)\right]\right\} \qquad (9.260)$$

where

$$p = \frac{\overline{\gamma}}{n^2+1}$$

$$a = n\left[\frac{1+2p}{2(1+p)} - \sqrt{\frac{p}{1+p}}\right]^{1/2} \qquad (9.261)$$

$$b = n\left[\frac{1+2p}{2(1+p)} + \sqrt{\frac{p}{1+p}}\right]^{1/2}$$

The optimum threshold can be found in a similar fashion for other modulation scheme/fading models combination. In general, this optimal threshold will be a solution of an integral equation similar to (9.256), but explicit closed-form solutions similar to the ones presented in (9.258), (9.259), and (9.261) will not always be possible to obtain. For example, the optimum threshold γ_T^* with DQPSK is the solution of the integral equation

$$\frac{1}{4\pi}\int_{-\pi}^{\pi} M_\gamma\left(-\frac{b^2(1-\zeta^2)^2}{2(1+2\zeta\sin\phi+\zeta^2)}\right)d\phi$$

$$= \frac{1}{4\pi}\int_{-\pi}^{\pi}\exp\left(-\frac{b^2(1-\zeta^2)^2}{2(1+2\zeta\sin\phi+\zeta^2)}\gamma_T^*\right)d\phi \qquad (9.262)$$

where $a = \sqrt{2-\sqrt{2}}$, $b = \sqrt{2+\sqrt{2}}$, and $\zeta = a/b$. In this case, one has to rely on numerical root-finding techniques to get an accurate numerical solution for γ_T^* in (9.262).

Comparison with MRC and SC. In Figs. 9.35 through 9.38 we present some numerical results comparing the average error rate performance of several modulation schemes with SSC, SC, and MRC. The SSC curves are generated as per the average error rate expressions given in Section 9.8.1.4 and with the optimum switching thresholds derived previously. MRC curves are based on the expressions derived in Section 9.2 on the performance of MRC receivers. For SC, the Rayleigh and Nakagami-m curves are based on the results presented in Section 9.7. For Nakagami-n (Rice) fading channels, we need the MGF at the SC output to be able to get average error rate expressions. Using the integral given in Appendix 4.6 of Ref. 123 this MGF can be found in closed form in terms of the Marcum Q-function as

$$M_{\gamma_{\text{SC}}}(s) = 2\left(1 - \frac{\overline{\gamma}s}{1+n^2}\right)^{-1} \cdot \exp\left(\frac{s\overline{\gamma}n^2}{1+n^2-\overline{\gamma}s}\right)$$
$$\times \left\{1 - \frac{v^2}{1+v^2}\left[1 - Q_1\left(\frac{\lambda v}{\sqrt{1+v^2}}, \frac{\mu}{\sqrt{1+v^2}}\right)\right]\right.$$
$$\left. - \frac{1}{1+v^2} Q_1\left(\frac{\mu}{\sqrt{1+v^2}}, \frac{\lambda v}{\sqrt{1+v^2}}\right)\right\} \quad (9.263)$$

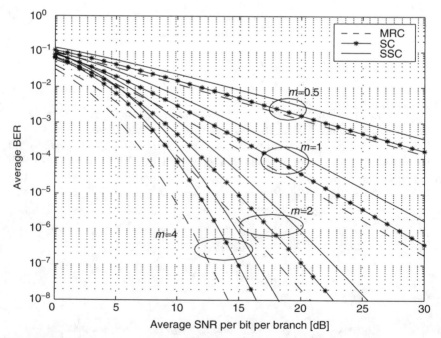

Figure 9.35. Comparison of the average BER of BPSK with MRC, SC, and SSC versus average SNR per bit per branch $\overline{\gamma}$ for Nakagami-m fading channel.

SWITCHED DIVERSITY 361

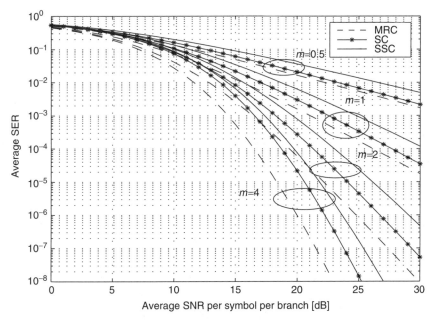

Figure 9.36. Comparison of the average SER of 8-PSK with MRC, SC, and SSC versus average SNR per symbol per branch $\bar{\gamma}$ for Nakagami-m fading channel.

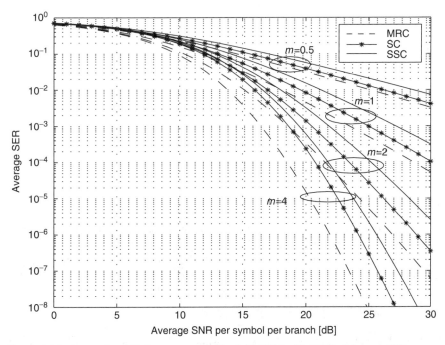

Figure 9.37. Comparison of the average SER of 16-QAM with MRC, SC, and SSC versus average SNR per symbol per branch $\bar{\gamma}$ for Nakagami-m fading channels.

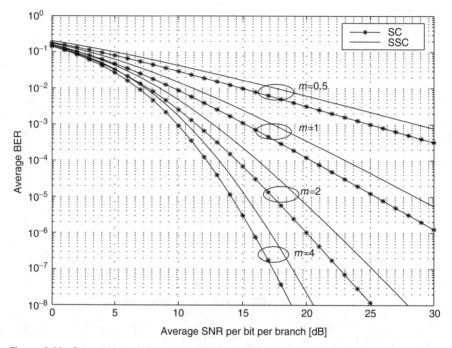

Figure 9.38. Comparison of the average BER of DQPSK with SC and SSC versus average SNR per bit per branch $\bar{\gamma}$ for Nakagami-m fading channels.

where

$$\mu = \sqrt{2}n$$

$$\lambda = \left(\frac{2(1+n^2)}{1+n^2-s\bar{\gamma}} \right)^{1/2}$$

$$v = \left(\frac{1+n^2}{1+n^2-s\bar{\gamma}} \right)^{1/2} \qquad (9.264)$$

9.8.2 Effect of Branch Unbalance

In the preceding section we analyzed SSC for the case of identically distributed branches. In this section we consider the performance of SSC systems in the more general case where the branches are still independent but not necessarily identically distributed. In particular, let us denote by $p_{\gamma_1}(\gamma_1)$ and $p_{\gamma_2}(\gamma_2)$ the PDF's of the two branches, by $P_{\gamma_1}(\gamma_1)$ and $P_{\gamma_2}(\gamma_2)$ their respective CDF's, and by $\bar{\gamma}_1$ and $\bar{\gamma}_2$ their respective average SNR's.

9.8.2.1 SSC Output Statistics.
Assuming a discrete time implementation of SSC let $\gamma_{1(n)}$ and $\gamma_{2(n)}$ be the instantaneous SNR of branch 1 and 2, respectively,

at time $t = nT$, and let γ_n denote the SSC output SNR at time $t = nT$. According to the mode of operation of SSC described above, we have

$$\gamma_n = \gamma_{1(n)} \quad \text{iff} \quad \begin{cases} \gamma_{n-1} = \gamma_{1(n-1)}, & \gamma_{1(n)} \geq \gamma_T \\ \gamma_{n-1} = \gamma_{2(n-1)}, & \gamma_{2(n)} < \gamma_T \end{cases} \qquad (9.265)$$

CDF. The CDF of γ_n, $P_{\gamma_{\text{SSC}}}(\gamma)$, can be written as

$$\begin{aligned}
P_{\gamma_{\text{SSC}}}(\gamma) &= \Pr\{\gamma_n \leq \gamma\} \\
&= \Pr\{\gamma_n = \gamma_{1(n)} \text{ and } \gamma_{1(n)} \leq \gamma\} + \Pr\{\gamma_n = \gamma_{2(n)} \text{ and } \gamma_{2(n)} \leq \gamma\} \\
&= \Pr\{\gamma_{1(n)} \geq \gamma_T \text{ and } \gamma_{1(n)} \leq \gamma\} \Pr\{\gamma_{n-1} = \gamma_{1(n-1)}\} \\
&\quad + \Pr\{\gamma_{2(n)} < \gamma_T \text{ and } \gamma_{1(n)} \leq \gamma\} \Pr\{\gamma_{n-1} = \gamma_{2(n-1)}\} \\
&\quad + \Pr\{\gamma_{2(n)} \geq \gamma_T \text{ and } \gamma_{2(n)} \leq \gamma\} \Pr\{\gamma_{n-1} = \gamma_{2(n-1)}\} \\
&\quad + \Pr\{\gamma_{1(n)} < \gamma_T \text{ and } \gamma_{2(n)} \leq \gamma\} \Pr\{\gamma_{n-1} = \gamma_{1(n-1)}\} \qquad (9.266)
\end{aligned}$$

which can be expressed in terms of the CDF of the individual branches as

$$P_{\gamma_{\text{SSC}}}(\gamma)$$
$$= \begin{cases} \Pr\{\gamma_{n-1} = \gamma_{1(n-1)}\} P_{\gamma_1}(\gamma_T) P_{\gamma_2}(\gamma) + \Pr\{\gamma_{n-1} = \gamma_{2(n-1)}\} P_{\gamma_2}(\gamma_T) P_{\gamma_1}(\gamma), \\ \qquad \gamma \leq \gamma_T \\ \Pr\{\gamma_{n-1} = \gamma_{1(n-1)}\}(P_{\gamma_1}(\gamma) - P_{\gamma_1}(\gamma_T) + P_{\gamma_1}(\gamma_T) P_{\gamma_2}(\gamma)) \\ \qquad + \Pr\{\gamma_{n-1} = \gamma_{2(n-1)}\}(P_{\gamma_2}(\gamma_T) P_{\gamma_1}(\gamma) + P_{\gamma_2}(\gamma) - P_{\gamma_2}(\gamma_T)), \\ \qquad \gamma > \gamma_T \end{cases}$$
$$(9.267)$$

To obtain the CDF of the SSC output, we need to find

$$\begin{aligned}
p_1 &\triangleq \Pr\{\gamma_{n-1} = \gamma_{1(n-1)}\} \\
&= \Pr\{(\gamma_{n-2} = \gamma_{1(n-2)} \text{ and } \gamma_{1(n-1)} \geq \gamma_T) \text{ or } (\gamma_{n-2} = \gamma_{2(n-2)} \text{ and } \gamma_{2(n-1)} < \gamma_T)\} \\
p_2 &\triangleq \Pr\{\gamma_{n-1} = \gamma_{2(n-1)}\} \\
&= \Pr\{(\gamma_{n-2} = \gamma_{2(n-2)} \text{ and } \gamma_{2(n-1)} \geq \gamma_T) \text{ or } (\gamma_{n-2} = \gamma_{1(n-2)} \text{ and } \gamma_{1(n-1)} < \gamma_T)\}
\end{aligned}$$
$$(9.268)$$

Assuming that the pairs of samples from each branch are i.i.d. (i.e., $\gamma_{1(n-1)}$ and $\gamma_{1(n)}$ are i.i.d. and $\gamma_{2(n-1)}$ and $\gamma_{2(n)}$ are i.i.d.), we can rewrite (9.268) as

$$\begin{aligned} p_1 &= p_1[1 - P_{\gamma_1}(\gamma_T)] + p_2 P_{\gamma_2}(\gamma_T) \\ p_2 &= p_2[1 - P_{\gamma_2}(\gamma_T)] + p_1 P_{\gamma_1}(\gamma_T) \end{aligned} \qquad (9.269)$$

Using the fact that the events $\gamma_n = \gamma_{1(n)}$ and $\gamma_n = \gamma_{2(n)}$ are mutually exclusive (i.e., $p_1 + p_2 = 1$), we can solve for p_1 and p_2 to get

$$p_1 = \frac{P_{\gamma_2}(\gamma_T)}{P_{\gamma_1}(\gamma_T) + P_{\gamma_2}(\gamma_T)}$$
$$p_2 = \frac{P_{\gamma_1}(\gamma_T)}{P_{\gamma_1}(\gamma_T) + P_{\gamma_2}(\gamma_T)}$$
(9.270)

Substituting (9.270) in (9.267) the CDF of the SSC output can be written solely in terms of the individual branch CDF's as

$$P_{\gamma_{\text{ssc}}}(\gamma) = \begin{cases} \dfrac{P_{\gamma_1}(\gamma_T) P_{\gamma_2}(\gamma_T)}{P_{\gamma_1}(\gamma_T) + P_{\gamma_2}(\gamma_T)}[P_{\gamma_1}(\gamma) + P_{\gamma_2}(\gamma)], & \gamma \leq \gamma_T \\ \dfrac{P_{\gamma_1}(\gamma_T) P_{\gamma_2}(\gamma_T)}{P_{\gamma_1}(\gamma_T) + P_{\gamma_2}(\gamma_T)}[P_{\gamma_1}(\gamma) + P_{\gamma_2}(\gamma) - 2] \\ + \dfrac{P_{\gamma_1}(\gamma) P_{\gamma_2}(\gamma_T) + P_{\gamma_1}(\gamma_T) P_{\gamma_2}(\gamma)}{P_{\gamma_1}(\gamma_T) + P_{\gamma_2}(\gamma_T)}, & \gamma > \gamma_T \end{cases}$$
(9.271)

PDF. Differentiating the expression of $P_{\gamma_{\text{ssc}}}(\gamma)$ as given by (9.271) with respect to γ, we get the PDF at the SSC output as

$$p_{\gamma_{\text{ssc}}}(\gamma) = \begin{cases} \dfrac{P_{\gamma_1}(\gamma_T) P_{\gamma_2}(\gamma_T)}{P_{\gamma_1}(\gamma_T) + P_{\gamma_2}(\gamma_T)}[p_{\gamma_1}(\gamma) + p_{\gamma_2}(\gamma)], & \gamma \leq \gamma_T \\ \dfrac{P_{\gamma_1}(\gamma_T) P_{\gamma_2}(\gamma_T)}{P_{\gamma_1}(\gamma_T) + P_{\gamma_2}(\gamma_T)}[p_{\gamma_1}(\gamma) + p_{\gamma_2}(\gamma)] \\ + \dfrac{p_{\gamma_1}(\gamma) P_{\gamma_2}(\gamma_T) + P_{\gamma_1}(\gamma_T) p_{\gamma_2}(\gamma)}{P_{\gamma_1}(\gamma_T) + P_{\gamma_2}(\gamma_T)}, & \gamma > \gamma_T \end{cases}$$
(9.272)

MGF. Taking the Laplace transform of the PDF as given by (9.272), it can be shown that the MGF of the SSC output $M_{\gamma_{\text{ssc}}}(s)$ when the branches are not necessarily identically distributed can be expressed with the help of (9.223) as

$$M_{\gamma_{\text{ssc}}}(s) = \frac{P_{\gamma_2}(\gamma_T)}{P_{\gamma_1}(\gamma_T) + P_{\gamma_2}(\gamma_T)} M^{(1)}_{\gamma_{\text{ssc}}}(s)$$
$$+ \frac{P_{\gamma_1}(\gamma_T)}{P_{\gamma_1}(\gamma_T) + P_{\gamma_2}(\gamma_T)} M^{(2)}_{\gamma_{\text{ssc}}}(s)$$
(9.273)

where $M^{(i)}_{\gamma_{\text{ssc}}}(s)$ is the MGF given by (9.224), (9.225), or (9.226). The superscript (i) in $M^{(i)}_{\gamma_{\text{ssc}}}(s)$ refers to the fact that in this analysis the two combined branches are allowed to have different average SNR's $\overline{\gamma}_1$ and $\overline{\gamma}_2$, different fading parameters such as m_1 and m_2 in the case of Nakagami-m fading, or (even) to be distributed according to two different families of distributions such as Nakagami-m and

Nakagami-n (Rice). As a check, note that (9.273) reduces to (9.224), (9.225), or (9.226) in the case of i.i.d. branches.

9.8.2.2 Average Output SNR.

The closed-form expression of the PDF as given by (9.272) readily allows obtaining the SSC output average SNR in the case of unbalanced branches. For instance, let us consider the average SNR at the SSC output of the Nakagami-m fading branches with the same fading parameter m but different average SNR's $\overline{\gamma}_1$ and $\overline{\gamma}_2$. In this case, averaging γ over (9.272), it can be shown after some manipulations that the average output SNR is given by

$$\overline{\gamma}_{\text{SSC}} = \frac{P_{\gamma_2}(\gamma_T)}{P_{\gamma_1}(\gamma_T) + P_{\gamma_2}(\gamma_T)} \left(1 + \frac{m^{m-1}(\gamma_T/\overline{\gamma}_1)^m e^{-(m/\overline{\gamma}_1)\gamma_T}}{\Gamma(m)} \right) \overline{\gamma}_1$$
$$+ \frac{P_{\gamma_1}(\gamma_T)}{P_{\gamma_1}(\gamma_T) + P_{\gamma_2}(\gamma_T)} \left(1 + \frac{m^{m-1}(\gamma_T/\overline{\gamma}_2)^m e^{-(m/\overline{\gamma}_2)\gamma_T}}{\Gamma(m)} \right) \overline{\gamma}_2 \quad (9.274)$$

where $P_{\gamma_i}(\cdot)$ is given in Table 9.5 for the Nakagami-m case.

As a comparison, recall that the average SNR at the output of a dual-branch MRC receiver is given by Eq. (5.102) of Ref. 3 as $\overline{\gamma}_{\text{MRC}} = \overline{\gamma}_1 + \overline{\gamma}_2$ regardless of the fading model, whereas the average SNR at the output of a dual-branch SC over Nakagami-m fading is given by (9.194) as

$$\overline{\gamma}_{\text{SC}} = \overline{\gamma}_1 + \overline{\gamma}_2 - \frac{(\overline{\gamma}_1 \overline{\gamma}_2)^{m+1} \Gamma(2m+1)}{m(m+1)\Gamma(m)^2 (\overline{\gamma}_1 + \overline{\gamma}_2)^{2m+1}}$$
$$\times \left[{}_2F_1 \left(1, 2m+1; m+2; \frac{\overline{\gamma}_1}{\overline{\gamma}_1 + \overline{\gamma}_2} \right) \right.$$
$$\left. + {}_2F_1 \left(1, 2m+1; m+2; \frac{\overline{\gamma}_2}{\overline{\gamma}_1 + \overline{\gamma}_2} \right) \right] \quad (9.275)$$

where ${}_2F_1[\cdot,\cdot;\cdot;\cdot]$ is the Gauss hypergeometric function [52, Chap. 15].

9.8.2.3 Average Probability of Error.

Using the closed-form MGF expression (9.273), we are in a position to derive the average probability of error for a wide variety of modulation schemes, as explained in Section 9.8.1.4. For example, the average BER of BPSK over Nakagami-m fading channels with unequal average SNR's is given by

$$P_b(E) = \frac{P_{\gamma_2}(\gamma_T)}{P_{\gamma_1}(\gamma_T) + P_{\gamma_2}(\gamma_T)} \frac{1}{\pi} \int_0^{\pi/2} M_{\gamma_{\text{SSC}}}^{(1)} \left(-\frac{1}{\sin^2 \phi} \right) d\phi$$
$$+ \frac{P_{\gamma_1}(\gamma_T)}{P_{\gamma_1}(\gamma_T) + P_{\gamma_2}(\gamma_T)} \frac{1}{\pi} \int_0^{\pi/2} M_{\gamma_{\text{SSC}}}^{(2)} \left(-\frac{1}{\sin^2 \phi} \right) d\phi \quad (9.276)$$

where $M_{\gamma_{\text{SSC}}}^{(i)}(\cdot)$ is given by (9.225).

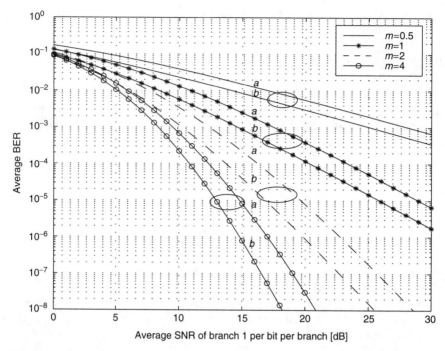

Figure 9.39. Average BER of BPSK versus average SNR of the first branch $\bar{\gamma}_1$ (a) for unequal average branch SNRs $\bar{\gamma}_1 = \bar{\gamma}_2/5$ and (b) for equal average branch SNRs $\bar{\gamma}_1 = \bar{\gamma}_2$ over Nakagami-m fading channels.

Figure 9.39 shows the effect of average SNR unbalance on the average BER of BPSK with SSC over Nakagami-m fading channels. We used the optimum switching threshold to generate these curves, and these optimum thresholds were found numerically by minimizing (9.276) with respect to γ_T. Note, for example, that in the case of $m = 2$, the unbalanced system under consideration suffers about a 3-dB penalty, for an average BER of 10^{-6} compared to a balanced system.

9.8.3 Effect of Branch Correlation

Recall that in Section 9.8.1 we considered the performance of SSC over i.i.d. branches, whereas in the preceding section we addressed the problem of branch unbalance. We assess in this section the effect of fading correlation on the performance of SSC receivers.

9.8.3.1 SSC Output Statistics

PDF. In the case of correlated Nakagami-m fading envelopes, the joint PDF $p_{\gamma_1,\gamma_2}(\gamma_1, \gamma_2)$ of the instantaneous SNR's γ_1 and γ_2 is given by [9, Eq. (19)]

$$p_{\gamma_1\gamma_2}(\gamma_1,\gamma_2) = \left(\frac{m}{\bar{\gamma}}\right)^{m+1} \frac{\gamma_1^{(m-1)/2}\gamma_2^{(m-1)/2}}{\Gamma(m)\rho^{(m-1)/2}(1-\rho)} e^{-[m/\bar{\gamma}(1-\rho)](\gamma_1+\gamma_2)}$$

$$\times I_{m-1}\left(\frac{2m\sqrt{\rho}}{(1-\rho)\bar{\gamma}}\sqrt{\gamma_1\gamma_2}\right), \quad \gamma_1 \geq 0, \quad \gamma_2 \geq 0 \quad (9.277)$$

Under these conditions and following the mode of operation of SSC systems as described above, Abu-Dayya and Beaulieu [9] showed that the PDF of the SSC output is given by [9, Eq. (21a)]

$$p_{\gamma_{\text{SSC}}}(\gamma) = \begin{cases} A(\gamma), & \gamma \leq \gamma_T \\ \left(\frac{m}{\bar{\gamma}}\right)^m e^{-(m/\bar{\gamma})\gamma} \frac{\gamma^{m-1}}{\Gamma(m)} + A(\gamma), & \gamma > \gamma_T, \end{cases} \quad (9.278)$$

where $A(\gamma)$ can be written with the help of Eq. (11) of Ref. 124 as [115]

$$A(\gamma) = \frac{(m/\bar{\gamma})^m \gamma^{m-1} e^{-m\gamma/\bar{\gamma}}}{\Gamma(m)}$$

$$\times \left[1 - Q_m\left(\sqrt{\frac{2m\rho\gamma}{(1-\rho)\bar{\gamma}}}, \sqrt{\frac{2m\gamma_T}{(1-\rho)\bar{\gamma}}}\right)\right] \quad (9.279)$$

MGF. Taking the Laplace transform of (9.278), the MGF of the SSC output SNR can be expressed in closed form in terms of the incomplete gamma function, with the help of Eq. (11) of Ref. 124, as

$$M_{\gamma_{\text{SSC}}}(\gamma) = \left(1 - \frac{\bar{\gamma}}{m}s\right)^{-m}$$

$$\times \left[1 + \frac{\Gamma\left(m, \frac{m\gamma_T}{\bar{\gamma}}\left(1 - \frac{\bar{\gamma}}{m}s\right)\right)}{\Gamma(m)} - \frac{\Gamma\left(m, \frac{m\gamma_T}{\bar{\gamma}}\frac{1-\bar{\gamma}s/m}{1-(1-\rho)\bar{\gamma}s/m}\right)}{\Gamma(m)}\right]$$

$$(9.280)$$

9.8.3.2 Average Output SNR. The average output SNR of SSC with correlated branches is obtained by averaging γ over the PDF of (9.278), yielding with the help of Eq. (28) of Ref. 125 and Eq. (8.356.2) of Ref. 36, and after some manipulation,

$$\bar{\gamma}_{\text{SSC}} = \bar{\gamma}\left(1 + \frac{(1-\rho)[(m/\bar{\gamma})\gamma_T]^m e^{-m\gamma_T/\bar{\gamma}}}{m\Gamma(m)}\right) \quad (9.281)$$

As a check, note that (9.281) reduces to the average SNR of a single-branch $\bar{\gamma}$ for fully correlated branches with $\rho = 1$. On the other hand, (9.281) reduces to the average output SNR over i.i.d. branches as given by (9.230) when $\rho = 0$. Differentiating (9.281) with respect to γ_T and setting the result equal to zero, it can easily be shown that the optimum threshold for maximum average output SNR is $\gamma_T^* = \overline{\gamma}$, which is the same as in the uncorrelated fading case. Hence the maximum average output SNR for SSC over correlated Nakagami-m fading channels is given by

$$\overline{\gamma}_{SSC} = \overline{\gamma}\left(1 + \frac{(1-\rho)m^{m-1}e^{-m}}{\Gamma(m)}\right) \quad (9.282)$$

For the purpose of comparison, recall that the average output SNR for dual-branch MRC is unaffected by fading correlation, while we showed that the average output SNR for dual-branch SC over equal average SNR correlated Nakagami-m fading paths is given by (9.192). Figure 9.40 compares the effect of fading correlation on the average output SNR of SSC and SC receivers. SC outperforms SSC, as expected, but has a slightly higher sensitivity to fading correlation.

9.8.3.3 Average Probability of Error.
Using the MGF of the SSC output SNR given by (9.280), we can determine the average probability of error

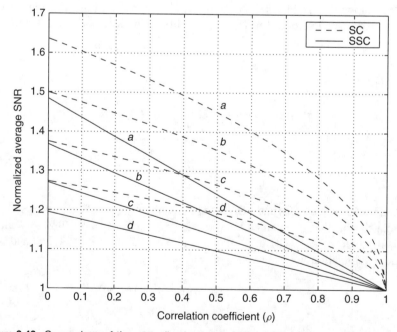

Figure 9.40. Comparison of the normalized average SNR of SC ($\overline{\gamma}_{SC}/\overline{\gamma}$) and SSC ($\overline{\gamma}_{SSC}/\overline{\gamma}$) versus correlation coefficients(ρ) over Nakagami fading with parameter (a) $m = 0.5$, (b) $m = 1$, (c) $m = 2$, and (d) $m = 4$.

of several modulation schemes, as explained in Section 9.8.1.4. For example, Fig. 9.41 depicts the effect of correlation on the BER performance of BPSK with SSC. From Fig. 9.41 one can conclude that correlation coefficients up to 0.6 do not seriously degrade the BER performance. Note that the curves in Fig. 9.41 used the optimum threshold in the minimum BER sense. In general, this optimum threshold cannot be expressed in an explicit closed form but can be found numerically by solving an integral equation. For example, in the particular BPSK case substituting (9.253) in (9.254) (with $M_\gamma(s)$ given by (9.280)) and differentiating with respect to γ_T (with the help of [36, Eq. (8.356.4)]) leads to the following integral equation for γ_T^*

$$\int_0^{\pi/2} \exp\left(-\left(\frac{1}{\sin^2\phi} + \frac{m}{\bar{\gamma}}\right)\gamma_T^*\right) d\phi$$
$$= \int_0^{\pi/2} \left(1 + \frac{\bar{\gamma}(1-\rho)}{m\sin^2\phi}\right)^{-m} \cdot \exp\left(-\frac{m\gamma_T^*}{\bar{\gamma}}\left(\frac{1 + \frac{\bar{\gamma}}{m\sin^2\phi}}{1 + (1-\rho)\frac{\bar{\gamma}}{m\sin^2\phi}}\right)\right) d\phi$$
(9.283)

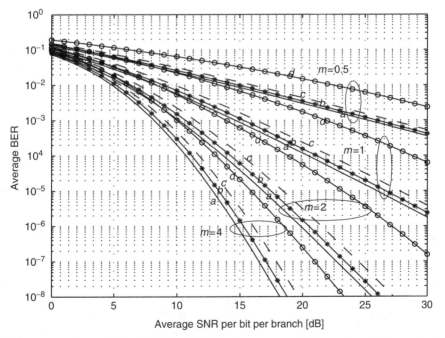

Figure 9.41. Average BER of BPSK with SSC versus average SNR per branch $\bar{\gamma}$ for various values of the correlation coefficient [ρ = (a) 0, (b) 0.3, (c) 0.6, and (d) 0.9] over Nakagami-m fading.

9.9 PERFORMANCE IN THE PRESENCE OF OUTDATED OR IMPERFECT CHANNEL ESTIMATES

In general, diversity combining techniques rely, to a large extent, on accurate channel estimation. As a typical first step in performance analysis, perfect channel estimation is assumed as was done in previous sections. However, in practice these estimates must be obtained in the presence of noise and time delay. For example, one common way to estimate the channel uses pilot symbol-assisted modulation (PSAM) [126; 127, Sec. 10.3.2], which periodically inserts pilots into the stream of data symbols to extract the channel-induced fading. Estimation error (due to additive noise as well as wireless channel variation/decorrelation over time) will cause the channel fading extracted from the pilot symbols to differ from the actual fading affecting the data symbols, thereby inducing a performance degradation.

The effects of channel estimation error or channel decorrelation on the performance of diversity systems has long been of interest. These previous studies focused on MRC receivers over Rayleigh fading channels [32, 128–131], postdetection EGC receivers over fast Rician fading channels [132,133], SC receivers over Rayleigh [134] and Nakagami-m [135] fading channels, and SSC receivers over Nakagami-m fading channels [115]. In this section we summarize briefly the work on the impact of channel estimation error or channel decorrelation on the performance of diversity systems.

9.9.1 Maximal-Ratio Combining

Gans [129] studied the effect of Gaussian-distributed weighting errors on the performance of MRC receivers. In particular, he showed that if the combined branches are subject to i.i.d. Rayleigh fading, the PDF of the combined SNR is given by

$$p_{\gamma_{\text{MRC}}}(\gamma) = \frac{(1-\rho)^{L-1} \exp(-\gamma/\overline{\gamma})}{\overline{\gamma}} \sum_{l=1}^{L} \frac{\binom{L-1}{l-1}}{(l-1)!} \left[\frac{\rho \gamma}{(1-\rho)\overline{\gamma}}\right]^{l-1} \quad (9.284)$$

where $\rho \in [0, 1]$ is the power correlation coefficient between the estimated and actual fadings. This coefficient can be viewed as a measure of the channel's rate of fluctuation and can be related solely to the time delay τ and to the maximum Doppler frequency shift f_d [e.g., for land mobile communication $\rho = J_0^2(2\pi f_d \tau)$, where $J_0(\cdot)$ is the zero-order Bessel function of the first kind]. The parameter ρ can also be viewed as a measure of the quality of the channel estimation and can be expressed, for example, in terms of PSAM parameters, such as the rate of pilot symbol insertion and SNR [136]. It is interesting to note that (9.284) can be rewritten as a weighted sum of L ideal MRC PDF's [131, Eq. (7)]

$$p_{\gamma_{\text{MRC}}}(\gamma) = \sum_{l=1}^{L} A(l) \frac{1}{(l-1)!\,\overline{\gamma}^l} \gamma^{l-1} e^{-\gamma/\overline{\gamma}} \quad (9.285)$$

where the weight coefficients

$$A(l) = \binom{L-1}{l-1}(1-\rho)^{L-l}\rho^{l-1} \tag{9.286}$$

are Bernstein polynomials. As a check, when $\rho = 1$ (perfect correlation between the pilot-extracted fading and the actual fading), perfect MRC combining is achieved and (9.284) or equivalently, (9.285) reduces to (9.5), as expected. As ρ decreases, the correlation between the pilot-extracted fading and the actual fading diminishes and performance degrades. In the limit as $\rho \to 0$ the fading estimate and its actual value are completely uncorrelated and (9.284) or equivalently, (9.285) approaches the PDF without diversity (i.e., $L = 1$) given by (9.4).

Taking the Laplace transform of (9.285) and using Eq. (3.351.3) of Ref. 36, the corresponding MGF $M(s)$ can easily be shown to be given by

$$M(s) = \sum_{l=1}^{L} \frac{A(l)}{(1-s\bar{\gamma})^l} \tag{9.287}$$

With (9.287) in hand and using the integrals derived in Appendix 5A, the average probability of error of several linear coherent modulations with imperfect MRC can be computed in closed form.

9.9.2 Noncoherent EGC over Rician Fast Fading

In this section we extend the results presented in Section 8.2.5.2 and consider the effect of fast Rician fading on binary DPSK when used in conjunction with L-branch postdetection EGC. Using the same notation as in Section 8.2.5.2, the EGC output decision variable corresponding to transmission of a $+1$ information bit during the kth bit time becomes

$$z_k = \sum_{l}^{L} w_{k_l}^* w_{k-1_l} + w_{k_l} w_{k-1_l}^* \tag{9.288}$$

where the subscript l refers to the lth branch. The probability of error is given by

$$P_b(E) = \Pr\{z_k < 0\} \tag{9.289}$$

Because the decision variable is a quadratic form of complex Gaussian random variables, we rely again on Appendix B of Ref. 7. Specifically, letting $A = B = 0$, $C = 1$, $D = z_k$, $X_k = w_{k-1_l}$, and $Y_k = w_{k_l}$, the decision variable (9.288) is identical to that in (9A.1) (or equivalently, Appendix B of Ref. 7) for any arbitrary L. Evaluating the various coefficients required in (9A.10) after much simplification produces the following results:

$$\eta = \frac{v_2}{v_1} = \frac{1+K+\bar{\gamma}(1+\rho)}{1+K+\bar{\gamma}(1-\rho)}$$

$$a = 0$$

$$b = \sqrt{\frac{2LK\bar{\gamma}}{1+K+\bar{\gamma}}} \qquad (9.290)$$

where K is the Rice factor and ρ is again the fading correlation whose value depends on the fast-fading channel model that is assumed, as mentioned earlier. Finally, substituting (9.290) in (9A.9) and recalling that $Q_m(b,0) = 1$ and (4.46), we obtain the desired average BER as

$$P_b(E) = \exp\left(-\frac{LK\bar{\gamma}}{1+K+\bar{\gamma}}\right)\left(\frac{1+K+\bar{\gamma}(1-\rho)}{2(1+K+\bar{\gamma})}\right)^{2L-1}$$

$$\times \left[\sum_{l=0}^{L-1}\binom{2L-1}{l}\left(\frac{1+K+\bar{\gamma}(1+\rho)}{1+K+\bar{\gamma}(1-\rho)}\right)^l\right.$$

$$\left.+\sum_{l=0}^{L-2}\binom{2L-1}{l}\left(\frac{1+K+\bar{\gamma}(1+\rho)}{1+K+\bar{\gamma}(1-\rho)}\right)^l\sum_{n=1}^{L-l-1}\frac{1}{n!}\left(\frac{LK\bar{\gamma}}{1+K+\bar{\gamma}}\right)^n\right] \qquad (9.291)$$

which can be shown to agree numerically with Eq. (76) of Ref. 133. The corresponding result for the Rayleigh ($K = 0$) channel is

$$P_b(E) = \left(\frac{1+\bar{\gamma}(1-\rho)}{2(1+\bar{\gamma})}\right)^{2L-1}\sum_{l=0}^{L-1}\binom{2L-1}{l}\left(\frac{1+\bar{\gamma}(1+\rho)}{1+\bar{\gamma}(1-\rho)}\right)^l \qquad (9.292)$$

in agreement with Eq. (3a) of Ref. 132 and Eq. (79) of Ref. 133. Similar to the no-diversity case, these expressions exhibit an irreducible bit error probability floor for any $\rho \neq 1$. Letting $\bar{\gamma}$ approach infinity in (9.291) and (9.292) yields

$$P_b(E) = e^{-LK}\left(\frac{1-\rho}{2}\right)^{2L-1}\left[\sum_{l=0}^{L}\binom{2L-1}{l}\left(\frac{1+\rho}{1-\rho}\right)^l\right.$$

$$\left.+\sum_{l=0}^{L-2}\binom{2L-1}{l}\left(\frac{1+\rho}{1-\rho}\right)^l\sum_{n=1}^{L-l-1}\frac{1}{n!}(LK)^n\right] \qquad (9.293)$$

$$P_b(E) = \left(\frac{1-\rho}{2}\right)^{2L-1}\sum_{l=0}^{L-1}\binom{2L-1}{l}\left(\frac{1+\rho}{1-\rho}\right)^l \qquad (9.294)$$

for Rician and Rayleigh channels, respectively. Figure 9.42 illustrates this bit error floor for a Rician factor $K = 10$ and assuming a correlation model

PERFORMANCE IN THE PRESENCE OF OUTDATED OR IMPERFECT CHANNEL ESTIMATES

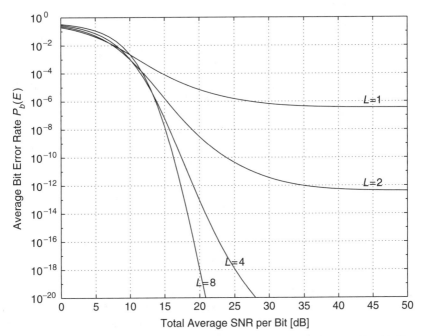

Figure 9.42. Average BER of binary DPSK over fast Rician channels; ($K = 10$ dB and $f_d T = 0.04$).

with $\rho = J_0^2(2\pi f_d T)$. Before concluding this section we should mention that Chow et al. [133] extend the analysis presented here to study the effect of fast Rician fading on M-DPSK ($M \leq 4$) when used in conjunction with post detection EGC.

9.9.3 Selection Combining

In this section we adopt the approach introduced by Ritcey and Azizoğlu [135] and study the impact of imperfect channel estimation or/and decorrelation on the performance of SC systems over Nakagami-m fading channels. This requires the second-order statistics of the channel variation which are fortunately known for Nakagami-m fading. Let α and α_τ denote the channel fading amplitudes at times t and $t + \tau$, respectively.[11] For a slowly varying channel we can assume that the *average* fading power remains constant over the time delay τ [i.e., $\Omega = E(\alpha^2) = E(\alpha_\tau^2)$]. Under these conditions the joint PDF $p_{\alpha,\alpha_\tau}(\alpha, \alpha_\tau)$ of these two correlated Nakagami-m distributed channel fading amplitudes is given by

---

[11] Equivalently, as mentioned above, α and α_τ can be viewed as the actual fading amplitude and the imperfectly estimated one, respectively.

[53, Eq. (126)]

$$p_{\alpha,\alpha_\tau}(\alpha, \alpha_\tau) = \frac{4(\alpha\alpha_\tau)^m}{(1-\rho)\Gamma(m)\rho^{(m-1)/2}} \left(\frac{m}{\Omega}\right)^{m+1} I_{m-1}\left(\frac{2m\sqrt{\rho}\alpha\alpha_\tau}{(1-\rho)\Omega}\right)$$
$$\times \exp\left(-\frac{m(\alpha^2 + \alpha_\tau^2)}{(1-\rho)\Omega}\right) \quad (9.295)$$

where $\rho \in [0, 1]$ again denotes the power correlation factor between α and α_τ. Denoting the instantaneous SNR per symbol at times t and $t + \tau$ by γ and γ_τ, respectively, the joint PDF of γ and γ_τ can be written as

$$p_{\gamma,\gamma_\tau}(\gamma, \gamma_\tau) = \left(\frac{m}{\overline{\gamma}}\right)^{m+1} \frac{\gamma^{(m-1)/2}\gamma_\tau^{(m-1)/2}}{(1-\rho)\Gamma(m)\rho^{(m-1)/2}} \exp\left(-\frac{m(\gamma + \gamma_\tau)}{(1-\rho)\overline{\gamma}}\right)$$
$$\times I_{m-1}\left(\frac{2m\sqrt{\rho\gamma\gamma_\tau}}{(1-\rho)\overline{\gamma}}\right) \quad (9.296)$$

where $\overline{\gamma}$ is the average SNR per symbol over the time delay τ.

Under these conditions it can be shown that for the dual-branch SC the MGF of the output of SC with an outdated/imperfect estimate can be obtained in closed form with the help of Eq. (6.455.2) of Ref. 36 as

$$M_{\gamma_{\text{SC}}}(s) = \frac{2\Gamma(2m)}{m\Gamma^2(m)} \left(2 - (2-\rho)\frac{\overline{\gamma}}{m}s\right)^{-2m} \left(1 - (1-\rho)\frac{\overline{\gamma}}{m}s\right)^m$$
$$\times {}_2F_1\left(1, 2m; m+1; \frac{1-(1-\rho)(\overline{\gamma}/m)s}{2-(2-\rho)(\overline{\gamma}/m)s}\right) \quad (9.297)$$

Using the well known result that the first moment of γ_{SC} is equal to its statistical average (9.188) we obtain the closed-form expression for the average output SNR of SC with an outdated/imperfect estimate as

$$\overline{\gamma}_{\text{SC}} = \frac{2\Gamma(2m)}{m\Gamma^2(m)} \left[\frac{{}_2F_1\left(1, 2m, m+1; \frac{1}{2}\right)}{2^{2m}} + \rho\frac{{}_2F_1\left(2, 2m+1, m+2; \frac{1}{2}\right)}{2^{2m+1}(m+1)}\right]\overline{\gamma} \quad (9.298)$$

With the MGF (9.297) in hand, the average probability of error can be found for a wide variety of modulation schemes, as explained in Chapter 8.

9.9.4 Switched Diversity

We now study the effect of channel decorrelation or imperfect channel estimation on the performance of SSC systems operating over Nakagami-m fading channels [115].

9.9.4.1 SSC Output Statistics.
The PDF of the SSC output at time $t + \tau$, $q_{\gamma_{SSC}}(\gamma_\tau)$, can be expressed in terms of the PDF of the SSC output at time t, $p_{\gamma_{SSC}}(\gamma)$, as

$$q_{\gamma_{SSC}}(\gamma_\tau) = \int_0^\infty p_{\gamma_{SSC}}(\gamma) p_\gamma(\gamma_\tau|\gamma) \, d\gamma \tag{9.299}$$

where the PDF of γ_τ conditioned on γ, $p_\gamma(\gamma_\tau|\gamma)$, is given by

$$p_{\gamma_\tau|\gamma}(\gamma_\tau|\gamma) = \frac{p_{\gamma,\gamma_\tau}(\gamma, \gamma_\tau)}{p_\gamma(\gamma)} \tag{9.300}$$

For simplicity, let us consider the case of i.i.d. branches. Inserting (9.219) and (9.300) in (9.299), we can write $q_{\gamma_{SSC}}(\gamma_\tau)$ as

$$q_{\gamma_{SSC}}(\gamma_\tau) = P_\gamma(\gamma_T) \int_0^{\gamma_T} p_{\gamma,\gamma_\tau}(\gamma, \gamma_\tau) \, d\gamma + [1 + P_\gamma(\gamma_T)] \int_{\gamma_T}^\infty p_{\gamma,\gamma_\tau}(\gamma, \gamma_\tau) \, d\gamma$$

$$= P_\gamma(\gamma_T) \int_0^\infty p_{\gamma,\gamma_\tau}(\gamma, \gamma_\tau) \, d\gamma + \int_{\gamma_T}^\infty p_{\gamma,\gamma_\tau}(\gamma, \gamma_\tau) \, d\gamma$$

$$= P_\gamma(\gamma_T) p_{\gamma_\tau}(\gamma_\tau) + \int_{\gamma_T}^\infty p_{\gamma,\gamma_\tau}(\gamma, \gamma_\tau) \, d\gamma \tag{9.301}$$

As a check, when γ and γ_τ are fully correlated [i.e., $p_{\gamma,\gamma_\tau}(\gamma, \gamma_\tau) = \delta(\gamma - \gamma_\tau) p_\gamma(\gamma)$], it can easily be shown that $q_{\gamma_{SSC}}(\gamma_\tau)$ reduces to $p_{\gamma_{SSC}}(\gamma_\tau)$ as given in (9.219), and full SSC diversity gain is achieved. On the other hand, when γ and γ_τ are uncorrelated [i.e., $p_{\gamma,\gamma_\tau}(\gamma, \gamma_\tau) = p_\gamma(\gamma) p_{\gamma_\tau}(\gamma_\tau)$], it is straightforward to show that $q_{\gamma_{SSC}}(\gamma_\tau)$ reduces to $p_{\gamma_\tau}(\gamma_\tau)$, which is the single-branch PDF, and hence no diversity gain is obtained.

For Nakagami-m fading, inserting the single-branch PDF and CDF as given in Table 9.5 as well as (9.296) in (9.301), $q_{\gamma_{SSC}}(\gamma_\tau)$ can be expressed in closed form as

$$q_{\gamma_{SSC}}(\gamma_\tau) = \left(\frac{m}{\overline{\gamma}}\right)^m \frac{\gamma_\tau^{m-1}}{\Gamma(m)} \exp\left(-m\frac{\gamma_\tau}{\overline{\gamma}}\right)$$

$$\times \left[1 - \frac{\Gamma(m, m\gamma_T/\overline{\gamma})}{\Gamma(m)} + Q_m\left(\sqrt{\frac{2m\rho\gamma_\tau}{(1-\rho)\overline{\gamma}}}, \sqrt{\frac{2m\gamma_T}{(1-\rho)\overline{\gamma}}}\right)\right] \tag{9.302}$$

The MGF at time $t + \tau$ is given by

$$M_{\gamma_{SSC}}(s) = \int_0^\infty e^{s\gamma_\tau} q_{\gamma_{SSC}}(\gamma_\tau) \, d\gamma_\tau \tag{9.303}$$

Substituting (9.302) in (9.303), and using the change of variable $x = \sqrt{2m\rho\gamma_\tau/(1-\rho)\bar{\gamma}}$, $M_{\gamma_{\text{SSC}}}(s)$ can be expressed in closed form with the help of Eq. (11) of Ref. 124 as

$$M_{\gamma_{\text{SSC}}}(s) = \left(1 - \frac{s\bar{\gamma}}{m}\right)^{-m} \left[1 + \frac{\Gamma\left(m, \frac{m\gamma_T}{\bar{\gamma}} \frac{1 - s\bar{\gamma}/m}{1-(1-\rho)s\bar{\gamma}/m}\right) - \Gamma\left(m, \frac{m\gamma_T}{\bar{\gamma}}\right)}{\Gamma(m)}\right]$$

(9.304)

When $\rho = 1$ and hence γ and γ_τ are perfectly correlated, it is easy to see that (9.304) reduces to the MGF with perfect SSC as given by (9.225). On the other hand, when γ and γ_τ are uncorrelated (i.e., $\rho = 0$), it is easy to see that (9.304) reduces to the MGF of single Nakagami-m channel reception as given in Table 9.5.

9.9.4.2 Average SNR. Averaging γ_τ over the PDF $q_{\gamma_{\text{SSC}}}(\gamma_\tau)$ as given by (9.302) yields the average output SNR as

$$\bar{\gamma}_{\text{SSC}} = \bar{\gamma}\left[1 + \frac{\rho(m\gamma_T/\bar{\gamma})^m e^{-(m\gamma_T/\bar{\gamma})}}{\Gamma(m+1)}\right]$$

(9.305)

Differentiating (9.305) with respect to γ_T and setting the result to zero, we find that the optimal threshold γ_T^* is given by $\gamma_T^* = \bar{\gamma}$ and results in a maximum average output SNR $\bar{\gamma}_{\text{SSC}}^*$ given by

$$\bar{\gamma}_{\text{SSC}}^* = \bar{\gamma}\left[1 + \rho \frac{m^{m-1}e^{-m}}{\Gamma(m)}\right]$$

(9.306)

9.9.4.3 Average Probability of Error. Consider as an example the average BER of binary DPSK or noncoherent FSK. In this case the average BER is given by

$$P_b(E) = \tfrac{1}{2} M_{\gamma_{\text{SSC}}}(-g)$$

(9.307)

where $g = 1$ for DPSK and $g = \tfrac{1}{2}$ for orthogonal FSK. Substituting (9.304) in (9.307), then differentiating with respect to γ_T, yields the optimal threshold as

$$\gamma_T^* = \frac{m + (1-\rho)g\bar{\gamma}}{g\rho} \ln\left(\frac{m + g\bar{\gamma}}{m + (1-\rho)g\bar{\gamma}}\right)$$

(9.308)

As a check, when $\rho = 0$ it is easy to see that (9.308) reduces to zero since no diversity gain can be achieved in this case and there is therefore no need to switch to the other branch. On the other hand, when $\rho = 1$, (9.308) reduces to the optimal threshold for perfect SSC as given by [9, Eq. (14)]

$$\gamma_T^* = \frac{m}{g} \ln\left(1 + \frac{g\bar{\gamma}}{m}\right) \quad (9.309)$$

Note that the optimum threshold γ_T^* in (9.309) is dependent on the correlation coefficient ρ. In the case where ρ is viewed as a measure of the channel estimation quality, γ_T^* can be set to (9.309) once ρ is known for minimum BER performance. However, in the case of outdated estimates, ρ is a function of the time delay τ, but γ_T^* (which is set to a particular value during the entire slot time[12]) cannot be changed as a function of τ. In this case, to minimize the degradation due to channel decorrelation, one may want to find a "global" optimum threshold independent of the delay τ. This may be achieved by finding the value, γ_T^*, that minimizes a cost function, C, similar to Eq. (29) of Ref. 9 over an average SNR range,

$$C = \int_{\rho_1}^{1} \log\left(\frac{P_b^*(\rho)}{P_b^*(\rho_0)}\right) d\rho \quad (9.310)$$

where $P_b^*(\rho)$ denotes the average BER, as given by (9.307), evaluated with the optimal threshold γ_T^*, as given by (9.309). In (9.310), ρ_1 denotes the minimal correlation coefficient experienced over the slot time, whereas ρ_0 can be chosen equal to $\rho_0 = (\rho_1 + 1)/2$.

9.9.5 Numerical Results

Figure 9.43 illustrates the analyses presented in previous sections by showing the dependence of the average BER of BPSK on the correlation coefficient $\rho \in [0, 1]$ between the estimated and actual fading for dual-branch MRC, SC, and SSC receivers and for Rayleigh type of fading. When $\rho = 1$ (i.e., perfect correlation between the pilot-extracted fading and the actual fading, or equivalently perfect channel estimation), MRC outperforms SC, which in turn outperforms SSC. As ρ decreases, the correlation between the pilot-extracted fading and the actual fading diminishes and performance degrades. In the limit as $\rho \to 0$, the fading estimate and its actual value are uncorrelated, and performance of all combining schemes approaches the performance without diversity. Figure 9.44 compares the effect of the correlation coefficient ρ on the average BER of binary DPSK with SC and SSC for various values of the Nakagami-m parameter. In this case, ρ is viewed as a measure of the channel estimation quality and the optimum switching threshold is set according to (9.308) for the SSC curves. We can see from these curves that the diversity gain offered by SC over SSC decreases as ρ decreases and tends eventually to zero as ρ tends to zero, as expected. Figure 9.45 shows the effect of channel decorrelation on the average BER of binary DPSK with SC and SSC. For the SSC curves the optimum switching threshold is fixed at the beginning

---

[12] By slot time we mean the time interval T between two consecutive switching instants.

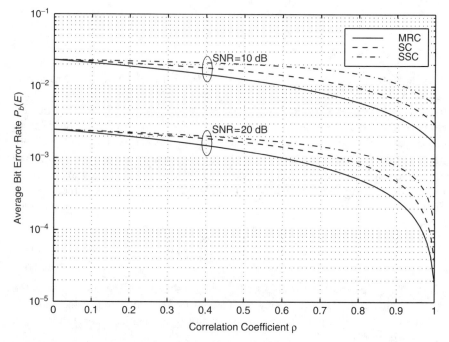

Figure 9.43. Average BER of BPSK with dual-branch MRC, SC, and SSC (with optimum threshold) versus the correlation coefficient ρ.

of the slot time. For the dashed SSC curves the optimum switching threshold is fixed to the optimum value for $\rho = 1$ or equivalently, for $\tau = 0$. On the other hand, the solid SSC curves are generated using optimum thresholds which were optimized over the entire slot time (according to the cost function as explained above), where we assumed that ρ is bounded between 0.8 and 1. From these figures we can see that at low average SNRs (10 dB), the two procedures yield nearly indistinguishable performance results. However, for higher average SNR (20 dB), the first procedure clearly yields better performance for ρ close to 1 (or equivalently, for small values of τ) before the two curves cross and the global optimization procedure starts to pay off.

9.10 HYBRID DIVERSITY SCHEMES

9.10.1 Generalized Selection Combining

In the context of spread-spectrum communication with RAKE reception, the complexity of MRC and EGC receivers depends on the number of resolvable paths available, which can be quite high, especially for multipath diversity of wideband spread-spectrum signals. In addition, MRC is sensitive to channel estimation errors, and these errors tend to be more important when the

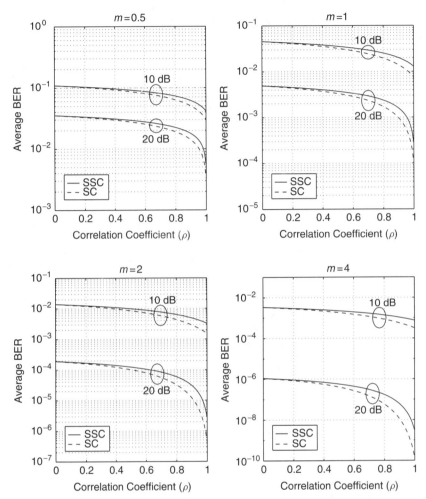

Figure 9.44. Average BER of binary DPSK with SC and SSC versus correlation coefficient (ρ) with $\bar{\gamma} = 10$ dB and $\bar{\gamma} = 20$ dB over Nakagami-m fading channel. The optimum threshold is set according to (9.308) and is thus a function of ρ and $\bar{\gamma}$.

instantaneous SNR is low. On the other hand, SC and SSC use only one path out of the L available (resolvable) multipaths [137] and hence do not fully exploit the amount of diversity offered by the channel. Recently, there has been an interest in bridging the gap between these two extremes (MRC/EGC and SC) by proposing *generalized selection combining (GSC)*, which adaptively combines (as per the rules of MRC or EGC) the L_c strongest (highest SNR) resolvable paths among the L available ones [11,138–142]. We denote such hybrid schemes as SC/MRC-L_c/L and SC/EGC-L_c/L. In the context of wideband spread-spectrum systems, these schemes offer less complex receivers than the conventional MRC RAKE receivers since they have a fixed number of fingers independent of the number

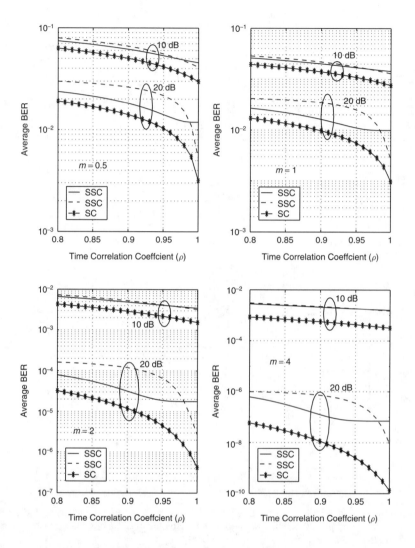

Figure 9.45. Average BER of binary DPSK with SC and SSC versus correlation coeffient (ρ) with $\bar{\gamma} = 10$ dB and $\bar{\gamma} = 20$ dB. The dashed line corresponds to the case where the optimum threshold is set for $\rho = 1$ or equivalently $\tau = 0$. The solid line corresponds to the case where the optimum threshold is optimized over $\rho = 0.8$ to 1 range.

of multipaths. In addition, SC/MRC receivers are expected to be more robust toward channel estimation errors since the weakest SNR paths (and hence the ones that are the most exposed to these errors) are excluded from the combining process. Finally, SC/MRC was shown to approach the performance of MRC [11], while SC/EGC was shown to outperform in certain cases conventional postdetection EGC since it is less sensitive to the "combining loss" of the very noisy (low-SNR) paths [11].

In Refs. 11 and 139 Eng, Kong, and Milstein present an error rate analysis of binary signals with the GSC scheme over Rayleigh fading channels with both i.i.d. distributions and an exponentially decaying power delay profile for $L_c = 2$ and $L_c = 3$. In Ref. 141, Kong and Milstein derived a simple, neat closed-form expression for the average combined SNR at the output of GSC diversity systems operating over Rayleigh fading channels with a constant (uniform) power delay profile. More recently, they extended their result to non-i.i.d. diversity paths [142]. In this section we show that by starting with the MGF of the GSC output SNR, we are able to analyze the performance of GSC receivers over i.i.d Rayleigh paths in terms of average combined SNR, outage probability, and average error rate for a wide variety of modulation schemes and for arbitrary L_c and L [143]. Extension to non-i.i.d. Rayleigh diversity paths is omitted here but can be found in Ref. 143. Finally, work on the performance analysis of GSC receivers using the virtual branch technique can also be found in Refs. 144 through 146.

9.10.1.1 GSC Statistics

GSC Input Joint PDF. Let $\alpha_1, \alpha_2, \ldots, \alpha_L$ denote the set of i.i.d. Rayleigh random fading amplitudes associated with the SC inputs, each of which has average power Ω. For RAKE reception with a matched filter receiver for each diversity path, we define as before the instantaneous SNR per symbol of the lth path as $\gamma_l = \alpha_l^2 E_s / N_0$, $l = 1, 2, \ldots, L$, and the corresponding average SNR per symbol for each path as $\overline{\gamma}_l = \overline{\alpha_l^2} E_s / N_0 = \Omega E_s / N_0$. Let $\gamma_{1:L} \geq \gamma_{2:L} \geq \cdots \geq \gamma_{L:L} \geq 0$ be the order statistics obtained by arranging the $\{\gamma_l\}_{l=1}^L$ in decreasing order of magnitude. Since the $\{\gamma_l\}_{l=1}^L$ are i.i.d., the joint PDF $p_{\gamma_{1:L},\ldots,\gamma_{L_c:L}}(\gamma_{1:L},\ldots,\gamma_{L_c:L})$ of the $\{\gamma_{l:L}\}_{l=1}^{L_c}$ ($L_c \leq L$) is given by [112, p. 185; 139, Eq. (9)]

$$p_{\gamma_{1:L},\ldots,\gamma_{L_c:L}}(\gamma_{1:L},\ldots,\gamma_{L_c:L}) = L_c! \binom{L}{L_c} [P_\gamma(\gamma_{L_c:L})]^{L-L_c} \prod_{l=1}^{L_c} p_\gamma(\gamma_{l:L});$$

$$\gamma_{1:L} \geq \gamma_{2:L} \geq \cdots \geq \gamma_{L_c:L} \quad (9.311)$$

where

$$\binom{L}{L_c} = \frac{L!}{L_c!(L-L_c)!}$$

denotes the binomial coefficient, $p_\gamma(\gamma)$ is the PDF of the $\{\gamma_l\}_{l=1}^L$, such as

$$p_\gamma(\gamma) = \frac{1}{\overline{\gamma}} e^{-\gamma/\overline{\gamma}} \quad (9.312)$$

and $P_\gamma(\gamma) = \int_0^\gamma p_\gamma(y)\,dy$ is the corresponding CDF, given by

$$P_\gamma(\gamma) = 1 - e^{-\gamma/\overline{\gamma}} \quad (9.313)$$

It is important to note that although the $\{\gamma_l\}_{l=1}^L$ are independent, the $\{\gamma_{l:L}\}_{l=1}^L$ are not, as can be seen from (9.311).

MGF of the Output SNR. The MGF of the total combined SNR $\gamma_{\text{GSC}} = \sum_{l=1}^{L_c} \gamma_{l:L}$ is defined by

$$M_{\gamma_{\text{GSC}}}(s) = E_{\gamma_{\text{GSC}}}[e^{s\gamma_{\text{GSC}}}] = E_{\gamma_{1:L},\gamma_{2:L},\ldots,\gamma_{L_c:L}}\left(e^{s\sum_{l=1}^{L_c}\gamma_{l:L}}\right) \quad (9.314)$$

with $E[\cdot]$ denoting the expectation operator. Substituting (9.311) in (9.314), we get

$$M_{\gamma_{\text{GSC}}}(s) = \underbrace{\int_0^\infty \int_{\gamma_{L_c:L}}^\infty \cdots \int_{\gamma_{2:L}}^\infty}_{L_c\text{-fold}} p_{\gamma_{1:L},\ldots,\gamma_{L_c:L}}(\gamma_{1:L},\ldots,\gamma_{L_c:L}) e^{s\sum_{l=1}^{L_c}\gamma_{l:L}}$$

$$\times d\gamma_{1:L}\cdots d\gamma_{L_c-1:L}\, d\gamma_{L_c:L} \quad (9.315)$$

Although the integrand is in a desirable separable form in the $\gamma_{l:L}$'s, we cannot partition the L_c-fold integral into a product of one-dimensional integrals as was possible for MRC in Section 9.2 and postdetection EGC in Section 9.4 because of the $\gamma_{l:L}$'s in the lower limits of the semifinite range (improper) integrals. To get around this difficulty, we take advantage of the following classical result, which is originally due to Sukhatme [147] and which subsequently played an important role in many order statistics problems [148,149], including, for example, radar detection analysis problems [150,151].

Theorem 1 (Sukhatme [147]). Consider the following transformation of random variables[13] by defining the "spacings"

$$\begin{aligned} x_l &\stackrel{\Delta}{=} \gamma_{l:L} - \gamma_{l+1:L}, \quad l = 1, 2, \ldots, L-1 \\ x_L &\stackrel{\Delta}{=} \gamma_{L:L} \end{aligned} \quad (9.316)$$

Then it can be shown that the $\{x_l\}_{l=1}^L$ are independent and distributed according to the exponential distribution $p_{x_l}(x_l)$, given by

$$p_{x_l}(x_l) = \frac{l}{\bar{\gamma}} e^{-lx_l/\bar{\gamma}}, \quad x_l \geq 0, \quad l = 1, 2, \ldots, L \quad (9.317)$$

A proof of this theorem is given in Appendix 9C.

---

[13] It should be pointed out that Kong and Milstein considered a very similar transformation [141, App.] in their derivation of the combined average SNR of GSC and hence implicitly used Theorem 1.

We now use Theorem 1 to derive a simple expression for the MGF of the total combined SNR γ_{GSC}, which can be expressed in terms of the x_l's as

$$\gamma_{GSC} = \sum_{l=1}^{L_c} \gamma_{l:L} = \sum_{l=1}^{L_c} \sum_{k=l}^{L} x_k$$
$$= x_1 + 2x_2 + \cdots + L_c x_{L_c} + L_c x_{L_c+1} + \cdots + L_c x_L \quad (9.318)$$

Hence the MGF of γ_{GSC} as defined in (9.314) can be expressed in terms of the x_l values as

$$M_{\gamma_{GSC}}(s) = \underbrace{\int_0^\infty \cdots \int_0^\infty}_{L\text{-fold}} p_{x_1,\ldots,x_L}(x_1,\ldots,x_L) e^{s(x_1+2x_2+\cdots+L_c x_{L_c}+L_c x_{L_c+1}+\cdots+L_c x_L)}$$
$$\times dx_1 \cdots dx_L \quad (9.319)$$

Since the x_l's are independent [i.e., $p_{x_1,\ldots,x_L}(x_1,\ldots,x_L) = \prod_{l=1}^L p_{x_l}(x_l)$], we can put the integrand in the *desired* product form, resulting in

$$M_{\gamma_{GSC}}(s) = \underbrace{\int_0^\infty \cdots \int_0^\infty}_{L\text{-fold}} \left[\prod_{l=1}^L p_{x_l}(x_l)\right] e^{sx_1} e^{2sx_2} \cdots e^{L_c s x_{L_c}} e^{L_c s x_{L_c+1}} \cdots e^{L_c s x_L}$$
$$\times dx_1 \cdots dx_L \quad (9.320)$$

Grouping terms of index l thereby partitioning the L-fold integral of (9.320) into a product of L one-dimensional integrals, then using the fact that the x_l values are exponentially distributed, we get the desired closed-form result

$$M_{\gamma_{GSC}}(s) = (1 - s\bar{\gamma})^{-L_c+1} \prod_{l=L_c}^{L} \left(1 - \frac{s\bar{\gamma}L_c}{l}\right)^{-1} \quad (9.321)$$

Using a partial fraction expansion of the product in (9.321), it can be shown that the MGF of γ_{GSC} can be rewritten in the following equivalent form:

$$M_{\gamma_{GSC}}(s) = (1 - s\bar{\gamma})^{-L_c+1} \sum_{l=0}^{L-L_c} \frac{(-1)^l \binom{L}{L_c} \binom{L-L_c}{l}}{1 + l/L_c - s\bar{\gamma}} \quad (9.322)$$

PDF of the Output SNR. Having a simple expression for the MGF as given by (9.322), we are now in a position to derive the PDF of the GSC output combined SNR γ_{GSC} for an arbitrary L_c and L. Letting $s = -p$, the Laplace transform of the PDF of γ_{GSC}, $\mathcal{L}_{\gamma_{\text{GSC}}}(p)$, is related to the MGF of γ_{GSC} by

$$\mathcal{L}_{\gamma_{\text{GSC}}}(p) = M_{\gamma_{\text{GSC}}}(-p) \tag{9.323}$$

which, by using (9.322), can be written as

$$\mathcal{L}_{\gamma_{\text{GSC}}}(p) = \frac{\binom{L}{L_c}}{\bar{\gamma}^{L_c}} \left[\frac{1}{(p+1/\bar{\gamma})^{L_c}} + \sum_{l=1}^{L-L_c} \frac{(-1)^l \binom{L-L_c}{l}}{(p+1/\bar{\gamma})^{L_c-1}[p+(1+l/L_c)/\bar{\gamma}]} \right] \tag{9.324}$$

Using the inverse Laplace transforms [152, Eqs. (1) and (3)] as well as the identity [36, Eq. (8.352.1)], we obtain the PDF of γ_{GSC} in closed form as the inverse Laplace transform of (9.324):

$$p_{\gamma_{\text{GSC}}}(\gamma) = \binom{L}{L_c} \left[\frac{\gamma^{L_c-1} e^{-\gamma/\bar{\gamma}}}{\bar{\gamma}^{L_c}(L_c-1)!} + \frac{1}{\bar{\gamma}} \sum_{l=1}^{L-L_c} (-1)^{L_c+l-1} \binom{L-L_c}{l} \left(\frac{L_c}{l} \right)^{L_c-1} \right.$$
$$\left. \times e^{-\gamma/\bar{\gamma}} \left(e^{-l\gamma/L_c \bar{\gamma}} - \sum_{m=0}^{L_c-2} \frac{1}{m!} \left(\frac{-l\gamma}{L_c \bar{\gamma}} \right)^m \right) \right] \tag{9.325}$$

As a check, note that (9.325) reduces to the well-known PDF of the SNR at an MRC ($L = L_c$) output as given by (9.5) and SC ($L_c = 1$) output

$$p_{\gamma_{\text{SC}}}(\gamma) = \frac{L}{\bar{\gamma}} \sum_{l=0}^{L-1} (-1)^l \binom{L-1}{l} \exp\left(-\frac{1+l}{\bar{\gamma}} \gamma \right) \tag{9.326}$$

9.10.1.2 Average Output SNR

Analysis. In this section, starting from the MGF of γ_{GSC}, we obtain the average combined SNR $\bar{\gamma}_{\text{GSC}}$ at the GSC output. For this purpose we first introduce the "second" MGF of γ_{GSC} (using the terminology of Papoulis [112, Sec. 5.5]) or equivalently, the cumulant generating function defined by

$$\Psi_{\gamma_{\text{GSC}}}(s) = \ln(M_{\gamma_{\text{GSC}}}(s)) \tag{9.327}$$

which for the GSC after substitution of (9.321) in (9.327) is given as

$$\Psi_{\gamma_{\text{GSC}}}(s) = -L_c \ln(1 - s\bar{\gamma}) - \sum_{l=L_c+1}^{L} \ln\left(1 - \frac{s\bar{\gamma}L_c}{l} \right) \tag{9.328}$$

We now use the well-known result that the first cumulant of γ_{GSC} is equal to its statistical average [112, Eq. (5-73)]:

$$\bar{\gamma}_{\text{GSC}} = \left.\frac{d\Psi_{\gamma_{\text{GSC}}}(s)}{ds}\right|_{s=0} \tag{9.329}$$

giving after substituting (9.328) in (9.329),

$$\bar{\gamma}_{\text{GSC}} = L_c \bar{\gamma} + \sum_{l=L_c+1}^{L} \frac{L_c \bar{\gamma}}{l}$$

$$= \left(1 + \sum_{l=L_c+1}^{L} \frac{1}{l}\right) L_c \bar{\gamma} \tag{9.330}$$

which is the beautifully simple closed-form result originally obtained by Kong and Milstein [141, Eq. (7)]. In Appendix 9D we give an alternative simple and direct proof[14] of the result in (9.330). It should be noted that since the MGF contains information about all the statistical moments of the underlying RV (and similarly for the cumulant generating function), it is then straightforward to obtain simple closed-form expressions for the higher-order moments and cumulants directly from higher-order derivatives of (9.321) and (9.328), respectively. For example, the variance of γ_{GSC}, which is equal to the second cumulant of γ_{GSC} [112, Eq. (5-73)], is given by

$$\text{var}(\gamma_{\text{GSC}}) = \left.\frac{d^2\Psi_{\gamma_{\text{GSC}}}(s)}{ds^2}\right|_{s=0} = \left(1 + \sum_{l=L_c+1}^{L} \frac{L_c}{l^2}\right) L_c \bar{\gamma}^2 \tag{9.331}$$

in agreement with Eq. (18) of Ref. 145, derived independently by Win and Winters.

We note, as pointed out in Ref. 141, that the result (9.330) generalizes the average SNR results for conventional SC and MRC. In particular, for the specific case of $L = L_c$ (i.e., conventional MRC) it is easy to see that (9.330) reduces to the classical result given in (9.50). Similarly, for the particular case of $L_c = 1$ (i.e., conventional SC), it is straightforward to see that (9.330) reduces to the well-known result $\bar{\gamma}_{\text{SC}} = \sum_{l=1}^{L}(1/l)\bar{\gamma}$ [4, Eq. (6.62)].

Numerical Examples. Figure 9.46 shows the normalized average combined SNR $\bar{\gamma}_{\text{GSC}}/\bar{\gamma}$ as a function of the number of available resolvable paths L for various values of the number of the L_c strongest combined paths. These results show that for a fixed number L of available diversity paths, diminishing diversity gain is obtained as the number of combined paths L_c increases. On the other hand, Fig. 9.47 shows the normalized average combined SNR $\bar{\gamma}_{\text{GSC}}/\bar{\gamma}$ as a function of the number of strongest combined paths L_c for various values of the number of

---

[14] By direct proof we mean a proof that does not rely on Theorem 1 or equivalently, on the transformation used in the Appendix of Ref. 141.

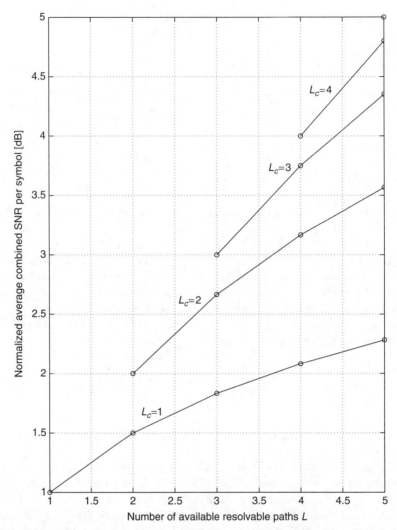

Figure 9.46. Normalized average combined SNR $\bar{\gamma}_{GSC}/\bar{\gamma}$ versus the number of available resolvable paths L for various values of the strongest combined paths L_c.

resolvable paths L. These curves indicate that for a fixed number of combined paths, a nonnegligible performance improvement can be gained by increasing the number of available diversity paths.

9.10.1.3 Outage Probability

Analysis. The outage probability, P_{out}, is defined as the probability that the GSC output SNR falls below a certain predetermined threshold SNR, γ_{th}, and hence can be obtained by integrating the PDF of γ_{GSC}, which can be obtained in closed

HYBRID DIVERSITY SCHEMES **387**

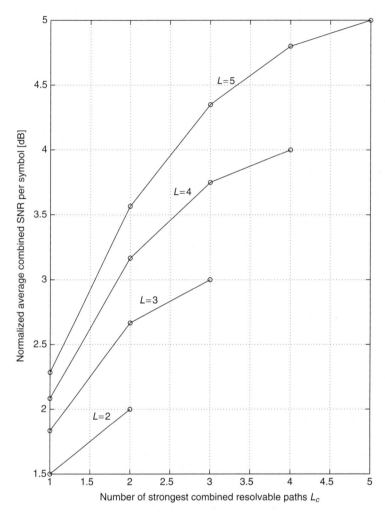

Figure 9.47. Normalized average combined SNR $\bar{\gamma}_{\text{GSC}}/\bar{\gamma}$ versus the number of the strongest combined paths L_c for various values of the number of the resolvable paths L.

form with the help of Eq. (3.351.1) of Ref. 36 as

$$P_{\text{out}}^{\text{GSC}} = \binom{L}{L_c} \left\{ 1 - e^{-\gamma_{\text{th}}/\bar{\gamma}} \sum_{l=0}^{L_c-1} \frac{(\gamma_{\text{th}}/\bar{\gamma})^l}{l!} \right.$$
$$+ \sum_{l=1}^{L-L_c} (-1)^{L_c+l-1} \binom{L-L_c}{l} \left(\frac{L_c}{l}\right)^{L_c-1} \left[\frac{1 - e^{-(1+l/L_c)(\gamma_{\text{th}}/\bar{\gamma})}}{1 + l/L_c} \right.$$
$$\left. \left. - \sum_{m=0}^{L_c-2} \left(\frac{-l}{L_c}\right)^m \left(1 - e^{-\gamma_{\text{th}}/\bar{\gamma}} \sum_{k=0}^{m} \frac{(\gamma_{\text{th}}/\bar{\gamma})^k}{k!} \right) \right] \right\} \quad (9.332)$$

As a check, it is easy to see that when $L_c = L$, (9.332) reduces to the well-known outage probability result for MRC [4, Eq. (6.69)]:

$$P_{\text{out}}^{\text{MRC}} = 1 - e^{-\gamma_{\text{th}}/\bar{\gamma}} \sum_{l=0}^{L-1} \frac{(\gamma_{\text{th}}/\bar{\gamma})^l}{l!} \tag{9.333}$$

In addition, for conventional SC ($L_c = 1$), (9.332) reduces to

$$P_{\text{out}}^{\text{SC}} = L \left[1 - e^{-\gamma_{\text{th}}/\bar{\gamma}} + \sum_{l=1}^{L-1} (-1)^l \binom{L-1}{l} \frac{1 - e^{-(1+l)(\gamma_{\text{th}}/\bar{\gamma})}}{1+l} \right]$$

$$= L \sum_{l=0}^{L-1} (-1)^l \binom{L-1}{l} \frac{1 - e^{-(1+l)(\gamma_{\text{th}}/\bar{\gamma})}}{1+l} \tag{9.334}$$

which can easily be shown using the identity [153, p. 171] and the binomial series expansion to be in agreement with the previously known result [4, Eq. (6.58)],

$$P_{\text{out}}^{\text{SC}} = (1 - e^{-\gamma_{\text{th}}/\bar{\gamma}})^L \tag{9.335}$$

as expected.

Numerical Examples. Figure 9.48 shows the outage probability $P_{\text{out}}^{\text{GSC}}$ as a function of the normalized average SNR per path $\bar{\gamma}/\gamma_{\text{th}}$ for various values of the available diversity paths L and strongest combined paths L_c. Notice again the diminishing returns as the number of combined paths increases. Figure 9.49 shows $P_{\text{out}}^{\text{GSC}}$ as function of $\bar{\gamma}/\gamma_{\text{th}}$ for a fixed $L_c = 3$ and $L = 3, 4,$ and 5. Clearly, these curves show that for fixed L_c a significant decrease in the outage probability is obtained as the number of available diversity paths increases.

9.10.1.4 Average Error Rate

Binary Signals. Using the closed-form expression for SNR at the GSC output (9.322), we get the average BER as a single finite-range integral given by

$$P_b(E) = \frac{\binom{L}{L_c}}{\pi} \int_0^{\pi/2} \frac{\sum_{l=0}^{L-L_c} (-1)^l \binom{L-L_c}{l} (1 + l/L_c + g\bar{\gamma}/\sin^2\phi)^{-1}}{(1 + g\bar{\gamma}/\sin^2\phi)^{L_c-1}} d\phi \tag{9.336}$$

Switching the order of summation and integration, and defining the integral $I_n(\theta; c_1, c_2)$ as in (5A.42), where in general c_1 and c_2 are two constants (independent of ϕ) that might be different, we can rewrite the average BER as

$$P_b(E) = \binom{L}{L_c} \sum_{l=0}^{L-L_c} \frac{(-1)^l \binom{L-L_c}{l}}{1 + l/L_c} I_{L_c-1}\left(\frac{\pi}{2}; g\bar{\gamma}, \frac{g\bar{\gamma}}{1 + l/L_c}\right) \tag{9.337}$$

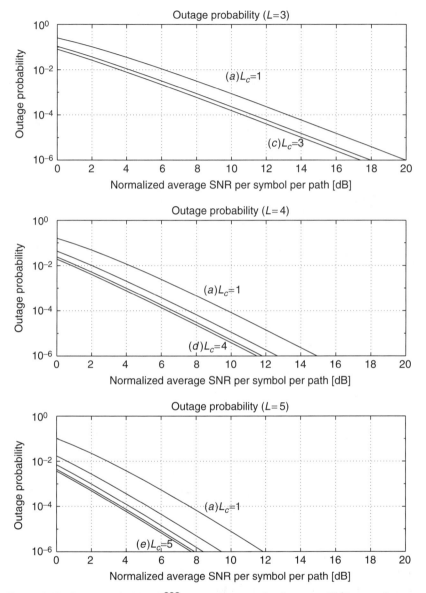

Figure 9.48. Outage probability $P_{\text{out}}^{\text{GSC}}$ versus the normalized average SNR per path $\bar{\gamma}/\gamma_{\text{th}}$.

Since the integrals $I_n(\theta; c_1, c_2)$ can be found in closed form (see Appendix 5A), (9.337) presents the final desired closed-form result. This result yields the same numerical results as Eqs. (9) and (12) of Ref. 11, for the average BER of BPSK ($g = 1$) with $L_c = 2$ and $L_c = 3$, respectively. Hence, (9.337) [or equivalently, (9.336)] is a generic expression valid for any $L_c \leq L$.

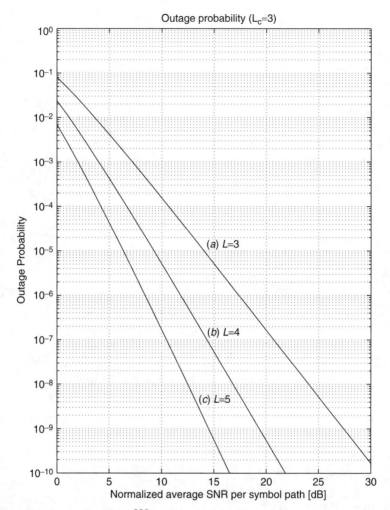

Figure 9.49. Outage probability P_{out}^{GSC} versus the normalized average SNR per path $\bar{\gamma}/\gamma_{th}$ for $L_c = 3$.

Average SER of M-PSK. Similarly, following the same steps as in (9.336)–(9.337), we obtain the average SER of M-PSK as

$$P_s(E) = \binom{L}{L_c} \sum_{l=0}^{L-L_c} \frac{(-1)^l \binom{L-L_c}{l}}{1+l/L_c} I_{L_c-1}\left(\frac{(M-1)\pi}{M}; g_{PSK}\bar{\gamma}, \frac{g_{PSK}\bar{\gamma}}{1+l/L_c}\right) \quad (9.338)$$

The result (9.338) generalizes the M-PSK average SER results of Chennakeshu and Anderson [33] with MRC and conventional SC. For instance, for the particular case of $L = L_c$ (i.e., MRC), it can easily be shown that (9.338) agrees with Eq. (21) of Ref. 33. Similarly, for the particular case of $L_c = 1$ (i.e., conventional SC), it can also be shown that (9.338) reduces to Eq. (26) of Ref. 33.

Average SER of M-QAM. Using the same steps as in (9.336)–(9.337), we obtain the average SER of square M-QAM as

$$P_s(E) = \binom{L}{L_c} \left[4\left(1 - \frac{1}{\sqrt{M}}\right) \sum_{l=0}^{L-L_c} \frac{(-1)^l \binom{L-L_c}{l}}{1 + l/L_c} \right.$$

$$\times I_{L_c-1}\left(\frac{\pi}{2}; g_{\text{QAM}}\bar{\gamma}, \frac{g_{\text{QAM}}\bar{\gamma}}{1+l/L_c}\right) - 4\left(1 - \frac{1}{\sqrt{M}}\right)^2$$

$$\left. \times \sum_{l=0}^{L-L_c} \frac{(-1)^l \binom{L-L_c}{l}}{1 + l/L_c} I_{L_c-1}\left(\frac{\pi}{4}; g_{\text{QAM}}\bar{\gamma}, \frac{g_{\text{QAM}}\bar{\gamma}}{1+l/L_c}\right) \right] \quad (9.339)$$

The result (9.339) generalizes the square M-QAM SER result given in (9.23) as well as those of Kim et al. [42,154,155] and Lu et al. [43] with conventional MRC and SC. For instance, for the particular case of $L = L_c$ (i.e., MRC), it can be shown that (9.339) yields the same numerical results as Eq. (6) of Ref. 155 or Eq. (15) of Ref. 42, or equivalently, Eq. (12) of Ref. 43. Similarly, for the particular case of $L_c = 1$ (i.e., conventional SC), it can also be shown that (9.339) reduces to Eq. (23) of Ref. 43 or equivalently, to Eq. (13) of Ref. 155.

Numerical Examples. Figures 9.50 through 9.55 show the effect of L_c and L on the average error rate of BPSK, 8-PSK, and 16-QAM. These curves confirm previous trends in the sense that diminishing returns are obtained as the number of strongest combined paths increases, but a significant performance improvement can be gained by increasing the number of available diversity paths.

9.10.1.5 Performance of GSC over Nakagami-m Channels. The "spacing" technique for ordered exponential RV's which allows the necessary partitioning of the integrand for the MGF-based approach to be applied successfully does not carry over to gamma-distributed variables, which are characteristic of the instantaneous SNR per path for Nakagami-*m* fading. Thus, an alternative approach is required to obtain analogous generic results for such channels.

A partial solution to this problem was provided by the authors in a recent paper [156] presenting a performance analysis of two specific hybrid SC/MRC receivers: SC/MRC-2/3 and SC/MRC-2/4. The final result for the average BER was shown to be expressible in terms of infinite series of hypergeometric functions suitable for numerical evaluation. However, the method used in Ref. 156

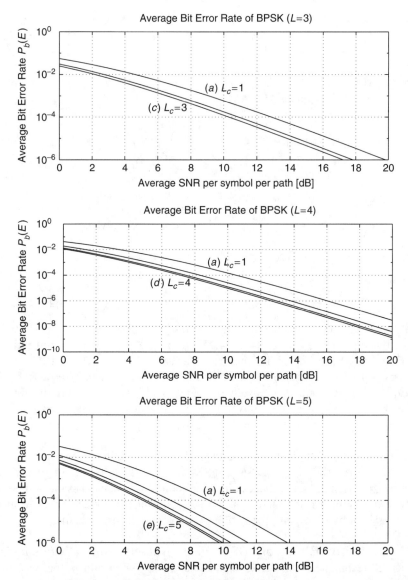

Figure 9.50. Average BER of BPSK versus the average SNR per path $\bar{\gamma}$.

does not allow for similar simplifications for the case of $L_c > 2$, and hence analysis is not amenable to application to other SC/MRC receivers. Furthermore, it is limited to binary coherent modulations and as such does not apply to the performance of SC/EGC or M-ary modulations such as M-ary phase-shift-keying (M-PSK) and M-ary quadrature amplitude modulation (M-QAM). In this section, applying the Dirichlet transformation [157] (a well-known technique found in classical textbooks on integral calculus [158, p. 492; 159, Chap. XXV]

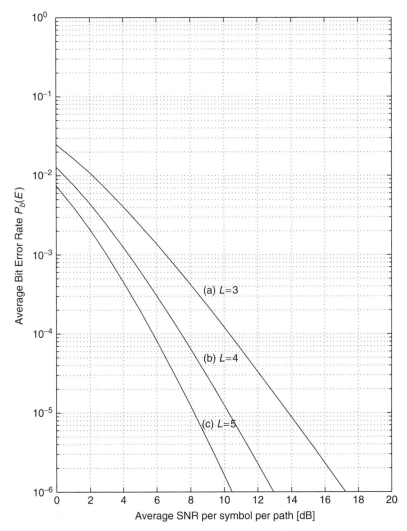

Figure 9.51. Average BER of BPSK versus the average SNR per path $\bar{\gamma}$ for $L_c = 3$.

to simplify certain multiple integrals), we develop a well-structured procedure that allows obtaining the average error rate for arbitrary L and L_c, and which is applicable for the performance analysis of not only SC/MRC with M-ary modulations but also SC/EGC receivers [160]. Specific results are presented for a number of examples both numerically and as simple closed-form expressions. These results are compared with the particular results corresponding to Rayleigh fading. As a by-product of the general results, some interesting closed-form expressions are presented in Appendix 9E for certain single and multiple definite integrals which heretofore appear not to have been reported in standard tabulations such as Refs. 36 and 52.

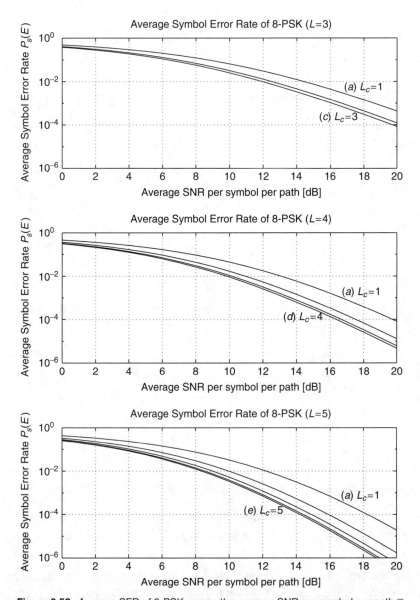

Figure 9.52. Average SER of 8-PSK versus the average SNR per symbol per path $\overline{\gamma}$.

GSC Input Joint Statistics. Let $\beta_l \stackrel{\Delta}{=} m\gamma_l/\overline{\gamma}$ denote the *normalized* instantaneous SNR for the lth channel with PDF

$$p_{\beta_l}(\beta_l) = \frac{\beta_l^{m-1}}{\Gamma(m)} \exp(-\beta_l), \qquad \beta_l \geq 0 \qquad (9.340)$$

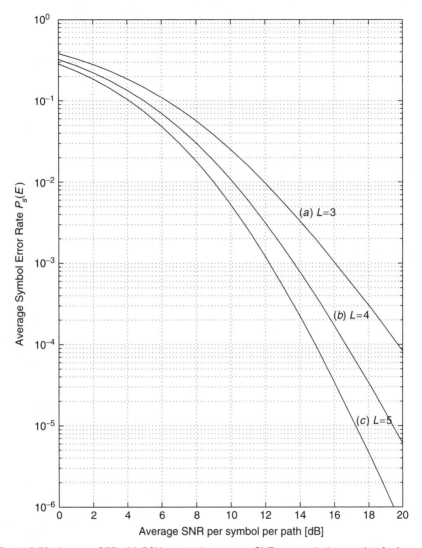

Figure 9.53. Average SER of 8-PSK versus the average SNR per symbol per path $\bar{\gamma}$ for $L_c = 3$.

Consider $\beta_{1:L} \geq \beta_{2:L} \geq \cdots \geq \beta_{L:L}$: the order statistics obtained by arranging the set $\{\beta_l\}_{l=1}^L$. Since the $\{\alpha_l\}_{l=1}^L$ are assumed to be i.i.d. RV's, so are the $\{\gamma_l\}_{l=1}^L$ and the $\{\beta_l\}_{l=1}^L$, and the joint PDF of the $\{\beta_{l:L}\}_{l=1}^L$ is then given by [112, p. 185; 139, Eq. (9)]

$$p_{\beta_{1:L},\ldots,\beta_{L:L}}(\beta_{1:L},\ldots,\beta_{L:L}) = L! \prod_{l=1}^{L} p_\beta(\beta_{l:L}),$$

$$\beta_{1:L} \geq \beta_{2:L} \geq \cdots \geq \beta_{L:L} \geq 0 \tag{9.341}$$

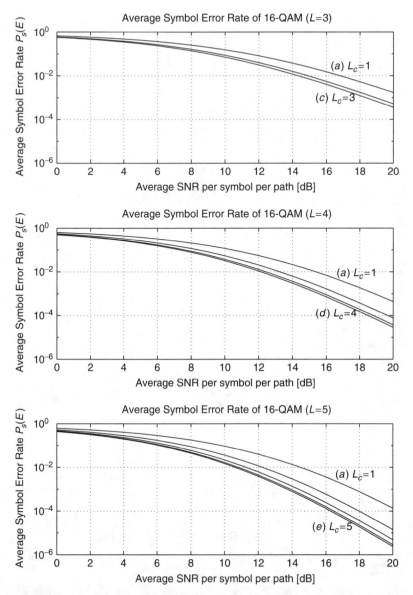

Figure 9.54. Average SER of 16-QAM versus the average SNR per symbol per path $\bar{\gamma}$.

which can be rewritten after substitution of (9.340) in (9.341) as

$$p_{\beta_{1:L},\ldots,\beta_{L:L}}(\beta_{1:L},\ldots,\beta_{L:L}) = \frac{L!}{[\Gamma(m)]^L} \left(\prod_{l=1}^{L} \beta_{l:L}\right)^{m-1} \exp\left(-\sum_{l=1}^{L} \beta_{l:L}\right),$$

$$\beta_{1:L} \geq \beta_{2:L} \geq \cdots \geq \beta_{L:L} \geq 0 \qquad (9.342)$$

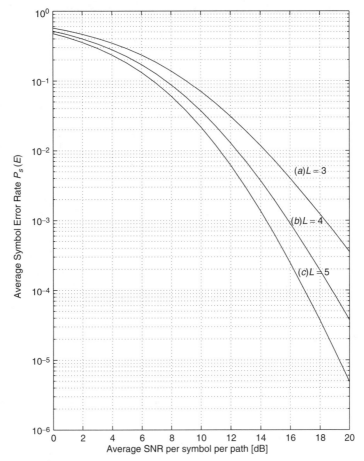

Figure 9.55. Average SER of 16-QAM versus the average SNR per symbol per path $\bar{\gamma}$ for $L_c = 3$.

MGF of GSC Output. In this section we first summarize the procedure proposed by Kabe [157] to obtain the MGF of any linear function of ordered gamma variates. We then show how this result can be used to get the MGF of the GSC output for arbitrary L_c and L.

To take advantage of the results of Kabe, let us first follow his notation and reorder the RV's from weakest to strongest by defining $\{x_{i:L}\}_{i=1}^{L}$ as the reverse-ordered set corresponding to $\{\beta_l\}_{l=1}^{L}$, namely, $x_{i:L} \triangleq \beta_{L-i+1:L}$, $i = 1, 2, \ldots, L$, with the joint PDF obtained from (9.342) as

$$p_{x_{1:L},\ldots,x_{L:L}}(x_{1:L}, \ldots, x_{L:L}) = \frac{L!}{[\Gamma(m)]^L} \left(\prod_{i=1}^{L} x_{i:L}\right)^{m-1} \exp\left(-\sum_{i=1}^{L} x_{i:L}\right),$$
$$0 \leq x_{1:L} \leq x_{2:L} \leq \cdots \leq x_{L:L} \qquad (9.343)$$

The MGF of $\sum_{i=1}^{L} u_i x_{i:L}$, where the $\{u_i\}_{i=1}^{L}$ are constants, is given by

$$E_{x_{1:L},\ldots,x_{L:L}}\left[e^{s\sum_{i=1}^{L} u_i x_{i:L}}\right] = \underbrace{\int_0^\infty \cdots \int_0^{x_{3:L}} \int_0^{x_{2:L}}}_{L-\text{fold}} \exp\left(s\sum_{i=1}^{L} u_i x_{i:L}\right)$$

$$\times p_{x_{1:L},\ldots,x_{L:L}}(x_{1:L},\ldots,x_{L:L})\,dx_{1:L}\,dx_{2:L}\cdots dx_{L:L}$$

$$= \frac{L!}{[\Gamma(m)]^L} \underbrace{\int_0^\infty \cdots \int_0^{x_{3:L}} \int_0^{x_{2:L}}}_{L-\text{fold}} \left(\prod_{i=1}^{L} x_{i:L}\right)^{m-1}$$

$$\times \exp\left[-\sum_{i=1}^{L}(1-u_i s)x_{i:L}\right] dx_{1:L}\,dx_{2:L}\cdots dx_{L:L}$$
(9.344)

which can be viewed as a multiple integral of the Dirichlet–Louiville type [159, Chap. XXV]. The difficulty in evaluating this L-fold integral is that the *integration limits are functions of the integration variables themselves*. This is where the Dirichlet transformation steps in to simplify the problem. In particular, using the transformation [157, Eq. (2.2)]

$$x_{i:L} = \prod_{l=i}^{L} \theta_l \tag{9.345}$$

so that $0 < \theta_l < 1$ for $l = 1, 2, \ldots, L-1$ and $\theta_L > 0$, (9.344) can be rewritten as

$$E_{x_{1:L},\ldots,x_{L:L}}\left[e^{s\sum_{i=1}^{L} u_i x_{i:L}}\right] = \frac{L!}{[\Gamma(m)]^L} \underbrace{\int_0^1 \cdots \int_0^1 \int_0^\infty}_{(L-1)-\text{fold}} \left(\prod_{l=1}^{L} \theta_l^{lm-1}\right)$$

$$\times \exp[-\theta_L D(s;\theta_1,\theta_2,\ldots,\theta_{L-1})]\,d\theta_L\,d\theta\,d\theta_{L-1}\cdots d\theta_1$$
(9.346)

where each integral now has limits that are independent of the integration variables and the function $D(s;\theta_1,\theta_2,\ldots,\theta_{L-1})$ is defined by

$$D(s;\theta_1,\theta_2,\ldots,\theta_{L-1}) = (1-u_L s) + \theta_{L-1}(1-u_{L-1}s) + \theta_{L-1}\theta_{L-2}(1-u_{L-2}s)$$
$$+ \cdots + \theta_{L-1}\theta_{L-2}\cdots\theta_1(1-u_1 s) \tag{9.347}$$

Note that $D(s;\theta_1,\theta_2,\ldots,\theta_{L-1})$ is independent of θ_L; thus, the first integration in (9.346) (i.e., the one on θ_L) is of the form

$$\int_0^\infty \theta_L^{Lm-1}\exp[-\theta_L D(s;\theta_1,\theta_2,\ldots,\theta_{L-1})]\,d\theta_L \tag{9.348}$$

which has the closed-form result [36, Eq. (3.381.4)]

$$\int_0^\infty \theta_L^{Lm-1} \exp[-D(s;\theta_1,\theta_2,\ldots,\theta_{L-1})\theta_L]\,d\theta_L = \frac{\Gamma(Lm)}{[D(s;\theta_1,\theta_2,\ldots,\theta_{L-1})]^{Lm}} \quad (9.349)$$

Thus, applying (9.349) to (9.346) immediately reduces the L-fold integral to an $(L-1)$-fold integral given by

$$E_{x_{1:L},\ldots,x_{L:L}}\left[e^{s\sum_{i=1}^L u_i x_{i:L}}\right] = \frac{L!\,\Gamma(Lm)}{[\Gamma(m)]^L}\underbrace{\int_0^1 \cdots \int_0^1}_{(L-1)\text{-fold}} \left(\prod_{l=1}^{L-1}\theta_l^{lm-1}\right)$$

$$\times [D(s;\theta_1,\theta_2,\ldots,\theta_{L-1})]^{-Lm}\,d\theta_{L-1}\cdots d\theta_1 \quad (9.350)$$

where each integral has finite limits that are independent of the integration variables.

We now use Kabe's procedure to derive the MGF of the total output SNR γ_{GSC}, which can be expressed in terms of the $\{x_{i:L}\}_{i=1}^L$ as

$$\gamma_{\text{GSC}} \overset{\Delta}{=} \sum_{l=1}^{L_c}\gamma_{l:L} = \frac{\overline{\gamma}}{m}\sum_{l=1}^{L_c}\beta_{l:L} = \sum_{i=1}^L u_i x_{i:L} \quad (9.351)$$

where the weights $\{u_i\}_{i=1}^L$ are defined by

$$\begin{aligned}u_1 = u_2 = \cdots = u_{L-L_c} &= 0 \\ u_{L-L_c+1} = u_{L-L_c+2} = \cdots u_L &= \frac{\overline{\gamma}}{m}\end{aligned} \quad (9.352)$$

Hence, using the result (9.350), the MGF of the total combined SNR can be written as

$$M_{\gamma_{\text{GSC}}}(s) = \frac{L!\,\Gamma(Lm)}{(\Gamma(m))^L}\underbrace{\int_0^1 \cdots \int_0^1}_{(L-1)\text{-fold}}\left(\prod_{l=1}^{L-1}\theta_l^{lm-1}\right)(D(s;\theta_1,\theta_2,\ldots,\theta_{L-1}))^{-Lm}$$

$$\times d\theta_{L-1}\cdots d\theta_1 \quad (9.353)$$

where $D(s;\theta_1,\theta_2,\ldots,\theta_{L-1})$ is found after substitution of the weights of (9.352) in (9.347), giving

$$D(s;\theta_1,\theta_2,\ldots,\theta_{L-1})$$
$$= \left(1 - \frac{\overline{\gamma}}{m}s\right)(1 + \theta_{L-1} + \theta_{L-1}\theta_{L-2} + \cdots + \theta_{L-1}\theta_{L-2}\cdots\theta_{L-L_c+2}\theta_{L-L_c+1})$$
$$+ \theta_{L-1}\theta_{L-2}\cdots\theta_{L-L_c+1}\theta_{L-L_c} + \cdots + \theta_{L-1}\theta_{L-2}\cdots\theta_2 + \theta_{L-1}\theta_{L-2}\cdots\theta_1 \quad (9.354)$$

In view of the form of $D(s; \theta_1, \theta_2, \ldots, \theta_{L-1})$ in (9.354) the next integrations (i.e., the ones on $\theta_{L-1}, \theta_{L-2}, \ldots$) can be computed in a recursive fashion [160].

Numerical Examples. With the MGF of the SNR output in hand, we can compute the average SER for a wide variety of modulation schemes. As examples, Figs 9.56 through 9.61 show the effect of L_c and L on the average error rate of

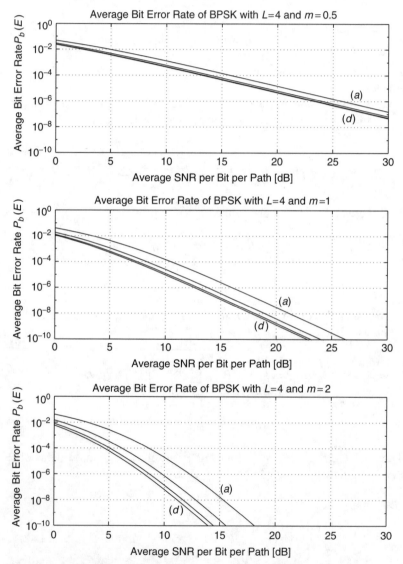

Figure 9.56. Average BER of BPSK versus the average SNR per bit path $\bar{\gamma}$ for $L=4$ and (a) $L_c = 1$ (SC), (b) $L_c = 2$, (c) $L_c = 3$, and (d) $L_c = 4$ (MRC).

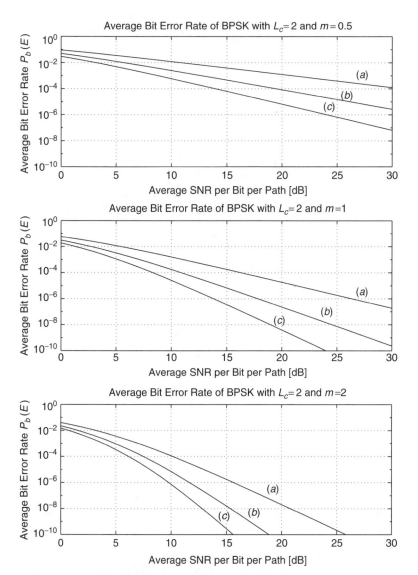

Figure 9.57. Average BER of BPSK versus the average SNR per bit per path $\bar{\gamma}$ for $L_c = 2$ and (a) $L = 2$, (b) $L = 3$, and (c) $L = 4$.

BPSK, 8-PSK, and 16-QAM, for various values of the fading parameter m. The curves for $m = 1$ are in agreement with the Rayleigh fading results, as expected. Furthermore, results for BPSK with SC/MRC-2/3 and SC/MRC-2/4 match the results reported in Ref. 156. These numerical results confirm trends observed for Rayleigh fading in the sense that diminishing returns are obtained as the number of combined paths increases, but a significant performance improvement

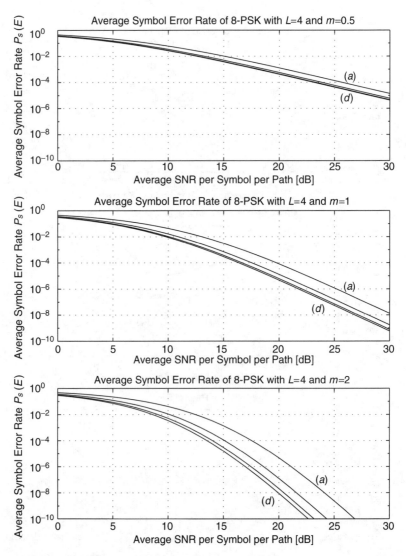

Figure 9.58. Average SER of 8-PSK versus the average SNR per symbol per path $\bar{\gamma}$ for $L = 4$ and (a) $L_c = 1$ (SC), (b) $L_c = 2$, (c) $L_c = 3$, and (d) $L_c = 4$ (MRC).

can be gained by increasing the number of available diversity paths. In addition, Figs. 9.56, 9.58, and 9.60 (in which the number of available diversity paths L is fixed at 4 and the number of combined paths L_c is varied from 1 to 4) indicate that the more severe the fading (i.e., the lower the fading parameter m), the more diminishing are the returns obtained for an increasing number of combined paths. On the other hand, Figs. 9.57, 9.59 and 9.61 (in which the number of combined paths L_c is fixed at 2 and the number of available diversity paths L is varied from 2 to 4) show that the performance improvement gained by increasing the

HYBRID DIVERSITY SCHEMES 403

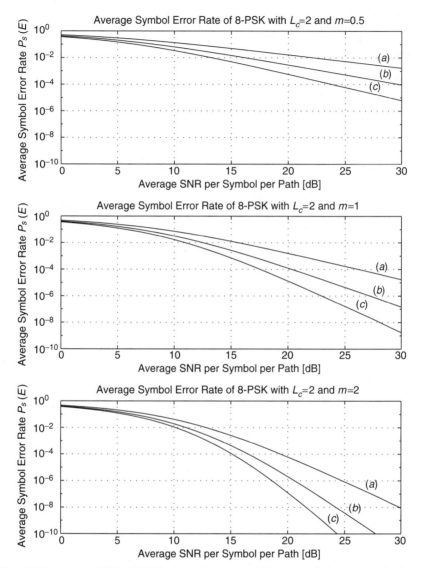

Figure 9.59. Average SER of 8-PSK versus the average SNR per symbol per path $\bar{\gamma}$ for $L_c = 2$ and (a) $L = 2$, (b) $L = 3$, and (c) $L = 4$.

number of available diversity paths is more important for channels subject to a low amount of fading.

9.10.2 Generalized Switched Diversity

We now focus on the performance of a GSSC scheme [115]. This scheme involves first SSC, followed by MRC or EGC, the operation of which is as

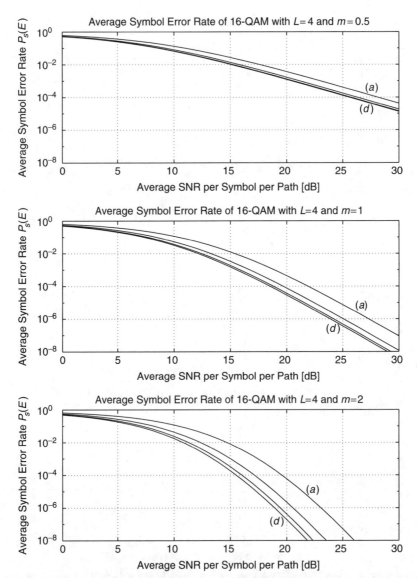

Figure 9.60. Average SER of 16-QAM versus the average SNR per symbol per path $\bar{\gamma}$ for $L = 4$ and (a) $L_c = 1$ (SC), (b) $L_c = 2$, (c) $L_c = 3$, and (d) $L_c = 4$ (MRC).

follows: The incoming signal is received over an even number $2L$ of diversity branches that are grouped in pairs. Every pair of signals is fed to a switching unit that operates according to the rules of SSC. The output from the L switching units are connected to an MRC or EGC combiner. This scheme is motivated by the GSC scheme that was analyzed in the preceding section, which inherits one

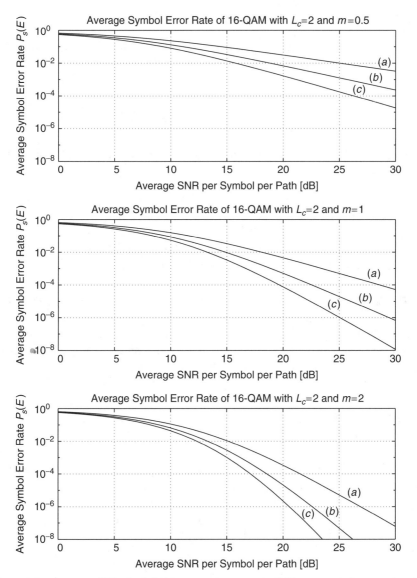

Figure 9.61. Average SER of 16-QAM versus the average SNR per symbol per path $\bar{\gamma}$ for $L_c = 2$ and (a) $L = 2$, (b) $L = 3$, and (c) $L = 4$.

of the main disadvantages of SC: the necessity of a "centralized," continuous, and simultaneous monitoring of all the diversity branches. On the other hand, the GSSC scheme offers a decentralized, simpler (although less efficient) solution and can be viewed as a more practical implementation of GSC. In what follows we evaluate the performance of the GSSC scheme and then compare it to GSC.

9.10.2.1 GSSC Output Statistics

Joint PDF. For simplicity let us assume that all the pairs of signals at the SSC unit inputs are i.i.d. Then the joint PDF at the MRC input is given by

$$p_{\gamma_1,\gamma_2,\ldots,\gamma_L}(\gamma_1, \gamma_2, \ldots, \gamma_L) = \prod_{l=1}^{L} p_{\gamma_{\text{SSC}}}(\gamma_l) \quad (9.355)$$

where $p_{\gamma_{\text{SSC}}}(\gamma_l)$ denotes the SNR PDF at the output of the lth SSC unit and is given by

$$p_{\gamma_{\text{SSC}}}(\gamma_l) = \begin{cases} P_{\gamma_l}(\gamma_l) p_{\gamma_l}(\gamma_l) & \text{if } \gamma_{T_l} < \gamma_l \\ [1 + P_{\gamma_l}(\gamma_{T_l})] p_{\gamma_l}(\gamma_l) & \text{if } \gamma_{T_l} \geq \gamma_l \end{cases} \quad (9.356)$$

where $P_{\gamma_l}(\gamma_l)$ and $p_{\gamma_l}(\gamma_l)$ are the CDF and PDF of the SSC unit's individual branches, respectively, and are given in Table 9.5.

MGF. Since the L MRC inputs are combined according to the rules of MRC, $\gamma_{\text{GSSC}} = \sum_{l=1}^{L} \gamma_l$, then assuming independent fading across the SSC units, the MGF of the GSSC SNR output is just the product of the L MGF's of the SNR's at the L SSC outputs. In the particular case of Nakagami-m fading, using (9.225), the MGF of the GSSC output can be obtained as

$$M_{\gamma_{\text{GSSC}}}(s) = \prod_{l=1}^{L} \left(1 - \frac{s\overline{\gamma}_l}{m_l}\right)^{-m_l} \left(1 + \frac{\Gamma(m_l, (1 - (\overline{\gamma}_l/m_l)s)(m_l \gamma_T/\overline{\gamma}_l)) - \Gamma(m_l, (m_l/\overline{\gamma}_l)\gamma_T)}{\Gamma(m_l)}\right) \quad (9.357)$$

Similarly for the Nakagami-n (Rician) fading case, using (9.226), the MGF of the GSSC output can be obtained as

$$M_{\gamma_{\text{GSSC}}}(s) = \prod_{l=1}^{L} \left(1 - \frac{s\overline{\gamma}_l}{1 + n_l^2}\right)^{-1} \exp\left(\frac{s\overline{\gamma}_l n_l^2}{1 + n_l^2 - s\overline{\gamma}_l}\right)$$

$$\times \left[1 - Q_1\left(n_l\sqrt{2}, \sqrt{\frac{2(1 + n_l^2)\gamma_T}{\overline{\gamma}_l}}\right)\right.$$

$$\left. + Q_1\left(n_l\sqrt{\frac{2(1 + n_l^2)}{1 + n_l^2 - s\overline{\gamma}_l}}, \sqrt{2\left(\frac{1 + n_l^2}{\overline{\gamma}_l} - s\right)\gamma_T}\right)\right] \quad (9.358)$$

9.10.2.2 Average Probability of Error.
Using the MGF of the GSSC output SNR determined in the preceding section, we can determine the average probability of error of several modulation schemes via the MGF-based approach.

Differentiating the resulting expressions with respect to the L switching thresholds, it can easily be shown (because of the product form of the integrand) that the optimal thresholds of the individual switching units will yield the overall optimum performance. This means that GSSC performance can be optimized without much more computational complexity than conventional SSC. As an example, Fig. 9.62 compares the average BER performance of SC/MRC-2/4 with the performance of SSC/MRC-2/4 (using the optimum switching thresholds in all the switching units) for $m = 0.5$, 1, and 2. Note that SSC/MRC-2/4 suffers about a 1-dB penalty compared to SC/MRC-2/4 in the medium- to high-average SNR region.

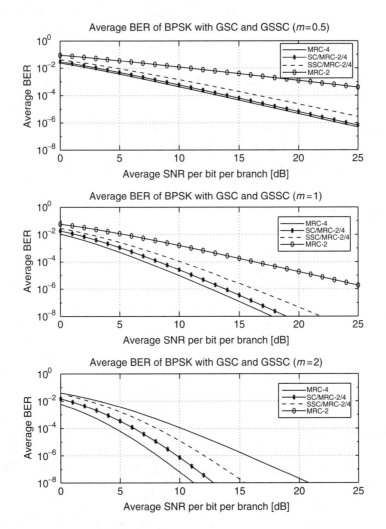

Figure 9.62. Comparison of the average BER of BPSK with MRC-2, SSC/MRC-2/4, SC/MRC-2/4, and MRC-4 over Nakagami-m fading channels.

9.10.3 Two-Dimensional Diversity Schemes

In this section we analyze the performance of two-dimensional diversity systems that we described in Section 9.1.3.2. The aim is to accurately quantify the effect of the fading severity and correlation as well as the power delay profile on the error rate and outage probability performance. As in the preceding section, we again rely on the MGF-based approach, which is going to be particularly handy in this case. In particular, when MRC or postdetection EGC is used for both dimensions, finding the PDF of the combined SNR in a simple form (as required by the classical approach to tackle these problems) is particularly difficult in the presence of fading correlation, a nonuniform power delay profile and/or when the fading tends to follow other than Rayleigh statistics, whereas the MGF-based approach will circumvent much of the tedium and intractability in the classical approach, as we show next.

9.10.3.1 Performance Analysis.
Consider a two-dimensional diversity system consisting, for example, of D antennas, each followed by an L_c-finger RAKE receiver. As an example of practical channel conditions of interest, let us assume that for a fixed antenna index d the $\{\gamma_{l,d}\}_{l=1}^{L_c}$ are independent but nonidentically distributed. On the other hand, let us assume that for a fixed multipath index l, the $\{\gamma_{l,d}\}_{d=1}^{D}$ are correlated (in space) according to model A, B, C, or D (as described in Section 9.6). When MRC or postdetection EGC combining is applied for both space and multipath diversity, we have a conditional combined SNR/symbol given by

$$\gamma_t = \sum_{d=1}^{D} \sum_{l=1}^{L_c} \gamma_{l,d}$$

$$= \sum_{d=1}^{D} \gamma_d \quad \left(\text{where } \gamma_d = \sum_{l=1}^{L_c} \gamma_{l,d}\right)$$

$$= \sum_{l=1}^{L_c} \gamma_l \quad \left(\text{where } \gamma_l = \sum_{d=1}^{D} \gamma_{l,d}\right) \qquad (9.359)$$

Finding the average error rate or outage probability performance of such systems with the classical PDF-based approach is difficult since the PDF of γ_t cannot be found in a simple form. However, using the MGF-based approach for the average BER, for example of BPSK, we have after switching the order of integration

$$P_b(E) = \frac{1}{\pi} \int_0^{\pi/2} E_{\gamma_1, \gamma_2, \ldots, \gamma_{L_c}} \left[\exp\left(-\frac{\sum_{l=1}^{L_c} \gamma_l}{\sin^2 \phi}\right)\right] d\phi \qquad (9.360)$$

Since the $\{\gamma_l\}_{l=1}^{L_c}$ are assumed to be independent, then

$$P_b(E) = \frac{1}{\pi} \int_0^{\pi/2} \prod_{l=1}^{L_c} E_{\gamma_l} \left[\exp\left(-\frac{\gamma_l}{\sin^2 \phi}\right) \right] d\phi$$

$$= \frac{1}{\pi} \int_0^{\pi/2} \prod_{l=1}^{L_c} M_{i_l}\left(-\frac{1}{\sin^2 \phi}\right) d\phi \quad (9.361)$$

where $M_{i_l}(s)$ is given by (9.156), (9.163), (9.167), or (9.169), depending on the space fading correlation model under consideration.

Application. As an application, let us consider a two-dimensional RAKE receiver operating over a Nakagami-m fading channel characterized by a spatial correlation coefficient ρ_l along the path of index l ($l = 1, 2, \ldots, L_c$) and the same exponential PDP for the D RAKE receivers:

$$\overline{\gamma}_{l,d} = \overline{\gamma}_{1,1} e^{-(l-1)\delta}, \qquad l = 1, 2, \ldots, L_c \quad (9.362)$$

where δ denotes the average fading power decay factor. Substituting (9.163) in (9.361), we obtain the average BER for BPSK with a constant spatial fading correlation profile as

$$P_b(E) = \frac{1}{\pi} \int_0^{\pi/2} \prod_{l=1}^{L_c} \left(1 + \frac{\overline{\gamma}_{l,d}(1 - \sqrt{\rho_l} + D\sqrt{\rho_l})}{m_l \sin^2 \phi}\right)^{-m_l}$$

$$\times \left(1 + \frac{\overline{\gamma}_{l,d}(1 - \sqrt{\rho_l})}{m_l \sin^2 \phi}\right)^{-m_l(D-1)} d\phi \quad (9.363)$$

Similarly substituting (9.178) in (9.361), we obtain the average BER for BPSK with a tridiagonal spatial fading correlation profile as

$$P_b(E) = \frac{1}{\pi} \int_0^{\pi/2} \prod_{l=1}^{L_c} \prod_{d=1}^{D} \left[1 - \frac{\overline{\gamma}_{l,d}}{m_l \sin^2 \phi}\left(1 + 2\sqrt{\rho_l} \cos \frac{d\pi}{D+1}\right)\right]^{-m_l} d\phi$$

$$(9.364)$$

9.10.3.2 Numerical Examples. As an example, the average BER performance curves of BPSK with two-dimensional MRC RAKE reception over an exponentially decaying power delay profile and with constant or tridiagonal correlation between the antenna elements of the array [as given by (9.363) and (9.364), respectively] are shown in Fig. 9.63. The corresponding outage probability curves obtained by using the numerical technique presented in Section 9.5 are given in Fig. 9.64. Again notice the relatively important effect of

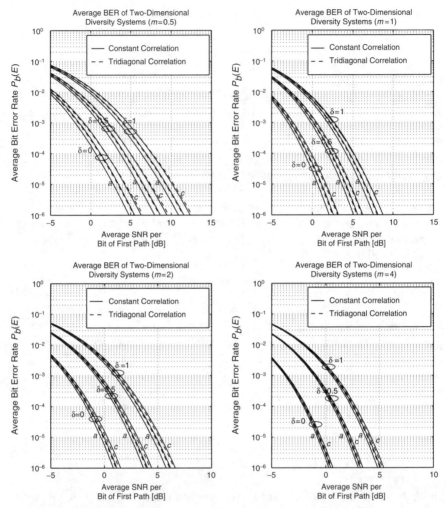

Figure 9.63. Average BER of BPSK with two-dimensional MRC RAKE reception ($L_c = 4$ and $D = 3$) over an exponentially decaying power delay profile and constant or tridiagonal spatial correlation between the antennas for various values of the correlation coefficient: (a) $\rho = 0$; (b) $\rho = 0.2$; (c) $\rho = 0.4$.

the power delay profile. Also, diversity systems subject to tridiagonal correlation have a slightly better performance than those subject to constant correlation in most cases. However, the opposite occurs at high-average SNR for channels with a high amount of fading ($m = 0.5$) and a relatively strong correlation between the paths ($\rho = 0.4$).

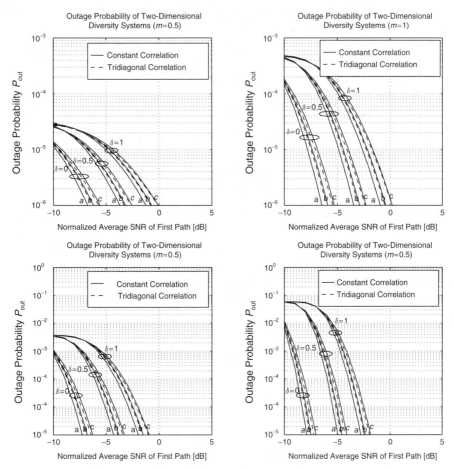

Figure 9.64. Outage probability with two-dimensional MRC or postdetection EGC RAKE reception ($L_c = 4$ and $D = 3$) over an exponentially decaying PDP and constant or tridiagonal spatial correlation between the antennas for various values of the correlation coefficient: (a) $\rho = 0$; (b) $\rho = 0.2$; (c) $\rho = 0.4$.

REFERENCES

1. D. Brennan, "Linear diversity combining techniques," *Proc. IRE*, vol. 47, June 1959, pp. 1075–1102.
2. W. C. Jakes, *Microwave Mobile Communication*, 2nd ed. Piscataway, NJ: IEEE Press, 1994.
3. G. L. Stüber, *Principles of Mobile Communications*. Norwell, MA: Kluwer Academic Publishers, 1996.

4. T. S. Rappaport, *Wireless Communications: Principles and Practice*. Upper Saddle River, NJ: Prentice Hall, 1996.
5. T. Sunaga and S. Sampei, "Performance of multi-level QAM with post-detection maximal ratio combining space diversity for digital land-mobile radio communications," *IEEE Trans. Veh. Technol.*, vol. VT-42, August 1993, pp. 294–301.
6. M. K. Simon, S. M. Hinedi, and W. C. Lindsey, *Digital Communication Techniques: Signal Design and Detection*. Upper Saddle River, NJ: Prentice Hall, 1995.
7. J. G. Proakis, *Digital Communications*, 3rd ed. New York: McGraw-Hill, 1995.
8. M. A. Blanco and K. J. Zdunek, "Performance and optimization of switched diversity systems for the detection of signals with Rayleigh fading," *IEEE Trans. Commun.*, vol. COM-27, December 1979, pp. 1887–1895.
9. A. A. Abu-Dayya and N. C. Beaulieu, "Analysis of switched diversity systems on generalized-fading channels," *IEEE Trans. Commun.*, vol. COM-42, November 1994, pp. 2959–2966.
10. A. A. Abu-Dayya and N. C. Beaulieu, "Switched diversity on microcellular Ricean channels," *IEEE Trans. Veh. Technol.*, vol. VT-43, November 1994, pp. 970–976.
11. T. Eng, N. Kong, and L. B. Milstein, "Comparison of diversity combining techniques for Rayleigh-fading channels," *IEEE Trans. Commun.*, vol. COM-44, September 1996, pp. 1117–1129. See also "Correction to 'Comparison of diversity combining techniques for Rayleigh-fading channels,' " *IEEE Trans. Commun.*, vol. COM-46, September 1998, p. 1111.
12. W. Xu and L. B. Milstein, "Performance of multicarrier DS CDMA systems in the presence of correlated fading," *Proc. IEEE Veh. Technol. Conf. (VTC'97)*, Phoenix, AZ, May 1997, pp. 2050–2054.
13. A. M. Sayeed and B. Aazhang, "Joint multipath-Doppler diversity in mobile wireless communications," *IEEE Trans. Commun.*, vol. COM-47, January 1999, pp. 123–132.
14. D. C. Cox, H. W. Arnold, and P. T. Porter, "Universal digital portable communications: a system perspective," *IEEE J. Sel. Areas Commun.*, vol. SAC-5, June 1987, pp. 764–773.
15. D. Molkdar, "Review on radio propagation into and within buildings," *IEE Proc. H*, vol. 138, February 1991, pp. 61–73.
16. COST 207 TD(86)51-REV 3 (WG1), "Proposal on channel transfer functions to be used in GSM test late 1986," *Tech. Rep.*, Office Official Publications European Communities, September 1986.
17. W. R. Braun and U. Dersch, "A physical mobile radio channel model," *IEEE Trans. Veh. Technol.*, vol. VT-40, May 1991, pp. 472–482.
18. K. A. Stewart, G. P. Labedz, and K. Sohrabi, "Wideband channel measurements at 900 MHz," *Proc. IEEE Veh. Technol. Conf. (VTC'95)*, Chicago, July 1995, pp. 236–240.
19. S. A. Abbas and A. U. Sheikh, "A geometric theory of Nakagami fading multipath mobile radio channel with physical interpretations," *Proc. IEEE Veh. Technol. Conf. (VTC'96)*, Atlanta, GA, April 1996, pp. 637–641.
20. M. K. Simon and M.-S. Alouini, "A unified approach to the performance analysis of digital communications over generalized fading channels," *Proc. IEEE*, vol. 86, September 1998, pp. 1860–1877.

21. M.-S. Alouini and A. J. Goldsmith, "A unified approach for calculating the error rates of linearly modulated signals over generalized fading channels," *IEEE Trans. Commun.*, vol. COM-47, September 1999, pp. 1324–1334. See also *Proc. IEEE Int. Conf. Commun. (ICC'98)*, Atlanta, GA, June 1998.
22. R. F. Pawula, S. O. Rice, and J. H. Roberts, "Distribution of the phase angle between two vectors perturbed by Gaussian noise," *IEEE Trans. Commun.*, vol. COM-30, August 1982, pp. 1828–1841.
23. J. W. Craig, "A new, simple, and exact result for calculating the probability of error for two-dimensional signal constellations," *Proc. IEEE Mil. Commun. Conf. (MILCOM'91)*, McLean, VA, October 1991, pp. 571–575.
24. C. Tellambura, A. J. Mueller, and V. K. Bhargava, "Analysis of M-ary phase-shift keying with diversity reception for land-mobile satellite channels," *IEEE Trans. Veh. Technol.*, vol. VT-46, November 1997, pp. 910–922.
25. A. Annamalai, C. Tellambura, and V. K. Bhargava, "Exact evaluation of maximal-ratio and equal-gain diversity receivers for M-ary QAM on Nakagami fading channels," *IEEE Trans. Commun.*, vol. COM-47, September 1999, pp. 1335–1344.
26. X. Dong, N. C. Beaulieu, and P. H. Wittke, "Signaling constellations for fading channels," *IEEE Trans. Commun.*, vol. COM-47, May 1999, pp. 703–714. See also *Proc. Commun. Theory Mini-conf. (CTMC-VII)* in conjunction with IEEE Global Commun. Conf. (GLOBECOM'98), Sydney, Australia, November 1998, pp. 22–27.
27. G. Efthymoglou and V. Aalo, "Performance of RAKE receivers in Nakagami fading channel with arbitary fading parameters," *Electron. Lett.*, vol. 31, August 1995, pp. 1610–1612.
28. J. E. Mazo, "Exact matched filter bound for two-beam Rayleigh fading," *IEEE Trans. Commun.*, vol. COM-39, July 1991, pp. 1027–1030.
29. J. E. Mazo, private communication, September 1998.
30. J. E. Mazo, "Matched filter bounds for multi-beam Rician fading with diversity," unpublished technical memorandum.
31. R. D. Yates and D. J. Goodman, *Probability and Stochastic Processes*. New York: Wiley, 1998.
32. J. G. Proakis, "Probabilities of error for adaptive reception of M-phase signals," *IEEE Trans. Commun. Technol.*, vol. COM-16, February 1968, pp. 71–81.
33. S. Chennakeshu and J. B. Anderson, "Error rates for Rayleigh fading multichannel reception of MPSK signals," *IEEE Trans. Commun.*, vol. COM-43, February–March–April 1995, pp. 338–346.
34. T. L. Staley, R. C. North, W. H. Ku, and J. R. Zeidler, "Performance of coherent MPSK on frequency selective slowly fading channels," *Proc. IEEE Veh. Technol. Conf. (VTC'96)*, Atlanta, GA, April 1996, pp. 784–788.
35. T. L. Staley, R. C. North, W. H. Ku, and J. R. Zeidler, "Probability of error evaluation for multichannel reception of coherent MPSK over Ricean fading channels," *Proc. IEEE Int. Conf. Commun. (ICC'97)*, Montreal, Quebec, Canada, June 1997, pp. 30–35.
36. I. S. Gradshteyn and I. M. Ryzhik, *Table of Integrals, Series, and Products*, 5th ed. San Diego, CA: Academic Press, 1994.
37. C. K. Pauw and D. L. Schilling, "Probability of error M-ary PSK and DPSK on a Rayleigh fading channel," *IEEE Trans. Commun.*, vol. COM-36, June 1988, pp. 755–758.

38. N. Ekanayake, "Performance of M-ary PSK signals in slow Rayleigh fading channels," *Electron. Lett.*, vol. 26, May 1990, pp. 618–619.
39. M. G. Shayesteh and A. Aghamohammadi, "On the error probability of linearly modulated signals on frequency-flat Ricean, Rayleigh, and AWGN channels," *IEEE Trans. Commun.*, vol. COM-43, February–March–April 1995, pp. 1454–1466.
40. Y. Miyagaki, N. Morinaga, and T. Namekawa, "Error probability characteristics for CPSK signal through m-distributed fading channel," *IEEE Trans. Commun.*, vol. COM-26, January 1978, pp. 88–100.
41. V. Aalo and S. Pattaramalai, "Average error rate for coherent MPSK signals in Nakagami fading channels," *Electron. Lett.*, vol. 32, August 1996, pp. 1538–1539.
42. C.-J. Kim, Y.-S. Kim, G.-Y. Jeong, and D.-D. Lee, "Matched filter bound of square QAM in multipath Rayleigh fading channels," *Electron. Lett.*, vol. 33, January 1997, pp. 20–21.
43. J. Lu, T. T. Tjhung, and C. C. Chai, "Error probability performance of L-branch diversity reception of MQAM in Rayleigh fading," *IEEE Trans. Commun.*, vol. 46, February 1998, pp. 179–181.
44. H. S. Abdel-Ghaffar and S. Pasupathy, "Asymptotical performance of M-ary and binary signals over multipath/multichannel Rayleigh and Ricain fading," *IEEE Trans. Commun.*, vol. COM-43, November 1995, pp. 2721–2731.
45. N. C. Beaulieu and A. A. Abu-Dayya, "Analysis of equal gain diversity on Nakagami fading channels," *IEEE Trans. Commun.*, vol. COM-39, February 1991, pp. 225–234.
46. A. A. Abu-Dayya and N. C. Beaulieu, "Microdiversity on Rician fading channels," *IEEE Trans. Commun.*, vol. COM-42, June 1994, pp. 2258–2267.
47. Q. T. Zhang, "Probability of error for equal-gain combiners over Rayleigh channels: some closed-form solutions," *IEEE Trans. Commun.*, vol. COM-45, March 1997, pp. 270–273.
48. N. C. Beaulieu, "An infinite series for the computation of the complementary probability distribution function of a sum of independent random variables and its application to the sum of Rayleigh random variables," *IEEE Trans. Commun.*, vol. COM-26, September 1990, pp. 1463–1474.
49. J. Gil-Pelaez, "Note on the inversion theorem," *Biometrika*, vol. 38, 1951, pp. 481–482.
50. Q. T. Zhang, "A simple approach to probability of error for equal gain combiners over Rayleigh channels," *IEEE Trans. Veh. Technol.*, vol. VT-48, July 1999, pp. 1151–1154.
51. M.-S. Alouini and M. K. Simon, "Error rate analysis of M-PSK with equal gain combining over Nakagami fading channels," *IEEE Veh. Technol. Conf. (VTC'99)*, Houston, TX, May 1999, pp. 2378–2382.
52. M. Abramowitz and I. A. Stegun, *Handbook of Mathematical Functions with Formulas, Graphs, and Mathematical Tables*, 9th ed. New York: Dover Publications, 1970.
53. M. Nakagami, "The m-distribution: A general formula of intensity distribution of rapid fading," in *Statistical Methods in Radio Wave Propagation*. Oxford: Pergamon Press, 1960, pp. 3–36.
54. T. Eng and L. B. Milstein, "Coherent DS-CDMA performance in Nakagami multipath fading," *IEEE Trans. Commun.*, vol. COM-43, February–March–April 1995, pp. 1134–1143.

55. M. K. Simon and M.-S. Alouini, "A unified approach to the probability of error for noncoherent and differentially coherent modulations over generalized fading channels," *IEEE Trans. Commun.*, vol. COM-46, December 1998, pp. 1625–1638.
56. M. K. Simon and M.-S. Alouini, "Bit error probability of noncoherent M-ary orthogonal modulation over generalized fading channels," *Int. J. Commun. Networks*, vol. 1, June 1999, pp. 111–117.
57. J. G. Proakis, "On the probability of error for multichannel reception of binary signals," *IEEE Trans. Commun. Technol.*, vol. COM-16, February 1968, pp. 68–71.
58. W. C. Lindsey, "Error probabilities for Ricean fading multichannel reception of binary and N-ary signals," *IEEE Trans. Inf. Theory*, vol. IT-10, October 1964, pp. 339–350.
59. U. Charash, "Reception through Nakagami fading multipath channels with random delays," *IEEE Trans. Commun.*, vol. COM-27, April 1979, pp. 657–670.
60. J. F. Weng and S. H. Leung, "Analysis of DPSK with equal gain combining in Nakagami fading channels," *Electron. Lett.*, vol. 33, April 1997, pp. 654–656.
61. F. Patenaude, J. H. Lodge, and J.-Y. Chouinard, "Noncoherent diversity reception over Nakagami-fading channels," *IEEE Trans. Commun.*, vol. COM-46, August 1998, pp. 985–991. See also *Proc. IEEE Veh. Technol. Conf. (VTC'97)*, Phoenix, AZ, May 1997, pp. 1484–1487.
62. S. Stein, "Unified analysis of certain coherent and noncoherent binary communication systems," *IEEE Trans. Inf. Theory*, vol. IT-10, January 1964, pp. 43–51.
63. G. M. Dillard, "Recursive computation of the generalized Q-function," *IEEE Trans. Aerosp. Electron. Syst.*, vol. AES-9, July 1973, pp. 614–615.
64. T. Eng and L. B. Milstein, "Comparison of hybrid FDMA/CDMA systems in frequency selective Rayleigh fading," *IEEE J. Sel. Areas Commun.*, vol. SAC-12, June 1994, pp. 938–951.
65. G. P. Efthymoglou, V. A. Aalo, and H. Helmken, "Performance analysis of noncoherent binary DS/CDMA systems in a Nakagami multipath channel with arbitrary parameters," *Proc. IEEE Global Commun. Conf. (GLOBECOM'96)*, London, November 1996, pp. 1296–1300. Full paper published in *IEE Proc. Commun.*, vol. 144, June 1997, pp. 166–172.
66. P. M. Hahn, "Theoretical diversity improvement in multiple frequency shift keying," *IRE Trans. Commun. Syst.*, vol. CS-10, June 1962, pp. 177–184.
67. P. J. Crepeau, "Uncoded and coded performance of MFSK and DPSK in Nakagami fading channels," *IEEE Trans. Commun.*, vol. COM-40, March 1992, pp. 487–493.
68. J. F. Weng and S. H. Leung, "Analysis of M-ary FSK square law combiner under Nakagami fading conditions," *Electron. Lett.*, vol. 33, September 1997, pp. 1671–1673.
69. Q. Bi, "Performance analysis of a CDMA system in the multipath fading environment," *Proc. IEEE Int. Symp. Personal Indoor Mobile Commun. (PIMRC'92)*, Boston, October 1992, pp. 108–111.
70. L. M. Jalloul and J. M. Holtzman, "Performance analysis of DS/CDMA with noncoherent M-ary orthogonal modulation in multipath fading channels," *IEEE J. Sel. Areas Commun.*, vol. SAC-12, June 1994, pp. 862–870.
71. P. I. Dallas and F. N. Pavlidou, "Cluster design of M-ary orthogonal DS/CDMA cellular system with Rayleigh fading and lognormal shadowing," *IEE Proc. Commun.*, vol. 144, August 1997, pp. 265–273.

72. V. Aalo, O. Ugweje, and R. Sudhakar, "Performance analysis of a DS/CDMA system with noncoherent M-ary orthogonal modulation in Nakagami fading," *IEEE Trans. Veh. Technol.*, vol. VT-47, February 1998, pp. 20–29.
73. D. Torrieri, "Performance of direct-sequence systems with long pseudonoise sequences," *IEEE J. Sel. Areas Commun.*, vol. SAC-10, May 1992, pp. 770–807.
74. Y.-C. Ko, M.-S. Alouini, and M. K. Simon, "Outage probability of diversity systems over generalized fading channels," to appear in *IEEE Trans. Commun.*
75. J. Abate and W. Whitt, "Numerical inversion of Laplace transforms of probability distributions," *ORSA J. Comput.*, vol. 7, no. 1, 1995, pp. 36–43.
76. J. N. Pierce and S. Stein, "Multiple diversity with nonindependent fading," *Proc. IRE*, vol. 48, January 1960, pp. 89–104.
77. V. A. Aalo, "Performance of maximal-ratio diversity systems in a correlated Nakagami-fading environment," *IEEE Trans. Commun.*, vol. COM-43, August 1995, pp. 2360–2369.
78. A. M. D. Turkmani, A. A. Arowojolu, P. A. Jefford, and C. J. Kellett, "An experimental evaluation of the performance of two-branch space and polarization diversity schemes at 1800 MHz," *IEEE Trans. Veh. Technol.*, vol. VT-44, May 1995, pp. 318–326.
79. S. Ruiz-Boque, M. Vilades, and J. Rodriguez, "Performance of two branch space and polarization diversity at 900 MHz," *Proc. IEEE Veh. Technol. Conf. (VTC'97)*, Phoenix, AZ, May 1997, pp. 1493–1497.
80. G. L. Turin, F. D. Clapp, T. L. Johnston, S. B. Fine, and D. Lavry, "A statistical model of urban multipath propagation," *IEEE Trans. Veh. Technol.*, vol. VT-21, February 1972, pp. 1–9.
81. A. S. Bajwa, "UHF wideband statistical model and simulation of mobile radio multipath propagation effects," *IEE Proc.*, vol. 132, August 1985, pp. 327–333.
82. F. Patenaude, J. Lodge, and J.-Y. Chouinard, "Temporal correlation analysis of frequency selective fading channels," *6th Annu. Int. Conf. Wireless Commun. (Wireless'94)*, Calgary, Alberta, Canada, July 1994, pp. 134–139.
83. F. Patenaude, J. Lodge, and J.-Y. Chouinard, "Eigen-analysis of wideband fading channel impulse responses," *IEEE Trans. Veh. Technol.*, vol. VT-48, March 1999, pp. 593–606.
84. E. K. Al-Hussaini and A. M. Al-Bassiouni, "Performance of MRC diversity systems for the detection of signals with Nakagami fading," *IEEE Trans. Commun.*, vol. COM-33, December 1985, pp. 1315–1319.
85. Q. T. Zhang, "Exact analysis of postdetection combining for DPSK and NFSK systems over arbitrary correlated Nakagami channels," *IEEE Trans. Commun.*, vol. COM-46, November 1998, pp. 1459–1467.
86. Q. T. Zhang, "Maximal-ratio combining over Nakagami fading channels with an arbitrary branch covariance matrix," *IEEE Trans. Veh. Technol.*, vol. VT-48, July 1999, pp. 1141–1150.
87. P. Lombardo, G. Fedele, and M. M. Rao, "MRC performance for binary signals in Nakagami fading with general branch correlation," *IEEE Trans. Commun.*, vol. COM-47, January 1999, pp. 44–52.
88. P. Lombardo and G. Fedele, "Post-detection diversity in Nakagami fading channels with correlated branches," *IEEE Commun. Lett.*, vol. 3, May 1999, pp. 132–135.

89. Y. Cho and J. H. Lee, "Effect of fading correlation on the SER performance of M-ary PSK with maximal ratio combining," *IEEE Commun. Lett.*, vol. 3, July 1999, pp. 199–201.

90. M. Z. Win, R. K. Mallik, G. Chrisikos, and J. H. Winters, "Canonical expressions for the error probability performance for M-ary modulation with hybrid selection/maximal-ratio combining in Rayleigh fading," *Proc. IEEE Wireless Commun. Network. Conf. (WCNC'99)*, New Orleans, LA, September 1999, pp. 266–270.

91. V. V. Veeravalli and A. Mantravadi, "Performance analysis for diversity reception of digital signals over correlated fading channels," *IEEE Veh. Technol. Conf. (VTC'99)*, Houston, TX, May 1999, pp. 1291–1295.

92. Y. Ma, C. C. Chai, and T. J. Lim, "Unified analysis of error probability for MRC in correlated fading channels," *Electron. Lett.*, vol. 35, August 1999, pp. 1314–1315.

93. M.-S. Alouini and M. K. Simon, "Multichannel reception of digital signals over correlated Nakagami fading channels," *Proc. 36th Allerton Conf. Commun. Control Comput. (Allerton'98)*, Allerton Park, IL, September 1998, pp. 146–155.

94. K. S. Packard, "Effect of correlation on combiner diversity," *Proc. IRE*, vol. 46, January 1958, pp. 362–363.

95. J. Gurland, "Distribution of the maximum of the arithmetic mean of correlated random variables," *Ann. Math. Stat.*, vol. 26, 1955, pp. 294–300.

96. J. A. Lawson and G. E. Uhlenbeck, *Threshold Signals*. New York: McGraw-Hill, 1952.

97. S. Kotz and J. Adams, "Distribution of sum of identically distributed exponentially correlated gamma variables," *Ann. Math. Stat.*, vol. 35, June 1964, pp. 277–283.

98. A. S. Krishnamoorthy and M. Parthasarathy, "A multivariate gamma-type distribution," *Ann. Math Stat.*, vol. 22, 1951, pp. 549–557.

99. S. J. Press, *Applied Multivariate Analysis*. New York: Holt, Rinehart and Winston, 1972.

100. M. Schwartz, W. R. Bennett, and S. Stein, *Communication Systems and Techniques*. New York: McGraw-Hill, 1966.

101. B. B. Barrow, "Diversity combination of fading signals with unequal mean strengths," *IEEE Trans. Commun. Syst.*, vol. CS-11, March 1963, pp. 73–78.

102. S. W. Halpern, "The effect of having unequal branch gains in practical predetection diversity systems for mobile radio," *IEEE Trans. Veh. Technol.*, vol. VT-26, February 1977, pp. 94–105.

103. M. A. Blanco, "Diversity receiver performance in Nakagami fading," *Proc. IEEE Southeast. Conf.*, Orlando, FL, April 1983, pp. 529–532.

104. F. Adachi, K. Ohno, and M. Ikura, "Postdetection selection diversity reception with correlated, unequal average power Rayleigh fading signals for $\pi/4$-shift QDPSK mobile radio," *IEEE Trans. Veh. Technol.*, vol. VT-41, May 1992, pp. 199–210.

105. S. Okui, "Probability of co-channel interference for selection diversity reception in the Nakagami m-fading channel," *IEE Proc. I*, vol. 139, February 1992, pp. 91–94.

106. W. C. Wan and J. C. Chen, "Fading distribution of diversity techniques with correlated channels," *Proc. IEEE Int. Symp. Personal Indoor Mobile Commun. (PIMRC'95)*, Toronto, Ontario, Canada, September 1995, pp. 1202–1206.

107. O. C. Ugweje and V. A. Aalo, "Performance of selection diversity system in correlated Nakagami fading," *Proc. IEEE Veh. Technol. Conf. (VTC'97)*, Phoenix, AZ, May 1997, pp. 1488–1492.

108. E. K. Al-Hussaini and A. M. Al-Bassiouni, "Performance of an ideal switched diversity receiver for NCFSK signals with Nakagami fading," *Trans. IECE Japan*, vol. E65, December 1982, pp. 750–751.

109. G. Fedele, I. Izzo, and M. Tanda, "Dual diversity reception of M-ary DPSK signals over Nakagami fading channels," *Proc. IEEE Int. Symp. Personal Indoor Mobile Commun. (PIMRC'95)*, Toronto, Ontario, Canada, September 1995, pp. 1195–1201.

110. M. K. Simon and M.-S. Alouini, "A unified performance analysis of digital communications with dual selective combining diversity over correlated Rayleigh and Nakagami-m fading channels," *IEEE Trans. Commun.*, vol. COM-47, January 1999, pp. 33–43. See also *Proc. Commun. Theory Mini-conf. (CTMC-VII)* in conjunction with IEEE Global Commun. Conf. (GLOBECOM'98), Sydney, Australia, November 1998, pp. 28–33.

111. Y.-C. Ko, M.-S. Alouini, and M. K. Simon, "Average SNR of dual selection combining over correlated Nakagami-m fading channels," *IEEE Commun. Lett.*, vol. 4, no. 1, January 2000, pp. 12–14.

112. A. Papoulis, *Probability, Random Variables, and Stochastic Processes*, 3rd ed. New York: McGraw-Hill, 1991.

113. C. C. Tan and N. C. Beaulieu, "Infinite series representation of the bivariate Rayleigh and Nakagami-m distributions," *IEEE Trans. Commun.*, vol. COM-45, October 1997, pp. 1159–1161.

114. H. Staras, "Diversity reception with correlated signals," *J. Appl. Phys.*, vol. 27, January 1956, pp. 93–94.

115. Y.-C. Ko, M.-S. Alouini, and M. K. Simon, "Performance analysis and optimization of switched diversity systems," *IEEE Veh. Technol. Conf. (VTC'99–Fall)*, Amsterdam, The Netherlands, September 1999, pp. 1366–1371. Full paper to appear in *IEEE Trans. Veh. Technol.*, vol. 49, September 2000.

116. A. H. Hausman, "An analysis of dual diversity receiving systems," *Proc. IRE*, vol. 42, June 1954, pp. 944–947.

117. E. Henze, "Theoretische Untersuchungen uber einige Diversity-verahren," *Arch. Elektrotech. Übertragung*, vol. 11, May 1957, pp. 183–194.

118. W. E. Shortall, "A switched diversity receiving system for mobile radio," *IEEE Trans. Commun.*, vol. COM-21, November 1973, pp. 1209–1275.

119. J. D. Parsons, M. Henze, P. A. Ratcliff, and M. J. Withers, "Diversity techniques for mobile radio reception," *IEEE Trans. Veh. Technol.*, vol. VT-25, August 1976, pp. 75–84.

120. G. Femenias and I. Furió, "Analysis of switched diversity TCM-MPSK systems on Nakagami fading channels," *IEEE Trans. Veh. Technol.*, vol. VT-46, February 1997, pp. 102–107.

121. Y. L. Luke, *Integrals of Bessel Functions*. New York: McGraw-Hill, 1962.

122. F. Vatalaro and G. E. Corazza, "Probability of error and outage in a Rice-lognormal channel for terrestrial and satellite personal communications," *IEEE Trans. Commun.*, vol. COM-44, August 1996, pp. 921–924.

123. J. C. Bic, D. Duponteil, and J. C. Imbeaux, *Elements of Digital Communication*. New York: Wiley, 1991.

124. A. H. Nuttall, "Some integrals involving the Q_M function," *IEEE Trans. Inf. Theory*, vol. IT-, January 1975, pp. 95–96.
125. A. H. Heatley, *Some Integrals, Differential Equations, and Series Related to the Modified Bessel Function of the First Kind*, University of Toronto Mathematical Series, no. 7, University of Toronto Press, Toronto, Ontario, Canada, 1939.
126. J. K. Cavers, "An analysis of pilot symbol assisted modulation for Rayleigh fading channels," *IEEE Trans. Veh. Technol.*, vol. VT-40, November 1991, pp. 686–693.
127. W. T. Webb and L. Hanzo, *Modern Quadrature Amplitude Modulation*. Piscataway, NJ; IEEE Press, 1994.
128. P. A. Bello and B. D. Nelin, "Predetection diversity combining with selectively fading channels," *IEEE Trans. Commun. Syst.*, vol. CS-10, 1962, pp. 32–42.
129. M. J. Gans, "The effect of Gaussian error in maximal ratio combiners," *IEEE Trans. Commun. Technol.*, vol. COM-19, August 1971, pp. 492–500.
130. M.-S. Alouini, S.-W. Kim, and A. Goldsmith, "RAKE reception with maximal-ratio and equal-gain combining for CDMA systems in Nakagami fading," *Proc. IEEE Int. Conf. Univ. Personal Commun. (ICUPC'97)*, San Diego, CA, October 1997, pp. 708–712.
131. B. R. Tomiuk, N. C. Beaulieu, and A. A. Abu-Dayya, "General forms for maximal ratio diversity with weighting errors," *IEEE Trans. Commun.*, vol. COM-47, April 1999, pp. 488–492. See also *Proc. IEEE Pacific Rim Conf. Commun. Comput. Signal Process. (PACRIM'95)*, Victoria, British Columbia, Canada, May 1995, pp. 363–368.
132. P. Y. Kam, "Bit error probabilities of MDPSK over the non-selective Rayleigh fading channel with diversity reception," *IEEE Trans. Commun.*, vol. COM-39, February 1991, pp. 220–224.
133. Y. C. Chow, J. P. McGeehan, and A. R. Nix, "Simplified error bound analysis for M-DPSK in fading channels with diversity reception," *IEE Proc. Commun.*, vol. 141, October 1994, pp. 341–350.
134. J. H. Barnard and C. K. Pauw, "Probability of error for selection diversity as a function of dwell time," *IEEE Trans. Commun.*, vol. COM-37, August 1989, pp. 800–803.
135. J. A. Ritcey and M. Azizoğlu, "Impact of switching constraints on selection diversity performance," *32rd Asilomar Conf. Signals Syst. Comput.*, Pacific Grove, CA, November 1998, pp. 795–799.
136. X. Tang, M.-S. Alouini, and A. J. Goldsmith, "Effect of channel estimate error on M-QAM performance in Rayleigh fading," *IEEE Trans. Commun.*, vol. COM-47, December 1999, pp. 1856–1864.
137. M. Kavehrad and P. J. McLane, "Performance of low-complexity channel coding and diversity for spread-spectrum in indoor, wireless communication," *AT & T Tech. J.*, vol. 64, October 1985, pp. 1927–1965.
138. E. Moriyama and H. Sasaoka, "Path diversity technology based on multiple path selection for a direct sequence spread spectrum communication system," *Proc. IEEE Veh. Technol. Conf. (VTC'95)*, Chicago, July 1995, pp. 424–428.
139. N. Kong, T. Eng, and L. B. Milstein, "A selection combining scheme for RAKE receivers," in *Proc. IEEE Int. Conf. Univ. Personal Commun. (ICUPC'95)*, Tokyo, November 1995, pp. 426–429.

140. K. J. Kim, S. Y. Kwon, E. K. Hong, and K. C. Whang, "Comments on 'Comparison of diversity combining techniques for Rayleigh-fading channels,'" *IEEE Trans. Commun.*, vol. COM-46, September 1998, pp. 1109–1110.

141. N. Kong and L. B. Milstein, "Average SNR of a generalized diversity selection combining scheme," *IEEE Commun. Lett.*, vol. 3, March 1999, p. 5759. See also *Proc. IEEE Int. Conf. Commun. (ICC'98)*, Atlanta, GA, June 1998, pp. 1556–1560.

142. N. Kong and L. B. Milstein, "A closed form expression for the average SNR when combining an arbitrary number of diversity branches with non-identical Rayleigh fading statistics," *Proc. IEEE Int. Conf. Commun. (ICC'99)*, Vancouver, British Columbia, Canada, June 1999, pp. 1864–1868.

143. M.-S. Alouini and M. K. Simon, "An MGF-based performance analysis of generalized selection combining over Rayleigh fading channels." *IEEE Trans. Commun.*, vol. COM-48, March 2000, pp. 401–415. See also *Proc. Commun. Theory Mini-conf.* in conjunction with IEEE Int. Conf. Commun. (ICC'99), Vancouver, British Columbia, Canada, June 1999, pp. 110–114.

144. M. Z. Win and J. H. Winters, "Analysis of hybrid selection/maximal-ratio combining of diversity branches with unequal SNR in Rayleigh fading," *IEEE Veh. Technol. Conf. (VTC'99)*, Houston, TX, May 1999, pp. 215–220.

145. M. Z. Win and J. H. Winters, "Analysis of hybrid selection/maximal-ratio combining in Rayleigh fading," *Proc. IEEE Int. Conf. Commun. (ICC'99)*, Vancouver, British Columbia, Canada, June 1999, pp. 6–10.

146. M. Z. Win and Z. Kostic, "Impact of spreading bandwidth on Rake reception in dense multipath channels," *Proc. Commun. Theory Mini-conf.* in conjunction with IEEE Int. Conf. Commun. (ICC'99), Vancouver, British Columbia, Canada, June 1999, pp. 78–82.

147. P. V. Sukhatme, "Tests of significance for samples of the χ^2 population with two degrees of freedom," *Ann. of Eugen.*, vol. 8, 1937, pp. 52–56.

148. H. A. David, *Order Statistics*. New York: Wiley, 1981.

149. B. C. Arnold, N. Balakrishnan, and H. N. Nagaraja, *A First Course in Order Statistics*. New York: Wiley, 1992.

150. J. A. Ritcey, "Performance analysis of the censored mean level detector," *IEEE Trans. Aerosp. Electron. Syst.*, vol. AES-22, July 1986, pp. 443–454.

151. J. A. Ritcey and J. L. Hines, "Performance of the MAX family of order statistic CFAR detectors," *IEEE Trans. Aerosp. Electron. Syst.*, vol. AES-27, January 1991, pp. 48–57.

152. A. Erdelyi, W. Magnus, F. Oberhettinger, and F. Tricomi, *Table of Integral Transforms*, vol. 2. New York: McGraw-Hill, 1954.

153. D. Zwillinger, *Standard Mathematical Tables and Formulae*, 30th ed. Boca Raton, FL: CRC Press, 1996.

154. C. Kim, Y. Kim, G. Jeong, and H. Lee, "BER analysis of QAM with MRC space diversity in Rayleigh fading channels," *Proc. IEEE Int. Symp. Personal Indoor Mobile Commun. (PIMRC'95)*, Toronto, Ontario, Canada, September 1995, pp. 482–485.

155. C.-J. Kim, Y.-S. Kim, G.-Y. Jeong, J.-K. Mun, and H.-J. Lee, "SER analysis of QAM with space diversity in Rayleigh fading channels," *ETRI J.*, vol. 17, January 1996, pp. 25–35.

156. M.-S. Alouini and M. K. Simon, "Performance of coherent receivers with hybrid SC/MRC over Nakagami-m fading channels," *IEEE Trans. Veh. Technol.*, vol. VT-48, July 1999, pp. 1155–1164.

157. D. G. Kabe, "Dirichlet's transformation and distributions of linear functions of ordered gamma variates," *Ann. Inst. Stat. Math.*, vol. 18, 1966, pp. 367–374.

158. G. A. Gibson, *Advanced Calculus*. London: Macmillan, 1931.

159. J. A. Edwards, *A Treatise on the Integral Calculus*, vol. II. London: Macmillan, 1922.

160. M.-S. Alouini and M. K. Simon, "Application of the Dirichlet transformation to the performance evaluation of generalized selection combining over Nakagami-m fading channels," *Int. J. Commun. Networks*, vol. 1, March 1999, pp. 5–13. See also *Proc. IEEE Veh. Technol. Conf. (VTC'99–Fall)*, Amsterdam, The Netherlands, September 1999, pp. 953–957.

161. E. Biglieri, G. Caire, G. Taricco, and J. Ventura-Traveset, "Simple method for evaluating error probabilities," *Electron. Lett.*, vol. 32, February 1996, pp. 191–192.

162. E. Biglieri, G. Caire, G. Taricco, and J. Ventura-Traveset, "Computing error probabilities over fading channels: a unified approach," *European Trans. Commun. Relat. Technol.*, vol. 9, January–February 1998, pp. 15–25.

APPENDIX 9A: ALTERNATIVE FORMS OF THE BIT ERROR PROBABILITY FOR A DECISION STATISTIC THAT IS A QUADRATIC FORM OF COMPLEX GAUSSIAN RANDOM VARIABLES

Appendix B of Ref. 7 considers a test against a zero threshold of a RV, D, which is a quadratic form of complex Gaussian RVs X_n, Y_n, that is,

$$D = \sum_{n=1}^{L} d_n \qquad (9A.1)$$

where

$$d_n = A|X_n|^2 + B|Y_n|^2 + CX_nY_n^* + C^*X_n^*Y_n \qquad (9A.2)$$

in which A, B, and C are constants. The pairs $\{X_n, Y_n\}$ are mutually independent, and thus the quadratic forms $\{d_n\}$ are likewise independent; however, for each n, X_n and Y_n can be correlated with each other. If D characterizes the decision variable at the output of the detector in a multichannel communication system transmitting binary signals over the AWGN channel, the probability $\Pr\{D < 0\}$ is representative of the BEP, $P_b(E)$, at the receiver. Using a characteristic function method, Proakis [7] finds an expression [see Eq. (B-21)] for this probability in terms of the first-order Marcum Q-function, modified Bessel functions of order $0, 1, \ldots, L-1$, and other elementary functions, namely,

$$P_b(E) = Q_1(a,b) - I_0(ab) \exp\left[-\tfrac{1}{2}(a^2+b^2)\right]$$
$$+ \frac{I_0(ab)\exp\left[-\tfrac{1}{2}(a^2+b^2)\right]}{(1+v_2/v_1)^{2L-1}} \sum_{k=0}^{L-1} \binom{2L-1}{k} \left(\frac{v_2}{v_1}\right)^k$$
$$+ \frac{\exp\left[-\tfrac{1}{2}(a^2+b^2)\right]}{(1+v_2/v_1)^{2L-1}} \sum_{n=1}^{L-1} I_n(ab) \sum_{k=0}^{L-1-n} \binom{2L-1}{k}$$
$$\times \left[\left(\frac{b}{a}\right)^n \left(\frac{v_2}{v_1}\right)^k - \left(\frac{a}{b}\right)^n \left(\frac{v_2}{v_1}\right)^{2L-1-k}\right] \tag{9A.3}$$

where

$$a = \left[\frac{2v_1^2 v_2(\xi_1 v_2 - \xi_2)}{(v_1+v_2)^2}\right]^{1/2}, \quad b = \left[\frac{2v_2^2 v_1(\xi_1 v_1 + \xi_2)}{(v_1+v_2)^2}\right]^{1/2} \tag{9A.4}$$

with

$$v_1 = \sqrt{w^2 + \frac{1}{4(\mu_{xx}\mu_{yy}-|\mu_{xy}|^2)(|C|^2-AB)}} - w$$

$$v_2 = \sqrt{w^2 + \frac{1}{4(\mu_{xx}\mu_{yy}-|\mu_{xy}|^2)(|C|^2-AB)}} + w$$

$$w = \frac{A\mu_{xx} + B\mu_{yy} + C\mu_{xy}^* + C^*\mu_{xy}}{4(\mu_{xx}\mu_{yy}-|\mu_{xy}|^2)(|C|^2-AB)}$$

$$\mu_{xx} = \tfrac{1}{2}E\{|X_n - \overline{X}_n|^2\}, \quad \mu_{yy} = \tfrac{1}{2}E\{|Y_n - \overline{Y}_n|^2\} \tag{9A.5}$$

$$\mu_{xy} = \tfrac{1}{2}E\{(X_n - \overline{X}_n)(Y_n - \overline{Y}_n)^*\}$$

$$\xi_1 = \sum_{n=1}^{L} \xi_{1n}, \quad \xi_2 = \sum_{n=1}^{L} \xi_{2n}$$

$$\xi_{1n} = 2(|C|^2 - AB)(|\overline{X}_n|^2 \mu_{yy} + |\overline{Y}_n|^2 \mu_{xx} - \overline{X}_n^* \overline{Y}_n \mu_{xy} - \overline{X}_n \overline{Y}_n^* \mu_{xy}^*)$$

$$\xi_{2n} = (A|\overline{X}_n|^2 + B|\overline{Y}_n|^2 + C\overline{X}_n^*\overline{Y}_n + C^*\overline{X}_n\overline{Y}_n^*)$$

Using the relation between the first- and mth-order Marcum Q-functions given by [see Eq. (B-17) and the associated footnote of Ref. 7]

$$Q_m(a,b) = Q_1(a,b) + \exp\left[-\tfrac{1}{2}(a^2+b^2)\right] \sum_{n=1}^{m-1} \left(\frac{b}{a}\right)^n I_n(ab) \tag{9A.6}$$

then after some simplification (9A.3) can be written as

$$P_b(E) = \frac{1}{(1+\eta)^{2L-1}} \left[\sum_{k=0}^{L-2} \binom{2L-1}{k} \eta^k Q_{L-k}(a,b) \right.$$

$$+ Q_1(a,b) \sum_{k=L-1}^{2L-1} \binom{2L-1}{k} \eta^k - \exp\left(-\frac{a^2+b^2}{2}\right) I_0(ab)$$

$$\left. \times \sum_{k=L}^{2L-1} \binom{2L-1}{k} \eta^k - \sum_{k=L+1}^{2L-1} \binom{2L-1}{k} \eta^k [Q_{k-L+1}(b,a) - Q_1(b,a)] \right]$$

(9A.7)

where $\eta \triangleq v_2/v_1$.

The advantage of the form in (9A.7) over that in (9A.3) is that when used to assess the performance in the presence of fading where additional statistical averaging over the arguments a and b is required, we shall be able to make use of the alternative forms of both the first-order and generalized (mth-order) Marcum Q-functions developed in Chapter 4. Specifically, we shall soon show that in principle, $P_b(E)$ can be expressed as a single integral with finite limits and an integrand which is a product of a trigonometric factor (complicated as it might be) and a single exponential factor of the form $\exp[-(b^2/2)(1+2\zeta \sin\theta + \zeta^2)]$, $0^+ \leq \zeta \triangleq a/b < 1$. Before proceeding with this development, however, we can first perform some further simplification of (9A.7). Recognizing that

$$\sum_{k=L-1}^{2L-1} \binom{2L-1}{k} \eta^k = \sum_{k=0}^{2L-1} \binom{2L-1}{k} \eta^k - \sum_{k=0}^{L-2} \binom{2L-1}{k} \eta^k$$

$$= (1+\eta)^{2L-1} - \sum_{k=0}^{L-2} \binom{2L-1}{k} \eta^k \qquad (9A.8)$$

then using this replacement in the second term of (9A.7) gives after some manipulation

$$P_b(E) = Q_1(a,b) - \left[1 - \frac{\sum_{k=0}^{L-1} \binom{2L-1}{k} \eta^k}{(1+\eta)^{2L-1}} \right] \exp\left(-\frac{a^2+b^2}{2}\right) I_0(ab)$$

$$+ \frac{1}{(1+\eta)^{2L-1}} \left[\sum_{k=0}^{L-2} \binom{2L-1}{k} \eta^k [Q_{L-k}(a,b) - Q_1(a,b)] \right.$$

$$\left. - \sum_{k=L+1}^{2L-1} \binom{2L-1}{k} \eta^k [Q_{k-L+1}(b,a) - Q_1(b,a)] \right] \qquad (9A.9)$$

For $L = 1$, the two latter two summations in (9A.9) do not contribute and the first sum reduces to the term corresponding to $k = 0$, whereupon one immediately obtains

$$P_b(E) = Q_1(a, b) - \frac{\eta}{1+\eta} \exp\left(-\frac{a^2+b^2}{2}\right) I_0(ab) \qquad (9\text{A}.10)$$

which agrees with the result in Eq. (B-21) of Ref. 7.

As a final simplification, in the second term of (9A.9) let $l = L - k$, and in the third term let $l = k - L + 1$. Then we can rewrite (9A.9) as

$$P_b(E) = Q_1(a, b) - \left[1 - \frac{\sum_{k=0}^{L-1} \binom{2L-1}{k} \eta^k}{(1+\eta)^{2L-1}}\right] \exp\left(-\frac{a^2+b^2}{2}\right) I_0(ab)$$

$$+ \frac{1}{(1+\eta)^{2L-1}} \left[\sum_{l=2}^{L} \binom{2L-1}{L-l} \eta^{L-l}[Q_l(a,b) - Q_1(a,b)]\right.$$

$$\left. - \sum_{l=2}^{L} \binom{2L-1}{L-l} \eta^{L-1+l}[Q_l(b,a) - Q_1(b,a)]\right] \qquad (9\text{A}.11)$$

The special case of $\eta = v_2/v_1 = 1$ (which occurs when $w = 0$) is of interest in many communication problems. For this case, (9A.11) simplifies to

$$P_b(E) = Q_1(a, b) - \frac{1}{2} \exp\left(-\frac{a^2+b^2}{2}\right) I_0(ab) + \frac{1}{2^{2L-1}} \sum_{k=0}^{L-1} \binom{2L-1}{k}$$

$$\times [(Q_{L-k}(a,b) - Q_{L-k}(b,a)) - (Q_1(a,b) - Q_1(b,a))] \qquad (9\text{A}.12)$$

where we have added back the $k = L - 1$ term in the sum since it contributes zero value anyway. Now recognizing that the first-order Marcum Q-function terms do not depend on the summation index k, we can simplify (9A.12) still further to

$$P_b(E) = Q_1(a, b) - \frac{1}{2} \exp\left(-\frac{a^2+b^2}{2}\right) I_0(ab) + \frac{1}{2^{2L-1}} \sum_{k=0}^{L-1} \binom{2L-1}{k}$$

$$\times [Q_{L-k}(a,b) - Q_{L-k}(b,a)] - \frac{1}{2}[Q_1(a,b) - Q_1(b,a)] \qquad (9\text{A}.13)$$

However, comparing Eqs. (40) and (42) of Ref. 62, we get

$$Q_1(a,b) - \frac{1}{2}\exp\left(-\frac{a^2+b^2}{2}\right)I_0(ab) = \frac{1}{2}[1 - Q_1(b,a) + Q_1(a,b)] \tag{9A.14}$$

Thus, combining (9A.13) and (9A.14) gives the final desired result:

$$P_b(E) = \frac{1}{2} + \frac{1}{2^{2L-1}}\sum_{k=0}^{L-1}\binom{2L-1}{k}[Q_{L-k}(a,b) - Q_{L-k}(b,a)]$$

$$= \frac{1}{2} + \frac{1}{2^{2L-1}}\sum_{l=1}^{L}\binom{2L-1}{L-l}[Q_l(a,b) - Q_l(b,a)] \tag{9A.15}$$

We are now prepared to justify an earlier statement: the ability to express the BEP in the form of a single integral with finite limits and an integrand composed of elementary functions. Using the alternative forms of the generalized Marcum Q-functions in (4.30) and (4.38), we immediately get

$$P_b(E) = \frac{1}{2\pi}\int_{-\pi}^{\pi}\frac{f(\theta;\zeta)}{1+2\zeta\sin\theta+\zeta^2}\exp\left[-\frac{b^2}{2}(1+2\zeta\sin\theta+\zeta^2)\right]d\theta,$$

$$0^+ \leq \zeta \triangleq a/b < 1 \tag{9A.16}$$

where

$$f(\theta;\zeta) = \frac{1}{2^{2L-1}}\sum_{l=1}^{L}\binom{2L-1}{L-l}\left[(\zeta^{-(l-1)} - \zeta^{l+1})\cos\left[(l-1)\left(\theta+\frac{\pi}{2}\right)\right]\right.$$

$$\left. + (\zeta^{-(l-2)} - \zeta^l)\cos\left[l\left(\theta+\frac{\pi}{2}\right)\right]\right] \tag{9A.17}$$

As a check, for $L = 1$, (9A.17) reduces to

$$f(\theta;\gamma) = \frac{1}{2}(1-\zeta^2) \tag{9A.18}$$

in which case (9A.16) becomes

$$P_b(E) = \frac{1}{4\pi}\int_{-\pi}^{\pi}\frac{1-\zeta^2}{1+2\zeta\sin\theta+\zeta^2}\exp\left[-\frac{b^2}{2}(1+2\zeta\sin\theta+\zeta^2)\right]d\theta,$$

$$0^+ \leq \zeta \triangleq a/b < 1 \tag{9A.19}$$

Comparing (9A.15) with (9A.19) agrees with a similar comparison of (8A.5) and (8A.12).

Finally, for the more general case where $\eta = v_2/v_1 \neq 1$, since the second term of (9A.11) is expressible in the desired single integral form [see (4.65)], it should be clear that the BEP of (9A.11) can be expressed in the form (9A.16), where the function $f(\theta; \zeta)$ is composed of the sum of the various trigonometric factors in the alternative representations of each term, each now weighted by terms that include a dependence on the ratio $\eta = v_2/v_1$. The final result can be written as

$$P_b(E) = \frac{\eta^L}{2\pi(1+\eta)^{2L-1}} \int_{-\pi}^{\pi} \frac{f(L;\zeta,\eta;\theta)}{1+2\zeta\sin\theta+\zeta^2} \exp\left[-\frac{b^2}{2}(1+2\zeta\sin\theta+\zeta^2)\right] d\theta,$$

$$0^+ \leq \zeta \stackrel{\Delta}{=} a/b < 1 \tag{9A.20}$$

where

$$f(L;\zeta,\eta;\theta) = f_0(L;\zeta,\eta;\theta) + f_1(L;\zeta,\eta;\theta) \tag{9A.21}$$

with[1]

$$f_0(L;\zeta,\eta;\theta) = \left[-\frac{(1+\eta)^{2L-1}}{\eta^L} + \sum_{l=1}^{L}\binom{2L-1}{L-l}(\eta^{-l}+\eta^{l-1})\right]\zeta(\zeta+\sin\phi)$$

$$f_1(L;\zeta,\eta;\theta) = \sum_{l=1}^{L}\binom{2L-1}{L-l}\left\{(\eta^{-l}\zeta^{-l+1} - \eta^{l-1}\zeta^{l+1})\cos\left[(l-1)\left(\theta+\frac{\pi}{2}\right)\right]\right.$$

$$\left. - (\eta^{-l}\zeta^{-l+2} - \eta^{l-1}\zeta^{l})\cos\left[l\left(\theta+\frac{\pi}{2}\right)\right]\right\} \tag{9A.22}$$

As a check, for the special case of $L=1$, we obtain

$$f_0(L;\zeta,\eta;\theta) = 0, \quad f_1(L;\zeta,\eta;\theta) = \frac{1-\eta\zeta^2+\zeta(1-\eta)\sin\theta}{\eta} \tag{9A.23}$$

and hence (9A.20) reduces to (9A.10), as expected.

Before concluding this appendix, we wish to point out that the upper and lower bounds found for the generalized Marcum Q-function in Chapter 4 are useful here in obtaining tight upper bounds on BEP. In applying these bounds on the Q-function itself, it is important to note that in the expressions for BEP developed in this appendix [e.g., (9A.11) and (9A.15)], the arguments of the second generalized Marcum Q-function are in reverse order to those in the first one. Thus, for example, for the case $a < b$ (the usual case of practical interest in communication problems), using the tight upper bound of (4.57a) for the first term and the tight lower bound of (4.64a) for the second would produce a simple tight upper bound on BEP. The specifics of these upper bounds on BEP are discussed in Section 9.4.1.3 paying most attention to the special case of $\eta = v_2/v_1 = 1$ and the further refinements of these bounds required by the necessity of averaging over the fading distributions.

---

[1] As the book was going to press, Mr. L.-F. Tsaur of Conexant Systems Inc, Newport Beach, CA pointed out to the authors that $f_0(L;\zeta,\eta;\phi)$ can be proven equal to zero for all values of L and η independent of ζ and ϕ.

APPENDIX 9B: SIMPLE NUMERICAL TECHNIQUES FOR THE INVERSION OF THE LAPLACE TRANSFORM OF CUMULATIVE DISTRIBUTION FUNCTIONS

9B.1 Euler Summation-Based Technique

Let X be a *positive* random variable (RV) with CDF $P_X(x)$ and let $\hat{P}_X(s)$ denote the Laplace transform of $P_X(x)$ so that

$$\hat{P}_X(s) = \int_0^\infty P_X(x) e^{-sx}\, dx$$

$$P_X(x) = \frac{1}{2\pi j} \int_{a-j\infty}^{a+j\infty} \hat{P}_X(s) e^{sx}\, ds \tag{9B.1}$$

$P_X(x)$ can be obtained from $\hat{P}_X(s)$ using the following three steps of a useful and simple numerical technique presented in Ref. 75.

Step 1. Using the change of variables $s = a + ju$ the inverse Laplace transform, as given in (9B.1), can be expressed as

$$P_X(x) = \frac{e^{ax}}{2\pi} \int_{-\infty}^\infty \operatorname{Re}\{\hat{P}_X(a+ju)\} \cos(ux) - \operatorname{Im}\{\hat{P}_X(a+ju)\} \sin(ux)\, du \tag{9B.2}$$

where $\operatorname{Re}\{\cdot\}$ and $\operatorname{Im}\{\cdot\}$ denote the real and imaginary parts, respectively. Since X is a positive RV, $P_X(-x) = 0$ for $x \geq 0$ and hence

$$\int_{-\infty}^\infty \operatorname{Re}\{\hat{P}_X(a+ju)\} \cos(ux)\, du = -\int_{-\infty}^\infty \operatorname{Im}\{\hat{P}_X(a+ju)\} \sin(ux)\, du \tag{9B.3}$$

Thus (9B.2) simplifies to

$$P_X(x) = \frac{e^{ax}}{\pi} \int_{-\infty}^\infty \operatorname{Re}\{\hat{P}_X(a+ju)\} \cos(ux)\, du \tag{9B.4}$$

In addition, using the fact that $P_X(x)$ is real, $\operatorname{Re}\{\hat{P}_X(a+ju)\}$ is even with respect to u. Thus, (9B.4) can be rewritten as

$$P_X(x) = \frac{2e^{ax}}{\pi} \int_0^\infty \operatorname{Re}\{\hat{P}_X(a+ju)\} \cos(ux)\, du \tag{9B.5}$$

Step 2. Letting $a = A/2x$, then evaluating the integral in (9B.5) by means of the trapezoidal rule with step size $h = \pi/(2x)$, we get

$$P_X(x) = \frac{e^{A/2}}{x} \sum_{n=0}^\infty \frac{(-1)^n}{\alpha_n} \operatorname{Re}\left\{\hat{P}_X\left(\frac{A + 2\pi jn}{2x}\right)\right\} + E(A) \tag{9B.6}$$

with

$$\alpha_n = \begin{cases} 2, & n = 0 \\ 1, & n = 1, 2, \ldots, N \end{cases}$$

and where $E(A)$ is a discretization error term that can be bounded, with the help of the Poisson summation formula, by

$$|E(A)| \leq \frac{e^{-A}}{1 - e^{-A}} \simeq e^{-A} \tag{9B.7}$$

Step 3. Truncating the infinite summation in (9B.6), we obtain

$$P_X(x) = \frac{e^{A/2}}{x} \sum_{n=0}^{N} \frac{(-1)^n}{\alpha_n} \operatorname{Re}\left\{\hat{P}_X\left(\frac{A + 2\pi jn}{2x}\right)\right\} + E(A) + E(N) \tag{9B.8}$$

where $E(N)$ is a truncation error. Since (9B.8) is in the form of an alternating series, its convergence can be accelerated by the Euler summation technique, which can be viewed as the binomial average of K partial series of length $N, N+1, \ldots, N+K$, respectively. This leads to the final desired result:

$$P_X(x) = \sum_{k=0}^{K} 2^{-K} \binom{K}{k} \left[\frac{e^{A/2}}{x} \sum_{n=0}^{N+k} \frac{(-1)^n}{\alpha_n} \operatorname{Re}\left\{\hat{P}_X\left(\frac{A + 2\pi jn}{2x}\right)\right\}\right]$$
$$+ E(A) + E(N, K) \tag{9B.9}$$

where the overall truncation error term $E(N, K)$ can be estimated by

$$E(N, K) \simeq \frac{e^{A/2}}{x} \sum_{k=0}^{K} 2^{-K} (-1)^{N+1+k} \binom{K}{k} \operatorname{Re}\left\{\hat{P}_X\left(\frac{A + 2\pi j(N + k + 1)}{2x}\right)\right\} \tag{9B.10}$$

9B.2 Gauss–Chebyshev Quadrature-Based Technique

Let Δ be a real RV with PDF $p_\Delta(x)$ and let $\hat{p}_\Delta(s)$ denote the Laplace transform of $p_\Delta(x)$, so that

$$\hat{p}_\Delta(s) = \int_{-\infty}^{\infty} p_\Delta(x) e^{-sx} \, dx \tag{9B.11}$$

The CDF of Δ evaluated at zero (i.e., $\Pr\{\Delta < 0\}$) can be computed from $\hat{p}_\Delta(s)$ using a Gauss–Chebyshev quadrature-based numerical technique given in Refs. 161 and 162, which we describe briefly next.

Using the Laplace inversion formula, we can write

$$\Pr\{\Delta < 0\} = \frac{1}{2\pi j} \int_{c-j\infty}^{c+j\infty} \hat{p}_\Delta(s) \frac{ds}{s} \quad (9B.12)$$

Since a probability is always a real quantity, (9B.12) can be rewritten as

$$\Pr\{\Delta < 0\} = \frac{1}{2\pi} \int_{-\infty}^{\infty} \frac{\hat{p}_\Delta(c + j\omega)}{c + j\omega} d\omega$$

$$= \frac{1}{2\pi} \int_{-\infty}^{\infty} \frac{c \cdot \text{Re}\{\hat{p}_\Delta(c + j\omega)\} + \omega \cdot \text{Im}\{\hat{p}_\Delta(c + j\omega)\}}{c^2 + \omega^2} d\omega \quad (9B.13)$$

Using the change of variable $\omega = c(\sqrt{1 - x^2}/x)$ in (9B.13) yields

$$\Pr\{\Delta < 0\} = \frac{1}{2\pi} \int_{-1}^{1} \left[\text{Re}\left\{ \hat{p}_\Delta\left(c + jc\frac{\sqrt{1-x^2}}{x}\right) \right\} \right.$$

$$\left. + \frac{\sqrt{1-x^2}}{x} \text{Im}\left\{ \hat{p}_\Delta\left(c + jc\frac{\sqrt{1-x^2}}{x}\right) \right\} \right] \frac{dx}{\sqrt{1-x^2}} \quad (9B.14)$$

which is of the form $\int_{-1}^{1} f(x)(dx/\sqrt{1-x^2})$ and thus can be computed efficiently using the Gauss–Chebyshev quadrature rule [52, Eq. (25.4.38)] with an even number n of nodes as[2]

$$\Pr\{\Delta < 0\} = \frac{1}{n} \sum_{k=1}^{n/2} [\text{Re}\{\hat{p}_\Delta(c + jc\tau_k)\} + \tau_k \text{Im}\{\hat{p}_\Delta(c + jc\tau_k)\}] + E_n \quad (9B.15)$$

where $\tau_k = \tan[(2k - 1)\pi/(2n)]$ and E_n is an error term that tends to zero as n goes to infinity. In practice, n can be chosen by computing (9B.15) for increasing values of n and stopping when the resulting numerical value does not change significantly. In addition, in (9B.15) c is a positive real number that should be chosen to guarantee quick convergence of (9B.15). Methods to choose proper values of c are discussed in Refs. 161 and 162. In particular, according to Refs. 161 and 162, the best value for c is the value for which $\hat{p}_\Delta(c)$ is minimum. Alternatively, if that minimum value cannot be found easily, c may be set equal to one-half the smallest real part of the poles of $\hat{p}_\Delta(s)$ without significant increase in n for a predetermined accuracy.

---

[2] Note that the form of the result in (9B.15) relies on the fact that $f(x)$ is an even function of x.

APPENDIX 9C: PROOF OF THEOREM 1

Starting with the transformation (9.3.17), note that

$$\frac{\partial x_l}{\partial \gamma_{l:L}} = 1$$
$$\frac{\partial x_l}{\partial \gamma_{l+1:L}} = -1 \qquad (9C.1)$$

Hence, the Jacobian of the transformation

$$|J(x_1, x_2, \ldots, x_L)| = \left\| \begin{bmatrix} 1 & -1 & 0 & \cdot & \cdot & 0 \\ 0 & 1 & -1 & 0 & \cdot & 0 \\ \cdot & \cdot & \cdot & \cdot & \cdot & \cdot \\ 0 & \cdot & \cdot & \cdot & 1 & -1 \\ 0 & \cdot & \cdot & 0 & 0 & 1 \end{bmatrix}_{L \times L} \right\| = 1 \qquad (9C.2)$$

Thus,

$$p_{x_1,x_2,\ldots,x_L}(x_1, x_2, \ldots, x_L) = \frac{p_{\gamma_{1:L},\gamma_{2:L},\ldots,\gamma_{L:L}}(\gamma_{1:L}, \gamma_{2:L}, \ldots, \gamma_{L:L})}{|J(x_1, x_2, \ldots, x_L)|}$$
$$= p_{\gamma_{1:L},\gamma_{2:L},\ldots,\gamma_{L:L}}(\gamma_{1:L}, \gamma_{2:L}, \ldots, \gamma_{L:L}) \qquad (9C.3)$$

Substituting (9.311) for $L_c = L$ in the right-hand side of (9C.3), we get

$$p_{x_1,x_2,\ldots,x_L}(x_1, x_2, \ldots, x_L) = \frac{L!}{\bar{\gamma}^L} \exp\left(-\frac{\sum_{l=1}^L \gamma_{l:L}}{\bar{\gamma}}\right) \qquad (9C.5)$$

Since the $\gamma_{l:L}$'s can be expressed in terms of the x_l values as

$$\gamma_{l:L} = \sum_{k=l}^L x_k$$

we can rewrite $p_{x_1,x_2,\ldots,x_L}(x_1, x_2, \ldots, x_L)$ of (9C.4) as

$$p_{x_1,\ldots,x_L}(x_1, \ldots, x_L) = \frac{L!}{\bar{\gamma}^L} \exp\left(-\frac{\sum_{l=1}^L \sum_{k=l}^L x_k}{\bar{\gamma}}\right)$$
$$= \frac{L(L-1)\cdots 1}{\bar{\gamma}\bar{\gamma}\cdots\bar{\gamma}} \exp\left(-\frac{x_1 + 2x_2 + \cdots + Lx_L}{\bar{\gamma}}\right)$$
$$= \prod_{l=1}^L \frac{l}{\bar{\gamma}} \exp\left(-\frac{lx_l}{\bar{\gamma}}\right) = \prod_{l=1}^L p_{x_l}(x_l) \qquad (9C.6)$$

which concludes the proof.

APPENDIX 9D: DIRECT PROOF OF EQ. (9.330)

Since the $\{\gamma_l\}_{l=1}^{L}$ are i.i.d., the PDF of the ordered RV $\gamma_{l:L}$, $p_{\gamma_{l:L}}(\gamma_{l:L})$, can be expressed in terms of the PDF, $p_\gamma(\gamma)$, and CDF, $P_\gamma(\gamma)$, of the original unordered RV γ as [112, Eq. (185); 148, Eq. (2.1.6)]

$$p_{\gamma_{l:L}}(\gamma) = \frac{L!}{(L-l)!(l-1)!}[P_\gamma(\gamma)]^{L-l}[1-P_\gamma(\gamma)]^{l-1}p_\gamma(\gamma), \qquad \gamma \geq 0 \tag{9D.1}$$

For Rayleigh fading, substituting (9.312) and (9.313) in (9D.1), we obtain the PDF of $\gamma_{l:L}$ as

$$p_{\gamma_{l:L}}(\gamma) = \frac{L!}{(L-l)!(l-1)!}(1-e^{-\gamma/\overline{\gamma}})^{L-l}\frac{e^{-l\gamma/\overline{\gamma}}}{\overline{\gamma}}, \qquad \gamma \geq 0 \tag{9D.2}$$

Hence, the average of $\gamma_{l:L}$ is given by

$$\overline{\gamma}_{l:L} = \frac{L!}{(L-l)!(l-1)!}\int_0^\infty \gamma(1-e^{-\gamma/\overline{\gamma}})^{L-l}\frac{e^{-l\gamma/\overline{\gamma}}}{\overline{\gamma}}\,d\gamma \tag{9D.3}$$

Making the change of variable $u = e^{-\gamma/\overline{\gamma}}$ in (9D.3), we get

$$\overline{\gamma}_{l:L} = -\frac{L!\overline{\gamma}}{(L-l)!(l-1)!}\int_0^1 \ln u(1-u)^{L-l}u^{l-1}\,du \tag{9D.4}$$

which can be evaluated in terms of the beta function, $B(\cdot,\cdot)$ [36, Sec. 8.38] and the psi function, $\psi(\cdot)$ [36, Sec. 8.36], with the help of Eq. (4.253.1) of Ref. 36 as

$$\overline{\gamma}_{l:L} = -\frac{L!\overline{\gamma}}{(L-l)!(l-1)!}B(l,L-l+1)[\psi(l)-\psi(L+1)] \tag{9D.5}$$

Using the beta function property [36, Eq. (8.384.1)]

$$B(l,L-l+1) = \frac{\Gamma(l)\Gamma(L-l+1)}{\Gamma(L+1)} = \frac{(l-1)!(L-l)!}{L!} \tag{9D.6}$$

as well as the psi function property [36, Eq. (8.365.3)]

$$\psi(L+1) - \psi(l) = \sum_{k=0}^{L-l}\frac{1}{l+k} \tag{9D.7}$$

in (9D.5) yields

$$\overline{\gamma}_{l:L} = \overline{\gamma}\sum_{k=0}^{L-l}\frac{1}{l+k} \tag{9D.8}$$

Since the combined SNR γ_{GSC} at the GSC output is $\gamma_{\text{GSC}} = \sum_{l=1}^{L_c} \gamma_{l:L}$, the average combined SNR $\overline{\gamma}_{\text{GSC}}$ is given by

$$\overline{\gamma}_{\text{GSC}} = \sum_{l=1}^{L_c} \overline{\gamma}_{l:L} \qquad (9\text{D}.9)$$

Using (9D.8) in (9D.9), we get

$$\begin{aligned}
\overline{\gamma}_{\text{GSC}} &= \overline{\gamma} \sum_{l=1}^{L_c} \sum_{k=0}^{L-l} \frac{1}{l+k} \\
&= \overline{\gamma} \left[\left(1 + \frac{1}{2} + \cdots + \frac{1}{L}\right) + \left(\frac{1}{2} + \frac{1}{3} + \cdots + \frac{1}{L}\right) \right. \\
&\qquad \left. + \cdots + \left(\frac{1}{L_c} + \frac{1}{L_c+1} + \cdots + \frac{1}{L}\right) \right] \\
&= \overline{\gamma} \left(1 + 2 \times \frac{1}{2} + 3 \times \frac{1}{3} + \cdots + L_c \times \frac{1}{L_c} + L_c \times \frac{1}{L_c+1} + \cdots + L_c \times \frac{1}{L}\right) \\
&= \overline{\gamma} \left(L_c + \frac{L_c}{L_c+1} + \frac{L_c}{L_c+2} + \cdots + \frac{L_c}{L}\right) \\
&= \left(1 + \sum_{l=L_c+1}^{L} \frac{1}{l}\right) L_c \overline{\gamma} \qquad (9\text{D}.10)
\end{aligned}$$

which is the final desired result in agreement with (9.330) or equivalently, Eq. (7) of Ref. 141. Note that the second moment of $\overline{\gamma^2}_{\text{GSC}} = \int_0^\infty \gamma^2 p_{\gamma_{l:L}}(\gamma)\,d\gamma$ [and hence the variance $\text{var}[\gamma_{\text{GSC}}] = \overline{\gamma^2}_{\text{GSC}} - (\overline{\gamma}_{\text{GSC}})^2$ as given in (9.331)] can also be obtained directly by using Eq. (4.261.21) of Ref. 36.

APPENDIX 9E: SPECIAL DEFINITE INTEGRALS

In this appendix we present some interesting closed-form expressions for certain single and multiple definite integrals which heretofore appear not to have been reported in standard tabulations such as Refs. 36 and 52. Consider the special case $L_c = L$ in (9.353), corresponding to combining all branches. For this case the function $D(s; \theta_1, \theta_2, \ldots, \theta_{L-1})$ of (9.354) reduces to

$$\begin{aligned}
D(s; \theta_1, \theta_2, \ldots, \theta_{L-1}) &= \left(1 - \frac{\overline{\gamma}}{m} s\right) (1 + \theta_{L-1} + \theta_{L-1}\theta_{L-2} + \theta_{L-1}\theta_{L-2}\theta_{L-3} \\
&\qquad + \cdots + \theta_{L-1}\theta_{L-2}\theta_{L-3} \cdots \theta_2\theta_1)
\end{aligned} \qquad (9\text{E}.1)$$

APPENDIX 9E: SPECIAL DEFINITE INTEGRALS

which results in the MGF of (9.353) becoming

$$M_{\gamma_{\text{GSC}}}(s) = L![\Gamma(m)]^{-L}\Gamma(Lm)\left(1-\frac{\bar{\gamma}}{m}s\right)^{-Lm}\underbrace{\int_0^1\cdots\int_0^1}_{(L-1)\text{-fold}} \theta_{L-1}^{(L-1)m-1}\cdots\theta_2^{2m-1}\theta_1^{m-1}$$

$$\times (1+\theta_{L-1}+\theta_{L-1}\theta_{L-2}+\theta_{L-1}\theta_{L-2}\theta_{L-3}$$
$$+\cdots+\theta_{L-1}\theta_{L-2}\theta_{L-3}\cdots\theta_2\theta_1)^{-Lm}\,d\theta_{L-1}\cdots d\theta_1 \quad (9\text{E}.2)$$

However, if all branches are combined, then whether or not the branches are ordered according to their SNR should have no bearing on the overall result; that is, the MGF of (9E.2) should be identical to that of true MRC of L i.i.d. Nakagami-m branches, for which the MGF is given by [21]

$$M_{\gamma_{\text{MRC}}}(s) = \left(1-\frac{\bar{\gamma}}{m}s\right)^{-Lm} \quad (9\text{E}.3)$$

Thus, equating (9E.2) and (9E.3) gives the following closed-form result for the $(L-1)$-fold integral:

$$\underbrace{\int_0^1\cdots\int_0^1}_{(L-1)\text{-fold}}\left(\prod_{l=1}^{L-1}\theta_l^{lm-1}\right)(1+\theta_{L-1}+\theta_{L-1}\theta_{L-2}+\cdots+\theta_{L-1}\theta_{L-2}\cdots\theta_1)^{-Lm}$$

$$\times\,d\theta_{L-1}\cdots d\theta_1 = \frac{[\Gamma(m)]^L}{L!\Gamma(Lm)} \quad (9\text{E}.4)$$

Equation (9E.4) is easily verified for $L=1$ since both the left- and right-hand sides evaluate to unity. For $L=2$ we obtain the single definite integral

$$\int_0^1 \theta_1^{m-1}(1+\theta_1)^{-2m}\,d\theta_1 = \frac{(\Gamma(m))^2}{2\Gamma(2m)} \quad (9\text{E}.5)$$

Neither the multiple definite integral of (9E.4) nor the single definite integral of (9E.5) appear in standard tabulations such as [36,52].[3] Both are valid for m integer or noninteger and have been numerically checked for their validity.

---

[3] A closed-form solution equivalent to (9E.5) is possible for m integer using the recursive form of Eq. (2.111.2) of Ref. 36 and at the end applying Eq. (2.117.2) of Ref. 36. However, to get the solution in the simple form equivalent to (9E.5) requires a good deal of effort and manipulation and again is restricted to m integer.

PART 4

APPLICATION IN PRACTICAL COMMUNICATION SYSTEMS

10

OPTIMUM COMBINING: A DIVERSITY TECHNIQUE FOR COMMUNICATION OVER FADING CHANNELS IN THE PRESENCE OF INTERFERENCE

Thus far in the book we have discussed the performance of digital communication systems perturbed only by a combination of AWGN and multipath fading. Under such noise-limited conditions, coherent diversity reception takes the form of a maximum ratio combiner (MRC), which as discussed in Chapter 9 is optimum from the standpoint of maximizing the signal-to-noise ratio (SNR) at the combiner output. In applications such as digital mobile radio, space diversity provided by an adaptive antenna array is an attractive means for providing such diversity [1]. In addition to combating multipath fading, space diversity can also be used in cellular radio systems to reduce the relative power of co-channel interferers (CCI's) that are present at each element of the array. When operating in this scenario, the appropriate diversity scheme to employ is one that combines the branch outputs in such a way as to maximize the signal-to-interference plus noise (SINR) ratio at the combiner output. Under such conditions, this scheme, which is referred to as *optimum combining (OC)*, will achieve a larger output SINR than MRC and is thus highly desirable even when the number of interferers exceeds the number of antenna array elements. This improved SINR efficiency can manifest itself in the cellular mobile radio application as a reduction in the number of base stations and/or an increased channel capacity through greater frequency reuse.

The maximization of output SINR using adaptive antenna arrays techniques has been studied extensively in the early literature [1–4], primarily in a pure AWGN environment (i.e., in the absence of fading). The application of these principles to the slow-fading channel that is typical of digital mobile radio applications was first studied by Bogachev and Kiselev [5], who evaluated the bit error probability (BEP) performance of an optimum combiner for coherent binary orthogonal signals in the presence of a flat Rayleigh fading assumed to be independent between antenna elements and a single interferer with the same fading characteristics. Later, Winters [6] amplified on these analytical results and also provided computer (Monte Carlo) simulation results for the multiple interferer case. In both of these papers, comparisons were made with comparable systems employing MRC, and significant performance improvement was demonstrated for the OC case. Subsequent to these early papers, many studies of OC followed which considered in detail such issues as (1) the number of interferers relative to the number of antenna elements, (2) channel correlation due to nonideal space separation, (3) practical signal and interference models that allow for analytical results corresponding to the multiple interferer case, and (4) simple upper bounds (as opposed to complicated exact expressions) on average BEP performance.

In this chapter we bring together the foregoing analytical results under a single roof by applying the unifying framework of the moment generating function (MGF) approach discussed in earlier chapters. We shall again see that such an approach not only simplifies the analytical expressions for average BEP associated with the various cases previously treated in the literature but allows extension to a large variety of modulation schemes as well as slow-fading channels other than those that are Rayleigh distributed. To illustrate the approach, we shall first treat the simple case of a single co-channel interferer impinging on an L-element antenna array. As we shall see there are, in principle, two approaches (one exact and one approximate) that can be taken to evaluate the average BEP for this scenario. What is important to observe is that a comparison of the two for the Rayleigh fading channel, assumed to be independent and identically distributed (i.i.d.) across the array, reveals that the simple (but approximate) approach yields performance results that are extremely close to those provided by the more complicated (but exact) approach [7]. From this observation, which can also be readily justified by intuitive reasoning, we shall then draw the conclusion that for the remainder of the cases (i.e., other modulations and fading channel models, more than one interferer), it is sufficient to evaluate average BEP performance using the simpler approach. Indeed this is the assumption made in many of the above-cited references without, however, the mathematical and numerical justification offered here and presented originally in Ref. 7.

10.1 PERFORMANCE OF OPTIMUM COMBINING RECEIVERS

10.1.1 Single Interferer, Independent Identically Distributed Fading

Consider a communication receiver [typical of the base station of the reverse link (mobile to base) of a digital mobile radio system] that provides space diversity via an L-element antenna array. Assume that the antenna elements of the array are placed sufficiently far apart so as to provide independent fading paths. Furthermore, assume that the fading is sufficiently slow as to allow coherent detection to be employed. Then the received signal vector $\mathbf{r}(t)$ at the outputs of the array elements may be expressed as[1]

$$\mathbf{r}(t) = \sqrt{P_d}\mathbf{c}_d s_d(t) + \sqrt{P_I}\mathbf{c}_I s_I(t) + \mathbf{n}(t) \quad (10.1)$$

where $s_d(t)$ and $s_I(t)$ are the desired and interfering signals normalized such that P_d and P_I represent their respective powers, $\mathbf{c}_d$, and $\mathbf{c}_I$ are the corresponding channel propagation vectors with components $c_{dl}|_{l=1}^{L}$, $c_{Il}|_{l=1}^{L}$, respectively, and $\mathbf{n}(t)$ is the AWGN vector each element of which has zero mean and variance σ^2. Each vector is of dimension L. In the absence of fading, the elements of $\mathbf{c}_d$ and $\mathbf{c}_I$ are constant complex quantities each with unit magnitude and a phase determined by the relative distance of its associated antenna element from the reference antenna element (often taken as the center element of the array). In the presence of fading, the elements of $\mathbf{c}_d$ and $\mathbf{c}_I$ become complex random variables (RVs) with statistics dependent on the fading channel model assumed. For example, for Rayleigh fading, the elements of $\mathbf{c}_d$ and $\mathbf{c}_I$ would be i.i.d. complex Gaussian RVs with zero means and unit mean-square value.[2] Finally, the desired signal, interference signal, and additive noise are assumed mutually independent, as would be the case in a practical system.

As in a conventional RAKE receiver, the components of $\mathbf{r}(t)$ are appropriately (complex) weighted and combined (summed) to form a decision statistic. The difference between the RAKE receiver for MRC and that for OC lies in selection of the weight vector $\mathbf{w}$. Specifically, for MRC the weights are selected for maximum instantaneous SNR at the combiner output, and thus $\mathbf{w} = \mathbf{c}_d/\sigma^2$. For OC the weights are selected for maximum instantaneous SINR at the same location, and thus $\mathbf{w} = R_{ni}^{-1}\mathbf{c}_d$, where R_{ni} is the noise plus interference covariance matrix defined below. As such, implementation of the RAKE receiver for OC requires complete knowledge of the channel corresponding to *both* the desired signal and the interferer. For this receiver, the maximum instantaneous SINR at the combiner output is given by [6]

$$\gamma_t = P_d \mathbf{c}_d^H R_{ni}^{-1} \mathbf{c}_d \quad (10.2)$$

where the superscript H stands for the Hermitian (transpose complex conjugate) operation and the noise plus interference covariance matrix is given by

$$R_{ni} = E\{[\sqrt{P_I}\mathbf{c}_I s_I(t) + \mathbf{n}(t)][\sqrt{P_I}\mathbf{c}_I s_I(t) + \mathbf{n}(t)]^H\}$$
$$= P_I \mathbf{c}_I \mathbf{c}_I^H + \sigma^2 I \quad (10.3)$$

where I is the $L \times L$ identity matrix. In the absence of interference, (10.3) becomes a purely diagonal matrix and thus (10.2) simplifies, as it should, to

---

[1] We assume for simplicity a baseband model corresponding to ideal coherent demodulation.
[2] The assumption of equal unit mean-square value for both the desired signal and interference propagation vector fading components results in no loss of generality and is made for consistency with the no-fading case.

the instantaneous SNR of the combiner output for MRC, namely,

$$\gamma_t = \frac{P_d}{\sigma^2}\mathbf{c}_d^H\mathbf{c}_d = \frac{P_d}{\sigma^2}\sum_{l=1}^{L}\alpha_{dl}^2 = \sum_{l=1}^{L}\gamma_l \quad (10.4)$$

where α_{dl} is the fading amplitude of the lth element of $\mathbf{c}_d$ and $\gamma_l \triangleq \alpha_{dl}^2 P_d/\sigma^2$ is its corresponding instantaneous SNR.

Since as we have seen in Chapter 9, the form of (10.4) is highly desirable from the standpoint of evaluating average BEP since for the i.i.d fading assumption it allows the MGF of γ_t to be expressed in product form, we shall find it expedient for the OC case to diagonalize the covariance matrix of (10.3) with the hope of applying the same MGF-based approach. The diagonalization of R_{ni} is accomplished by applying a unitary matrix transformation U to it (i.e., $\Lambda \triangleq U^H R_{ni} U$ is a diagonal matrix with elements $\lambda_1, \lambda_2, \ldots, \lambda_L$ corresponding to the eigenvalues³ of R_{ni}). The rows of U are the corresponding complex eigenvectors, which from the properties of a unitary matrix are orthonormal. From the definition of Λ, the inverse of R_{ni} is easily found to be $R_{ni}^{-1} \triangleq U\Lambda^{-1}U^H$. Thus, substituting this result in (10.2) gives

$$\gamma_t = P_d \mathbf{c}_d^H U\Lambda^{-1}U^H \mathbf{c}_d \triangleq P_d \mathbf{s}^H \Lambda^{-1} \mathbf{s} = P_d \sum_{l=1}^{L}\frac{|s_l|^2}{\lambda_l} \quad (10.5)$$

where $\mathbf{s} = U^H \mathbf{c}_d$ is the transformed desired signal propagation vector with components s_l, $l = 1, 2, \ldots, L$. It is clear from (10.5) that *conditioned on the set of eigenvalues* λ_l, $l = 1, 2, \ldots, L$, the MGF γ_t will be expressible in product form if the transformed instantaneous signal powers $|s_l|^2$, $l = 1, 2, \ldots, L$, are mutually independent. To see how this condition can be satisfied, we proceed as follows. If the elements of $\mathbf{c}_d$ are modeled as complex Gaussian RVs (as for Rayleigh or Rician fading) or sums of complex Gaussian RVs (as for Nakagami-m fading), a linear operation (e.g., multiplication by U^H) on $\mathbf{c}_d$ results in a vector (i.e., $\mathbf{s}$) whose components are again complex Gaussian RVs. Furthermore, from the orthonormal property of the rows of U, the mean-square value of the components of $\mathbf{s}$ are all equal to unity (as is the case for $\mathbf{c}_d$). In addition, these components are mutually uncorrelated, and since they are Gaussian, they are mutually independent. Thus, we conclude that the transformed desired signal propagation vector $\mathbf{s}$ has statistics identical to those of its untransformed version $\mathbf{c}_d$, and therefore for analytical purposes we can replace $|s_l|^2$ by $|c_{dl}|^2 = \alpha_{dl}^2$ in (10.5). Doing so allows us to rewrite (10.5) as

$$\gamma_t = \frac{P_d}{\sigma^2}\sum_{l=1}^{L}\frac{\sigma^2}{\lambda_l}\alpha_{dl}^2 = \sum_{l=1}^{L}\frac{\sigma^2}{\lambda_l}\gamma_l \quad (10.6)$$

---

³ In general, these eigenvalues are RVs, although as we shall see shortly, for the single-interferer case, only one of them, say λ_1, is random.

and hence the conditional (on the eigenvalues) MGF of γ_t, $M_{\gamma_t|\lambda_1,\lambda_2,\ldots,\lambda_L}(s)$, is given by the product

$$M_{\gamma_t|\lambda_1,\lambda_2,\ldots,\lambda_L}(s) = \prod_{l=1}^{L} M_\gamma\left(s; \frac{\sigma^2}{\lambda_l}\overline{\gamma}_d\right) \tag{10.7}$$

where $M_\gamma(s; \overline{\gamma}_d)$ is the MGF of any of the γ_l's with mean value $\overline{\gamma}_d = P_d/\sigma^2$, which also represents the average SNR of the desired signal per antenna.

Before proceeding with the evaluation of average BEP, we first specify the eigenvalues for the single-interferer case. It has been shown in several places in the literature [5–8] that the eigenvalues of the covariance matrix of (10.3) are given by

$$\lambda_l = \begin{cases} P_I \sum_{n=1}^{L} |c_{In}|^2 + \sigma^2 = P_I \sum_{n=1}^{L} \alpha_{In}^2 + \sigma^2, & l = 1 \\ \sigma^2, & l = 2, 3, \ldots, L \end{cases} \tag{10.8}$$

That is, $L-1$ of them are constant and one of them, λ_1, is a RV. Making this substitution in (10.7) gives

$$M_{\gamma_t|\lambda_1}(s) = [M_\gamma(s; \overline{\gamma}_d)]^{L-1} M_\gamma\left(s; \frac{P_d}{\lambda_1}\right) \tag{10.9}$$

To evaluate the average BEP of coherent BPSK exactly using an OC receiver, we must average the conditional (on the fading) BEP over the fading distribution of the combiner output statistic. In particular,

$$\begin{aligned} P_b(E) &= \int_0^\infty Q(\sqrt{2\gamma_t}) p_{\gamma_t}(\gamma_t) d\gamma_t \\ &= \int_{\sigma^2}^\infty \int_0^\infty Q(\sqrt{2\gamma_t}) p_{\gamma_t}(\gamma_t|\lambda_1) d\gamma_t \, p_{\lambda_1}(\lambda_1) d\lambda_1 \end{aligned} \tag{10.10}$$

where $Q(x)$ is, as before, the Gaussian Q-function and $p_{\gamma_t}(\gamma_t|\lambda_1)$ is the probability density function (PDF) of the combiner output SINR conditioned on the single random eigenvalue λ_1 [with PDF $p_{\lambda_1}(\lambda_1)$] and is ordinarily found by first evaluating the conditional MGF, $M_{\gamma_t|\lambda_1}(s)$, and then taking its inverse Laplace transform. Although direct evaluation of (10.9) may be possible, it typically involves complicated analysis, which includes first determining the PDF $p_{\gamma_t}(\gamma_t|\lambda_1)$ in closed form and then successively performing the remaining integrations over the Gaussian Q-function and the eigenvalue probability distribution. Quite often, the closed-form expressions obtained at any stage in the process are given in terms of functions not readily available in standard software packages such as Mathematica and are in a form that provides little insight into their dependence on such system parameters as P_d, P_I, and σ^2.

10.1.1.1 Rayleigh Fading: Exact Evaluation of Average Bit Error Probability.
To illustrate the foregoing point, Shah et al. [7] evaluate (10.10) for the Rayleigh channel. First, the conditional MGF of γ_t is found from (10.9) to be

$$M_{\gamma_t|\lambda_1}(s) = \frac{1}{(1 - s\bar{\gamma}_d)^{L-1}[1 - s(P_d/\lambda_1)]} \quad (10.11)$$

which has the inverse Laplace transform [9, p. 410]

$$p_{\gamma_t}(\gamma_t|\lambda_1) = \frac{\lambda_1 \gamma_t^{L-1} \exp(-\lambda_1 \gamma_t/P_d) {}_1F_1(L-1; L; (\lambda_1 - \sigma^2)\gamma_t/P_d)}{\sigma^2 \Gamma(L)(\bar{\gamma}_d)^L},$$

$$\gamma_t \geq 0, \quad \lambda_1 \geq \sigma^2, \quad L \geq 1 \quad (10.12)$$

where ${}_1F_1(\bullet; \bullet; \bullet)$ is Kummer's confluent hypergeometric function [10, Eq. (9.210)]. Performing the first integration (on γ_t) in (10.10) gives

$$P_b(E|\lambda_1)$$

$$= \int_0^\infty Q(\sqrt{2\gamma_t}) p_{\gamma_t}(\gamma_t|\lambda_1) d\gamma_t$$

$$= \frac{1}{2} - \frac{\lambda_1 \Gamma\left(L + \frac{1}{2}\right) P_d^{L+1/2} F_2\left(L + \frac{1}{2}, 1, L-1; \frac{3}{2}, L; \frac{P_d}{\lambda_1 + P_d}, \frac{\lambda_1 - \sigma^2}{\lambda_1 + P_d}\right)}{\sqrt{\pi} \sigma^2 \Gamma(L)(\bar{\gamma}_d)^L (\lambda_1 + P_d)^{L+1/2}} \quad (10.13)$$

where $F_2(\bullet, \bullet, \bullet; \bullet, \bullet; \bullet, \bullet)$ is Appell's hypergeometric function of two variables [10, Eq. (9.180.2)]. Since each α_{In}^2 in (10.8) has a chi-square distribution, λ_1 has the PDF

$$p_{\lambda_1}(\lambda_1) = \frac{1}{\Gamma(L) P_I^L} (\lambda_1 - \sigma^2)^{L-1} \exp\left(-\frac{\lambda_1 - \sigma^2}{P_I}\right), \quad \lambda_1 \geq \sigma^2 \quad (10.14)$$

Finally, the average BEP is obtained by averaging (10.13) over the PDF in (10.14) in accordance with (10.10), namely,

$$P_b(E) = \frac{1}{2} - \frac{\Gamma\left(L + \frac{1}{2}\right) P_d^{L+1/2}}{\sqrt{\pi} \sigma^2 \Gamma(L)(\bar{\gamma}_d)^L} \frac{1}{\Gamma(L) P_I^L}$$

$$\times \int_{\sigma^2}^\infty \frac{\lambda_1}{(\lambda_1 + P_d)^{L+1/2}} (\lambda_1 - \sigma^2)^{L-1} \exp\left(-\frac{\lambda_1 - \sigma^2}{P_I}\right)$$

$$\times F_2\left(L + \frac{1}{2}, 1, L-1; \frac{3}{2}, L; \frac{P_d}{\lambda_1 + P_d}, \frac{\lambda_1 - \sigma^2}{\lambda_1 + P_d}\right) d\lambda_1 \quad (10.15)$$

which cannot be obtained in closed form except for the special case of $L = 1$, which has the result

$$P_b(E) = \frac{1}{2} - \sqrt{\pi \frac{P_d}{P_I}} \exp\left(\frac{P_d + \sigma^2}{P_I}\right) Q\left(\sqrt{2\frac{P_d + \sigma^2}{P_I}}\right)$$

$$= \frac{1}{2} - \sqrt{\pi \frac{\overline{\gamma}_d}{\overline{\gamma}_I}} \exp\left(\frac{1 + \overline{\gamma}_d}{\overline{\gamma}_I}\right) Q\left(\sqrt{2\frac{1 + \overline{\gamma}_d}{\overline{\gamma}_I}}\right) \quad (10.16)$$

where $\overline{\gamma}_I \stackrel{\Delta}{=} P_I/\sigma^2$ is the average interference-to-noise ratio per antenna.

We recall from Chapter 9 that using the alternative form of the Gaussian Q-function allows expressing the BEP directly in terms of the MGF of γ_t. Specifically, conditioned on λ_1, we have

$$P_b(E|\lambda_1) = \frac{1}{\pi} \int_0^{\pi/2} M_{\gamma_t|\lambda_1}\left(-\frac{1}{\sin^2\theta}\right) d\theta \quad (10.17)$$

Then, the average BEP is given by

$$P_b(E) = \frac{1}{\pi} \int_{\sigma^2}^{\infty} \int_0^{\pi/2} M_{\gamma_t|\lambda_1}\left(-\frac{1}{\sin^2\theta}\right) d\theta \, p_{\lambda_1}(\lambda_1) \, d\lambda_1$$

$$= \frac{1}{\pi} \int_{\sigma^2}^{\infty} \int_0^{\pi/2} \left[M_\gamma\left(-\frac{1}{\sin^2\theta}; \overline{\gamma}_d\right)\right]^{L-1}$$

$$\times M_\gamma\left(-\frac{1}{\sin^2\theta}; \frac{\sigma^2}{\lambda_1}\overline{\gamma}_d\right) d\theta \, p_{\lambda_1}(\lambda_1) \, d\lambda_1 \quad (10.18)$$

which for the Rayleigh channel becomes

$$P_b(E) = \frac{1}{\pi} \int_{\sigma^2}^{\infty} \int_0^{\pi/2} \left(\frac{\sin^2\theta}{\sin^2\theta + \overline{\gamma}_d}\right)^{L-1} \left[\frac{\sin^2\theta}{\sin^2\theta + (\sigma^2/\lambda_1)\overline{\gamma}_d}\right] d\theta \, p_{\lambda_1}(\lambda_1) \, d\lambda_1 \quad (10.19)$$

Performing the integral on λ_1 first, we obtain after much manipulation

$$P_b(E) = \frac{1}{\pi}\left(\frac{1}{\overline{\gamma}_I}\right)^L \int_0^{\pi/2} \exp\left[\left(1 + \frac{\overline{\gamma}_d}{\sin^2\theta}\right)\frac{1}{\overline{\gamma}_I}\right] \left[\Gamma\left(1 - L, \left(1 + \frac{\overline{\gamma}_d}{\sin^2\theta}\right)\frac{1}{\overline{\gamma}_I}\right)\right.$$

$$\left. + L\left(1 + \frac{\overline{\gamma}_d}{\sin^2\theta}\right) \Gamma\left(-L, \left(1 + \frac{\overline{\gamma}_d}{\sin^2\theta}\right)\frac{1}{\overline{\gamma}_I}\right)\right] d\theta \quad (10.20)$$

where $\Gamma(a, x)$ is the complementary incomplete gamma function [10, Eq. (8.353.3)]. If one wants to simplify the notation a bit (which will be convenient when extending the results to other modulations), define

$$f(\theta) \triangleq \left(1 + \frac{\overline{\gamma}_d}{\sin^2\theta}\right)\frac{1}{\overline{\gamma}_I} \tag{10.21}$$

in which case (10.20) simplifies to[4]

$$P_b(E) = \frac{1}{\pi}\left(\frac{1}{\overline{\gamma}_I}\right)^L \int_0^{\pi/2} \exp[f(\theta)][\Gamma(1-L, f(\theta)) + L\overline{\gamma}_I f(\theta)\Gamma(-L, f(\theta))]\, d\theta \tag{10.22}$$

Clearly, (10.19) or (10.22) together with (10.22) is considerably simpler in form than (10.15) and as we shall see momentarily, allows us to draw conclusions immediately for certain special cases.

For the special case of no interferer [i.e., $P_I = 0$ ($\lambda_1 = \sigma^2$)], we immediately see from (10.19) that

$$P_b(E) = \int_0^{\pi/2} \left(\frac{\sin^2\theta}{\sin^2\theta + \overline{\gamma}_d}\right)^L d\theta \tag{10.23}$$

which corresponds to the performance of coherent PSK with MRC and an L-element array.

For the special case of a single infinite power interferer [i.e., $P_I = \infty$ ($\lambda_1 = \infty$)], Eq. (10.19) simplifies to

$$P_b(E) = \int_0^{\pi/2} \left(\frac{\sin^2\theta}{\sin^2\theta + \overline{\gamma}_d}\right)^{L-1} d\theta \tag{10.24}$$

Thus, based on *exact* expressions for BEP, we observe that for the infinite power interferer, the array uses up one entire order of diversity in its attempt to cancel

---

[4] Using an analogous approach based on an alternative form of the complementary error function (equivalently, the Gaussian Q-function), Eq. (4A.1), Aalo and Zhang [11] were able to arrive at a closed-form expression for average BEP, namely,

$$P_b(E) = \frac{1}{2}\left[1 - \sqrt{\frac{\overline{\gamma}_d}{\overline{\gamma}_d+1}}\sum_{k=0}^{L-2}\binom{2k}{k}\left(\frac{1}{4(\overline{\gamma}_d+1)}\right)^k\right]$$
$$- \frac{1}{2\Gamma(L)(-L\overline{\gamma}_I)^{L-1}}\left[\sqrt{\frac{\pi\overline{\gamma}_d}{L\overline{\gamma}_I}}\exp\left(\frac{\overline{\gamma}_d+1}{L\overline{\gamma}_I}\right)\text{erfc}\left(\sqrt{\frac{\overline{\gamma}_d+1}{L\overline{\gamma}_I}}\right)\right.$$
$$\left. - \sqrt{\frac{\overline{\gamma}_d}{\overline{\gamma}_d+1}}\sum_{k=0}^{L-2}\frac{(2k)!}{k!}\left(\frac{-L\overline{\gamma}_I}{4(\overline{\gamma}_d+1)}\right)^k\right]$$

which checks numerically with (10.22).

it. This same conclusion was reached in Ref. 7 based on "upper" bounds on conditional BEP and in Ref. 6 based on approximate expressions for average BEP (obtained by replacing λ_1 by its mean value $\overline{\lambda}_1 = \sigma^2 + LP_I$). We shall ourselves pursue the legitimacy of this approximation momentarily.

Before proceeding, we wish to point out that the foregoing conclusion, reached from a comparison of the no interferer and infinite power interferer could have been reached earlier from the context of a generalized slow-fading channel. This can easily be seen by evaluating (10.9) for the cases of $P_I = 0$ ($\lambda_1 = \sigma^2$) and $P_I = \infty$ ($\lambda_1 = \infty$), resulting in $M_{\gamma_t|\lambda_1=\sigma^2}(s) = M_\gamma(s, \overline{\gamma}_d)^L$ and $M_{\gamma_t|\lambda_1=\infty}(s) = M_\gamma(s, \overline{\gamma}_d)^{L-1}$, respectively, which from (10.17) establishes the desired conclusion.

10.1.1.2 Rayleigh Fading: Approximate Evaluation of Average Bit Error Probability.
To simplify the analysis, several authors [5–8] have made the assumption of replacing λ_1 by its mean value $\overline{\lambda}_1 = \sigma^2 + LP_I$ in the conditional MGF of γ_t (and similarly, the same assumption in the conditional PDF of γ_t) and then computing the average BEP from

$$P_b(E) \simeq \int_0^\infty Q(\sqrt{2\gamma_t}) p_{\gamma_t}(\gamma_t|\lambda_1)|_{\lambda_1=\overline{\lambda}_1} \, d\gamma_t \tag{10.25}$$

Clearly, this avoids evaluating the PDF of the eigenvalue λ_1 and the integration over this RV in (10.10) and also makes the expression for average BEP independent of the probability statistics of the interferer (i.e., it is only necessary to know its average power). When this substitution is made in the conditional PDF of (10.12), the following closed-form expression results for Rayleigh fading [6, Eq. (25)]:[5]

$$P_b(E) = \frac{1 + L\overline{\gamma}_I}{2(-L\overline{\gamma}_I)^{L-1}} \left\{ -\frac{L\overline{\gamma}_I}{1 + L\overline{\gamma}_I} + \sqrt{\frac{\overline{\gamma}_d}{1 + \overline{\gamma}_d}} - \frac{1}{1 + L\overline{\gamma}_I} \sqrt{\frac{\overline{\gamma}_d}{1 + L\overline{\gamma}_I + \overline{\gamma}_d}} \right.$$
$$\left. - \sum_{k=1}^{L-2} (-L\overline{\gamma}_I)^k \left[1 - \sqrt{\frac{\overline{\gamma}_d}{1 + \overline{\gamma}_d}} \left(1 + \sum_{i=1}^{k} \frac{(2i-1)!!}{i!(2 + 2\overline{\gamma}_d)^i} \right) \right] \right\} \tag{10.26}$$

This result should also agree with that obtained by substituting $\overline{\lambda}_1 = \sigma^2 + LP_I$ for λ_1 in (10.19), that is,

$$P_b(E) = \frac{1}{\pi} \int_0^{\pi/2} \left(\frac{\sin^2\theta}{\sin^2\theta + \overline{\gamma}_d/(1 + L\overline{\gamma}_I)} \right) \left(\frac{\sin^2\theta}{\sin^2\theta + \overline{\gamma}_d} \right)^{L-1} d\theta \tag{10.27}$$

---

[5] The result in (10.26) was obtained from the expression in Ref. 5 for the BEP of coherent binary *orthogonal* signaling by replacing $\overline{\gamma}_d$ by $2\overline{\gamma}_d$, reflecting the 3-dB difference between orthogonal and antipodal signaling in Gaussian noise. Also, although not stated explicitly in either reference, the result in (10.26) is valid only for $L \geq 2$. The lack of validity for $L = 1$ (i.e., no diversity) can easily be seen by observing that for $\overline{\gamma}_I = 0$ (i.e., no interference), (10.26) would yield $P_b(E) = 0$, which is an incorrect result.

The result in (10.27) can be also be evaluated in closed form using (5A.56), that is,

$$P_b(E) = \frac{1}{2}\left\{1 - \sqrt{\frac{\overline{\gamma}_d}{1+\overline{\gamma}_d}} \sum_{k=0}^{L-2} \binom{2k}{k} \frac{1}{[4(1+\overline{\gamma}_d)]^k}\left[1 - \left(-\frac{1}{L\overline{\gamma}_I}\right)^{L-1-k}\right] \right.$$
$$\left. - \sqrt{\frac{\overline{\gamma}_d}{1+L\overline{\gamma}_I+\overline{\gamma}_d}}\left(-\frac{1}{L\overline{\gamma}_I}\right)^{L-1}\right\} \qquad (10.28)$$

The equivalence between (10.28) and (10.26) can be established mathematically (after much tedious manipulation). Furthermore, (10.28) is valid for $L = 1$ and is thus the more general result.

Figures 10.1 through 10.3 illustrate the evaluation of $P_b(E)$ from (10.20) (the exact result) and from (10.28) (the approximate result) as a function of average total SNR $L\overline{\gamma}_d$ for two- and four-element arrays and for the special case of equal average desired signal and interference powers (or equivalently, $\overline{\gamma}_d = \overline{\gamma}_I$.) We observe that the difference between the curves using (10.20) and (10.28) is small. There is a simple, intuitive explanation for why the average BEP values are as close as they appear. For either $L = 2$ or $L = 4$, the array has a sufficient number of degrees of freedom to suppress the single interferer, regardless of whether or not it is degraded by fading. Thus, the performance is affected predominantly by the interferer's average power (which is all that is needed in the approximate

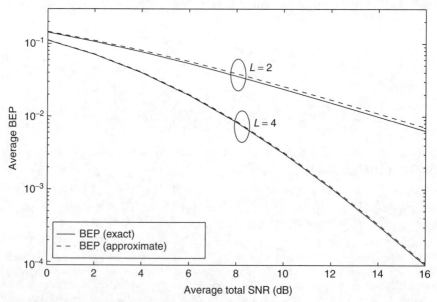

Figure 10.1. Average BEP of the optimum combiner in Rayleigh fading with a single fading co-channel interferer; $\overline{\gamma}_d = \overline{\gamma}_I$.

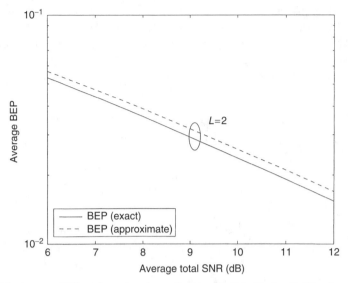

Figure 10.2. Average BEP performance (magnified scale); $\bar{\gamma}_d = \bar{\gamma}_I$, $L = 2$. (Courtesy of Shah, Haimovich, Simon, Alouini [7].)

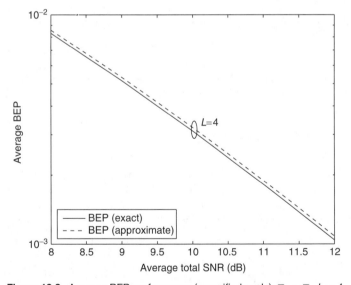

Figure 10.3. Average BEP performance (magnified scale); $\bar{\gamma}_d = \bar{\gamma}_I$, $L = 4$.

evaluation case) rather than its complete statistical description (which is what is needed in the exact evaluation case).

As we shall soon see, obtaining exact results in useful form for more complex fading channels is difficult. Thus, in view of the observation above, we shall resort to using the approximate approach for evaluating average BEP in these cases based on the same intuitive argument as that made for the simpler Rayleigh channel. To illustrate the difficulty of the exact evaluation method and the relative simplicity of the approximate approach, we present next the average BEP results for the Rician and Nakagami-m fading channels. Before doing this, however, we first present results for the Rayleigh channel corresponding to coherent M-PSK and QAM.

10.1.1.3 Extension to Other Modulations.
From the treatment of coherent M-PSK in the presence of fading discussed in Chapter 8, the conditional SEP can be written as

$$P_s(E|\lambda_1) = \frac{1}{\pi} \int_0^{(M-1)\pi/M} M_{\gamma_t|\lambda_1}\left(-\frac{g_{\text{PSK}}}{\sin^2\theta}\right) d\theta \tag{10.29}$$

where $g_{\text{PSK}} = \sin^2 \pi/M$. Thus, following the same procedure as for binary PSK, then after averaging over the PDF of λ_1, we obtain [analogous to (10.22)]

$$P_s(E) = \frac{1}{\pi}\left(\frac{1}{\bar{\gamma}_I}\right)^L \int_0^{\pi/2} \exp[f_{\text{PSK}}(\theta)][\Gamma(1-L, f_{\text{PSK}}(\theta))$$
$$+ L\bar{\gamma}_I f_{\text{PSK}}(\theta)\Gamma(-L, f_{\text{PSK}}(\theta))] d\theta \tag{10.30}$$

where now

$$f_{\text{PSK}}(\theta) \triangleq \left(1 + \frac{\bar{\gamma}_s g_{\text{PSK}}}{\sin^2\theta}\right)\frac{1}{\bar{\gamma}_I} \tag{10.31}$$

and $\bar{\gamma}_s$ denotes the average desired signal SNR per *symbol* per antenna.

For QAM with $M = 2^k$ signal points, the conditional SER is given by

$$P_s(E|\lambda_1) = \frac{4}{\pi}\left(1 - \frac{1}{\sqrt{M}}\right) \int_0^{\pi/2} M_{\gamma_t|\lambda_1}\left(-\frac{g_{\text{QAM}}}{\sin^2\theta}\right) d\theta$$
$$- \frac{4}{\pi}\left(1 - \frac{1}{\sqrt{M}}\right)^2 \int_0^{\pi/4} M_{\gamma_t|\lambda_1}\left(-\frac{g_{\text{QAM}}}{\sin^2\theta}\right) d\theta \tag{10.32}$$

where $g_{\text{QAM}} \triangleq 3/[2(M-1)]$. Thus, analogous to (10.22), the average SER is given by

$$P_s(E) = \frac{4}{\pi}\left(1 - \frac{1}{\sqrt{M}}\right)\left(\frac{1}{\bar{\gamma}_I}\right)^L \int_0^{\pi/2} \exp\{f_{\text{QAM}}(\theta)\}[\Gamma(1-L, f_{\text{QAM}}(\theta))$$
$$+ L\bar{\gamma}_I f_{\text{QAM}}(\theta)\Gamma(-L, f_{\text{QAM}}(\theta))] d\theta - \frac{4}{\pi}\left(1 - \frac{1}{\sqrt{M}}\right)^2 \left(\frac{1}{\bar{\gamma}_I}\right)^L$$
$$\times \int_0^{\pi/4} \exp\{f_{\text{QAM}}(\theta)\}[\Gamma(1-L, f_{\text{QAM}}(\theta))$$
$$+ L\bar{\gamma}_I f_{\text{QAM}}(\theta)\Gamma(-L, f_{\text{QAM}}(\theta))]d\theta \tag{10.33}$$

where now

$$f_{\text{QAM}}(\theta) \triangleq \left(1 + \frac{\bar{\gamma}_s g_{\text{QAM}}}{\sin^2\theta}\right)\frac{1}{\bar{\gamma}_I} \qquad (10.34)$$

and again $\bar{\gamma}_s$ denotes the average desired signal symbol SNR per antenna.

10.1.1.4 Rician Fading: Evaluation of Average Bit Error Probability.
Analogous to (10.11), the conditional MGF is now

$$M_{\gamma_t|\lambda_1}(s) = \left[\frac{1}{\left(1 - s\frac{1}{1+K}\frac{P_d}{\lambda_1}\right)\left(1 - s\frac{1}{1+K}\bar{\gamma}_d\right)^{L-1}}\right]$$

$$\times \exp\left\{\frac{K}{1+K}s\left[\frac{P_d/\lambda_1}{1 - s\frac{1}{1+K}\frac{P_d}{\lambda_1}} + (L-1)\frac{\bar{\gamma}_d}{1 - s\frac{1}{1+K}\bar{\gamma}_d}\right]\right\} \qquad (10.35)$$

Also, the largest eigenvalue has a noncentral chi-square PDF given by

$$p_{\lambda_1}(\lambda_1) = \left(\frac{1+K}{P_I}\right)\left(\frac{1+K}{LK}\right)^{(L-1)/2}\left(\frac{\lambda_1 - \sigma^2}{P_I}\right)^{(L-1)/2}$$

$$\times \exp\left\{-\left[LK + (1+K)\left(\frac{\lambda_1 - \sigma^2}{P_I}\right)\right]\right\}$$

$$\times I_{L-1}\left(2\sqrt{LK(1+K)\left(\frac{\lambda_1 - \sigma^2}{P_I}\right)}\right), \qquad \lambda_1 \geq \sigma^2 \qquad (10.36)$$

For $L = 1$ and $y = (\lambda_1 - \sigma^2)/P_I$, Eq. (10.36) reduces to

$$p_y(y) = (1+K)\exp\{-[K+(1+K)y]\}I_0(2\sqrt{K(1+K)y}), \qquad y \geq 0 \qquad (10.37)$$

which is the standard PDF for the square of a Rician RV. Also, for $K = 0$, using the asymptotic (small argument) form of $I_{L-1}(x)$ in (10.36), namely,

$$I_{L-1}(x) \simeq \frac{1}{\Gamma(L)}\left(\frac{x}{2}\right)^{L-1} \qquad (10.38)$$

the PDF of (10.36) can be shown to reduce to that in (10.14), as it should.

The conditional BEP for coherent PSK is from (10.17) and (10.35) given by

$$P_b(E|\lambda_1) = \frac{1}{\pi} \int_0^{\pi/2} \frac{1}{\left(1 + \frac{1}{1+K}\frac{P_d}{\lambda_1 \sin^2\theta}\right)\left(1 + \frac{1}{1+K}\frac{\overline{\gamma}_d}{\sin^2\theta}\right)^{L-1}}$$

$$\times \exp\left\{-\frac{K}{1+K}\frac{1}{\sin^2\theta}\left[\frac{P_d/\lambda_1}{1 + \frac{1}{1+K}\frac{P_d}{\lambda_1 \sin^2\theta}}\right.\right.$$

$$\left.\left. + (L-1)\frac{\overline{\gamma}_d}{1 + \frac{1}{1+K}\frac{\overline{\gamma}_d}{\sin^2\theta}}\right]\right\} d\theta \quad (10.39)$$

To evaluate the average BEP exactly, we must average (10.39) over the PDF in (10.36). After some algebraic manipulation, we get the following result:

$$P_b(E) = \exp(-LK)(1+K)\left(\frac{1+K}{LK}\right)^{(L-1)/2} \frac{1}{\pi}\int_0^{\pi/2}\left(\frac{\sin^2\theta}{\sin^2\theta + \frac{1}{1+K}\overline{\gamma}_d}\right)^{L-1}$$

$$\times \exp\left[-\frac{(L-1)K}{1+K}\left(\frac{\overline{\gamma}_d}{\sin^2\theta + \frac{1}{1+K}\overline{\gamma}_d}\right)\right]\int_0^\infty \frac{\sin^2\theta}{\sin^2\theta + \frac{1}{1+K}\frac{\overline{\gamma}_d}{1+\overline{\gamma}_I y}}$$

$$\times \exp\left[-\frac{K}{1+K}\left(\frac{\overline{\gamma}_d/(1+\overline{\gamma}_I y)}{\sin^2\theta + \frac{1}{1+K}\frac{\overline{\gamma}_d}{1+\overline{\gamma}_I y}}\right)\right] y^{(L-1)/2}$$

$$\times \exp[-(1+Ky)]I_{L-1}(2\sqrt{LK(1+K)y})\,dy\,d\theta \quad (10.40)$$

Unfortunately, the integral on y cannot be obtained in closed form. Thus, we now resort to the approximate approach, corresponding to $\overline{\lambda}_1 = LP_I + \sigma^2$ for λ_1 in the MGF of (10.35). When this is done, the approximate average BEP from (10.39) becomes

$$P_b(E) = \frac{1}{\pi} \int_0^{\pi/2} \left(\frac{\sin^2\theta}{\sin^2\theta + \frac{1}{1+K}\frac{\bar{\gamma}_d}{1+L\bar{\gamma}_I}} \right) \left(\frac{\sin^2\theta}{\sin^2\theta + \frac{1}{1+K}\bar{\gamma}_d} \right)^{L-1}$$

$$\times \exp\left\{ -\frac{K}{1+K} \left[\frac{\bar{\gamma}_d/(1+L\bar{\gamma}_I)}{\sin^2\theta + \frac{1}{1+K}\frac{\bar{\gamma}_d}{1+L\bar{\gamma}_I}} + \frac{(L-1)\bar{\gamma}_d}{\sin^2\theta + \frac{1}{1+K}\bar{\gamma}_d} \right] \right\} d\theta \quad (10.41)$$

For $K = 0$, (10.41) reduces to

$$P_b(E) = \frac{1}{\pi} \int_0^{\pi/2} \left(\frac{\sin^2\theta}{\sin^2\theta + \frac{\bar{\gamma}_d}{1+L\bar{\gamma}_I}} \right) \left(\frac{\sin^2\theta}{\sin^2\theta + \bar{\gamma}_d} \right)^{L-1} d\theta \quad (10.42)$$

which is identical to (10.27).

Figures 10.4 and 10.5 illustrate the approximate evaluation of $P_b(E)$ from (10.41) as a function of average total SNR $L\bar{\gamma}_d$ with Rician factor K as a parameter for two- and four-element arrays, again assuming that $\bar{\gamma}_d = \bar{\gamma}_I$.

10.1.1.5 Nakagami-m Fading: Evaluation of Average Bit Error Probability.
For this fading channel we can model each of the interference vector components as a sum of m i.i.d. complex Gaussian RVs each with zero mean and variance $1/m$. In this way, each component of the interference vector (e.g., c_{In}) still has unity mean-square value. Then, after transformation by the unitary matrix, the new vector $\mathbf{s}$, which corresponds to a weighted sum of the components $\mathbf{c}_I$, will still have i.i.d. complex Gaussian components, and its properties are thus preserved as in the Rayleigh and Rician cases. Also, the RV $\sum_{l=1}^{L} |c_{Il}|^2$ is now central chi-square distributed with $2mL$ degrees of freedom, each of which has variance $1/2m$. Thus, the PDF of $\lambda_1 = P_I \sum_{l=1}^{L} |c_{Il}|^2 + \sigma^2$ is now given analogous to (10.14) by

$$p_{\lambda_1}(\lambda_1) = \frac{m^{mL}}{\Gamma(mL)(P_I)^{mL}} (\lambda_1 - \sigma^2)^{mL-1} \exp\left(-m\frac{\lambda_1 - \sigma^2}{P_I} \right), \quad \lambda_1 \geq \sigma^2 \quad (10.43)$$

Similarly, analogous to (10.11), the conditional MGF of the combiner output SNR is now

$$M_{\gamma_t|\lambda_1}(s) = \frac{1}{[1 - s(\bar{\gamma}_d/m)]^{m(L-1)}[1 - s(P_d/m\lambda_1)]} \quad (10.44)$$

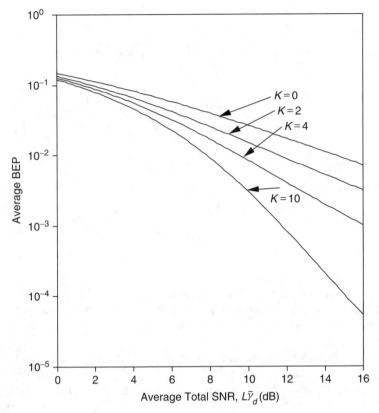

Figure 10.4. Average BEP of the optimum combiner in Rician fading with a single fading co-channel interferer; $\bar{\gamma}_d = \bar{\gamma}_I$, $L = 2$.

Using (10.44), the conditional BEP of (10.17) is given by

$$P_b(E|\lambda_1) = \frac{1}{\pi} \int_0^{\pi/2} \frac{1}{(1+\bar{\gamma}_d/m \sin^2 \theta)^{m(L-1)}} \frac{1}{(1+P_d/m\lambda_1 \sin^2 \theta)^m} d\theta \quad (10.45)$$

For no interferer ($\lambda_1 = \sigma^2$), (10.45) becomes

$$P_b(E) = \frac{1}{\pi} \int_0^{\pi/2} \frac{1}{(1+\bar{\gamma}_d/m \sin^2 \theta)^{mL}} d\theta \quad (10.46)$$

which is equivalent to MRC combining with mL orders of diversity and channels each with $(1/m)$th of the power. For an infinite power interferer ($\lambda_1 = \infty$), we get

$$P_b(E) = \frac{1}{\pi} \int_0^{\pi/2} \frac{1}{(1+\bar{\gamma}_d/m \sin^2 \theta)^{m(L-1)}} d\theta \quad (10.47)$$

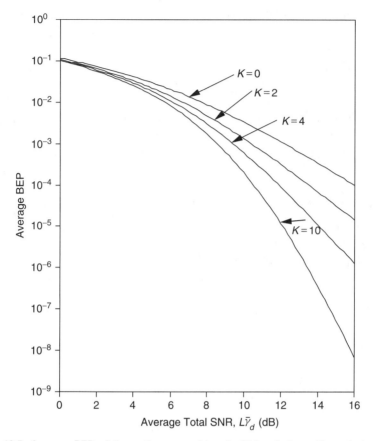

Figure 10.5. Average BEP of the optimum combiner in Rician fading with a single fading co-channel interferer; $\bar{\gamma}_d = \bar{\gamma}_I$, $L = 4$.

Thus, analogous to the Rayleigh and Rician cases, the diversity is reduced by m orders of magnitude in attempting to cancel the interferer, where each channel now has only $(1/m)$th of the power.

If we attempt to get the exact expression for average BEP by averaging (10.45) over the PDF in (10.43), we obtain after some simplification

$$P_b(E) = \frac{1}{\pi \Gamma(mN)} \left(\frac{m}{\bar{\gamma}_I}\right)^{mN} \int_0^{\pi/2} \frac{1}{(1+\bar{\gamma}_d/m \sin^2 \theta)^{L-1}}$$
$$\times \int_0^\infty \left(\frac{1+z}{1+\bar{\gamma}_d/m \sin^2 \theta + z}\right)^m z^{mL-1} \exp\left(-\frac{m}{\bar{\gamma}_I} z\right) dz\, d\theta$$
(10.48)

Unfortunately, the integral on z cannot be obtained in closed form so once again we must resort to the approximate approach. Substituting $\bar{\lambda}_1 = LP_I + \sigma^2$ for λ_1 in the MGF of (10.44) and then applying (10.17) results in the approximate average BEP

$$P_b(E) = \frac{1}{\pi} \int_0^{\pi/2} \left(\frac{\sin^2 \theta}{\sin^2 \theta + (\bar{\gamma}_d/m)/(1 + L\bar{\gamma}_I)} \right)^m \left(\frac{\sin^2 \theta}{\sin^2 \theta + \bar{\gamma}_d/m} \right)^{m(L-1)} d\theta \qquad (10.49)$$

Analogous to Figs. 10.4 and 10.5, Figs. 10.6 and 10.7 illustrate the approximate evaluation of $P_b(E)$ from (10.49) as a function of $L\bar{\gamma}_d$ with Nakagami-m factor m as a parameter for two- and four-element arrays, again assuming that $\bar{\gamma}_d = \bar{\gamma}_I$.

10.1.2 Multiple Interferers, Independent Identically Distributed Fading

Consider now a scenario where more than a single co-channel interferer exists at each antenna element. On the one hand, to increase network capacity, the cellular system might be operating in an interference-limited environment, in which case the number of co-channel interferers, N_I, could typically exceed the number of

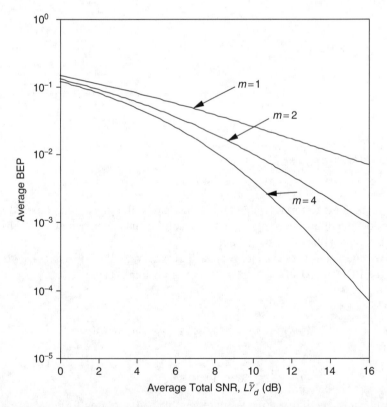

Figure 10.6. Average BEP of the optimum combiner in Nakagami-m fading with a single fading co-channel interferer; $\bar{\gamma}_d = \bar{\gamma}_I$, $L = 2$.

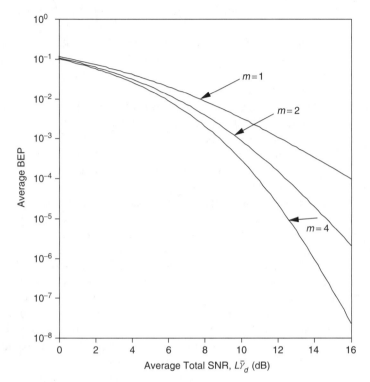

Figure 10.7. Average BEP of the optimum combiner in Nakagami-m fading with a single fading co-channel interferer; $\bar{\gamma}_d = \bar{\gamma}_I$, $L = 4$.

antenna elements, L. Under such conditions, which is typical of CDMA, the number of degrees of freedom provided by the array is insufficient to allow a partitioning of the observation space into distinct noise and interferer subspaces (i.e., the optimum combiner processes both the noise and interference together as a single entity and cannot cancel all the interfering signals). On the other hand, in certain other practical cellular mobile applications, for instance in a TDMA system such as the Global System for Mobile (GSM), most of the interference can be due to only a limited number of dominant interferers. Here the array has sufficient degrees of freedom to partition the observation space into distinct noise and interferer subspaces and combat them as separate entities (much like the case of the single interferer discussed earlier). As we shall soon see, the treatment of this scenario is a natural extension of the single-interferer case and therefore will be the first to be discussed.

Exact and approximate evaluation of average BER for optimum combining receivers in the presence of multiple narrowband interferers has been considered by several authors. Specifically, Cui, Falconer, and Sheikh [12,13][6] considered

---

[6] Also considered in these papers is the average BEP performance of MRC receivers in the presence of an arbitrary number of interferers.

optimum combining receivers for DPSK modulation with $N_I = 2$ and Rayleigh fading. Their results also yield an upper bound on the performance of coherent BPSK and QAM based on the approach taken in Ref. 14. Optimum combining for coherent BPSK systems with arbitrary $N_I < L$ and Rayleigh fading was considered by Winters and Salz [15] and Winters et al. [16]. However, their results were presented only in the form of an upper bound on average BEP, which unfortunately is not accurate for low BEP and a large number of antenna elements. Most recently, Aalo and Zhang [11] obtained exact average BEP results (not in closed form, however) for optimum combining coherent BPSK systems with $N_I = 2$ and again Rayleigh fading. The approach taken there was the same as that used to derive closed-form results for the single-interferer case, as mentioned in footnote 4 of this chapter.

A more generic approach to the problem of spatial combining in the presence of multiple equal-power narrowband interferers was considered by Haimovich and Shah [17] and Shah and Haimovich [18,19],[7] who assumed that the number of interferers is sufficiently large as to justify an interference-limited environment (i.e., AWGN was ignored). The results of this work are presented in Section 10.1.2.2 for the case where the number of interferers is equal to or greater than the number of array elements. For the case where the number of interferers is less than the number of array elements, Villier [8] assumed that the CCI is dominated by the N_I strongest interferers[8] and that the approximate approach for evaluating error probability (i.e., replacing the random eigenvalues by their mean value is once again valid). (We shall say more about this assumption momentarily.) As we shall see, this approach allows for consideration of generalized fading channels as well as a variety of modulation types. Before considering this approach and the performance results derived from it, we first present the exact average BEP results for the two-interferer coherent BPSK Rayleigh channel case as given in Ref. 11, starting with a generalization of the system model to the multiple-interferer scenario.

10.1.2.1 Number of Interferers Less Than the Number of Array Elements. Analogous to (10.1), the received signal vector $\mathbf{r}(t)$ at the outputs of the array elements may be expressed as

$$\mathbf{r}(t) = \sqrt{P_d}\mathbf{c}_d s_d(t) + \sum_{n=1}^{N_I} \sqrt{P_I}\mathbf{c}_{I_n} s_{I_n}(t) + \mathbf{n}(t) \qquad (10.50)$$

where $s_{I_n}(t)$ and $\mathbf{c}_{I_n}$ are now, respectively, the signal and propagation vectors associated with the nth interfering signal. Assuming that the N_I interference signals are mutually independent, then analogous to (10.3), the noise plus

---

[7] In Ref. 19, EGC and MRC are compared with OC, and in addition, allowance is made for Rayleigh, Rician, or no fading on the desired signal when the fading on the interference is Rayleigh distributed.
[8] Any other interfering signal is assumed to have a power level significantly lower than that of the N_I strongest ones and as such is included in the additive noise term $\mathbf{n}(t)$.

interference covariance matrix R_{ni} is given by

$$R_{ni} = E\left\{\left[\sum_{n=1}^{N_I}\sqrt{P_I}\mathbf{c}_{I_n}s_{I_n}(t) + \mathbf{n}(t)\right]\left[\sum_{n=1}^{N_I}\sqrt{P_I}\mathbf{c}_{I_n}s_{I_n}(t) + \mathbf{n}(t)\right]^H\right\}$$

$$= P_I\sum_{n=1}^{N_I}\mathbf{c}_{I_n}\mathbf{c}_{I_n}^H + \sigma^2 I \qquad (10.51)$$

Since the maximum instantaneous SINR at the combiner output is still given by the relation in (10.2), where R_{ni}^{-1} is now the inverse of (10.51), then once again the eigenvalue decomposition of R_{ni} yields the result in (10.5), where now there exist $L - N_I$ nonrandom eigenvalues $\lambda_{N_I+1}, \lambda_{N_I+2}, \ldots, \lambda_L$, each with value σ^2 and N_I random eigenvalues $\lambda_1, \lambda_2, \ldots, \lambda_{N_I}$. Although we still obtain a product form for the MGF of the combiner output SINR, namely,

$$M_{\gamma_t|\lambda_1,\lambda_2,\ldots,\lambda_{N_I}}(s) = [M_\gamma(s;\overline{\gamma}_d)]^{L-N_I}\prod_{n=1}^{N_I}M_\gamma\left(s;\frac{\sigma^2}{\lambda_n}\overline{\gamma}_d\right) \qquad (10.52)$$

the difficulty now lies in the determination of the N_I random eigenvalues which are related in a complex manner to the total interference power received by the array. For the case where the interference sources are represented by mutually orthogonal propagation vectors, the random eigenvalues $\lambda_1, \lambda_2, \ldots, \lambda_{N_I}$ become equal in form to that of (10.8), with PDFs given by [see (10.14)]

$$p_{\lambda_n}(\lambda_n) = \frac{1}{\Gamma(L)P_I^L}(\lambda_n - \sigma^2)^{L-1}\exp\left(-\frac{\lambda_n - \sigma^2}{P_I}\right), \qquad \lambda_n \geq \sigma^2,$$
$$n = 1, 2, \ldots, N_I \qquad (10.53)$$

When the number of interferers is restricted to two and the fading is Rayleigh distributed, the two random eigenvalues can be found exactly. Using the results given in Refs. 12 and 13, and later in Ref. 11, we have

$$\lambda_1 = \sigma^2\left[1 + \frac{\gamma_1 + \gamma_2}{2} + \frac{1}{2}\sqrt{(\gamma_1 - \gamma_2)^2 + 4\rho_{12}\gamma_1\gamma_2}\right]$$
$$\lambda_2 = \sigma^2\left[1 + \frac{\gamma_1 + \gamma_2}{2} - \frac{1}{2}\sqrt{(\gamma_1 - \gamma_2)^2 + 4\rho_{12}\gamma_1\gamma_2}\right] \qquad (10.54)$$

where $\gamma_n \overset{\Delta}{=} \sum_{l=1}^{L}\alpha_{I_nl}^2 P_I/\sigma^2 = \gamma_I\sum_{l=1}^{L}\alpha_{I_nl}^2$, $n = 1, 2$, is the instantaneous SNR for the nth interferer, with α_{I_nl} the Rayleigh fading amplitude on its lth branch and has the PDF [analogous to (10.14)]

$$p_{\gamma_n}(\gamma_n) = \begin{cases} \dfrac{1}{\Gamma(L)(\overline{\gamma}_I)^L} \gamma_n^{L-1} \exp\left(-\dfrac{\gamma_n}{\overline{\gamma}_I}\right), & \gamma_n \geq 0, n = 1, 2 \\ 0, & \text{otherwise} \end{cases} \quad (10.55)$$

Also, in (10.54), ρ_{12} is the normalized correlation between the two interference propagation vectors, $\mathbf{c}_{I_1}$ and $\mathbf{c}_{I_2}$, which is a beta-distributed RV with PDF given by

$$p_{\rho_{12}}(\rho_{12}) = \begin{cases} (L-1)(1-\rho_{12})^{L-2}, & 0 \leq \rho_{12} \leq 1 \\ 0, & \text{otherwise} \end{cases} \quad (10.56)$$

Note that for uncorrelated interference vectors (i.e., $\rho_{12} = 0$), (10.54) simplifies to $\lambda_n = \sigma^2(1 + \gamma_n) = P_I \sum_{l=1}^{L} \alpha_{I_n l}^2 + \sigma^2$, $n = 1, 2$, which as alluded to above, has the form of (10.8).

To evaluate the average BEP for coherent BPSK, we proceed analogous to (10.17) and (10.18). Specifically, from (10.52) the conditional MGF for Rayleigh fading is now

$$M_{\gamma_t | \lambda_1, \lambda_2}(s) = \frac{1}{(1 - s\overline{\gamma}_d)^{L-2}[1 - s(P_d/\lambda_1)][1 - s(P_d/\lambda_2)]}$$

$$= \frac{1}{(1 - s\overline{\gamma}_d)^{L-2}[1 - s\overline{\gamma}_d(\sigma^2/\lambda_1)][1 - s\overline{\gamma}_d(\sigma^2/\lambda_2)]} \quad (10.57)$$

Then the average BEP can be written as

$$P_b(E) = \frac{1}{\pi} \int_{\sigma^2}^{\infty} \int_{\sigma^2}^{\infty} \int_0^{\pi/2} M_{\gamma_t | \lambda_1, \lambda_2}\left(-\frac{1}{\sin^2 \theta}\right) d\theta \, p_{\lambda_1}(\lambda_1) p_{\lambda_2}(\lambda_2) \, d\lambda_1 \, d\lambda_2$$

$$= \frac{1}{\pi} \int_0^{\pi/2} \left(\frac{\sin^2 \theta}{\sin^2 \theta + \overline{\gamma}_d}\right)^{L-2} d\theta$$

$$\times \int_0^{\infty} \int_0^{\infty} \int_0^1 \frac{1}{\left(1 + \dfrac{\overline{\gamma}_d}{\sin^2 \theta} \dfrac{\sigma^2}{\lambda_1}\right)\left(1 + \dfrac{\overline{\gamma}_d}{\sin^2 \theta} \dfrac{\sigma^2}{\lambda_2}\right)}$$

$$\times p_{\rho_{12}}(\rho_{12}) \, d\rho_{12} \, p_{\gamma_1}(\gamma_1) p_{\gamma_2}(\gamma_2) \, d\gamma_1 \, d\gamma_2 \quad (10.58)$$

where λ_1, and λ_2 are in turn expressed in terms of the RVs γ_1, γ_2, and ρ_{12}, as in (10.54). The statistical average over ρ_{12} [inner integral of the triple integral in (10.58)] can be evaluated in closed form. Using results from Refs. 12 and 13, it can be shown that this average can be put in the form

$$\int_0^1 \frac{1}{\left(1 + \dfrac{\overline{\gamma}_d}{\sin^2\theta} \dfrac{\sigma^2}{\lambda_1}\right)\left(1 + \dfrac{\overline{\gamma}_d}{\sin^2\theta} \dfrac{\sigma^2}{\lambda_2}\right)} p_{\rho_{12}}(\rho_{12}) d\rho_{12}$$

$$= \frac{(L-1)(1+\gamma_1+\gamma_2)}{\left(1+\dfrac{\overline{\gamma}_d}{\sin^2\theta}\right)\left(1+\dfrac{\overline{\gamma}_d}{\sin^2\theta}+\gamma_1+\gamma_2\right)} \int_0^1 \frac{1+au}{1+bu} u^{L-2} du \quad (10.59)$$

where we have made the change of variables $u = 1 - \rho_{12}$ and

$$a \triangleq \frac{\gamma_1\gamma_2}{1+\gamma_1+\gamma_2}, \quad b \triangleq \frac{\gamma_1\gamma_2}{(1+\overline{\gamma}_d/\sin^2\theta)(1+\overline{\gamma}_d/\sin^2\theta+\gamma_1+\gamma_2)} \quad (10.60)$$

Evaluating the integral in (10.59) as in [Eq. (49) of Ref. 13, then after some manipulation we arrive at the closed-form result (also see Ref. 11)

$$\int_0^1 \frac{1}{\left(1 + \dfrac{\overline{\gamma}_d}{\sin^2\theta} \dfrac{\sigma^2}{\lambda_1}\right)\left(1 + \dfrac{\overline{\gamma}_d}{\sin^2\theta} \dfrac{\sigma^2}{\lambda_2}\right)} p_{\rho_{12}}(\rho_{12}) d\rho_{12}$$

$$= 1 + (L-1)\left(\frac{b-a}{ab}\right)\left[\sum_{k=0}^{L-3} \frac{1}{(L-2-k)(-b)^k} + \frac{1}{(-b)^{L-2}}\ln(b+1)\right] \quad (10.61)$$

Substituting (10.61) into (10.58) leaves a triple integral for evaluating $P_b(E)$, which according to Aalo and Zhang [11], "can be easily evaluated numerically."

Motivated by the desire to obtain a simple expression for assessing the average BEP of the OC in the presence of multiple interferers with $N_I < L$, Villier [8] proposes to use an approximate approach analogous to that for the single interference case, wherein each random eigenvalue is replaced by its average over the fading distribution and all of them are made equal in accordance with what would be obtained from λ_1 of (10.8) (i.e., $\overline{\lambda}_i = LP_I + \sigma^2, i = 1, 2, \ldots, N_I$). The validity of this approximate approach is justified for the case where the interference power level is high (relative to the desired signal power) and the number of antenna elements is considerably greater than the number of dominant interferers, which translates mathematically to $\overline{\lambda}_i \gg P_d$ and $L \gg N_I$. In such instances, the product $\prod_{n=1}^{N_I} M_\gamma(s; (\sigma^2/\lambda_n)\overline{\gamma}_d)$ in (10.52) tends to become insignificant compared to the remainder of the product $(M_\gamma(s; \overline{\gamma}_d))^{L-N_I}$, which then dominates the MGF. A further justification corresponds to the scenario where the interferers are indeed orthogonal, in which case, based on our previous discussion, the mean values of the random eigenvalues would become equal and given precisely by $\overline{\lambda}_i = LP_I + \sigma^2, i = 1, 2, \ldots, N_I$.

Proceeding under the assumption of the approximate approach above, then analogous to (10.11), the MGF for the Rayleigh fading channel becomes

$$M_{\gamma_t|\bar{\lambda}_1}(s) = \frac{1}{(1-s\bar{\gamma}_d)^{L-N_I}[1-s(P_d/\bar{\lambda}_1)]^{N_I}}$$

$$= \frac{1}{(1-s\bar{\gamma}_d)^{L-N_I}(1-s[\bar{\gamma}_d/(1+L\bar{\gamma}_I)])^{N_I}} \quad (10.62)$$

and similarly, analogous to (10.27), the average BEP of coherent BPSK with OC is given by

$$P_b(E) = \frac{1}{\pi}\int_0^{\pi/2}\left(\frac{\sin^2\theta}{\sin^2\theta+\bar{\gamma}_d/(1+L\bar{\gamma}_I)}\right)^{N_I}\left(\frac{\sin^2\theta}{\sin^2\theta+\bar{\gamma}_d}\right)^{L-N_I}d\theta \quad (10.63)$$

Note the similarity in the form of (10.63) (N_I independent interferers with Rayleigh fading) with that in (10.49) (one interferer in Nakagami-m fading.) Furthermore, Villier [8] gives a closed-form result for the N_I independent interferer, Rayleigh fading case which from the similarity of the integrals mentioned above would therefore imply that the single interferer in Nakagami-m fading performance [see (10.49)] would have a similar closed-form result (this is left as an exercise for the reader). For the former, we have [8, Eq. (20)]

$$P_b(E) = \frac{(1+L\bar{\gamma}_I)^{N_I-1}}{2(-L\bar{\gamma}_I)^{L-1}}\left[\sum_{k=0}^{N_I-1}\left(\frac{-L\bar{\gamma}_I}{1+L\bar{\gamma}_I}\right)^k B_k I_k\left(\frac{\bar{\gamma}_d}{1+L\bar{\gamma}_I}\right)\right.$$
$$\left. -(1+L\bar{\gamma}_I)\sum_{k=0}^{L-N_I-1}(-L\bar{\gamma}_I)^k C_k I_k(\bar{\gamma}_d)\right] \quad (10.64)$$

where the coefficients B_k and C_k are given by[9]

$$B_k \triangleq \frac{A_k}{\binom{L-1}{k}}, \quad C_k \triangleq \sum_{n=0}^{N_I-1}\frac{\binom{k}{n}}{\binom{L-1}{n}}A_n,$$

$$A_k \triangleq (-1)^{N_I-1+k}\frac{\binom{N_I-1}{k}}{(N_I-1)!}\prod_{\substack{n=1\\n\neq k+1}}^{N_I}(L-n) \quad (10.65)$$

---

[9] Note that by convention $\binom{k}{n}=0$ for $n>k$. Also, for $N_I=1$, by convention the product $\prod_{n=1,n\neq k+1}^{N_I}(L-n)=1$ and the only nonzero-valued coefficients are $A_0=B_0=1$ and $C_k=1$, $k=0,1,\ldots,L-2$. For $N_I>1$, the coefficients A_k, B_k, and C_k clearly depend on both N_I and L.

and

$$I_k(c) \triangleq \frac{1}{k!} \int_0^\infty x^k \, \mathrm{erfc}(\sqrt{cx}) e^{-x} \, dx \qquad (10.66)$$

which has the series form

$$I_k(c) = 1 - \sqrt{\frac{c}{1+c}} \left[1 + \sum_{n=1}^{k} \frac{(2n-1)!!}{n! \, 2^n (1+c)^n} \right] \qquad (10.67)$$

with the double-factorial notation denoting the product of only odd integers from 1 to $2k - 1$.

An alternative closed-form expression for the average BEP can be obtained from the results in Ref. 17 using the approach in (10.12) and (10.13) but with λ_1 replaced by $\bar{\lambda}_1$. In particular, the inverse Laplace transform of (10.62) is [17, Eq. (71)]

$$p_{\gamma_t}(\gamma_t | \bar{\lambda}_1) = \left(\frac{\bar{\lambda}_1}{\sigma^2} \right)^{N_I} \frac{\gamma_t^{L-1} \exp(-\bar{\lambda}_1 \gamma_t / P_d) \, {}_1F_1(L - N_I; L; (\bar{\lambda}_1 - \sigma^2)\gamma_t / P_d)}{\Gamma(L)(\bar{\gamma}_d)^L},$$

$$\gamma_t \geq 0, \quad N_I < L \qquad (10.68)$$

which clearly reduces to (10.12) with λ_1 replaced by $\bar{\lambda}_1$ when $N_I = 1$. Integrating the Gaussian Q-function over the PDF in (10.64) as in (10.13) gives [17, Eq. (83)][10]

$$\begin{aligned} P_b(E) &= \int_0^\infty Q(\sqrt{2\gamma_t}) p_{\gamma_t}(\gamma_t | \bar{\lambda}_1) \, d\gamma_t = \frac{1}{2} - \left(\frac{\bar{\lambda}_1}{\sigma^2} \right)^{N_I} \\ &\times \frac{\Gamma\left(L + \frac{1}{2}\right) \left(\frac{P_d}{\bar{\lambda}_1} \right)^{L+1/2} F_2\left(L + \frac{1}{2}, \frac{1}{2}, L - N_I; \frac{3}{2}, L; -\frac{P_d}{\bar{\lambda}_1}, \frac{\bar{\lambda}_1 - \sigma^2}{\bar{\lambda}_1}\right)}{\sqrt{\pi} \Gamma(L)(\bar{\gamma}_d)^L} \\ &= \frac{1}{2} - (1 + L\bar{\gamma}_I)^{N_I} \frac{\Gamma\left(L + \frac{1}{2}\right)\left(\frac{\bar{\gamma}_d}{1 + L\bar{\gamma}_I} \right)^{L+1/2} \times F_2\left(L + \frac{1}{2}, \frac{1}{2}, L - N_I; \frac{3}{2}, L; -\frac{\bar{\gamma}_d}{1 + L\bar{\gamma}_I}, \frac{L\bar{\gamma}_I}{1 + L\bar{\gamma}_I}\right)}{\sqrt{\pi}\Gamma(L)(\bar{\gamma}_d)^L} \end{aligned}$$

$$(10.69)$$

---

[10] Equation (83) of Haimovich and Shah [17] has a typographical error. A factor of $\rho^{(N-r)}$ should appear in the denominator.

where again $F_2(\bullet, \bullet, \bullet; \bullet, \bullet; \bullet, \bullet)$ is Appell's hypergeometric function of two variables [10, Eq. (9.180.2)]. Using the functional relation between hypergeometric functions of two variables in Eq. (9.183.2) of Ref. 10, the BEP of (10.69) can be rewritten as

$$P_b(E) = \frac{1}{2} - \left(\frac{\bar{\lambda}_1}{\sigma^2}\right)^{N_I}$$

$$\times \frac{\Gamma\left(L + \frac{1}{2}\right)(P_d)^{L+1/2} F_2\left(L + \frac{1}{2}, 1, L - N_I; \frac{3}{2}, L; \frac{P_d}{\bar{\lambda}_1 + P_d}, \frac{\bar{\lambda}_1 - \sigma^2}{\bar{\lambda}_1 + P_d}\right)}{\sqrt{\pi}\Gamma(L)(\bar{\gamma}_d)^L (\bar{\lambda}_1 + P_d)^{L+1/2}}$$

$$= \frac{1}{2} - (1 + L\bar{\gamma}_I)^{N_I}$$

$$\times \frac{\Gamma\left(L + \frac{1}{2}\right)\left(\frac{\bar{\gamma}_d}{1 + \bar{\gamma}_d + L\bar{\gamma}_I}\right)^{L+1/2}}{\sqrt{\pi}\Gamma(L)(\bar{\gamma}_d)^L}$$

$$\times F_2\left(L + \frac{1}{2}, 1, L - N_I; \frac{3}{2}, L; \frac{\bar{\gamma}_d}{1 + \bar{\gamma}_d + L\bar{\gamma}_I}, \frac{L\bar{\gamma}_I}{1 + \bar{\gamma}_d + L\bar{\gamma}_I}\right)$$

(10.70)

which reduces to (10.13) with λ_1 replaced by $\bar{\lambda}_1$ when $N_I = 1$.

Figures 10.8 and 10.9 illustrate $P_b(E)$ as computed from (10.64) [or equivalently, from (10.70)] versus average total SNR, $L\bar{\gamma}_d$, for four- and eight-element arrays with multiple interferers and equal desired signal and interference powers. Included for comparison are the corresponding results for the single-interferer case, which for the four-element array are obtained from Fig. 10.3. The numerical results clearly show the increased performance penalty produced by the additional CCI.

10.1.2.2 Number of Interferers Equal to or Greater Than the Number of Array Elements.
The scenario where the number of interfering signals is no less than the number of array elements is treated in Ref. 18 for the case of Rayleigh fading on both the desired signal and the CCI. In particular, assuming that the system is interference limited (hence, thermal noise can be neglected), it is shown there that[11]

$$p_{\gamma_t}(\gamma_t) = \frac{\Gamma(N_I + 1)}{\Gamma(L)\Gamma(N_I + 1 - L)} \left(\frac{P_d}{P_I}\right)^{N_I + 1 - L} \frac{\gamma_t^{L-1}}{(P_d/P_I + \gamma_t)^{N_I + 1}}, \quad \gamma_t \geq 0$$

(10.71)

---

[11] Shah and Haimovich [18] also note that the PDF in (10.71) does not depend on the form of the covariance matrix of the interference propagation vectors. Thus, the performance of the optimum combiner obtained from using this PDF is the same regardless of whether or not the fading at the L receiver elements is independent. It is to be emphasized, however, that this statement is true only for the equal power interferer and $N_I \geq L$ case, as is being considered here.

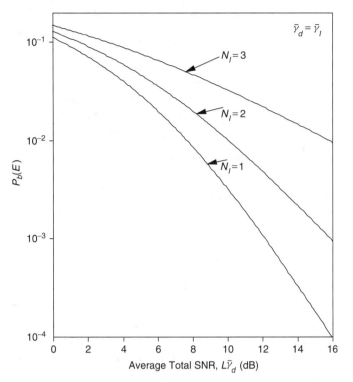

Figure 10.8. Average BEP versus average total SNR for optimum combining with Rayleigh fading and multiple interferers; $L = 4$.

where γ_t now denotes the signal-to-interference ratio (SIR) at the combiner output. This PDF is an example of one that has a finite number of finite moments, with all remaining higher-order moments being infinite. We will show this explicitly after evaluating the MGF.

To compute the MGF associated with (10.71), we make use of a result in Eq. (12) of Ref. 20, for the generalized Stieltjes integral

$$\int_0^\infty \frac{x^\lambda}{(x+y)^\rho} e^{-ax} \, dx = \Gamma(\lambda+1) a^{[(\rho-\lambda)/2]-1} y^{(\lambda-\rho)/2} e^{(1/2)ay} W_{k,m}(ay),$$

$$k = \frac{-\lambda - \rho}{2}, \quad m = \frac{\lambda - \rho + 1}{2} \tag{10.72}$$

where $W_{k,m}(z)$ is the Whittaker function defined in Eq. (9.222) of Ref. 10.[12]

---

[12] Note that the condition on the integral definition of $W_{k,m}(z)$ in Eq. (9.222) of Ref. 10, namely, $m - k > -\frac{1}{2}$, is satisfied since here

$$m - k = \frac{L - N_I - 1}{2} - \frac{-L - N_I}{2} = L - \frac{1}{2}$$

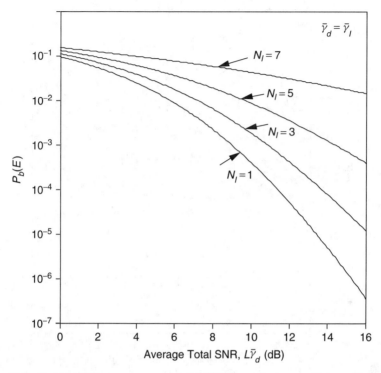

Figure 10.9. Average BEP versus average total SNR for optimum combining with Rayleigh fading and multiple interferers; $L = 8$.

Then, letting $x = \gamma_t$, $a = s$, $\lambda = L - 1$, $\rho = N_I + 1$, and $y = P_d/P_I$, we obtain, after simplification,

$$M_{\gamma_t}(-s) = \int_0^\infty p_{\gamma_t}(\gamma_t) e^{-s\gamma_t} \, d\gamma_t$$

$$= \frac{\Gamma(N_I + 1)}{\Gamma(L)\Gamma(N_I + 1 - L)} \left(\frac{P_d}{P_I}\right)^{N_I+1-L} \int_0^\infty \frac{\gamma_t^{L-1}}{(P_d/P_I + \gamma_t)^{N_I+1}} e^{-s\gamma_t} \, d\gamma_t$$

$$= \frac{\Gamma(N_I + 1)}{\Gamma(N_I + 1 - L)} \left(s\frac{P_d}{P_I}\right)^{(N_I-L)/2} \exp\left(\frac{1}{2} s \frac{P_d}{P_I}\right)$$

$$\times W_{(-L-N_I)/2, (L-N_I-1)/2} \left(s\frac{P_d}{P_I}\right) \tag{10.73}$$

To obtain the moments of γ_t, we must evaluate the derivatives of the MGF in (10.73) at $s = 0$. To obtain these we first rewrite (10.73) using the integral definition of the Whittaker function in Ref. 10 [Eq. (9.222)], to give

$$M_{\gamma_t}(-s) = \frac{\Gamma(N_I+1)}{\Gamma(N_I+1-L)\Gamma(L)} \int_0^\infty \exp\left(-s\frac{P_d}{P_I}t\right) \frac{t^{L-1}}{(1+t)^{N_I+1}} dt \quad (10.74)$$

Then

$$E\{\gamma_t^n\} = (-1)^n \frac{d^n}{ds^n} M_{\gamma_t}(-s)|_{s=0}$$

$$= \frac{\Gamma(N_I+1)}{\Gamma(N_I+1-L)\Gamma(L)} \left(\frac{P_d}{P_I}\right)^n \int_0^\infty \frac{t^{n+L-1}}{(1+t)^{N_I+1}} dt \quad (10.75)$$

The integral in (10.75) can be evaluated using [10, Eq. (3.241.4)], which is restricted to $n < N_I - L$. When this done, the following results:

$$E\{\gamma_t^n\} = \frac{\Gamma(L+n)\Gamma(N_I+1-L-n)}{\Gamma(N_I+1-L)\Gamma(L)} \left(\frac{P_d}{P_I}\right)^n$$

$$= \frac{L(L+1)\cdots(L+n-1)}{(N_I-L)(N_I-L-1)\cdots(N_I+1-L-n)} \left(\frac{P_d}{P_I}\right)^n, \quad n < N_I - L$$

$$(10.76)$$

The first moment corresponding to $n=1$ is simply given by

$$E\{\gamma_t\} = \frac{L}{N_I - L} \left(\frac{P_d}{P_I}\right) \quad (10.77)$$

which agrees with Eq. (14) of Ref. 18 and varies linearly with L when the number of interferers is large compared to the number of antenna elements. For $n \geq N_I - L$, the integral in (10.75) diverges, and thus as stated above, only a finite number $(N_I - L - 1)$ of moments have finite value.

Assuming that the number of interferers is large and that the total interference can be modeled as a Gaussian RV [18], the conditional error probability for coherent detection is still described by a Gaussian Q-function, and as before, the average BEP of OC can be computed based entirely knowledge of the MGF; for example, for BPSK we have, analogous to (10.17),

$$P_b(E) = \frac{1}{\pi} \int_0^{\pi/2} M_{\gamma_t}\left(-\frac{1}{\sin^2\theta}\right) d\theta \quad (10.78)$$

Alternatively, a closed-form expression for the average BEP corresponding to coherent BPSK is obtained in Ref. 18 by direct integration of the conditional

BEP over the PDF in (10.71), with the result

$$P_b(E) = \frac{1}{2} + \frac{1}{2\sqrt{\pi}\Gamma(L)\Gamma(N_I+1-L)} \left[\left(\frac{P_d}{P_I}\right)^{N_I+1-L} \frac{\Gamma\left(L-N_I-\frac{1}{2}\right)\Gamma(L+1)}{\Gamma(L-N_I-1)} \right.$$

$$\times {}_2F_2\left(N_I+1, N_I+1-L; N_I-L+\frac{3}{2}, N_I-L+2; \frac{P_d}{P_I}\right)$$

$$-2\left(\frac{P_d}{P_I}\right)^{1/2} \Gamma\left(N_I-L+\frac{1}{2}\right) \Gamma\left(L+\frac{1}{2}\right)$$

$$\left. \times {}_2F_2\left(L+\frac{1}{2},\frac{1}{2}; L-N_I+\frac{1}{2},\frac{3}{2}; \frac{P_d}{P_I}\right) \right] \quad (10.79)$$

where ${}_pF_q(\overbrace{\bullet,\cdots,\bullet}^{p}; \overbrace{\bullet,\cdots,\bullet}^{q}; \bullet)$ is the generalized hypergeometric series defined in Eq. (9.14.1) of Ref. 10. Again the simplicity of (10.78) combined with (10.65) compared with (10.79) is to be observed.

10.1.3 Comparison with Results for MRC in the Presence of Interference

As mentioned previously, using MRC in the presence of interference results in suboptimum performance in that it produces a smaller SINR at the combiner output than OC. As such, it is of interest to evaluate the performance of MRC with interference and then compare the amount by which it suffers relative to that obtained with OC. Based on the approximate approach (replacing the eigenvalues by their statistical means), the analytical results describing the performance of MRC in the presence of interference can be obtained directly from the results for the performance of MRC in the absence of interference (see Chapter 9) by replacing the average SNR with the average SINR. For example, for Rayleigh fading, analogous to (9.6), the appropriate expression for average BEP would be

$$P_b(E) = \frac{1}{2^L}\left(1 - \sqrt{\frac{\bar{\gamma}}{1+\bar{\gamma}}}\right)^L \sum_{k=0}^{L-1} \frac{1}{2^k} \binom{L-1+k}{k} \left(1 + \sqrt{\frac{\bar{\gamma}}{1+\bar{\gamma}}}\right)^k \quad (10.80)$$

where now

$$\bar{\gamma} \triangleq \frac{\bar{\gamma}_d}{1+N_I\bar{\gamma}_I} \quad (10.81)$$

The result in (10.80) agrees with that obtained by Winters [6] for the single-interferer case and that obtained by Villier [8] for the multiple-interferer case. We further note that (10.80) is also obtained as the limit of (10.26) (for $L \geq 2$) or (10.28) (for $L \geq 1$) when $\bar{\gamma}_I$ approaches zero, in which case $\bar{\gamma}$ of (10.81) becomes equal to $\bar{\gamma}_d$.

For the case where the system is interference limited and the interference is assumed to be modeled by a Gaussian distribution (Shah and Haimovich [19] justify this assumption using histograms obtained from computer simulations), the exact approach discussed in Section 10.1.2.2 is applied to derive results for MRC performance in the presence of CCI. Specifically, the PDF of the SIR γ_t is shown to be[13]

$$p_{\gamma_t}(\gamma_t) = \frac{\Gamma(N_I+L)}{\Gamma(L)\Gamma(N_I)} \left(\frac{P_d}{P_I}\right)^{N_I} \frac{\gamma_t^{L-1}}{(P_d/P_I + \gamma_t)^{N_I+L}}, \quad \gamma_t \geq 0 \quad (10.82)$$

Comparing (10.82) with (10.71) we observe that they are of similar form and thus the approach taken to compute the MGF in (10.73) can also be used here. Specifically using (10.72) with now $x = \gamma_t$, $a = s$, $\lambda = L-1$, $\rho = N_I + L$, $y = P_d/P_I$, we obtain after simplification

$$M_{\gamma_t}(-s) = \frac{\Gamma(N_I+L)}{\Gamma(N_I)} \left(s\frac{P_d}{P_I}\right)^{(N_I-1)/2}$$

$$\times \exp\left(\frac{1}{2}s\frac{P_d}{P_I}\right) W_{(-2L-N_I+1)/2, -N_I/2}\left(s\frac{P_d}{P_I}\right) \quad (10.83)$$

Hence, the average BEP can be computed from (10.78) with $M_{\gamma_t}(-s)$ as in (10.83).

Analogous to (10.79), a closed-form expression for $P_b(E)$ was found by Shah and Haimovich [19] by direct integration of the conditional BEP over the PDF in (10.71) with the result

$$P_b(E) = \frac{1}{2\sqrt{\pi}\Gamma(N_I)\Gamma(L)} \left[\left(\frac{P_d}{P_I}\right)^{N_I} \frac{\Gamma(\frac{1}{2}-N_I)\Gamma(N_I+L)}{\Gamma(-N_I)} \right.$$

$$\times {}_2F_2\left(N_I+L, N_I; N_I+\frac{1}{2}, N_I+1; \frac{P_d}{P_I}\right) - 2\left(\frac{P_d}{P_I}\right)^{1/2} \Gamma\left(N_I-\frac{1}{2}\right)$$

$$\left. \times \Gamma\left(L+\frac{1}{2}\right) {}_2F_2\left(L+\frac{1}{2}, \frac{1}{2}; \frac{3}{2}-N_I, \frac{3}{2}; \frac{P_d}{P_I}\right) + \sqrt{\pi}\Gamma(N_I)\Gamma(L) \right]$$
(10.84)

Borrowing on numerical results presented in Ref. 19, Fig. 10.10 is a plot of average BEP versus SIR per channel, P_d/P_I, as computed from (10.79) for OC and (10.84) for MRC for two values of the number of antennas, namely, $L=3$ and $L=6$, and also $N_I=18$ interfering sources. It is observed that OC can provide improved performance over MRC even when the number of interfering sources is much larger than the number of antennas. For example, for $P_b(E) = 10^{-3}$ and $L=6$, OC requires 1 dB less SIR than MRC. Figure 10.11

---

[13] Note that the PDF for MRC in (10.82) is not restricted to $L \geq N_I$ as was the case for (10.71) corresponding to OC.

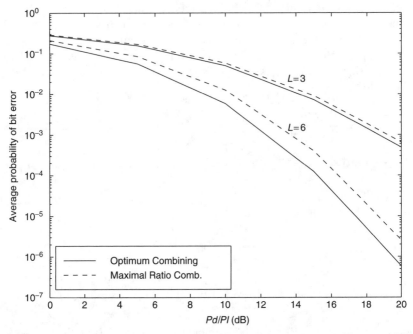

Figure 10.10. Average BEP versus channel SIR for Rayleigh fading; $N_I = 18$. (Courtesy of Shah and Haimovich [19].)

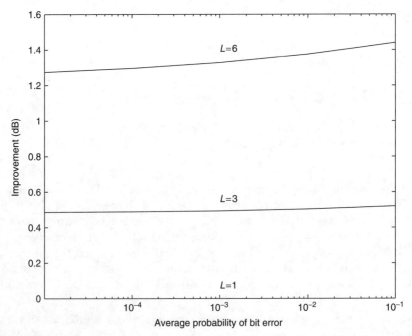

Figure 10.11. Improvement of OC over MRC in Rayleigh fading; $N_I = 18$. (Courtesy of Shah and Haimovich [19].)

explicitly shows the improvement of OC over MRC as a function of the average BEP for the same numbers of antennas as in Fig. 10.10. (Clearly, for a single antenna, the two combining schemes produce identical performance.) By improvement in Fig. 10.11 is meant the reduction in required SIR per channel to obtain a given average BEP using OC as compared with MRC. We observe from this figure that for a fixed number of antenna elements, the improvement is quite insensitive to the value of average BEP over a range of four decades. On the other hand, for a fixed average BEP, a noticeable improvement is obtained as the number of antenna elements increases.

The extension of the above results to the case of Nakagami-m fading has recently been considered by Aalo and Zhang [21]. In particular, letting m_d and m_I denote the Nakagami-m parameters on the desired signal and interference, respectively, from Ref. 21 the result for the average BEP of coherent BPSK with MRC is in our notation given by

$$P_b(E) = \frac{1}{2} - \frac{\Gamma(m_d L + m_I N_I)\Gamma\left(m_d L + \frac{1}{2}\right)\sqrt{\pi}}{\Gamma(m_d L)\Gamma(m_I N_I)\cos(\pi m_I N_I)}$$

$$\times \left[\frac{1}{2m_I N_I} \left(\frac{P_d}{P_I}\right)^{m_I N_I} \frac{1}{\Gamma\left(m_d L + \frac{1}{2}\right)\Gamma\left(m_I N_I + \frac{1}{2}\right)} \right.$$

$$\times {}_2F_2\left(m_I N_I, m_d L + m_I N_I; m_I N_I + 1, m_I N_I + \frac{1}{2}; \frac{P_d}{P_I}\right)$$

$$- \left(\frac{P_d}{P_I}\right)^{1/2} \frac{1}{\Gamma(m_d L + m_I N_I)\Gamma\left(\frac{3}{2} - m_I N_I\right)}$$

$$\left. \times {}_2F_2\left(\frac{1}{2}, m_d L + \frac{1}{2}; \frac{3}{2}, \frac{3}{2} - m_I N_I; \frac{P_d}{P_I}\right) \right] \quad (10.85)$$

An identically equivalent result can be obtained from (10.84) by replacing L with $m_d L$ and N_I with $m_I N_I$. Using this same replacement in (10.79) gives the analogous result for OC, namely,

$$P_b(E)$$
$$= \frac{1}{2} + \frac{1}{2\sqrt{\pi}\Gamma(m_d L)\Gamma(m_I N_I + 1 - m_d L)} \left[\left(\frac{P_d}{P_I}\right)^{m_I N_I + 1 - m_d L} \right.$$

$$\times \frac{\Gamma\left(m_d L - m_I N_I - \frac{1}{2}\right)\Gamma(L+1)}{\Gamma(m_d L - m_I N_I - 1)} {}_2F_2\left(m_I N_I + 1, m_I N_I + 1 - m_d L; m_I N_I\right.$$

$$\left. - m_d L + \frac{3}{2}, m_I N_I - m_d L + 2; \frac{P_d}{P_I}\right) - 2\left(\frac{P_d}{P_I}\right)^{1/2} \Gamma\left(m_I N_I - m_d L + \frac{1}{2}\right)$$

$$\left. \times \Gamma\left(m_d L + \frac{1}{2}\right) {}_2F_2\left(m_d L + \frac{1}{2}, \frac{1}{2}; m_d L - m_I N_I + \frac{1}{2}, \frac{3}{2}; \frac{P_d}{P_I}\right) \right] \quad (10.86)$$

REFERENCES

1. W. C. Jakes, *Microwave Mobile Communications*, New York: Wiley, 1974.
2. B. Widrow, P. E. Mantey, L. J. Griffiths, and B. B. Goode, "Adaptive antenna systems," *Proc. IEEE*, vol. 55, December 1967, p. 2143.
3. S. R. Applebaum, "Adaptive antenna systems," *IEEE Trans. Antenna Propag.*, vol. AP-24, September 1976, p. 585.
4. R. A. Monzingo and T. W. Miller, *Introduction to Adaptive Arrays*, New York: Wiley, 1980.
5. V. M. Bogachev and I. G. Kiselev, "Optimum combining of signals in space-diversity reception," *Telecommun. Radio Eng.*, vol. 34–35, October 1980, pp. 83–85.
6. J. H. Winters, "Optimum combining in digital mobile radio with co-channel interference, *IEEE Trans. Veh. Technol.*, vol. VT-33, August 1984, pp. 144–155.
7. A. Shah, A. M. Haimovich, M. K. Simon, and M.-S. Alouini, "Exact bit error probability for optimum combining with a Rayleigh fading Gaussian co-channel interferer," to appear in the June 2000 issue of the *IEEE Trans. Commun.*
8. E. Villier, "Performance analysis of optimum combining with multiple interferers in flat Rayleigh fading," *IEEE Trans. Commun.*, vol. 47, October 1999, pp. 1503–1510.
9. J. V. DiFranco and W. L. Rubin, *Radar Detection*. Upper Saddle River, NJ: Prentice Hall, 1968.
10. I. S. Gradshteyn and I. M. Ryzhik, *Table of Integrals, Series, and Products*, 5th ed., San Diego, CA: Academic Press, 1994.
11. V. A. Aalo and J. Zhang, "Performance of antenna array systems with optimum combining in a Rayleigh fading environment," to appear in *IEEE Comm. Letters*.
12. J. Cui, D. D. Falconer, and A. U. H. Sheikh, "Analysis of BER for optimum combining with two co-channel interferers and maximal ratio combining with arbitrary number of interferers," *Proc. IEEE Int. Symp. Personal indoor Mobile Commun. (PIMRC'96)*, October 1996, pp. 53–57.
13. J. Cui, D. D. Falconer, and A. U. H. Sheikh, "Performance evaluation of optimum combining and maximal ratio combining in the presence of co-channel interference and channel correlation for wireless communication systems," *Mobile Networks Appl.*, vol. 2, 1997, pp. 315–324.
14. G. J. Foschini and J. Salz, "Digital communications over fading radio channels," *Bell Syst. Tech. J.*, vol. 62, February 1983, pp. 429–456.
15. J. H. Winters and J. Salz, "Upper bounds on the bit error rate of optimum combining in wireless systems," *Proc. Veh. Technol. Conf.*, 1994, pp. 942–946. (See also *IEEE Trans. Commun.*, vol. 46, December 1998, pp. 1619–1624.)
16. J. H. Winters, J. Salz, and R. Gitlin, "The impact of antenna diversity on the capacity of wireless communication systems," *IEEE Trans. Commun.*, vol. 42, February/March/April 1994, pp. 1740–1751.
17. A. M. Haimovich and A. Shah, "The performance of space-time processing for suppressing narrowband interference in CDMA communications," *Wireless Personal Commun.*, vol. 7, August 1998, pp. 233–255.
18. A. Shah and A. M. Haimovich, "Performance analysis of optimum combining in wireless communications with Rayleigh fading and cochannel interference," *IEEE Trans. Commun.*, vol. 46, April 1998, pp. 473–479.

19. A. Shah and A. M. Haimovich, "Performance analysis of maximal ratio combining and comparison with optimum combining for mobile radio communications with cochannel interference," to appear in *IEEE Trans. Veh. Technol.*
20. A. Erdelyi et al., *Tables of Integral Transforms*, vol. 2, New York: McGraw-Hill, 1954.
21. V. A. Aalo and J. Zhang, "On the effect of cochannel interference on average error rates in Nakagami-fading channels," *IEEE Commun. Lett.*, vol. 3, May 1999, pp. 136–138.

11

DIRECT-SEQUENCE CODE-DIVISION MULTIPLE ACCESS

In its generic form, direct-sequence code-division multiple access (DS-CDMA) is a spread-spectrum (SS) technique for simultaneously transmitting a number of signals representing information messages from a multitude of users over a channel employing a common carrier.[1] The method by which the various users share the channel is the assignment of a unique pseudonoise (PN)-type code to each user (which accompanies the transmission of the information) with orthogonal-like properties that allows the composite received signal to be separated into its individual user components, each of which can then be demodulated and detected. The deployment of the code (assumed to be represented by a binary waveform with PN properties[2]) at the transmitter (*spreading process*, i.e., the superposition onto the binary information waveform) is accomplished by a simple multiplication [which is equivalent to modulo-2 addition of their (0,1) representations], hence the term *direct-sequence* modulation. Similarly, the removal of the code at the receiver (*despreading process*) is also accomplished by the identical multiplication operation. For our purposes in this chapter, we assume that the receiver is perfectly capable of regenerating the transmitted codes corresponding to each of the users' transmissions, and as such we shall ignore all synchronization issues dealing with the acquisition and tracking of these codes at the receiver. A complete discussion of techniques for accomplishing these functions and their impact on system performance can be found in Part 4, Chapters 1 and 2, of Ref. 1.

The DS-CDMA technique has its roots in the literature dealing with military network applications (for a complete historical perspective on early SS systems

---

[1] Later we address a multiple-carrier version of this modulation, which has become of interest in recent years.

[2] Depending on the particular application, the codes may or not be purely orthogonal; however, in either event they are chosen to have large autocorrelation and small cross-correlation. A detailed discussion of the design of codes for SS applications such as CDMA is beyond the scope of this book but can be found in Ref. 1, Part 1, Chapter 5.

employing such modulations, see Part 1, Chapter 2, of Ref. 1), where their use was primarily to combat intentional jamming introduced by an enemy. As such, the communication channel was typically modeled as additive white Gaussian noise (AWGN) combined with jamming of one sort or another (see Part 1, Chapters 3 and 4, and Part 2, Chapter 1, of Ref. 1). More recently, however, DS-CDMA has secured a strong foothold in the commercial market primarily because of its adoption as the IS-95 standard that governs digital cellular telephony in the United States and elsewhere. It is this and related wireless communication applications that motivate the results presented in this chapter, since here the primary channel of interest is the fading channel along with the possibility of narrowband interference. In fact, it is the inclusion of this possibility that has stimulated researchers to investigate a multiple-carrier form of DS-CDMA, since as we shall see later in the chapter, this particular form offers significant advantage over the traditional single-carrier version in combating such interference.

In view of the discussion above, it seems natural, therefore, to divide the chapter into two main sections: *single-* and *multiple-carrier DS-CDMA*. Each of these sections focuses on the average bit error rate (BER) performance of the corresponding DS-CDMA system when transmitted over a generalized fading channel such as those modeled in previous chapters. As before, the emphasis will be on using the alternative forms of the classic functions developed in Chapter 4 to simplify the resulting expressions. Although we shall restrict ourselves to binary DS modulation (spreading waveforms), we allow for the possibility of other than a rectangular pulse shape to represent the PN code chip waveform.

11.1 SINGLE-CARRIER DS-CDMA SYSTEMS

There are a large number of papers dealing with the performance of DS-CDMA systems over frequency-selective fading channels [2–13]. In this section we apply the MGF-based approach presented in Chapter 9 to derive the average BER performance of binary DS-CDMA systems operating over these channels [14]. The results presented in this section are applicable to systems that employ RAKE reception with coherent maximal-ratio combining (MRC). Extensions to other type of detection/diversity combining techniques is straightforward (in view of the results presented in Chapter 9) and are therefore left as an exercise for the reader. Also, although specifically developed for receiver type of diversity, the analysis presented in this section can be used with some modifications to assess the performance of transmit diversity CDMA systems [15,16].

11.1.1 System and Channel Models

11.1.1.1 Transmitted Signal. We consider a binary DS-CDMA system with K_u independent users sharing a channel simultaneously, each transmitting with power P at a common carrier frequency $f_c = \omega_c/2\pi$, using a data rate $R_b = 1/T_b$

and a chip rate $R_c = 1/T_c$. The kth user, $k = 1, 2, \ldots, K_u$, is assigned a unique code sequence $\{a_{k,j}\}$ of chip elements $(+1, -1)$, so that its code waveform is given by

$$a_k(t) = \sum_{j=-\infty}^{+\infty} a_{k,j} P_{T_c}(t - jT_c) \tag{11.1}$$

where the function $P_T(\cdot)$ denotes the chip pulse of duration T. In the single-carrier case we assume that $P_T(\cdot)$ is a unit rectangular pulse, whereas in the multicarrier case we will consider Nyquist pulses. The code sequence $\{a_{k,j}\}$ is assumed to be periodic, with period equal to the processing gain PG $= T_b/T_c$. The data signal waveform $b_k(t)$ given by

$$b_k(t) = \sum_{j=-\infty}^{+\infty} b_{k,j} P_{T_b}(t - jT_b) \tag{11.2}$$

is binary phase-shift-keyed (BPSK) onto the carrier at f_c, which is then spread by that user's code sequence and transmitted over the channel. The resulting kth user's transmitted signal $s_k(t)$ is thus given by

$$s_k(t) = \sqrt{2P} a_k(t) b_k(t) \cos(\omega_c t) \tag{11.3}$$

The composite transmitted signal $s(t)$ at the input of the channel can then be expressed as

$$s(t) = \sum_{k=1}^{K_u} \sqrt{2P} a_k(t) b_k(t) \cos(\omega_c t) \tag{11.4}$$

11.1.1.2 Channel Model. DS-CDMA systems involve a spreading process which results in a transmitted signal whose bandwidth is much wider then the channel coherence bandwidth, and therefore undergoes frequency-selective fading. Following our discussion in Chapter 2, this type of fading is typically modeled by a linear filter which for the kth user is characterized by a complex-valued lowpass equivalent impulse response [17–19]

$$h_k(t) = \sum_{l=1}^{L_p} \alpha_{k,l} e^{-j\theta_{k,l}} \delta(t - \tau_{k,l}) \tag{11.5}$$

where $\delta(\cdot)$ is the Dirac delta function, l the propagation path index, and $\{\alpha_{k,l}\}_{l=1}^{L_p}$, $\{\theta_{k,l}\}_{l=1}^{L_p}$, and $\{\tau_{k,l}\}_{l=1}^{L_p}$ the random path amplitudes, phases, and delays, respectively. We assume that the sets $\{\alpha_{k,l}\}_{l=1}^{L_p}$, $\{\theta_{k,l}\}_{l=1}^{L_p}$, and $\{\tau_{k,l}\}_{l=1}^{L_p}$ are mutually independent. In (11.5), L_p is the number of resolvable paths (the first path being the reference path whose delay $\tau_1 = 0$) and is related to the ratio of the maximum delay spread $\tau_{\max}$ to the chip time duration T_c.

We assume slow fading, so that L_p is constant over time, and $\{\alpha_{k,l}\}_{l=1}^{L_p}$, $\{\theta_{k,l}\}_{l=1}^{L_p}$, and $\{\tau_{k,l}\}_{l=1}^{L_p}$ are all constant over a symbol interval. If the different paths of a given impulse response are generated by different scatterers, they tend to exhibit negligible correlations [20,21]. In this case it is reasonable to assume that the $\{\alpha_{k,l}\}_{l=1}^{L_p}$ are statistically independent random variables (RV's). We make this assumption in our analysis but remind the reader that this assumption can be relaxed for Nakagami-m fading channels in view of the results presented in Section 9.6.4. We denote the fading amplitude of the kth-user lth resolved path by $\alpha_{k,l}$, which is a RV whose mean-square value $\overline{\alpha_{k,l}^2}$ is assumed to be independent of k and is denoted by Ω_l.

After passing through the fading channel, the signal is perturbed by additive white Gaussian noise (AWGN) with a one-sided power spectral density which is denoted by N_0 (W/Hz). The AWGN is assumed to be independent of the fading amplitudes $\{\alpha_{k,l}\}_{l=1}^{L_p}$. Hence, the instantaneous SNR per bit of the lth channel is given by $\gamma_{k,l} = \alpha_{k,l}^2 E_b/N_0$, where E_b (J) is the energy per bit, and the average SNR per bit of the lth channel is given by $\overline{\gamma}_l = \Omega_l E_b/N_0$.

11.1.1.3 Receiver. With multipath propagation, it follows from (11.5) and (11.4) that the received signal $r(t)$, whose signal component is the time convolution of $s(t)$ and $h(t)$, may be written as

$$r(t) = \sqrt{2P} \sum_{k=1}^{K_u} \sum_{l=1}^{L_p} \alpha_{k,l} a_k(t - \tau_{k,l}) b_k(t - \tau_{k,l}) \cos[\omega_c(t - \tau_{k,l}) + \theta_{k,l}] + n(t) \tag{11.6}$$

where $n(t)$ is the receiver AWGN random process.

We consider an L-branch (finger) MRC RAKE receiver, as shown in Fig. 9.2. The optimal value for L is L_p, but L may be chosen less than L_p, due to receiver complexity constraints. Let us consider the kth-user receiver. Each of the L paths to be combined is first coherently demodulated through multiplication by the unmodulated carrier $\cos[\omega_c(t - \tau_{k,l}) + \theta_{k,l}]$, then lowpass filtered to remove the second harmonics of the carrier. All these operations assume that the receiver is correctly time and phase synchronized at every branch (i.e., perfect carrier recovery, and perfect phase $\{\theta_{k,l}\}_{l=1}^{L}$ and time delay $\{\tau_{k,l}\}_{l=1}^{L}$ estimates). Using MRC (see Section 9.2) and assuming perfect knowledge of the fading amplitude on each finger, the L lowpass filter outputs $\{r_{m,l}\}_{l=1}^{L}$ are individually weighted by their respective fading amplitudes and then combined by a linear combiner yielding the decision variable

$$r_k = \sum_{l=1}^{L} \alpha_{k,l} r_{k,l}, \qquad k = 1, 2, \ldots, K_u \tag{11.7}$$

11.1.2 Performance Analysis

Without loss of generality, let us consider the kth user's performance. The decision variable r_k may be written as the sum of a desired signal component and three interference/noise components [10]:

$$r_k = \pm \left(\sum_{l=1}^{L} \alpha_{k,l}^2\right)\sqrt{E_b} + \sum_{l=1}^{L} \alpha_{k,l}(I_S + I_M + N) \tag{11.8}$$

where I_S is the self-interference component induced by the autocorrelation function of the kth user's spreading code, I_M is the multiple-access interference (MAI) component induced by the other $K_u - 1$ users on the desired user, and N is a zero-mean AWGN component with variance $\sigma_N^2 = N_0/2$. Eng and Milstein showed that I_S can be considered to be a zero-mean Gaussian RV with variance [10, Eq. (9)]

$$\sigma_S^2 = \frac{\Omega_T - 1}{2PG}\Omega_1 E_b \tag{11.9}$$

where $\Omega_T = \sum_{l=1}^{L_p} \Omega_l/\Omega_1$ can be interpreted as the normalized (to the first path) total average fading power. Similarly, under the standard Gaussian approximation (large numbers of users) [3,5,6,10], I_M can be modeled as a zero-mean Gaussian RV with variance [10, Eq. (6)]

$$\sigma_M^2 = \frac{2(K_u - 1)\Omega_T}{6PG}\Omega_1 E_b \tag{11.10}$$

Under these assumptions $\bar{r}_k$ may be considered to be a conditional Gaussian RV (conditioned on $\{\alpha_{k,l}\}_{l=1}^{L}$) with a conditional mean $\mathrm{E}[\bar{r}_k/\{\alpha_{k,l}\}_{l=1}^{L}]$ and a conditional variance $\mathrm{var}[\bar{r}_k/\{\alpha_{k,l}\}_{l=1}^{L}]$ given by

$$\mathrm{E}\left[r_k|\{\alpha_{k,l}\}_{l=1}^{L}\right] = \pm \left(\sum_{l=1}^{L} \alpha_{k,l}^2\right)\sqrt{E_b} \tag{11.11}$$

$$\mathrm{var}\left(r_k|\{\alpha_{k,l}\}_{l=1}^{L}\right) = \sum_{l=1}^{L} \alpha_{k,l}^2 \left(\sigma_N^2 + \sigma_S^2 + \sigma_M^2\right) \tag{11.12}$$

Assuming that the data bits $+1$ or -1 are equally probable, the kth-user conditional SNR, $\mathrm{SNR}(\{\alpha_{k,l}\}_{l=1}^{L})$, is given as

$$\mathrm{SNR}\left(\{\alpha_{k,l}\}_{l=1}^{L}\right) = \frac{(\mathrm{E}[r_i|\underline{\alpha}_i])^2}{2\,\mathrm{var}(r_i|\underline{\alpha}_i)}$$

$$= \left(\sum_{l=1}^{L} \alpha_{k,l}^2\right)\frac{E_b}{N_e} \tag{11.13}$$

where $N_e/2$ is the equivalent two-sided interference plus noise power spectral density defined as

$$\frac{N_e}{2} \triangleq \sigma_N^2 + \sigma_S^2 + \sigma_M^2$$

$$= \frac{(2K_u + 1)\Omega_T - 3}{6PG} \Omega_1 E_b + \frac{N_0}{2} \qquad (11.14)$$

with $\bar{\gamma}_1 = \Omega_1 E_b/N_0$ the average received SNR per bit corresponding to the first path.

11.1.2.1 General Case. In view of (11.13) and (11.14), and since we are assuming BPSK modulation, the average BER performance expression obtained in (9.11) applies here for DS-CDMA by replacing N_0 with N_e, or equivalently, replacing the average SNR per bit of the lth path $\bar{\gamma}_l = \Omega_l E_b/N_0$ by

$$\bar{\gamma}_{l,e} = \frac{\Omega_l E_b}{N_e} = \bar{\gamma}_l \left[1 + \frac{(2K_u + 1)\Omega_T - 3}{3PG} \bar{\gamma}_1 \right]^{-1} \qquad (11.15)$$

11.1.2.2 Application to Nakagami-m Fading Channels. The performance of coherent DS-CDMA systems over Nakagami-m frequency-selective fading channels with MRC [10,12] and postdetection EGC [9,13,22] RAKE reception has received considerable attention in the recent literature. In particular, Eng and Milstein [10] have provided a BER performance analysis of coherent DS-CDMA systems in a Nakagami-m fading environment with an equally spaced exponentially decaying power delay profile ($\Omega_l = \Omega_1 e^{\delta(l-1)}, l = 1, 2, \ldots, L_p$), where δ is the power decay factor. Their analysis relies on a classical PDF-based approach and uses a Nakagami approximation [23, Eq. (80)] to the PDF of the sum of squares of independent nonidentically distributed Nakagami-m RVs, which leads to a closed-form *approximation* to the BER in terms of the Gauss hypergeometric series, $_2F_1(\cdot,\cdot;\cdot;\cdot)$. The approximation is accurate for small values of the power decay factors δ but loses its accuracy as δ increases. More recently, Efthymoglou et al. [12] applied the Gil–Pelaez lemma [12, Eq. (26)] to obtain *exact* BER performance in the Nakagami-m fading environment with arbitrary fading parameters along the different resolvable paths. Applying the approach described above to the problem treated in Refs. 10 and 12 and using the MGF corresponding to Nakagami-m fading given in Table 2.1 yields after some manipulations the following expression for the average BER:

$$P_b(E) = \frac{1}{\pi} \int_0^{\pi/2} \prod_{l=1}^{L} \left(1 + \frac{\bar{\gamma}_e e^{-(l-1)\delta}}{m_l \sin^2 \phi} \right)^{-m_l} d\phi \qquad (11.16)$$

where $\bar{\gamma}_e = \Omega_1 E_b/N_e$ can be viewed as the equivalent signal-to-noise-plus-interference ratio per bit corresponding to the first path, and which may be expressed using (11.15) as

$$\overline{\gamma}_e = \left(\frac{(2K_u + 1)\Omega_T - 3}{3PG} + \frac{1}{\overline{\gamma}_1} \right)^{-1} \quad (11.17)$$

In this case, Ω_T reduces to $\Omega_T = \sum_{l=1}^{L_p} e^{-(l-1)\delta} = (1 - e^{-L_p \delta})/(1 - e^{-\delta})$. The exact and simple form of the average BER in (11.16) has advantage over previous equivalent forms which are either an approximation [10, Eq. (16a)] or are expressed in terms of an integral with an infinite upper limit and with a much more complicated integrand [12, Eq. (33)]. Many such DS-CDMA performance analysis problems (for other types of channel models [24] or other types of detection/diversity combining [9,13,22]) may be solved using the approach described above.

11.2 MULTICARRIER DS-CDMA SYSTEMS

Multicarrier code-division multiple-access (MC-CDMA) systems have recently received widespread interest due to their potential for high-speed transmission and their effectiveness in mitigating the effects of multipath fading and in rejecting narrowband interference [25]. Different classes of these systems have been presented and studied by various researchers [25]. Here we focus on one particular scheme proposed by Kondo and Milstein [26]. In the absence of partial band interference (PBI), this scheme was shown to achieve performance equivalent to RAKE reception of single-carrier DS-CDMA systems operating over frequency-selective Rayleigh channels with a flat multipath intensity profile. However, the MC-CDMA system has the advantage of not requiring a continuous spectrum since the various subbands over which the modulated signal is transmitted can be chosen in various parts of the "unused" spectrum. In addition, the proposed MC-CDMA scheme was shown to outperform single-carrier systems in the presence of PBI and can therefore be favorably used in an overlay CDMA/FDMA or CDMA/TDMA situation without the need for sophisticated adaptive notch filtering to excise the interference.

There have been several studies that built on the work of Kondo and Milstein. In particular, Rowitch and Milstein examined the performance of coded versions of the original scheme [27]. The effect of fading correlation between the various subbands was studied in Refs. 28 and 29. In this section we present generic expressions for the average BER performance of MC-CDMA systems operating over generalized fading channels with and without PBI [30]. The results are again applicable to systems that employ coherent detection with MRC but can easily be extended, in view of the results in Chapter 9, to other detection and combining technique combinations. Aside from providing equivalent forms for known expressions corresponding to the performance of MC-CDMA over Rayleigh fading channels [26], utilization of the MGF-based approach for this specific application yields a solution for many scenarios that would be otherwise difficult to analyze in simple form. More specifically, this approach is particularly useful for the performance of MC-CDMA systems operating in a microcellular environment where the fading tends to follow a Rice or Nakagami-m type of distribution. Indeed, as we shall see in the following sections, because of the PBI the variance of the subband correlator outputs may be different, and hence computing the PDF of the conditional SNR, which can be done for the Rayleigh fading case, becomes a difficult task for other types of fading. In fact, even in the absence of PBI, the variance of the subband correlator outputs may be different because of the variation in the average fading power across the band. This is particularly true when the various subbands are not contiguous (i.e., subcarriers are far apart) because of the dependence of the path loss on the carrier frequency [31, Sec. 2.5].

11.2.1 System and Channel Models

11.2.1.1 Transmitter.
Consider a BPSK multicarrier coherent DS-CDMA system with K_u independent users each transmitting with power P. The users are simultaneously sharing an available bandwidth $\text{BW} = (1+\alpha)/T_c$, where T_c is the chip duration of a corresponding single-carrier wideband DS-CDMA system, and α ($0 \leq \alpha \leq 1$) is the rolloff factor of the chip wave-shaping Nyquist filter. The available spectrum BW is divided into (not necessarily contiguous) M_f equal bandwidth subbands each of width BW_{M_f} approximately equal to the coherence bandwidth of the channel (see Fig. 11.1). Each subband is assigned a carrier (at frequency $f_l, l = 1, 2, \ldots, M_f$) which is DS-CDMA modulated with the same user information at bit rate $1/T_b$ and chip rate $1/(M_f T_c)$ (see Fig. 11.2). Each user is effectively assigned a specific periodic code sequence of chip elements $(+1,-1)$ and of processing gain per subband $\text{PG}' = \text{PG}/M_f$. We

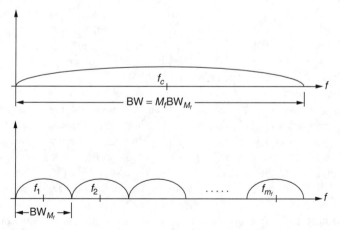

Figure 11.1. Comparison between single and multicarrier DS-CDMA systems power spectral densities.

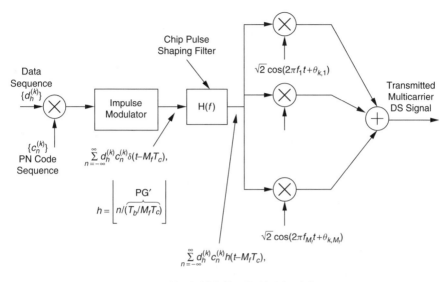

Figure 11.2. Transmitter of multicarrier DS-CDMA system for the kth user.

assume deterministic subband PN codes with ideal autocorrelation function. The use of bandlimited (Nyquist-shaped) spreading waveforms with wave-shaping filter transfer function denoted by $H(f)$ guarantees that the DS waveforms do not overlap.

11.2.1.2 Channel.

Following the system design and modeling assumptions of Kondo and Milstein [26], the number of subbands as well as their bandwidths are chosen so that the separate subbands fade slowly and nonselectively. Under these assumptions, the channel transfer function of the lth subband for the kth user is $\alpha'_{k,l} \exp(j\theta'_{k,l})$, where the $\{\alpha'_{k,l}\}_{l=1}^{M_f}$ are the fading amplitude RV's and $\{\theta'_{k,l}\}_{l=1}^{M_f}$ are independent uniformly distributed RV's over $[0, 2\pi]$. The average fading power of the lth subband is denoted by $\Omega'_l = \overline{(\alpha'_{k,l})^2}$ and is assumed to be independent of k.

11.2.1.3 Receiver.

The receiver consists of a bank of M_f matched filters followed by MRC (see Fig. 11.3). Each of the received modulated subband carriers is first passed through a bandpass chip-matched filter $H^*(f)$, then coherently demodulated, sampled, despread, and summed. All these operations assume that the receiver is correctly phase and time synchronized at every branch (i.e., perfect carrier recovery and bit synchronization). We denote by $X(f) = H(f)H^*(f) = |H(f)|^2$ the overall frequency response of the chip wave-shaping Nyquist filter and assume that $X(f)$ is a root raised-cosine frequency

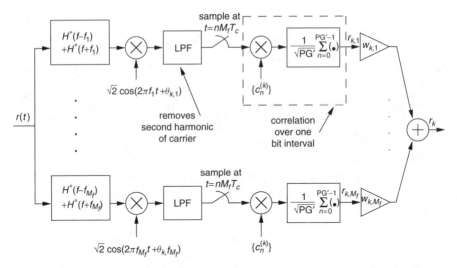

Figure 11.3. Receiver of multicarrier DS-CDMA system for *k*th user (ideal coherent demodulation).

response given by

$$X(f) = \begin{cases} \dfrac{1}{W}, & 0 \le |f| < \dfrac{W}{2}(1-\alpha) \\ \dfrac{1}{2W}\left[1-\sin\left(\dfrac{1}{2\alpha}\left(\dfrac{2\pi|f|}{W}-\pi\right)\right)\right], & \dfrac{W}{2}(1-\alpha) \le |f| \le \dfrac{W}{2}(1+\alpha) \\ 0, & \dfrac{W}{2}(1+\alpha) \le |f| \end{cases}$$

(11.18)

with $W = 1/T'_c = 1/(M_f T_c)$ for multicarrier and $W = 1/T_c$ for single carrier. In addition, we normalize the chip correlators by $1/\sqrt{PG'}$ (Kondo and Milstein [26] do not normalize). Finally, the M_f test statistics $r_{k,l}$ are individually weighted by the coefficients $w_{k,l}$ and then combined according to the rules of MRC to form the decision variable $r_k = \sum_{l=1}^{M_f} w_{k,l} r_{k,l}$.

11.2.1.4 Notation. For clarity we show in this section the equivalence between multicarrier and single-carrier system parameters. To distinguish between them we add a prime to the multicarrier system parameters.

- Subband chip time $T'_c = M_f T_c$.
- Subband processing gain $PG' = T_b/M_f T_c = PG/M_f$.
- Subband chip energy $E'_c = E_c [(P/M_f)M_f T_c = PT_c]$.
- Subband bit energy $E'_b = E_b/M_f = PG'E'_c = PG'E_c$.

11.2.2 Performance Analysis

Without loss of generality let us consider the kth user's performance. Two cases are of interest. In the first case the received signal is affected by fading, MAI, and AWGN. The second case corresponds to the scenario where the received signal is affected not only by the fading, MAI, and AWGN but also by Gaussian PBI.

11.2.2.1 Conditional SNR

Case 1: No Partial-Band Interference. The decision variable of the kth user may be written as the sum of a desired signal component and two interference/noise components [26]:

$$r_k = \sum_{l=1}^{M_f} w_{k,l} r_{k,l} = \pm \sum_{l=1}^{M_f} w_{k,l} \alpha'_{k,l} \sqrt{\frac{E_b}{M_f}} + \sum_{l=1}^{M_f} w_{k,l}(I_{M_l} + N) \quad (11.19)$$

where $E_b = PT_b$ is the energy per bit, I_{M_l} is the MAI term induced by the other $K_u - 1$ users in the lth subband, and N is the zero-mean AWGN component with variance $\sigma_N^2 = N_0/2$. Under the standard Gaussian approximation (valid for large number of users), I_{M_l} can be modeled as a zero-mean Gaussian RV with variance $\sigma_{M_l}^2$ given by (assuming Nyquist chip pulses)

$$\sigma_{M_l}^2 = \frac{(K_u - 1)E'_c \Omega'_l}{2T'_c} \int_{-\infty}^{\infty} X^2(f) df = \frac{(K_u - 1)E_b \Omega'_l}{2M_f PG'} \left(1 - \frac{\alpha}{4}\right) \quad (11.20)$$

Assuming that the MAI and AWGN are independent, we define an equivalent additive interference/noise with two-sided power spectral density

$$\begin{aligned}\frac{N_{e_l}}{2} = \sigma_N^2 + \sigma_{M_l}^2 &= \frac{(K_u - 1)E_b \Omega'_l}{2M_f PG'} \left(1 - \frac{\alpha}{4}\right) + \frac{N_0}{2} \\ &= \frac{N_0}{2} \left[1 + \frac{K_u - 1}{M_f PG'} \left(1 - \frac{\alpha}{4}\right) \overline{\gamma_l}\right]\end{aligned} \quad (11.21)$$

where $\overline{\gamma_l} = \Omega'_l E_b/N_0$ represents the average SNR/bit of the lth subband. Under these assumptions r_k is a conditionally Gaussian RV (conditioned on $\{\alpha'_{k,l}\}_{l=1}^{M_f}$) with conditional mean $E[r_k|\{\alpha'_{k,l}\}_{l=1}^{M_f}]$ and conditional variance $\mathrm{var}(r_k|\{\alpha'_{k,l}\}_{l=1}^{M_f})$ given by

$$E\left[r_k | \{\alpha'_{k,l}\}_{l=1}^{M_f}\right] = \pm \left(\sum_{l=1}^{M_f} w_{k,l} \alpha'_{k,l}\right) \sqrt{\frac{E_b}{M_f}}$$

$$\mathrm{var}\left(r_k | \{\alpha'_{k,l}\}_{l=1}^{M_f}\right) = \sum_{l=1}^{M_f} (w_{k,l})^2 \frac{N_{e_l}}{2}$$

(11.22)

Assuming that there is an equal probability that the data bits are $+$ or -1, we get the kth-user conditional SNR, SNR $(\{\alpha'_{k,l}\}_{l=1}^{M_f})$, as

$$\text{SNR}(\{\alpha'_{k,l}\}_{l=1}^{M_f}) = \frac{\left(\text{E}[r_k|\{\alpha'_{k,l}\}_{l=1}^{M_f}]\right)^2}{2\,\text{var}\left(r_k|\{\alpha'_{k,l}\}_{l=1}^{M_f}\right)}$$

$$= \frac{E_b}{M_f} \frac{\left(\sum_{l=1}^{M_f} w_{k,l}\alpha'_{k,l}\right)^2}{\sum_{l=1}^{M_f} (w_{k,l})^2 N_{e_l}} \quad (11.23)$$

The maximum conditional SNR is obtained when the various frequency diversity subbands are weighted as per the rule of MRC by the coefficients

$$w_{k,l} = \frac{\text{E}[r_{k,l}|\{\alpha'_{k,l}\}_{l=1}^{M_f}]}{\text{var}\left(r_{k,l}|\{\alpha'_{k,l}\}_{l=1}^{M_f}\right)} = \frac{2\alpha'_{k,l}}{N_{e_l}}\sqrt{\frac{E_b}{M_f}} \quad (11.24)$$

Substituting (11.24) in (11.23), we obtain the kth-user maximum conditional SNR, $\text{SNR}_{\max}(\{\alpha'_{k,l}\}_{l=1}^{M_f})$, at the MRC output as

$$\text{SNR}_{\max}(\{\alpha'_{k,l}\}_{l=1}^{M_f}) = \frac{E_b}{M_f}\sum_{l=1}^{M_f}\frac{(\alpha'_{k,l})^2}{N_{e_l}}$$

$$= \frac{E_b}{N_0 M_f}\sum_{l=1}^{M_f}\beta_l(\alpha'_{k,l})^2 \quad (11.25)$$

with

$$\beta_l = \left[1 + \frac{K_u - 1}{M_f PG'}\left(1 - \frac{\alpha}{4}\right)\overline{\gamma_l}\right]^{-1} \quad (11.26)$$

Kondo and Milstein [26] compared the performance of their proposed multicarrier system with a wideband single-carrier coherent CDMA system with L fingers RAKE reception [10]. For the special case of a single user ($K_u = 1$), the MAI term vanishes and the maximum conditional SNR of the multicarrier system, as given by (11.25), reduces to

$$\text{SNR}_{\max}(\{\alpha'_{k,l}\}_{l=1}^{M_f}) = \frac{E_b}{N_0 M_f}\sum_{l=1}^{M_f}(\alpha'_{k,l})^2 \quad (11.27)$$

On the other hand, it is well know that the maximum conditional SNR, $\text{SNR}_{\max}(\{\alpha_{k,l}\}_{l=1}^{L})$, of a single-carrier system with L fingers RAKE reception and in which the self-interference is negligible is given by [10]

$$\text{SNR}_{\max}\left(\{\alpha_{k,l}\}_{l=1}^{L}\right) = \frac{E_b}{N_0}\sum_{l=1}^{L}(\alpha_{k,l})^2 \tag{11.28}$$

Thus, for exact equivalence between the single-carrier and multicarrier systems, what we need is

$$L = M_f, \qquad \alpha_{k,l} = \frac{\alpha'_{k,l}}{\sqrt{M_f}} \tag{11.29}$$

For the multiuser case the maximum conditional SNR of the multicarrier system is given by (11.25), whereas the maximum conditional SNR of the single-carrier system is given by

$$\text{SNR}_{\max}\left(\{\alpha_{k,l}\}_{l=1}^{L}\right) = \frac{E_b}{N_0}\sum_{l=1}^{L}(\alpha_{k,l})^2 \left[1 + \frac{(K_u - 1)\Omega_T}{\text{PG}}\left(1 - \frac{\alpha}{4}\right)\bar{\gamma}_1\right]^{-1} \tag{11.30}$$

Hence, in the multiuser case even if $L = L_p$ and even if the conditions in (11.29) are met, comparing (11.25) and (11.30) we see that we do not have in general equivalence between the single- and multicarrier systems. However, equivalence can be achieved in the particular case of a uniform average fading power delay profile ($\Omega_l = \Omega, l = 1, 2, \ldots, L_p$) for single-carrier systems and a uniform average fading power across the frequency band for multicarrier systems ($\Omega'_l = \Omega', l = 1, 2, \ldots, M_f$). Indeed, in this special case if $L = L_p = M_f$ and if the single-user conditions (11.29) are met, we have $\Omega_T = L$ and $\Omega = \Omega'/M_f$, and thus

$$\Omega_T \Omega_1 = L\Omega = L\frac{\Omega'}{M_f} = \Omega' \tag{11.31}$$

which is the condition necessary for equivalence between (11.25) and (11.30).

Case 2: Partial-Band Interference. Consider now the presence of a PBI jammer modeled as a bandlimited white Gaussian noise with bandwidth $W_J = BW_{M_f}$ and power spectral density $S_{n_J}(f)$ such as

$$S_{n_J}(f) = \begin{cases} \dfrac{\eta_J}{2}, & f_J - \dfrac{W_J}{2} \leq |f| \leq f_J + \dfrac{W_J}{2} \\ 0, & \text{elsewhere} \end{cases} \tag{11.32}$$

where f_J denotes the jammer carrier frequency. The decision variable of the kth user may now be written as the sum of a desired signal component and three interference/noise components [26],

$$r_k = \sum_{l=1}^{M_f} w_{k,l} r_{k,l} = \pm \sum_{l=1}^{M_f} w_{k,l} \alpha'_{k,l} \sqrt{\frac{E_b}{M_f}} + \sum_{l=1}^{M_f} w_{k,l}(N + I_{M_l} + I_{J_l}) \quad (11.33)$$

where I_{J_l} is the Gaussian PBI present in the lth subband with variance

$$\sigma_{J_l}^2 \triangleq \frac{N_{J_l}}{2} = \frac{1}{2} \int_{-\infty}^{\infty} [S_{n_J}(f - f_l) + S_{n_J}(f + f_l)] X(f) \, df \quad (11.34)$$

Assuming that the PBI is independent of the MAI and AWGN, we can define a new equivalent additive interference/noise with two-sided power spectral density

$$\frac{N_{e_l}}{2} = \sigma_N^2 + \sigma_{M_l}^2 + \sigma_{J_l}^2$$

$$= \frac{N_0}{2} \left[1 + \frac{K_u - 1}{M_f \text{PG}'} \left(1 - \frac{\alpha}{4}\right) \overline{\gamma_l} + \frac{N_{J_l}}{N_0} \right] \quad (11.35)$$

Note that even if the average fading power is uniform across the subbands, some of the N_{e_l} values will still depend on l because of the presence of the PBI N_{J_l} in these subbands. Since the various frequency diversity subbands are weighted as per the rule of MRC by the weights (11.24) the kth-user maximum conditional SNR, $\text{SNR}_{\max}(\{\alpha'_{k,l}\}_{l=1}^{M_f})$, is still given by (11.25), where β_l is now given by

$$\beta_l = \left[1 + \frac{K_u - 1}{M_f \text{PG}'} \left(1 - \frac{\alpha}{4}\right) \overline{\gamma_l} + \frac{N_{J_l}}{N_0} \right]^{-1} \quad (11.36)$$

Because of the assumption that the bandwidth of the PBI is equal to the bandwidth of one subband, two subcases have to be considered. First, if the jammer carrier f_J coincides with one of the system subcarriers, say f_v, $v = 1, 2, \ldots, M_f$, the PBI completely overlaps the vth subband and

$$\frac{N_{J_v}}{N_0} = \frac{\text{JSR}_v \overline{\gamma}_v}{\text{PG}'(1 + \alpha)}, \quad v = 1, 2, \ldots \text{ or } M_f$$

$$\frac{N_{J_l}}{N_0} = 0, \quad l \neq v, \quad (11.37)$$

where $\text{JSR}_v = \eta_J W_J / \Omega_v E_b / T_b$ represents the interference (jamming) to average signal power ratio in the vth subband. Now if the jammer carrier f_J is between two of the system subcarriers, say $f_v < f_J < f_{v+1}$, the PBI partially overlaps the vth and $(v+1)$th subbands and it can be shown by substituting (11.18) in

(11.34) that

$$\frac{N_{J_v}}{N_0} = \frac{\text{JSR}_v \overline{\gamma}_v}{\text{PG'}(1+\alpha)} \left[1 - \frac{1+\alpha}{4} \frac{|\Delta f|}{W_J/2} + \frac{\alpha}{2\pi} \sin\left(\frac{\pi}{2} \frac{1+\alpha}{\alpha} \frac{|\Delta f|}{W_J/2}\right) \right]$$

$$\frac{N_{J_{v+1}}}{N_0} = \frac{\text{JSR}_{v+1} \overline{\gamma}_{v+1}}{\text{PG'}(1+\alpha)} \left[\frac{1+\alpha}{4} \frac{|\Delta f|}{W_J/2} - \frac{\alpha}{2\pi} \sin\left(\frac{\pi}{2} \frac{1+\alpha}{\alpha} \frac{|\Delta f|}{W_J/2}\right) \right]$$

$$\frac{N_{J_l}}{N_0} = 0, \quad l \neq v, v+1 \qquad (11.38)$$

where $\Delta f = f_J - f_v$ and $|\Delta f| \leq [\alpha/(\alpha+1)]W_J$.

For single-carrier systems affected by Gaussian PBI, the maximum conditional SNR becomes

$$\text{SNR}_{\max}\left(\{\alpha_{k,l}\}_{l=1}^L\right) = \frac{E_b}{N_0} \sum_{l=1}^L (\alpha_{k,l})^2 \left[1 + \frac{(K_u - 1)\Omega_T}{\text{PG}} \left(1 - \frac{\alpha}{4}\right) \overline{\gamma}_1 + \frac{N_J}{N_0} \right]^{-1}$$

(11.39)

where N_J is as in (11.34), so that

$$\frac{N_J}{N_0} = \frac{\text{JSR}\,\Omega_T \overline{\gamma}_1}{\text{PG}} \qquad (11.40)$$

where $\text{JSR} = \eta_J W_J / [(\sum_{l=1}^L \Omega_l) E_b / T_b]$ represents the interference to total average signal power ratio. Note that even in the special case of uniform power delay profile and uniform power distribution across the band, in the presence of PBI, we do *not* have equivalence between single- and multicarrier systems. In fact, if we employ the same equivalence conditions as for the no-interference case, *the multicarrier system has an SNR advantage in mitigating the PBI*, since, in general, some of N_{J_l}s will be equal to zero. This is confirmed in Section 11.3.3.

11.2.2.2 Average BER. Since the output of the MRC is a conditional (on the $\{\alpha_{k,l}\}_{l=1}^{M_f}$) Gaussian RV with a conditional SNR given by (11.25), the kth user's conditional BER, $P_b(E|\{\alpha'_{k,l}\}_{l=1}^{M_f})$, is given by

$$P_b\left(E\big|\{\alpha'_{k,l}\}_{l=1}^{M_f}\right) = Q\left(\sqrt{\frac{2E_b}{N_0 M_f} \sum_{l=1}^{M_f} \beta_l (\alpha'_{k,l})^2}\right) \qquad (11.41)$$

where $Q(\cdot)$ denotes the Gaussian Q-function. The goal is to evaluate the system performance in terms of the users' average BER, which requires the averaging of the conditional BER as given by (11.41) over the random fading amplitudes $\{\alpha'_l\}_{l=1}^{M_f}$. Recall that the classical PDF-based approach for solving this problem is to first find the PDF of $\gamma'_t \stackrel{\Delta}{=} \sum_{l=1}^{M_f} \beta_l(\alpha'_{k,l})/N_0$ and then to average (11.41) over that PDF. This is, in fact, the approach used by Kondo and Milstein, since for

Rayleigh fading this PDF can be found either in closed form or can be evaluated by residue calculations. However, because γ'_t is a weighted sum of RVs, finding its PDF for other fading conditions of interest (such as Rician or Nakagami-m with or without a uniform average fading power across the subbands) is a difficult task. To circumvent this difficulty we apply the alternative MGF-based approach.

Independent Fading Across the Subbands. Following the procedure in Section 9.2.3, we partition the conditional BER (11.41) in a separable product form, thereby obtaining the unconditional BER by independently averaging over the fading of each subband, resulting in an average BER expression given by

$$P_b(E) = \frac{1}{\pi} \int_0^{\pi/2} \prod_{l=1}^{M_f} M_{\gamma'_l}\left(-\frac{\beta_l}{M_f \sin^2 \phi}\right) d\phi \qquad (11.42)$$

where $M_{\gamma'_l}(s)$ denotes the MGF of the lth-subband SNR/bit, as given in Table 2.1.

Correlated Nakagami-m Fading. As discussed in Refs. 28 and 29, fading correlation among the various subband fading amplitudes induces a certain performance degradation. Under these conditions, using a procedure similar to the one adopted in Section 9.6.4.1, the average BER performance of MC-CDMA systems can be expressed as

$$P_b(E) = \frac{1}{\pi} \int_0^{\pi/2} M_{\gamma'_t}\left(-\frac{1}{\sin^2 \phi}\right) d\phi \qquad (11.43)$$

where $M_{\gamma'_t}(s)$ is the MGF of the combined output SNR with arbitrary correlated Nakagami-m faded subbands and which can be found in closed form based on Eq. (2.3) of Ref. 32 as

$$M_{\gamma'_t}(s) = E_{\gamma'_1, \gamma'_2, \ldots, \gamma'_{M_f}}\left[\exp\left(s \sum_{l=1}^{M_f} \frac{\beta_l}{M_f} \gamma'_l\right)\right]$$

$$= \prod_{l=1}^{M_f}\left(1 - \frac{s\beta_l}{mM_f}\right)^{-m} [\det[C_{ij}]_{M_f \times M_f}]^{-m} \qquad (11.44)$$

where

$$C_{ij} = \begin{cases} 1, & i = j \\ \sqrt{\rho_{ij}}\left(1 - \frac{mM_f}{s\beta_j}\right)^{-1}, & \text{otherwise} \end{cases} \qquad (11.45)$$

with ρ_{ij} the fading power correlation coefficient between subbands i and j.

11.2.3 Numerical Examples

We present in this section some numerical examples illustrating the effect of the severity of fading on the performance of MC-CDMA systems operating over a Nakagami-m fading channel with uniform average fading power profile across the band. Using the same system parameters as the ones in Kondo and Milstein [26] (i.e., $M_f = 4$, $\alpha = 0.5$, $K_u = 50$, PG$'$ = 128, $W_J = \text{BW}_{M_f}$, and $f_J = f_l, l = 1, 2, 3, 4$), we plot the BER performance of both systems in terms of $\overline{\gamma}' = M_f \overline{\gamma}$. Figures 11.4 through 11.6 compare the performance of an MC-CDMA system with its corresponding SC-CDMA system (with a flat power delay profile) for $m = 0.5$, $m = 1$, and $m = 2$, respectively. The results for the Rayleigh case ($m = 1$) check with the results published in Ref. 26. Note first that both MC-CDMA and SC-CDMA systems are more sensitive to the JSR for channels with a lower amount of fading since we observe a higher dynamic range in the BER performance for higher values of m. Furthermore, regardless of the severity of fading, the performance of SC-CDMA and MC-CDMA are almost the same for negligible JSR, but MC-CDMA outperforms SC-CDMA for high values of JSR. However, the relative difference between the MC-CDMA and SC-CDMA systems tends to increase as the amount of fading decreases (i.e., higher m), which means that MC-CDMA is even a better choice in a microcellular environment. Figures 11.7 through and 11.9 compare the performance of aligned MC-CDMA systems with the misaligned ones for

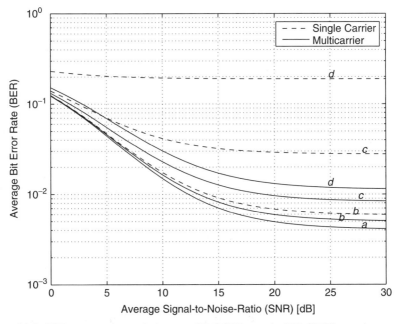

Figure 11.4. BER comparison between SC-CDMA and MC-CDMA systems over Nakagami-m fading channels, $m = 0.5$: (a) JSR $= -\infty$ dB; (b) JSR $= 10$ dB; (c) JSR $= 20$ dB; (d) JSR $= 30$ dB.

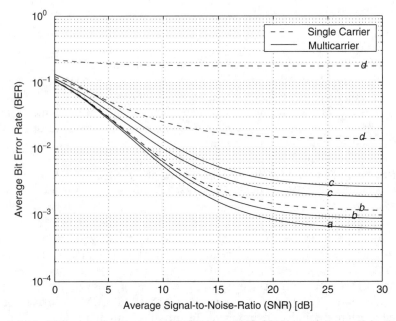

Figure 11.5. BER comparison between SC-CDMA and MC-CDMA systems over Nakagami-m fading channels, $m = 1$: (a) JSR $= -\infty$ dB; (b) JSR $= 10$ dB; (c) JSR $= 20$ dB; (d) JSR $= 30$ dB.

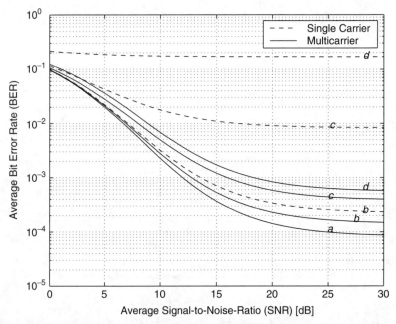

Figure 11.6. BER comparison between SC-CDMA and MC-CDMA systems over Nakagami-m fading channels, $m = 2$: (a) JSR $= -\infty$ dB; (b) JSR $= 10$ dB; (c) JSR $= 20$ dB; (d) JSR $= 30$ dB.

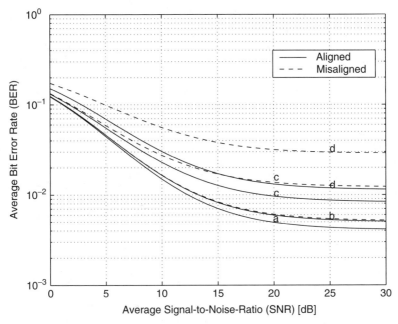

Figure 11.7. Average BER of MC-CDMA systems over Nakagami-m fading channels, $m = 0.5$: (a) JSR $= -\infty$ dB; (b) JSR $= 10$ dB; (c) JSR $= 20$ dB; (d) JSR $= 30$ dB.

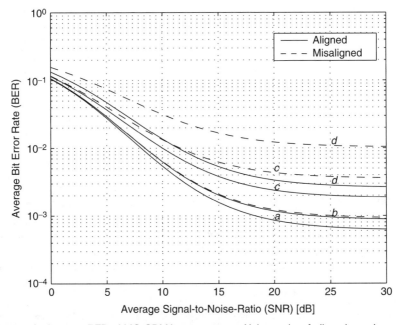

Figure 11.8. Average BER of MC-CDMA systems over Nakagami-m fading channels, $m = 1$: (a) JSR $= -\infty$ dB; (b) JSR $= 10$ dB; (c) JSR $= 20$ dB; (d) JSR $= 30$ dB.

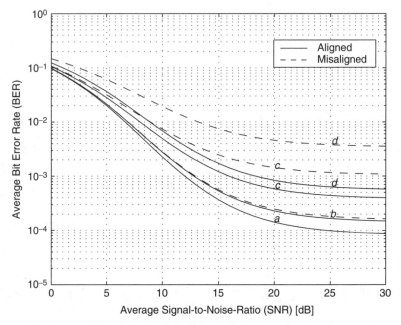

Figure 11.9. Average BER of MC-CDMA systems over Nakagami-m fading channels, $m = 2$: (a) JSR $= -\infty$ dB; (b) JSR $= 10$ dB; (c) JSR $= 20$ dB; (d) JSR $= 30$ dB.

$m = 0.5$, $m = 1$, and $m = 2$, respectively. Aligned systems correspond to the case of $f_J = f_l, l = 1, 2, 3, 4$, whereas misaligned systems correspond to the case $f_J = (f_l + f_{l+1})/2, l = 1, 2, 3$. For all values of m, the system performance is better when only one subband is affected by the interference. Furthermore, the relative difference between the two systems slightly increases for channels with a lower amount of fading.

REFERENCES

1. M. K. Simon, J. K. Omura, R. A. Scholtz, and B. K. Levitt, *Spread Spectrum Communications Handbook*, 2nd ed. New York: McGraw-Hill, 1994. Originally published in three parts as *Spread Spectrum Communications*, Computer Science Press, New York, 1984.
2. G. Turin, "The effects of multipath and fading on the performance of direct sequence CDMA systems," *IEEE J. Sel. Areas Commun.*, vol. SAC-2, April 1984, pp. 597–603.
3. E. A. Geraniotis and M. B. Pursley, "Performance of coherent direct-sequence spread-spectrum communications over specular multipath fading channels," *IEEE Trans. Commun.*, vol. COM-33, June 1985, pp. 502–508.
4. H. Xiang, "Binary code-division multiple-access systems operating in multipath fading, noisy channels," *IEEE Trans. Commun.*, vol. COM-33, August 1985, pp. 775–784.

5. E. A. Geraniotis, "Direct-sequence spread-spectrum communications multiple-access communications over nonselective and frequency-selective Rician fading channels," *IEEE Trans. Commun.*, vol. COM-34, August 1986, pp. 756–764.
6. J. S. Lehnert and M. B. Pursley, "Error probabilities for binary direct-sequence spread-spectrum communications with random signature sequences," *IEEE Trans. Commun.*, vol. COM-35, February 1987, pp. 87–98.
7. R. Prasad, H. S. Misser, and A. Kegel, "Performance analysis of direct-sequence spread-spectrum multiple access communication in an indoor Rician-fading channel with DPSK modulation," *Electron. Lett.*, vol. 26, August 1990, pp. 1366–1367.
8. R. D. J. van Nee, H. S. Misser, and R. Prasad, "Direct-sequence spread spectrum in a shadowed Rician fading land-mobile satellite channel," *IEEE J. Sel. Areas Commun.*, vol. SAC-10, February 1992, pp. 350–357.
9. T. Eng and L. B. Milstein, "Comparison of hybrid FDMA/CDMA systems in frequency selective Rayleigh fading," *IEEE J. Sel. Areas Commun.*, vol. SAC-12, June 1994, pp. 938–951.
10. T. Eng and L. B. Milstein, "Coherent DS-CDMA performance in Nakagami multipath fading," *IEEE Trans. Commun.*, vol. COM-43, February–March–April 1995, pp. 1134–1143.
11. R. Prasad, H. S. Misser, and A. Kegel, "Performance evaluation of direct-sequence spread spectrum multiple-access for indoor wireless communication in a Rician fading channel," *IEEE Trans. Commun.*, vol. COM-43, February–March–April 1995, pp. 581–592.
12. G. P. Efthymoglou, V. A. Aalo, and H. Helmken, "Performance analysis of coherent DS-CDMA systems in a Nakagami fading channel with arbitrary parameters," *IEEE Trans. Veh. Technol.*, vol. VT-46, May 1997, pp. 289–297.
13. G. P. Efthymoglou, V. A. Aalo, and H. Helmken, "Performance analysis of noncoherent binary DS/CDMA systems in a Nakagami multipath channel with arbitrary parameters," *Proc. IEEE Global Commun. Conf. (GLOBECOM'96)*, London, November 1996, pp. 1296–1300. Full paper published in *IEE Proc. Commun.*, vol. 144, June 1997, pp. 166–172.
14. M.-S. Alouini, M. K. Simon, and A. J. Goldsmith, "A unified performance analysis of DS-CDMA systems over generalized frequency-selective fading channels," *Proc. IEEE Int. Symp. Inf. Theory (ISIT'98)*, Cambridge, MA, August 1998, p. 8.
15. V. Weerackody, "Diversity for the direct-sequence spread spectrum system using multiple transmit antennas," *Proc. IEEE Int. Conf. Commun. (ICC'93)*, Geneva, June 1993, pp. 1775–1779.
16. R. Esmailzadeh and M. Nakagawa, "Pre-RAKE diversity combination for direct sequence spread spectrum communication systems," *Proc. IEEE Int. Conf. Commun. (ICC'93)*, Geneva, June 1993, pp. 463–467.
17. G. L. Turin, "Communication through noisy, random-multipath channels," *IRE Natl. Conv. Rec.*, March 1956, pp. 154–166.
18. G. L. Turin, F. D. Clapp, T. L. Johnston, S. B. Fine, and D. Lavry, "A statistical model of urban multipath propagation," *IEEE Trans. Veh. Technol.*, vol. VT-21, February 1972, pp. 1–9.
19. H. Suzuki, "A statistical model for urban multipath propagation," *IEEE Trans. Commun.*, vol. COM-25, July 1977, pp. 673–680.

20. H. Hashemi, "Impulse response modeling of indoor radio propagation channels," *IEEE J. Sel. Areas Commun.*, vol. SAC-11, September 1993, pp. 967–978.
21. S. A. Abbas and A. U. Sheikh, "A geometric theory of Nakagami fading multipath mobile radio channel with physical interpretations," *Proc. IEEE Veh. Technol. Conf. (VTC'96)*, Atlanta, GA, April 1996, pp. 637–641.
22. V. Aalo, O. Ugweje, and R. Sudhakar, "Performance analysis of a DS/CDMA system with noncoherent M-ary orthogonal modulation in Nakagami fading," *IEEE Trans. Veh. Technol.*, vol. VT-47, February 1998, pp. 20–29.
23. M. Nakagami, "The m-distribution: a general formula of intensity distribution of rapid fading," in *Statistical Methods in Radio Wave Propagation*, Oxford: Pergamon Press, 1960, pp. 3–36.
24. J. R. Foerster and L. B. Milstein, "Analysis of hybrid, coherent FDMA/CDMA systems in Ricean multipath fading," *IEEE Trans. Commun.*, vol. COM-45, January 1997, pp. 15–18.
25. K. Fazel and G. P. Fettweis, *Multi-carrier Spread-Spectrum*. Norwell, MA: Kluwer Academic Publishers, 1997.
26. S. Kondo and L. B. Milstein, "Performance of multicarrier DS CDMA systems," *IEEE Trans. Commun.*, vol. COM-44, February 1996, pp. 238–246.
27. D. N. Rowitch and L. B. Milstein, "Coded multicarrier DS-CDMA in the presence of partial band interference," *Proc. IEEE Mil. Commun. Conf. (MILCOM'96)*, McLean, VA, November 1996.
28. R. E. Ziemer and N. Nadgauda, "Effect of correlation between subcarriers of a MCM/DSSS communication system," *Proc. IEEE Veh. Technol. Conf. (VTC'96)*, Atlanta, GA, April 1996, pp. 146–150.
29. W. Xu and L. B. Milstein, "Performance of multicarrier DS CDMA systems in the presence of correlated fading," *Proc. IEEE Veh. Technol. Conf. (VTC'97)*, Phoenix, AZ, May 1997, pp. 2050–2054.
30. M. K. Simon and M.-S. Alouini, "BER performance of multicarrier DS-CDMA systems over generalized fading channels," *Proc. Commun. Theory Mini-conference* in conjunction with IEEE Int. Conf. on Commun. (ICC'99), Vancouver, British Columbia, Canada, June 1999, pp. 72–77.
31. G. L. Stüber, *Principles of Mobile Communications*. Norwell, MA: Kluwer Academic Publishers, 1996.
32. A. S. Krishnamoorthy and M. Parthasarathy, "A multivariate gamma-type distribution," *Ann. Math. Stat.*, vol. 22, 1951, pp. 549–557.

PART 5
FURTHER EXTENSIONS

12

CODED COMMUNICATION OVER FADING CHANNELS

Thus far we have considered the performance of uncoded digital communication systems over fading channels. As such, it has only been necessary to model the fading channel in a single symbol interval, T_s, since for uncoded transmission, decisions are made on a symbol-by-symbol basis. When error-correction coding is applied to the transmitted modulation and decisions are made based on an observation of the received signal much longer than T_s, it becomes necessary to consider the variation of the fading channel from symbol interval to symbol interval.

For the case of slow fading, the fading parameters (e.g., amplitude, phase) are typically treated as being constant over many symbol intervals, thereby introducing unintentional memory into the channel and an associated degradation of performance. A common method for breaking up these fading bursts without disturbing the intentional memory introduced by the coding is to employ *interleaving* at the transmitter and *deinterleaving* at the receiver. The combination of interleaving and deinterleaving acts in such a way as to produce fades that are independent from symbol to symbol, whereupon the fading channel once again becomes memoryless.

In studying the performance of coded communications over memoryless channels (with or without fading), the results are given as upper bounds on the average bit error probability (BEP). In principle, there are three different approaches to arriving at these bounds all of which employ obtaining the *pairwise error probability* (i.e., the probability of choosing one symbol sequence over another for a given pair of possible transmitted symbol sequences), followed by a weighted summation over all pairwise events. The first approach, which typically gives the weakest result but the simplest to evaluate, upper bounds the pairwise error probability by a Chernoff bound and then further upper bounds the summation over all pairwise events by a union bound (i.e., it treats the pairwise events as if they were independent where in reality they are correlated). Bounds on average error probability and average BEP arrived at in this fashion are referred to as *union–Chernoff bounds* and are typically the most common form found in the literature (e.g., for trellis-coded M-PSK modulation over Rayleigh and Rician fading channels see Refs. 1 and 2 (later documented in tutorial fashion in Refs. 3 through 5)). Furthermore, the evaluation of the union-bound portion of the overall upper bound is conveniently accomplished using the transfer function bound method originally proposed in Ref. 6 and later documented in tutorial form in Chapter 4 of Ref. 7.

The second approach evaluates the pairwise error probability exactly but considers only a finite number of pairwise events (i.e., those that are dominant) in place of the true union bound that considers *all* pairwise events. Examples of this approach can be found in Refs. 8 and 9. In the limiting case only the single dominant error event corresponding to the minimum distance between the correct and incorrect sequences is considered, which results in the simplest of this form of upper bound.[1]

The third approach also exactly evaluates the pairwise error probability but accounts for all pairwise events by using the transfer function bound to evaluate the true union bound. Clearly, of the three approaches this form will result in the tightest upper bound; however, it is, in general, more complex to evaluate analytically. Significant contributions using this approach can be found in Refs. 10 through 13, all of which focused on trellis-coded modulation (TCM) transmitted over fading channels. The degree to which these contributions differ from each other depends on the nature of the detection schemes (i.e., coherent versus differentially coherent), the statistics of the fading channel (e.g., Rayleigh, Rician), and the amount of knowledge concerning the state of the channel [i.e., the availability of *channel state information (CSI)*]. For example, the approach taken in Ref. 10 is not readily applicable to differential detection, and the approach taken by the authors of Refs. 11 and 12 is easily computed only for Rayleigh and Rician channels, where the difference of the decision metrics can be modeled as a quadratic form in complex Gaussian random variables (RVs). On the other hand, the method taken in Ref. 13 has the advantage that it can be extended to include a larger class of fading channels other than just Rayleigh and Rician, at the same time allowing for both coherent and differentially coherent detection of a variety of different modulation schemes of the form discussed previously. Unfortunately, however, as we shall demonstrate, this method is not practical when other-than-perfect CSI knowledge is available.

In this chapter we focus on the results obtained from the third approach since these provide the tightest upper bounds on the true performance. The first emphasis will be placed on evaluating the pairwise error probability with and without CSI, following which we deal with how the results of these evaluations can be used via the transfer-bound approach to evaluate the average BEP of coded modulation transmitted over the fading channel. For the method in Ref. 13, we shall soon show that the use of interleaving/deinterleaving to create a memoryless fading channel results in decision statistics akin to those obtained

---

[1] Upper bounds found by this approach are not *true* upper bounds but rather are *approximate* upper bounds because of the limited number of error events considered.

when diversity combining is employed to enhance system performance for the uncoded communication case (see Chapter 9). Because of this analogy, we therefore find it possible to apply the unified approach to coded communication over the memoryless fading channel in much the same way as it was used to simplify the evaluation of performance for multiple reception of uncoded communications. In particular, by using the alternative representations of the classic functions given in Chapter 4, we shall be able to exactly evaluate the pairwise error probability in the form of a product of integrals, each with finite limits and an integrand composed of tabulated functions. In the situations where the method in Ref. 13 is not practical, we shall turn to the method in Ref. 12, which requires evaluation of an integral with doubly infinite limits whose integrand is a product of characteristic functions.

Although, in principle, results can once again be obtained for a variety of fading channel models and modulation/coding types, to allow comparison with results obtained previously by, say, the first approach, we focus specifically on the combination of M-PSK modulation with trellis coding. We begin by considering the case of ideal coherent detection.

12.1 COHERENT DETECTION

12.1.1 System Model

Consider the block diagram of the trellis-coded M-PSK system illustrated in Fig. 12.1. Random binary i.i.d. information bits are inputted to a rate $n_c/(n_c + 1)$ trellis encoder whose output symbols are then block interleaved to break up fading bursts according to the discussion above.[2] Groups of $n_c + 1$ interleaved code symbols are mapped (in accordance with the *set partitioning method* of Ungerboeck [14][3] into M-PSK signals chosen from a set of $M = 2^{n_c+1}$ members. For example, a rate $-\frac{1}{2}$ encoder ($n_c = 1$) would be combined with a QPSK ($M = 4$) modulation. The in-phase (I) and quadrature (Q) components of the mapped signal point are then modulated onto quadrature carriers (with or without pulse shaping) for transmission over the fading channel. The usual additive white Gaussian noise (AWGN) is added at the input to the receiver, which first demodulates the I and Q signal components, soft quantizes the results of these demodulations, and then passes them through a block deinterleaver to recreate the codewords temporarily scrambled by the interleaver at the transmitter. The soft-quantized deinterleaved code symbols are then passed to the trellis decoder, which implements a Viterbi algorithm [15] with a metric depending on whether or not channel state information is available. For our purposes here, we assume that

---

[2] As was done in Ref. 2, we assume for the purpose of analysis an infinite interleaving depth, resulting in an ideal memoryless channel. In practice, the depth of interleaving would be finite and chosen in relation to the maximum duration of fade anticipated.

[3] A detailed discussion of trellis-coded modulation (TCM) and its application is beyond the scope of this book. The reader is referred to Ref. 3 for a thorough treatment of this subject.

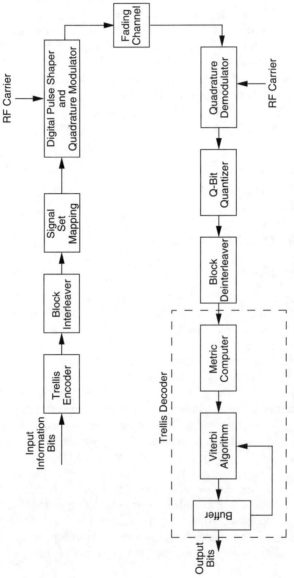

Figure 12.1. Block diagram of a trellis-coded *M*-PSK system.

such information is either perfectly known or totally unknown, without regard to the manner in which this information is obtained (e.g., via pilot tone techniques). Finally, the tentative soft decisions from the Viterbi decoder are stored in a buffer whose size is typically a design parameter (depending on the nature of the encoded information, e.g., speech), but for simplicity of analysis is assumed to be infinite.

A mathematical model for the foregoing system can be derived from the simple analysis block diagram in Fig. 12.2. The block labeled *encoder* includes both the binary input–binary output trellis encoder together with mapping onto the M-PSK signal set. Hence, the output of the encoder is a succession of coded M-PSK symbols, which for a sequence of length N is denoted by

$$\mathbf{x} = (x_1, x_2, \ldots, x_N) \qquad (12.1)$$

where the kth element of $\mathbf{x}$, namely, x_k, represents the normalized[4] transmitted M-PSK symbol (in complex form) at time k and, in general, is a nonlinear function of the state of the encoder, s_k, and the information symbol, u_k, representing the n_c i.i.d. information bits at its input [i.e., $x_k = f(s_k, u_k)$]. The transition from state to state is defined by a similar nonlinear relation, namely, $s_{k+1} = g(s_k, u_k)$. Corresponding to the transmission of $\mathbf{x}$, the channel outputs the sequence

$$\mathbf{y} = (y_1, y_2, \ldots, y_N) \qquad (12.2)$$

where the kth element of $\mathbf{y}$, namely, y_k, representing the channel output at time k is given by

$$y_k = \alpha_k \sqrt{2E_s} x_k + n_k \qquad (12.3)$$

Here α_k is the fading amplitude for the kth transmission and n_k is a zero-mean complex Gaussian RV with variance $\sigma^2 = N_0$ per dimension (i.e., $E\{|n_k|^2\} = 2N_0$). Based on an observation of $\mathbf{y}$, the maximum-likelihood (ML) receiver chooses as the transmitted information bit *sequence* the one that maximizes the a posteriori probability $p(\mathbf{u}_k|\mathbf{y})$ or equivalently (since the information bit sequences

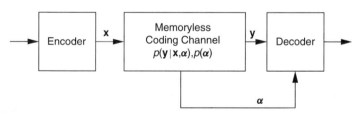

Figure 12.2. Simple analysis block diagram.

---

[4] We assume that the M-PSK symbols are normalized such that $|x_n| = 1$ (i.e., the signals lie on the perimeter of the unit circle). The actual transmitted M-PSK symbol in the nth interval is given by $\sqrt{2E_s} x_n$, where $E_s = n_c E_b$ is the symbol energy, with E_b denoting the information bit energy.

are equiprobable), the likelihood $p(\mathbf{y}|\mathbf{u}_k)$. Receivers that make decisions in this fashion are referred to as *maximum-likelihood sequence estimators* (MLSE)[5] and are practically implemented by the Viterbi algorithm [15].

12.1.2 Evaluation of Pairwise Error Probability

The first step in evaluating the average error probability is to compute the pairwise error probability associated with the transmitted M-PSK symbol sequences [i.e., the probability of choosing the sequence $\hat{\mathbf{x}} = (\hat{x}_1, \hat{x}_2, \ldots, \hat{x}_N)$ when in fact $\mathbf{x} = (x_1, x_2, \ldots, x_N)$ was transmitted given that these are the only two possible choices]. We refer to this occurrence as an *error event of length N*. The assumption of infinite interleaving/deinterleaving (i.e., an ideal memoryless channel) allows one to express the channel probabilities as

$$p(\mathbf{y}|\mathbf{x}, \boldsymbol{\alpha}) = \prod_{n=1}^{N} p(y_n|x_n, \alpha_n) \quad (12.4)$$

and

$$p(\boldsymbol{\alpha}) = \prod_{n=1}^{N} p(\alpha_n) \quad (12.5)$$

Since conditioned on α_n and x_n, y_n of (12.3) is a Gaussian RV with PDF,

$$p(y_n|x_n, \alpha_n) = \frac{1}{2\pi\sigma^2} \exp\left(-\frac{|y_n - \alpha_n\sqrt{2E_s}x_n|^2}{2\sigma^2}\right) \quad (12.6)$$

then for the case of perfectly known channel state information, substituting (12.6) in (12.4) and taking the natural logarithm of the result gives the decision metric (ignoring unnecessary scale factors)

$$m(\mathbf{y}, \mathbf{x}; \boldsymbol{\alpha}) \stackrel{\Delta}{=} \sum_{n=1}^{N} m(y_n, x_n; \alpha_n) = -\sum_{n=1}^{N} |y_n - \alpha_n\sqrt{2E_s}x_n|^2 \quad (12.7)$$

In the absence of channel state information, the optimum decision metric would be obtained by averaging (12.4) over (12.5) to obtain $p(\mathbf{y}|\mathbf{x})$ and then again taking the natural logarithm. This procedure would yield a decision metric whose form depends on the actual fading PDF assumed (see Chapter 7). To

---

[5] Strictly speaking, the term *maximum-likelihood sequence estimator* is reserved for sequences whose length N approaches infinity. However, it has become common practice to use this terminology even when the sequence length is finite, since practical implementations of the ML decision rule such as the Viterbi algorithm begin making decisions prior to an infinite observation time interval.

simplify matters, we propose (as was done in Ref. 1) a suboptimum decision metric that treats the channel as if it were purely Gaussian:

$$m(\mathbf{y}, \mathbf{x}) \stackrel{\Delta}{=} \sum_{n=1}^{N} m(y_n, x_n) = -\sum_{n=1}^{N} |y_n - \sqrt{2E_s} x_n|^2 \quad (12.8)$$

and which therefore is not a function of the fading sequence $\boldsymbol{\alpha} = (\alpha_1, \alpha_2, \ldots, \alpha_N)$.

We now proceed to evaluate the pairwise error probability for these two extreme cases of channel state information.

12.1.2.1 Known Channel State Information.
Since for the decision metric of (12.7), the sequence $\hat{\mathbf{x}} = (\hat{x}_1, \hat{x}_2, \ldots, \hat{x}_N)$ would be chosen over $\mathbf{x} = (x_1, x_2, \ldots, x_N)$ whenever $m(\mathbf{y}, \hat{\mathbf{x}}; \boldsymbol{\alpha}) \geq m(\mathbf{y}, \mathbf{x}; \boldsymbol{\alpha})$, the pairwise error probability (i.e., the probability of an error event of length N) for this case is given by

$$P(\mathbf{x} \to \hat{\mathbf{x}} | \boldsymbol{\alpha}) = \Pr\left\{ \sum_{n=1}^{N} m(y_n, \hat{x}_n; \alpha_n) \geq \sum_{n=1}^{N} m(y_n, x_n; \alpha_n) | \mathbf{x} \right\} \quad (12.9)$$

where the conditioning on $\mathbf{x}$ in the right-hand side of (12.9) indicates the fact that the components of the observation $\mathbf{y}$ are to be computed assuming that $\mathbf{x}$ was transmitted. Substituting (12.7) into (12.9) and recalling that both $\mathbf{x}$ and $\hat{\mathbf{x}}$ have components with unit squared magnitude, we obtain

$$P(\mathbf{x} \to \hat{\mathbf{x}} | \boldsymbol{\alpha}) = \Pr\left\{ \sum_{n=1}^{N} \alpha_n \operatorname{Re}\{y_n \hat{x}_n^*\} \geq \sum_{n=1}^{N} \alpha_n \operatorname{Re}\{y_n x_n^*\} | \mathbf{x} \right\}$$

$$= \Pr\left\{ \sum_{n \in \eta} \alpha_n \operatorname{Re}\{y_n (\hat{x}_n - x_n)^*\} \geq 0 | \mathbf{x} \right\} \quad (12.10)$$

where η is the set of all n for which $\hat{x}_n \neq x_n$. Based on (12.3), the decision variable to be compared with the zero threshold in (12.10) is Gaussian with mean $\sqrt{2E_s} \sum_{n \in \eta} \alpha_n^2 \operatorname{Re}\{x_n(\hat{x}_n - x_n)^*\}$ and variance $N_0 \sum_{n \in \eta} \alpha_n^2 |\hat{x}_n - x_n|^2$. Thus, since for constant envelope signal sets such as M-PSK, where $|x|^2 = |\hat{x}|^2$, it is straightforward to show that

$$|x - \hat{x}|^2 = 2\operatorname{Re}\{x(x - \hat{x})^*\} \quad (12.11)$$

then the pairwise error probability immediately evaluates to

$$P(\mathbf{x} \to \hat{\mathbf{x}} | \boldsymbol{\alpha}) = Q\left(\sqrt{\frac{E_s}{2N_0} \sum_{n \in \eta} \alpha_n^2 |\hat{x}_n - x_n|^2} \right) \quad (12.12)$$

The pairwise error probability in (12.12) has a form analogous to that obtained for the probability of error of uncoded BPSK transmitted over a multichannel with maximal-ratio combining (MRC) employed at the receiver (see Chapter 9). In fact, for convolutionally encoded BPSK where $|\hat{x}_n - x_n|^2 = 4$ for all $n \in \eta$, (12.12) would simplify to

$$P(\mathbf{x} \to \hat{\mathbf{x}}|\boldsymbol{\alpha}) = Q\left(\sqrt{\frac{2E_s}{N_0} \sum_{n \in \eta} \alpha_n^2}\right) \tag{12.13}$$

and hence the number of diversity channels for the uncoded application can be seen to be directly analogous to the number of symbols that are in error in the coded case.

In Ref. 1, the conditional (on the fading) pairwise error $P(\mathbf{x} \to \hat{\mathbf{x}}|\boldsymbol{\alpha})$ was upper (Chernoff) bounded to allow averaging over the fading statistics (a feat not possible using the classic definition of the Gaussian Q-function), thereby obtaining a closed-form upper bound on $P(\mathbf{x} \to \hat{\mathbf{x}})$. In particular, it was shown there that

$$P(\mathbf{x} \to \hat{\mathbf{x}}) \leq \overline{D^{d^2(\mathbf{x},\hat{\mathbf{x}})}}^{\boldsymbol{\alpha}} \tag{12.14}$$

where the overbar denotes statistical averaging over the random vector $\boldsymbol{\alpha}$,

$$d^2(\mathbf{x}, \hat{\mathbf{x}}) \triangleq \sum_{n \in \eta} \alpha_n^2 |\hat{x}_n - x_n|^2 \tag{12.15}$$

and

$$D \triangleq \exp\left(-\frac{E_s}{4N_0}\right) \tag{12.16}$$

is the *Bhattacharyya parameter* [7]. Since the fading amplitudes are independent, it was then possible to write (12.14) as the product

$$P(\mathbf{x} \to \hat{\mathbf{x}}) \leq \prod_{n \in \eta} \overline{D^{\alpha_n^2 |\hat{x}_n - x_n|^2}}^{\alpha_n} \tag{12.17}$$

which is in a form that lends itself to application of the transfer function bound approach for computing an upper bound on average BEP. The individual terms of the product in (12.17) were evaluated in Ref. 1 for Rayleigh and Rician fading amplitude statistics.

Since in Chapter 4 it was demonstrated that the alternative form of the Gaussian Q-function has the analytic advantages of the Chernoff bound without the disadvantage of being a bound, it is reasonable to apply that approach here to the conditional pairwise error probability in much the same manner that it was used in Chapter 9 to unify and simplify the analysis of multichannel reception of BPSK with MRC. In particular, using the alternative form of $Q(x)$ given in (4.2), we can express (12.13) as

$$P(\mathbf{x} \to \hat{\mathbf{x}}|\boldsymbol{\alpha}) = \frac{1}{\pi} \int_0^{\pi/2} \exp\left(-\frac{E_s}{4N_0 \sin^2\theta} \sum_{n \in \eta} \alpha_n^2 |\hat{x}_n - x_n|^2\right) d\theta$$

$$= \frac{1}{\pi} \int_0^{\pi/2} [D(\theta)]^{d^2(\mathbf{x},\hat{\mathbf{x}})} d\theta \qquad (12.18)$$

where, analogous to (12.15),

$$D(\theta) \stackrel{\Delta}{=} \exp\left(-\frac{E_s}{4N_0 \sin^2\theta}\right) \qquad (12.19)$$

Hence, the unconditional pairwise error probability is given by

$$P(\mathbf{x} \to \hat{\mathbf{x}}) = \overline{\frac{1}{\pi} \int_0^{\pi/2} [D(\theta)]^{d^2(\mathbf{x},\hat{\mathbf{x}})} d\theta}^{\boldsymbol{\alpha}} = \frac{1}{\pi} \int_0^{\pi/2} \overline{[D(\theta)]^{d^2(\mathbf{x},\hat{\mathbf{x}})}}^{\boldsymbol{\alpha}} d\theta$$

$$= \frac{1}{\pi} \int_0^{\pi/2} \prod_{n \in \eta} \overline{D(\theta)^{\alpha_n^2 |\hat{x}_n - x_n|^2}}^{\alpha_n} d\theta \qquad (12.20)$$

which is to be compared with the upper bound in (12.17).

The statistical average required by each term in the product of (12.20) can be written as

$$\overline{D(\theta)^{\alpha_n^2 |\hat{x}_n - x_n|^2}}^{\alpha_n} = \int_0^\infty \exp\left(-\frac{\alpha_n^2 E_s}{4N_0 \sin^2\theta} |\hat{x}_n - x_n|^2\right) p_{\alpha_n}(\alpha_n) d\alpha_n \qquad (12.21)$$

Since for a rate $n_c/(n_c+1)$ trellis encoder n_c information bits produce one $M(=2^{n_c+1})$-ary symbol, then in terms of the bit energy-to-noise ratio, E_b/N_0, (12.24) can be rewritten as

$$\overline{D(\theta)^{\alpha_n^2 |\hat{x}_n - x_n|^2}}^{\alpha_n} = \int_0^\infty \exp\left(-\frac{n_c \alpha^2 E_b}{4N_0 \sin^2\theta} |\hat{x}_n - x_n|^2\right) p_\alpha(\alpha) d\alpha \qquad (12.22)$$

where we dropped the n subscript on α since the fading variables are all identically distributed. Alternatively, in terms of the instantaneous SNR per bit $\gamma \stackrel{\Delta}{=} \alpha^2 E_b/N_0$, (12.22) becomes

$$\overline{D(\theta)^{\alpha_n^2 |\hat{x}_n - x_n|^2}}^{\alpha_n} = \int_0^\infty \exp\left(-\gamma \frac{n_c |\hat{x}_n - x_n|^2}{4 \sin^2\theta}\right) p_\gamma(\gamma) d\gamma \qquad (12.23)$$

Integrals of the form in (12.23) were considered in Chapter 5 for a wide variety of fading channel types. For example, for Rayleigh fading, using (5.5) we would obtain[6]

$$\overline{D(\theta)^{\alpha_n^2 |\hat{x}_n - x_n|^2 \alpha_n}} = \frac{1}{1 + n_c |\hat{x}_n - x_n|^2 \overline{\gamma}/4 \sin^2 \theta} = \frac{\sin^2 \theta}{\sin^2 \theta + n_c |\hat{x}_n - x_n|^2 \overline{\gamma}/4} \quad (12.24)$$

where $\overline{\gamma} \triangleq \overline{\alpha^2} E_s/N_0$ is the average SNR per bit. For Rician fading, using (5.11) we would obtain

$$\overline{D(\theta)^{\alpha_n^2 |\hat{x}_n - x_n|^2 \alpha_n}} = \frac{1 + K}{1 + K + n_c |\hat{x}_n - x_n|^2 \overline{\gamma}/4 \sin^2 \theta}$$

$$\times \exp\left(-\frac{K[n_c |\hat{x}_n - x_n|^2 \overline{\gamma}/4 \sin^2 \theta]}{1 + K + n_c |\hat{x}_n - x_n|^2 \overline{\gamma}/4 \sin^2 \theta}\right) \quad (12.25)$$

which clearly reduces to (12.24) for $K = 0$. Finally, for Nakagami-m fading, using (5.15), the analogous expression to (12.24) and (12.25) becomes

$$\overline{D(\theta)^{\alpha_n^2 |\hat{x}_n - x_n|^2 \alpha_n}} = \frac{1}{\left(1 + n_c |\hat{x}_n - x_n|^2 \overline{\gamma}/4m \sin^2 \theta\right)^m}$$

$$= \left(\frac{\sin^2 \theta}{\sin^2 \theta + n_c |\hat{x}_n - x_n|^2 \overline{\gamma}/4m}\right)^m \quad (12.26)$$

Note that had one chosen to use the upper bound on pairwise error probability (as was done in Ref. 1) of (12.17) rather than the exact result of (12.20), the terms that would be required for the product in the former equation would be obtained simply by setting $\theta = \pi/2$ ($\sin^2 \theta = 1$) in (12.24), (12.25), and (12.26), respectively, as discussed in Chapter 4, that is,

$$\overline{D^{\alpha_n^2 |\hat{x}_n - x_n|^2 \alpha_n}} = \overline{D(\theta)^{\alpha_n^2 |\hat{x}_n - x_n|^2 \alpha_n}}\Big|_{\theta = \pi/2} \quad (12.27)$$

12.1.2.2 Unknown Channel State Information.
When channel state information is not available, the expression for the pairwise error probability analogous to (12.9) becomes

$$P(\mathbf{x} \to \hat{\mathbf{x}} | \boldsymbol{\alpha}) = \Pr\left\{\sum_{n=1}^{N} m(y_n, \hat{x}_n) \geq \sum_{n=1}^{N} m(y_n, x_n) | \mathbf{x}\right\} \quad (12.28)$$

---

[6] For the special case of BPSK modulation wherein $|x_n - \hat{x}_n|^2 = 4$, a similar result for the pairwise error probability to (12.20) together with (12.24) was obtained by Hall and Wilson [16].

COHERENT DETECTION   507

Substituting (12.8) into (12.28) and recalling that both **x** and $\hat{\mathbf{x}}$ have components with unit squared magnitude, we obtain

$$P(\mathbf{x} \to \hat{\mathbf{x}}|\boldsymbol{\alpha}) = \Pr\left\{\sum_{n\in\eta} \text{Re}\left\{y_n(\hat{x}_n - x_n)^*\right\} \geq 0 | \mathbf{x}\right\} \quad (12.29)$$

where η is again the set of all n for which $\hat{x}_n \neq x_n$. Based on (12.3), the decision variable to be compared with the zero threshold in (12.29) is now Gaussian with mean $\sum_{n\in\eta} \alpha_n \sqrt{2E_s}\,\text{Re}\{x_n(\hat{x}_n - x_n)^*\}$ and variance $N_0 \sum_{n\in\eta} |\hat{x}_n - x_n|^2$. Once again using the relation in (12.11), which is valid for constant envelope signal sets such as M-PSK, the pairwise error probability immediately evaluates to

$$P(\mathbf{x} \to \hat{\mathbf{x}}|\boldsymbol{\alpha}) = Q\left(\sqrt{\frac{E_s}{2N_0}\frac{\left(\sum_{n\in\eta}\alpha_n|\hat{x}_n - x_n|^2\right)^2}{\sum_{n\in\eta}|\hat{x}_n - x_n|^2}}\right)$$

$$= Q\left(\sqrt{\left(\frac{E_s}{2N_0}\sum_{n\in\eta}\alpha_n\frac{|\hat{x}_n - x_n|^2}{\sqrt{\sum_{k\in\eta}|\hat{x}_k - x_k|^2}}\right)^2}\right) \quad (12.30)$$

The pairwise error probability in (12.30) has a form analogous to that obtained for the probability of error of uncoded BPSK transmitted over a multichannel with equal-gain combining (EGC) employed at the receiver (see Chapter 9). In fact, for convolutionally encoded BPSK, where $|\hat{x}_n - x_n|^2 = 4$ for all $n \in \eta$, (12.30) would simplify to

$$P(\mathbf{x} \to \hat{\mathbf{x}}|\boldsymbol{\alpha}) = Q\left(\sqrt{\frac{2E_s}{N_0}\frac{1}{L_\eta}\left(\sum_{n\in\eta}\alpha_n\right)^2}\right) \quad (12.31)$$

where L_η is the number of elements in the set η or equivalently, the Hamming distance between the correct and incorrect sequences. Hence, the number of diversity channels for the uncoded application can again be seen to be directly analogous to the number of symbols that are in error in the coded case.

To obtain the unconditional pairwise error probability, one must average (12.31) over the i.i.d. fading statistics of the α_n's. In particular, defining

$$\alpha_T \overset{\Delta}{=} \sum_{n\in\eta} \alpha_n \frac{|\hat{x}_n - x_n|^2}{\sqrt{\sum_{k\in\eta}|\hat{x}_k - x_k|^2}} \overset{\Delta}{=} \sum_{n\in\eta} \alpha_n d_n^2 \quad (12.32)$$

the L_η-fold integral obtained by averaging (12.31) over the fading amplitudes can be collapsed to a single integral, namely,

$$P(\mathbf{x} \to \hat{\mathbf{x}}) = \int_0^\infty Q\left(\sqrt{\frac{E_s}{2N_0}\alpha_T^2}\right) p_{\alpha_T}(\alpha_T)\,d\alpha_T \quad (12.33)$$

Applying the alternative representation of the Gaussian Q-function to (12.33) gives

$$P(\mathbf{x} \to \hat{\mathbf{x}}) = \frac{1}{\pi} \int_0^{\pi/2} \int_0^\infty \exp\left(-\frac{E_s}{4N_0 \sin^2\theta}\alpha_T^2\right) p_{\alpha_T}(\alpha_T)\, d\alpha_T\, d\theta \qquad (12.34)$$

Direct evaluation of (12.34) requires finding the PDF of α_T defined in (12.32), which, in general, is difficult even when the α_n's are i.i.d. Instead, we follow the procedure given in Chapter 9 by first representing $p_{\alpha_T}(\alpha_T)$ in terms of its characteristic function. Thus,

$$P(\mathbf{x} \to \hat{\mathbf{x}}) = \frac{1}{2\pi^2} \int_0^{\pi/2} \int_{-\infty}^\infty \left[\prod_{l=1}^{L_\eta} \psi_{\alpha_l}(jvd_l^2)\right]$$
$$\times \left[\overbrace{\int_0^\infty \exp\left(-\frac{E_s}{4N_0 \sin^2\theta}\alpha_T^2 - jv\alpha_T\right) d\alpha_T}^{J(v,\theta)}\right] dv\, d\theta \qquad (12.35)$$

Despite the apparent similarity of (12.35) with (9.61), the difficulty here lies in the fact that the weight d_l^2 of (12.32) that appears in the argument of the lth characteristic function does not depend only on the squared Euclidean distance for the lth branch of the error sequence but because of its normalization also depends on the *total* squared Euclidean distance of the entire error sequence. Even in the case of BPSK modulation, each of these weights would be normalized by the length L_η of the error sequence. Because of this, it will not be possible to obtain an integral form for $P(\mathbf{x} \to \hat{\mathbf{x}})$, where the integrand is a product of terms each of which depends only on the squared Euclidean distance associated with that term. Thus, we abandon this method for the case of unknown CSI and instead turn to the *inverse Laplace transform method* introduced for problems of this type by Cavers and Ho [9][7] with additional generalizations reported in Ref. 12, and additional methods for evaluation later explored by Biglieri, Caire, Taricco, and Ventura-Traveset [17,18].

Consider a RV z with PDF $p(z)$ and moment generating function (MGF)[8] $M_z(s) = \int_{-\infty}^\infty e^{sz} p_z(z)\, dz$. Then the CDF $P(z) = \int_{-\infty}^z p(y)\, dy$ is related to $M_z(s)$

---

[7] Cavers and Ho [9] refer to this method as the *characteristic function method*. As we shall see momentarily, in keeping with the distinction made in Chapter 5 between the moment generating function and the characteristic function of a RV in terms of their relations to the Laplace and Fourier transforms, respectively, the method in Ref. 9 is more appropriately called a *moment generating function method*. Also, in their definition of characteristic function, Cavers and Ho [9] do not reverse the sign of the exponent in the Laplace transform and as such are not consistent with the traditional definition of this function.

[8] The definition of moment generating function used here is the generalization of that introduced in Chapter 5 to the case where the underlying RV takes on both positive and negative values. Thus, the MGF is now the *bilateral* (as opposed to unilateral) Laplace transform of the PDF with the sign of the exponent reversed.

by (see Chapter 9)

$$P(z) = \frac{1}{2\pi j} \int_{\sigma-j\infty}^{\sigma+j\infty} \frac{M_z(-s)}{s} e^{sz}\, ds \qquad (12.36)$$

where σ is chosen such that the vertical line of integration lies in the region of convergence (ROC) of the bilateral Laplace transform. Since $P(a) = \Pr\{z \le a\}$,

$$\Pr\{z \le 0\} = P(0) = \frac{1}{2\pi j} \int_{\sigma-j\infty}^{\sigma+j\infty} \frac{M_z(-s)}{s}\, ds \qquad (12.37)$$

Methods for evaluating integrals of the type in (12.37) are discussed in detail in Appendix 9B. Hence, we shall draw upon these results only when needed to perform numerical evaluations. When applied to the difference decision metric

$$z = \sum_{n=1}^{N} \frac{1}{2}[m(y_n, x_n) - m(y_n, \hat{x}_n)] | \mathbf{x} \triangleq \sum_{n=1}^{N} z_n$$

(12.37) results in the evaluation of the pairwise error probability. Since as we have already discussed, the interleaving/deinterleaving operation makes the z_n's independent, then denoting the MGF of z_n by $M_{z_n}(s)$, we obtain from (12.37) that

$$P(\mathbf{x} \to \hat{\mathbf{x}}) = \frac{1}{2\pi j} \int_{\sigma-j\infty}^{\sigma+j\infty} \frac{1}{s} \prod_{n=1}^{N} M_{z_n}(-s)\, ds \qquad (12.38)$$

Using (12.3), the RV

$$z_n \triangleq \mathrm{Re}\left\{ y_n (x_n - \hat{x}_n)^* | \mathbf{x} \right\} = \mathrm{Re}\left\{ \alpha_n \sqrt{2E_s} x_n (x_n - \hat{x}_n)^* + n_n (x_n - \hat{x}_n)^* | \mathbf{x} \right\}$$

is conditionally (on the fading) Gaussian with mean $\mu_{z_n|\alpha_n} \triangleq \mathrm{Re}\{\alpha_n\sqrt{2E_s}x_n(x_n - \hat{x}_n)^*\}$ and variance $\sigma^2_{z_n|\alpha_n} \triangleq N_0|\hat{x}_n - x_n|^2$. Thus, using (12.11), the unconditional RV z_n has PDF

$$\begin{aligned} p_{z_n}(z_n) &= \int_0^\infty p_{z_n}(z_n|\alpha_n) p_{\alpha_n}(\alpha_n)\, d\alpha_n \\ &= \int_0^\infty \frac{1}{\sqrt{2\pi N_0|\hat{x}_n - x_n|^2}} \\ &\quad \times \exp\left(-\frac{z_n - \frac{1}{2}\alpha_n\sqrt{2E_s}|\hat{x}_n - x_n|^2}{2N_0|\hat{x}_n - x_n|^2} \right) p_{\alpha_n}(\alpha_n)\, d\alpha_n \end{aligned} \qquad (12.39)$$

Since for a Gaussian RV Y with mean μ_Y and variance σ_Y^2 the MGF is given by

$$M_Y(s) = e^{s\mu_Y + \sigma_Y^2 s^2/2} \qquad (12.40)$$

then applying (12.40) to the conditional Gaussian RV $z_n|\alpha_n$, we have from (12.39) that the MGF of z_n is

$$M_{z_n}(s) = \int_{-\infty}^{\infty} p_{z_n}(z_n) e^{sz_n} dz_n$$

$$= \int_0^{\infty} e^{(1/2)s\alpha_n \sqrt{2E_s}|\hat{x}_n - x_n|^2 + (1/2)N_0|\hat{x}_n - x_n|^2 s^2} p_{\alpha_n}(\alpha_n) d\alpha_n$$

$$= \overline{e^{(1/2)[s\alpha_n\sqrt{2E_s} + N_0 s^2]|\hat{x}_n - x_n|^2}}^{\alpha_n} \tag{12.41}$$

or in terms of the MGF of the fading random variable $\alpha_n = \alpha$,

$$M_{z_n}(s) = e^{(1/2)N_0 s^2|\hat{x}_n - x_n|^2} M_\alpha\left(s\frac{\sqrt{2E_s}|\hat{x}_n - x_n|^2}{2}\right) \tag{12.42}$$

Finally, substituting (12.42) into (12.38) gives the pairwise error probability in the desired product form

$$P(\mathbf{x} \to \hat{\mathbf{x}}) = \frac{1}{2\pi j} \int_{\sigma-j\infty}^{\sigma+j\infty} \frac{1}{s} \prod_{n \in \eta} \left[e^{(1/2)N_0 s^2|\hat{x}_n - x_n|^2} M_\alpha\left(-\frac{s\sqrt{2E_s}|\hat{x}_n - x_n|^2}{2}\right) \right] ds \tag{12.43}$$

where the nth term of the product depends only on the squared Euclidean distance for the nth branch of the sequence and not on the distance properties of the entire sequence as in (12.35). Also, in (12.43) it has again become possible to replace the product over all branches by the product over only those for which $\hat{x}_n \neq x_n$ (i.e., $n \in \eta$), since for the terms where $\hat{x}_n = x_n$, the MGF $M_\alpha(s\sqrt{2E_s}|\hat{x}_n - x_n|^2/2)$ is equal to unity.

12.1.3 Transfer Function Bound on Average Bit Error Probability

To compute the true upper (union) bound (TUB) on the average BEP, one sums the pairwise error probability over all error events (sequence pairs), corresponding to a given transmitted sequence weighting each term by the number of information bit errors associated with that event, then statistically averages this sum over all possible transmitted sequences, finally dividing by the number of input bits per transmission. In mathematical terms, if $P(\mathbf{x})$ denotes the probability that the sequence $\mathbf{x}$ is transmitted, $n(\mathbf{x}, \hat{\mathbf{x}})$ the number of information bit errors committed by choosing $\hat{\mathbf{x}}$ instead of $\mathbf{x}$, and n_c the number of information bits per transmission, the average BEP has the TUB

$$P_b(E) \leq \frac{1}{n_c} \sum_{\mathbf{x}} P(\mathbf{x}) \sum_{\mathbf{x} \neq \hat{\mathbf{x}}} n(\mathbf{x}, \hat{\mathbf{x}}) P(\mathbf{x} \to \hat{\mathbf{x}}) \tag{12.44}$$

An efficient method for computing this weighted sum was originally proposed by Viterbi [6] for convolutional codes transmitted over the AWGN channel and is referred to as the *transfer function bound approach*. To apply this approach, one must be able to write $P(\mathbf{x} \to \hat{\mathbf{x}})$ in a product form where the nth term of the product is associated with the nth branch of the particular path through the state diagram defined by the error event. Also, evaluation of this nth term must depend only on the distance between the nth branch of the correct and incorrect sequences and not on the distance properties of the entire sequence. Alternatively, one can write $P(\mathbf{x} \to \hat{\mathbf{x}})$ in the form of an integral whose *integrand* is a product of terms satisfying the above-mentioned condition. In this case it is possible to evaluate the TUB on average BER by first applying the transfer function bound approach to the integrand (conditioned on the integration variable) and then performing the required integration.

For trellis codes we have seen that the appropriate distance measure is Euclidean distance and thus for the AWGN channel $P(\mathbf{x} \to \hat{\mathbf{x}})$ would take the form [see (12.20) omitting the averaging on the fading][9]

$$P(\mathbf{x} \to \hat{\mathbf{x}}) = \frac{1}{\pi} \int_0^{\pi/2} \prod_{n \in \eta} D(\theta)^{|\hat{x}_n - x_n|^2} \, d\theta \qquad (12.45)$$

where $D(\theta)$ is still defined as in (12.19). To incorporate $n(\mathbf{x}, \hat{\mathbf{x}})$ into the product, we define $n(x_n, \hat{x}_n)$ as the number of bit errors in the nth interval of the error event, in which case $n(\mathbf{x}, \hat{\mathbf{x}}) = \sum_{n=1}^{N} n(x_n, \hat{x}_n)$. Then, defining an indicator variable I, we can rewrite the second summation in (12.44) as

$$\sum_{\mathbf{x} \neq \hat{\mathbf{x}}} n(\mathbf{x}, \hat{\mathbf{x}}) P(\mathbf{x} \to \hat{\mathbf{x}}) = \sum_{\mathbf{x} \neq \hat{\mathbf{x}}} \frac{1}{\pi} \int_0^{\pi/2} \frac{\partial}{\partial I} \prod_n D(\theta)^{|x_n - \hat{x}_n|^2} I^{n(x_n, \hat{x}_n)} \bigg|_{I=1} d\theta$$

$$= \frac{1}{\pi} \int_0^{\pi/2} \left[\frac{\partial}{\partial I} \sum_{\mathbf{x} \neq \hat{\mathbf{x}}} \prod_n D(\theta)^{|x_n - \hat{x}_n|^2} I^{n(x_n, \hat{x}_n)} \bigg|_{I=1} \right] d\theta$$

$$\triangleq \frac{1}{\pi} \int_0^{\pi/2} \left[\frac{\partial}{\partial I} T(D(\theta), I) \bigg|_{I=1} \right] d\theta \qquad (12.46)$$

where $T(D, I)$ is the transfer function associated with the error state diagram [3] of a particular TCM scheme and in general depends on the transmitted sequence $\mathbf{x}$. The form of $T(D, I)$ is typically a ratio of polynomials in D and I, as will become clear when we consider some examples. Finally, combining (12.44) and

---

[9] From here on, for simplicity of notation, we denote the product over the N branches of an error event by $\prod_n$, with the understanding that it need only be taken over those branches for which an error occurs.

(12.46) gives the TUB on average BER for the AWGN channel, namely,

$$P_b(E) \le \frac{1}{n_c} \sum_{\mathbf{x}} P(\mathbf{x}) \frac{1}{\pi} \int_0^{\pi/2} \left[\frac{\partial}{\partial I} T(D(\theta), I) \bigg|_{I=1} \right] d\theta \qquad (12.47)$$

For a large class of trellis codes, a symmetry property exists such that for the purpose of evaluation of the TUB, the correct sequence $\mathbf{x}$ can always be chosen as the all-zeros sequence, thus avoiding the necessity of averaging over all possible transmitted code sequences in (12.44). Codes of this type, referred to as *uniform error probability (UEP) codes*, are the only ones considered in this chapter, although, in principle, the generic methods discussed also apply when the UEP criterion is not valid. Thus, for UEP TCM schemes, (12.44) simplifies to

$$P_b(E) \le \frac{1}{\pi} \int_0^{\pi/2} \left[\frac{1}{n_c} \frac{\partial}{\partial I} T(D(\theta), I) \bigg|_{I=1} \right] d\theta \qquad (12.48)$$

Had we applied the Chernoff bound to the pairwise error probability rather than obtain its exact form, the equivalent result to (12.47) would become

$$P_b(E) \le \frac{1}{n_c} \sum_{\mathbf{x}} P(\mathbf{x}) \frac{\partial}{\partial I} T(D, I) \bigg|_{I=1, D=\exp(-E_s/4N_0)} \qquad (12.49)$$

or for UEP TCM schemes

$$P_b(E) \le \frac{1}{n_c} \frac{\partial}{\partial I} T(D, I) \bigg|_{I=1, D=\exp(-E_s/4N_0)} \qquad (12.50)$$

both of which are looser than the TUB.

To find the TUB on the average BEP for TCM transmitted over the fading channel, we simply substitute in (12.44) the expressions found in Sections 12.1.2.1 or 12.1.2.2 for the pairwise error probability corresponding to the cases of perfectly known CSI or unknown CSI, respectively. We now present the specific results for the two cases of CSI knowledge.

12.1.3.1 Known Channel State Information.

Comparing the integrand of (12.20) with (12.45), we observe the analogy between $\overline{D(\theta)^{\alpha_n^2 |\hat{x}_n - x_n|^2}}^{\alpha_n}$ of the former and $D(\theta)^{|x_n - \hat{x}_n|^2}$ of the latter. Thus, based on the discussion above, the average BEP for trellis-coded M-PSK transmitted over the slow-fading channel has a TUB analogous to (12.48), namely,[10]

$$P_b(E) \le \frac{1}{\pi} \int_0^{\pi/2} \frac{1}{n_c} \frac{\partial}{\partial I} T(\overline{D(\theta)}, I) \bigg|_{I=1, D(\theta)=e^{-E_s/4N_0 \sin^2\theta}} d\theta \qquad (12.51)$$

---

[10] The implication of the simple notation $\overline{D(\theta)}$ is that the label $D(\theta)^{|x_n - \hat{x}_n|^2}$ on each branch between transitions be replaced by $\overline{D(\theta)^{\alpha_n^2 |x_n - \hat{x}_n|^2}}^{\alpha_n}$.

The averages over the fading required in (12.51) have been evaluated in closed form in Section. 12.1.2.2 [e.g., see (12.24) for Rayleigh fading, (12.25) for Rician fading, and (12.26) for Nakagami-m fading].

12.1.3.2 Unknown Channel State Information.
For this case the appropriate product form of the pairwise error probability integrand is (12.43). Thus, by analogy with (12.48), the TUB on the average BEP is given by

$$P_b(E) \leq \frac{1}{2\pi j} \int_{\sigma-j\infty}^{\sigma+j\infty} \frac{1}{s} \left[\frac{1}{n_c} \frac{\partial}{\partial I} T(s, I) \Big|_{I=1} \right] ds \qquad (12.52)$$

where $T(s, I)$ is the transfer function computed from the state transition diagram for the AWGN channel with the label $D(\theta)^{|x_n - \hat{x}_n|^2}$ on each branch between transitions replaced by

$$D(s; |\hat{x}_n - x_n|^2) \triangleq e^{(1/2)N_0 s^2 |\hat{x}_n - x_n|^2} M_\alpha \left(-\frac{s\sqrt{2E_s}|\hat{x}_n - x_n|^2}{2} \right) \qquad (12.53)$$

12.1.4 Alternative Formulation of the Transfer Function Bound

A variation of the previous approach to computing the transfer function introduced by Divsalar [19] is referred to as the *pair-state method*. It is particularly useful for non-UEP codes since it circumvents averaging over the transmitted code sequences by incorporating it in the transfer function itself. In this method, a *pair-state transition diagram* is constructed wherein each pair state $S_k = (s_k, \hat{s}_k)$ corresponds to a pair of states, s_k and $\hat{s}_k$, in the trellis diagram. One is said to be in a correct pair-state when $s_k = \hat{s}_k$ and an incorrect pair-state when $s_k \neq \hat{s}_k$. A transition between pair-states $S_k = (s_k, \hat{s}_k)$ and $S_{k+1} = (s_{k+1}, \hat{s}_{k+1})$ in the transition diagram corresponds to a pair of transitions in the trellis diagram (i.e., s_k to $\hat{s}_k$ and s_{k+1} to $\hat{s}_{k+1}$). Since associated with each transition in the pair is an M-PSK symbol and a corresponding information symbol (a sequence of n_c information bits), the transition between two pair-states in the transition diagram is characterized by the squared Euclidean distance δ^2 between the corresponding M-PSK symbols and the Hamming distance d_H between the corresponding information bit sequences.

Based on the discussion above, in the absence of fading (i.e., the AWGN channel), each branch between pair-states in the transition diagram has a gain G of the form

$$G = \sum \frac{1}{2^{n_c}} I^{d_H} D^{\delta^2} \qquad (12.54)$$

where, as before, I is an indicator variable and D is the Bhattacharyya parameter defined in (12.19). The summation in (12.54) accounts for the possibility of parallel paths between states of the trellis diagram. Since the pair-state method accounts for all possible transmitted symbol sequences and their probability, the union–Chernoff bound would be given by (12.50) (which was formally restricted to UEP codes), where the transfer function $T(D, I)$ is now computed from the pair-state transition diagram based on the gains of (12.54). Extending this approach to exact evaluation of the pairwise error probability, if instead of (12.54) we were to label each branch with a gain

$$G(\theta) = \sum \frac{1}{2^{n_c}} I^{d_H} D(\theta)^{\delta^2} \qquad (12.55)$$

with $D(\theta)$ as in (12.19), the TUB would be given by (12.48), where the transfer function $T(D(\theta), I)$ is now computed from the pair-state transition diagram based on the gains of (12.55).

In the presence of fading, the appropriate substitutions for D and $D(\theta)$ as discussed in Section 12.1.3. would result in upper bounds on average BEP. Since our interest is in the TUB, we consider only the case where the pair-state gains are as in (12.55) for the AWGN and their equivalences for the fading channel.

12.1.5 Example

Consider the case of rate $-\frac{1}{2}$ coded QPSK using a two-state trellis. The signal constellation and appropriate set partitioning [14] are illustrated in Fig. 12.3, and

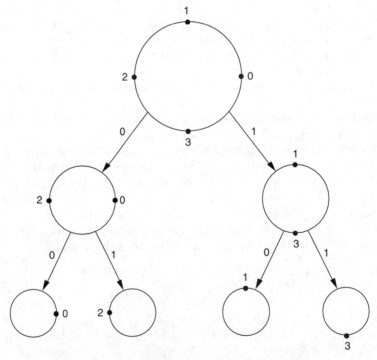

Figure 12.3. Set partitioning of QPSK signal constellation.

the corresponding trellis diagram and pair-state transition diagram are shown in Figs. 12.4 and 12.5, respectively. The dashed branch in the trellis diagram corresponds to a transition resulting from a "0" information bit, whereas the solid branch corresponds to a transition resulting from a "1" information bit. The branches are labeled with the M-PSK output symbol that is transmitted as a result of the information bit above being input to the encoder. The branches of the pair-state transition diagram are labeled with the gains computed from (12.56). For example, for the transition from the pair state 0,0 to the pair state 0,1 in Fig. 12.5 or equivalently, the pair of transitions from state 0 to state 0 and state 0 to state 1 in Fig. 12.4, the corresponding output M-PSK symbols are 0 and 2. Since the signal constellation is normalized to unit radius circle, then from Fig. 12.3 the squared Euclidean distance between symbols 0 and 2 is $\delta^2 = 4$. Similarly, the transition from state 0 to state 0 is the result of transmitting a single 0 information bit, whereas the transition from state 0 to state 1 is the result of transmitting a single 1 information bit. Thus, the Hamming distance between these two information bits is $d_H = 1$. Since there are no parallel branches in the trellis diagram and $n_c = 1$, then, in accordance with (12.55), the gain associated with the transition from pair state 0,0 to pair state 0,1 is $a = \frac{1}{2} I D^4(\theta)$ (see Fig. 12.5). The gains $b = \frac{1}{2} I D^2(\theta)$ and $c = \frac{1}{2} D^2(\theta)$ follow from similar considerations.

Defining the states of the pair-state diagram by ξ with the input state having value unity, the transfer function is obtained by solving the following set of state equations:

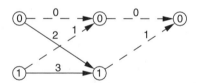

Figure 12.4. Trellis diagram and QPSK signal assignment.

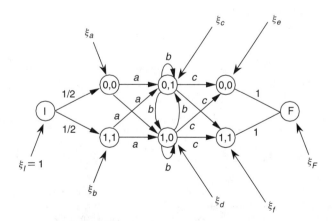

Figure 12.5. Pair-state transition diagram.

$$T(D(\theta), I) = \xi_e + \xi_f$$
$$\xi_e = \xi_f = c(\xi_c + \xi_d), \quad \xi_c = \xi_d = b(\xi_c + \xi_d) + a(\xi_a + \xi_b)$$
$$\xi_a = \xi_b = \tfrac{1}{2} \tag{12.56}$$

resulting in

$$T(D(\theta), I) = \frac{4ac}{1 - 2b} = \frac{ID^6(\theta)}{1 - ID^2(\theta)} \tag{12.57}$$

Thus, for the AWGN channel, the TUB on average BEP would be given by

$$P_b(E) \leq \frac{1}{\pi} \int_0^{\pi/2} \left[\frac{d}{dI} T(D(\theta), I) \Big|_{I=1} \right] d\theta = \frac{1}{\pi} \int_0^{\pi/2} \frac{D^6(\theta)}{[1 - D^2(\theta)]^2} d\theta$$
$$= \frac{1}{\pi} \int_0^{\pi/2} \frac{\exp(-3E_b/2N_0 \sin^2 \theta)}{[1 - \exp(-E_b/2N_0 \sin^2 \theta)]^2} d\theta \tag{12.58}$$

For the fading channel with known channel state information, the transfer function would become

$$T(\overline{D(\theta)}, I) = \frac{I \overline{D^{2\alpha^2}(\theta)}^\alpha \overline{D^{4\alpha^2}(\theta)}^\alpha}{1 - I \overline{D^{2\alpha^2}(\theta)}^\alpha} \tag{12.59}$$

from which we would obtain the TUB

$$P_b(E) \leq \frac{1}{\pi} \int_0^{\pi/2} \frac{\overline{D^{2\alpha^2}(\theta)}^\alpha \overline{D^{4\alpha^2}(\theta)}^\alpha}{[1 - \overline{D^{2\alpha^2}(\theta)}^\alpha]^2} d\theta \tag{12.60}$$

where the statistical averages are obtained from (12.24), (12.25), and (12.26) for Rayleigh, Rician, and Nakagami-m channels, respectively. For example, for Rayleigh fading, using (12.24) gives the simple TUB

$$P_b(E) \leq \frac{1}{\pi} \int_0^{\pi/2} \left(\frac{2}{\overline{\gamma}} \right)^2 \sin^4 \theta \left(\frac{\sin^2 \theta + \overline{\gamma}/2}{\sin^2 \theta + \overline{\gamma}} \right) d\theta \tag{12.61}$$

As a check on the results, the union–Chernoff bound would be obtained by replacing the integrand in (12.61) by its value at $\theta = \pi/2$, resulting in

$$P_b(E) \leq \frac{2}{\overline{\gamma}^2} \left(\frac{1 + \overline{\gamma}/2}{1 + \overline{\gamma}} \right) \tag{12.62}$$

which agrees[11] with Eq. (49) of Ref. 1.

---

[11] The result in (12.63) is actually one-half of the result in Eq. (49) of Ref. 1 since the bound on the alternative representation of the Gaussian Q-function obtained by replacing the integrand by its value at $\theta = \pi/2$ (see Chapter 4) is one-half of the result obtained from the conventional Chernoff bound.

For the case of no channel state information, the branch gains of Fig. 12.5 would become

$$a = \frac{I}{2} e^{2N_0 s^2} M_\alpha(-2s\sqrt{2E_s}), \qquad b = \frac{I}{2} e^{N_0 s^2} M_\alpha(-s\sqrt{2E_s}),$$

$$c = \frac{1}{2} e^{N_0 s^2} M_\alpha(-s\sqrt{2E_s}) \qquad (12.63)$$

and thus from (12.57), the transfer function is given by

$$T(s, I) = \frac{I[e^{N_0 s^2} M_\alpha(-s\sqrt{2E_s})][e^{2N_0 s^2} M_\alpha(-2s\sqrt{2E_s})]}{1 - I[e^{N_0 s^2} M_\alpha(-s\sqrt{2E_s})]}$$

$$= \frac{I[e^{3N_0 s^2} M_\alpha(-s\sqrt{2E_s}) M_\alpha(-2s\sqrt{2E_s})]}{1 - I[e^{N_0 s^2} M_\alpha(-s\sqrt{2E_s})]} \qquad (12.64)$$

with the corresponding TUB

$$P_b(E) \le \frac{1}{2\pi j} \int_{\sigma-j\infty}^{\sigma+j\infty} \frac{1}{s} \left[\frac{e^{3N_0 s^2} M_\alpha(-s\sqrt{2E_s}) M_\alpha(-2s\sqrt{2E_s})}{[1 - e^{N_0 s^2} M_\alpha(-s\sqrt{2E_s})]^2} \right] ds \qquad (12.65)$$

For even the simplest of fading channels (e.g., Rayleigh) the MGF is not available in a simple form involving elementary functions which lend themselves to integration. Nevertheless, as we shall see momentarily, for the Rayleigh fading channel it is still relatively straightforward to obtain results using the method of Gauss–quadrature (in particular, Gauss–Hermite) integration. The procedure is as follows.

The MGF of a Rayleigh RV with mean-square value $\overline{\alpha^2} = \Omega$ is given by

$$M_\alpha(s) = \exp\left(\frac{s^2 \Omega}{4}\right) \left[{}_1F_1\left(-\frac{1}{2}; \frac{1}{2}; -\frac{s^2\Omega}{4}\right) + s\sqrt{\Omega}\frac{\sqrt{\pi}}{2}\right] \qquad (12.66a)$$

where ${}_1F_1(\bullet; \bullet; \bullet)$ is the Kummer confluent hypergeometric function [20, Eq. (12.2)] or equivalently, using the relation between ${}_1F_1(\bullet; \bullet; \bullet)$ and the error function erf($\bullet$),

$$M_\alpha(s) = 1 + \frac{\sqrt{\pi}}{2} s\sqrt{\Omega} \exp\left(\frac{s^2\Omega}{4}\right) \text{erfc}\left(-\frac{s\sqrt{\Omega}}{2}\right) \qquad (12.66b)$$

Renormalizing the complex integration variable in (12.65) as $\xi = s\sqrt{N_0\overline{\gamma}}$, the TUB can be written as

$$P_b(E) \le \frac{1}{2\pi j} \int_{\sigma'-j\infty}^{\sigma'+j\infty} \frac{1}{\xi} \left[\frac{e^{3\xi^2/\overline{\gamma}} M'_\alpha(-\xi) M'_\alpha(-2\xi)}{[1 - e^{\xi^2/\overline{\gamma}} M'_\alpha(-\xi)]^2} \right] d\xi \qquad (12.67)$$

where now $\sigma' = \sigma\sqrt{N_0\bar{\gamma}}$ and

$$M'_\alpha(\xi) = \exp\left(\frac{\xi^2}{2}\right) A(\xi), \quad A(\xi) \triangleq {}_1F_1\left(-\frac{1}{2};\frac{1}{2};-\frac{\xi^2}{2}\right) + \xi\sqrt{\frac{\pi}{2}} \quad (12.68a)$$

or equivalently,

$$M'_\alpha(\xi) = 1 + \sqrt{\frac{\pi}{2}}\xi \exp\left(\frac{\xi^2}{2}\right) \text{erfc}\left(-\frac{\xi}{\sqrt{2}}\right) \quad (12.68b)$$

To evaluate the bound on average BEP in (12.67), one has the option of using either of two Gauss–quadrature methods. First, substituting (12.68a) into (12.67) and simplifying gives

$$P_b(E) \leq \frac{1}{2\pi j} \int_{\sigma'-j\infty}^{\sigma'+j\infty} \frac{1}{\xi} \left[\frac{e^{\xi^2[(\bar{\gamma}+3)/\bar{\gamma}]}A(-\xi)A(-2\xi)}{(1 - e^{\xi^2[(\bar{\gamma}+2)/2\bar{\gamma}]}A(-\xi))^2} \right] d\xi \quad (12.69)$$

To evaluate the integration along the vertical line in (12.69), we recognize that along this line the complex integration variable can be expressed as $\xi = \sigma' + j\omega$, where σ' is fixed and ω varies from $-\infty$ to ∞. Thus, making this change of variables in (12.69) gives

$$P_b(E) \leq \frac{1}{2\pi} \int_{-\infty}^{\infty} \frac{1}{\sigma' + j\omega} \left[\frac{e^{(\sigma'+j\omega)^2[(\bar{\gamma}+3)/\bar{\gamma}]}A(-\sigma'-j\omega)A(-2\sigma'-2j\omega)}{(1 - e^{(\sigma'+j\omega)^2[(\bar{\gamma}+2)/2\bar{\gamma}]}A(-\sigma'-j\omega))^2} \right] d\omega \quad (12.70)$$

which is of the form[12]

$$P_b(E) \leq \int_{-\infty}^{\infty} e^{-a\omega^2} f(\omega)\,d\omega = \frac{1}{\sqrt{a}} \int_{-\infty}^{\infty} e^{-x^2} f(x/\sqrt{a})\,dx \quad (12.71)$$

and thus can be evaluated by Gauss–Hermite integration [20, Sec. 25.4.46], namely,

$$\int_{-\infty}^{\infty} \exp(-x^2) f\left(\frac{x}{\sqrt{a}}\right) dx \simeq \sum_{n=1}^{N_p} H_{x_n} f\left(\frac{x_n}{\sqrt{a}}\right) \quad (12.72)$$

where $\{x_n\}$ are the N_p zeros of the N_p-order Hermite polynomial $H_{N_p}(x)$ and H_{x_n} are corresponding weight factors defined by

$$H_{x_n} \triangleq \frac{2^{N_p-1}N_p!\sqrt{\pi}}{(N_p)^2[H_{N_p-1}(x_n)]^2} \quad (12.73)$$

---

[12] Note that $f(\omega)$ is a complex function of ω, and thus it might appear that the upper bound is also complex. However, the imaginary part of $f(\omega)$ will be an odd function of ω, and thus since $e^{-a\omega^2}$ is an even function of ω, the imaginary part of the integral will evaluate to zero.

The zeros and the weight factors are both tabulated in Table 25.10 of Ref. 20, for various polynomial orders. Typically, $N_p = 20$ is sufficient for excellent accuracy.

The second approach to evaluating an upper bound on average BEP is to substitute (12.68b) into (12.67) and use the Gauss–Chebyshev method of Appendix 9B. In particular,

$$\frac{1}{2\pi j} \int_{\sigma'-j\infty}^{\sigma'+j\infty} \frac{1}{\xi} \left[\frac{e^{3\xi^2/\bar{\gamma}} M'_\alpha(-\xi) M'_\alpha(-2\xi)}{(1 - e^{\xi^2/\bar{\gamma}} M'_\alpha(-\xi))^2} \right] d\xi \stackrel{\Delta}{=} \frac{1}{2\pi j} \int_{\sigma'-j\infty}^{\sigma'+j\infty} \frac{1}{\xi} f(\xi) d\xi$$

$$\simeq \frac{1}{n} \sum_{k=1}^{n/2} [\text{Re}\{f(\sigma' + j\sigma'\tau_k)\} + j\,\text{Im}\{f(\sigma' + j\sigma'\tau_k)\}],$$

$$\tau_k = \tan\frac{(2k-1)\pi}{2n} \tag{12.74}$$

where the choice of n and σ' are discussed in Appendix 9B.[13] This approach is perhaps the simpler of the two in that it does not involve computation of the zeros and weight factors of the Hermite polynomials.

The MGF of a Nakagami-m RV with mean-square value $\overline{\alpha^2} = \Omega$ is given by

$$M_\alpha(s) = \exp\left(\frac{s^2\Omega}{4m}\right) \left[{}_1F_1\left(\frac{1}{2} - m; \frac{1}{2}; -\frac{s^2\Omega}{4m}\right) \right.$$

$$\left. + s\sqrt{\Omega}\frac{\Gamma\left(m + \frac{1}{2}\right)}{\sqrt{m}\,\Gamma(m)} {}_1F_1\left(1 - m; \frac{3}{2}; -\frac{s^2\Omega}{4m}\right) \right] \tag{12.75}$$

Again renormalizing the integration variable in (12.65) as $\xi = s\sqrt{N_0\bar{\gamma}}$, the TUB can be written as in (12.67), where now

$$M'_\alpha(\xi) = \exp\left(\frac{\xi^2}{2m}\right) A(\xi; m) \tag{12.76}$$

with

$$A(\xi; m) = {}_1F_1\left(\frac{1}{2} - m; \frac{1}{2}; -\frac{\xi^2}{2m}\right) + \xi\sqrt{\frac{2}{m}}\frac{\Gamma\left(m + \frac{1}{2}\right)}{\Gamma(m)} {}_1F_1\left(1 - m; \frac{3}{2}; -\frac{\xi^2}{2m}\right) \tag{12.77}$$

[Note that for $m = 1$ (i.e., Rayleigh fading), $A(\xi; 1)$ reduces to $A(\xi)$ of (12.68a) as it should.] It should now be obvious that the TUB on average BEP is given by (12.70) with the appropriate substitution of $A(\xi; 1)$ for $A(\xi)$, which again can be evaluated using the Gauss–Hermite integration method.

---

[13] Specifically, the function $f(\xi)$ in (12.74) should be numerically minimized and the resulting value of $\xi > 0$ at which this minimum occurs is then equated to σ'. Note that the minimization must be performed for each $\bar{\gamma}$ and thus the value of σ' used to evaluate (12.74) is a function of $\bar{\gamma}$.

12.2 DIFFERENTIALLY COHERENT DETECTION

12.2.1 System Model

Consider the block diagram of a trellis-coded M-DPSK system[14] illustrated in Fig. 12.6. The only difference between this block diagram and that of Fig. 12.1 is the inclusion of a differential (phase) encoder prior to the modulator and the replacement of the coherent demodulator by a differential (phase) detector. Thus, if x_k still denotes the kth trellis-coded M-PSK symbol corresponding to the information symbol u_k, the actual M-PSK symbol transmitted over the channel is[15]

$$v_k = v_{k-1} x_k \tag{12.78}$$

Analogous to (12.3), the fading channel output at time k is

$$w_k = \alpha_k \sqrt{2E_s} v_k + n_k \tag{12.79}$$

where the noise sample n_k has the same properties as for the coherent detection case, and the output of the differential detector is

$$y_k = w_k w_{k-1}^* \tag{12.80}$$

If we again assume that the fading is independent from symbol to symbol, we can represent the combination of the differential encoding/detection operations and the fading channel as a memoryless coding channel whose input is the information M-PSK symbol x_k and whose output is the decision variable y_k. As such, Fig. 12.2 also represents a simple block diagram of the trellis-coded M-DPSK system that is suitable for performance analysis the primary difference being that, conditioned on the fading, the memoryless coding channel is no longer AWGN. This is easily seen by combining Eqs. (12.78) through (12.80), which yields

$$y_k = w_k w_{k-1}^* = (\alpha_k \sqrt{2E_s} v_{k-1} x_k + n_k)(\alpha_{k-1} \sqrt{2E_s} v_{k-1} + n_{k-1})^*$$
$$= \alpha_k \alpha_{k-1} 2E_s x_k + \text{noise (non-Gaussian) terms} \tag{12.81}$$

The optimum decision metric still depends on the availability or lack thereof of CSI. Such metrics for multiple channel reception of differentially detected M-PSK were considered in Chapter 7 and also in Ref. 2. For the case of perfect CSI, the branch decision metric is complicated (involving the zero-order Bessel function) and thus theoretical analysis of the average BEP is difficult, if not impossible. Even for the case of no CSI, depending on the statistics of the fading amplitude, the optimum branch metric can also be quite complicated. [The one

---

[14] By M-DPSK, we refer in this chapter to the conventional (two-symbol observation) form of differentially detected M-PSK.
[15] Note that the transmitted M-PSK symbols are still normalized such that $|v_k| = 1$.

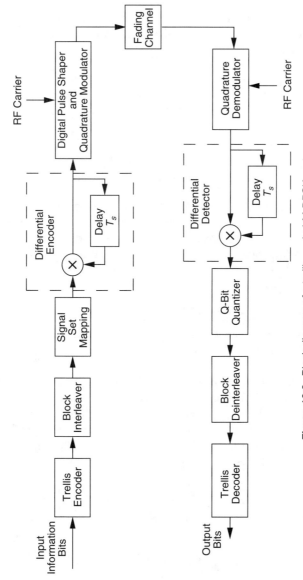

Figure 12.6. Block diagram of a trellis-coded *M*-DPSK system.

case that has a simple solution is the branch metric for Rayleigh fading with no CSI, which happens to be the same as that for the pure AWGN channel (see Chapter 7).] In view of this difficulty, we will follow the approach taken in Ref. 2 and assume the simpler Gaussian metric for both the known and unknown CSI cases. This approach was also taken by Johnston and Jones [21] in analyzing the performance of a block-coded M-DPSK system over a Rayleigh fading channel. Thus, analogous to (12.7) and (12.8) the decision metrics for the differentially detected M-PSK case become

$$m(\mathbf{y}, \mathbf{x}; \boldsymbol{\alpha}) \triangleq -\sum_{n=1}^{N} |y_n - \alpha_n \alpha_{n-1} 2E_s x_n|^2 \qquad (12.82)$$

and

$$m(\mathbf{y}, \mathbf{x}) \triangleq -\sum_{n=1}^{N} |y_n - 2E_s x_n|^2 \qquad (12.83)$$

corresponding, respectively, to perfectly known and unknown CSI. Since, as mentioned above, the Gaussian decision metric is (at least for Rayleigh fading) optimum for unknown CSI, we focus our attention first on this case since the suboptimality of the metric for known CSI will tend to reduce the performance improvement.

12.2.2 Performance Evaluation

In this section we evaluate the TUB on the average BEP starting first with evaluation of the pairwise error probability.

12.2.2.1 Unknown Channel State Information.
As for the coherent detection case, we must first find the pairwise error probability in a form suitable for application of the transfer function bound approach. The first approach to try is find the conditional pairwise error probability and then average this result over the fading PDF. Substituting (12.81) into (12.83), then, the conditional pairwise error probability of (12.28) becomes

$$P(\mathbf{x} \to \hat{\mathbf{x}}|\boldsymbol{\alpha}) = \Pr\left\{ \sum_{n \in \eta} \mathrm{Re}\left\{ w_n w_{n-1}^* (\hat{x}_n - x_n)^* \right\} \geq 0 | \mathbf{x} \right\}$$

$$= \Pr\left\{ \sum_{n \in \eta} \mathrm{Re}\left\{ w_n w_{n-1}^* (x_n - \hat{x}_n)^* \right\} < 0 | \mathbf{x} \right\} \qquad (12.84)$$

Since w_n and w_{n-1} are conditionally (on the fading) complex Gaussian, the probability required in (12.84) is an extension of the problem considered in Appendix 8A to a sequence of weighted RVs. In particular, because of the assumption of ideal interleaving/deinterleaving the decision variable in (12.84) is a sum of independent complex conjugate products of Gaussian RVs, which would be appear to be a special case of the quadratic form considered in Appendix B of Ref. 22. However, because each Gaussian product RV in the sum is weighted by $(\hat{x}_n - x_n)^*$, a constant that, in general, depends on the summation index n, the approach taken in Ref. 22, which would lead to a conditional pairwise error probability expressed in terms of the Marcum Q-function, does not apply here. Nevertheless, because of the independence of the terms in the sum, we can still apply the method used in Section 12.1.2.2, in particular, (12.38), where we must now find the MGF of $z_n \triangleq \mathrm{Re}\{w_n w_{n-1}^*(x_n - \hat{x}_n)^*|\mathbf{x}\}$. The solution is presented in Appendix 12A, which for the case of slow fading (i.e., $\alpha_{n-1} = \alpha_n = \alpha$) gives the result

$$M_{z_n}(s) = \overline{\frac{1}{1 - (sN_0|x_n - \hat{x}_n|)^2} \exp\left\{\frac{\alpha^2(E_s/N_0)|x_n - \hat{x}_n|^2[(2s^2N_0^2 + sN_0)]}{1 - (sN_0|x_n - \hat{x}_n|)^2}\right\}}^{\alpha} \quad (12.85)$$

where as before the overbar denotes statistical averaging with respect to the fading RV α. Defining as before the instantaneous SNR per bit $\gamma \triangleq \alpha^2 E_b/N_0$ and recalling that $E_s = n_c E_b$, then (12.85) can be rewritten as

$$M_{z_n}(s) = \overline{\frac{1}{1 - (sN_0|x_n - \hat{x}_n|)^2} \exp\left\{\gamma \frac{n_c|x_n - \hat{x}_n|^2[(2s^2N_0^2 + sN_0)]}{1 - (sN_0|x_n - \hat{x}_n|)^2}\right\}}^{\gamma} \quad (12.86)$$

Statistical averages of the type required in (12.86) are in fact Laplace transforms of the PDF $p_\gamma(\gamma)$ (with exponent reversed) and have been evaluated in Chapter 5 for a wide variety of fading channels. Finally, making the change of complex variables $\xi = sN_0$ in (12.86) and then substituting the result in (12.38) gives the pairwise error probability with an integrand in the desired product form, namely,

$$P(\mathbf{x} \to \hat{\mathbf{x}}) = \frac{1}{2\pi j} \int_{\sigma-j\infty}^{\sigma+j\infty} \frac{1}{\xi} \prod_{n \in \eta} M'_{z_n}(-\xi)\, d\xi \quad (12.87)$$

where

$$M'_{z_n}(\xi) = \overline{\frac{1}{1 - \xi^2|x_n - \hat{x}_n|^2} \exp\left\{\gamma \frac{n_c|x_n - \hat{x}_n|^2(2\xi^2 + \xi)}{1 - \xi^2|x_n - \hat{x}_n|^2}\right\}}^{\gamma} \quad (12.88)$$

For the fast-fading case where $\alpha_{n-1} \neq \alpha_n$ but because of the interleaving and deinterleaving the products $\alpha_{n-1}\alpha_n$ are still independent of each other, (12.87) is still valid, now with

$$M'_{z_n}(\xi) = \overline{\frac{1}{1 - \xi^2|x_n - \hat{x}_n|^2} \exp\left\{\frac{n_c|x_n - \hat{x}_n|^2[(\gamma_{n-1} + \gamma_n)\xi^2 + \sqrt{\gamma_{n-1}\gamma_n}\xi]}{1 - \xi^2|x_n - \hat{x}_n|^2}\right\}}^{\gamma_{n-1},\gamma_n} \quad (12.89)$$

Note that evaluation of (12.89) now requires second-order fading statistics.

12.2.2.2 Known Channel State Information.
Comparing the decision metrics of (12.82) with (12.83), then following the same approach as for the unknown CSI case, we immediately see that to evaluate the pairwise error probability as given by (12.38), we now need to obtain the MGF of the decision variable $z_n \triangleq \text{Re}\{\alpha_n \alpha_{n-1} w_n w_{n-1}^* (x_n - \hat{x}_n)^* | \mathbf{x}\}$ or, for the slow-fading assumption, $z_n \triangleq \text{Re}\{\alpha_n^2 w_n w_{n-1}^* (x_n - \hat{x}_n)^* | \mathbf{x}\}$. The approach taken in Appendix 12A is still appropriate, and with suitable redefinition of the constant C in (12A.7), we obtain the following result:

$$M_{z_n}(s) = \frac{\exp\left[\dfrac{\alpha_n^4 (E_s/N_0)|x_n - \hat{x}_n|^2 \left(2\alpha_n^2 s^2 N_0^2 + s N_0\right)}{1 - \left(sN_0 \alpha_n^2 |x_n - \hat{x}_n|\right)^2}\right]^{-\alpha_n}}{1 - \left(sN_0 \alpha_n^2 |x_n - \hat{x}_n|\right)^2} \quad (12.90)$$

Analytical evaluation of the statistical averages of (12.90) in closed form is not possible even for the simplest of fading channels such as the Rayleigh. Similar evaluation problems were noted in Ref. 2 in connection with trying to Chernoff bound this very same pairwise error probability. Furthermore, Divsalar and Simon [2] reported that computer simulation results corresponding to several examples revealed that for the Gaussian metric under consideration, little performance was gained from having knowledge of CSI available at the receiver. Thus, because of the difficulty associated with the analysis, we shall abandon pursuit of the known CSI case as was done there.

12.2.3 Example

Consider the same example as in Section 12.1.5, namely rate $-\frac{1}{2}$ coded QPSK using the two-state trellis illustrated in Fig. 12.4. Here we are interested in computing the TUB on average BER when differential detection is employed at the receiver and no CSI is available. We now develop the specific result for the case of slow Rayleigh fading. Results for other fading channel models follow directly from the Laplace transforms given in Chapter 5.

Using the Laplace transform of (5.5) to evaluate the statistical average over the fading, the MGF of (12.88) becomes

$$M'_{z_n}(\xi) = \frac{1}{1 - \xi^2 |x_n - \hat{x}_n|^2} \left(1 - \bar{\gamma} \frac{|x_n - \hat{x}_n|^2 (2\xi^2 + \xi)}{1 - \xi^2 |x_n - \hat{x}_n|^2}\right)^{-1}$$

$$= \frac{1}{1 - |x_n - \hat{x}_n|^2 [(2\bar{\gamma} + 1)\xi^2 + \bar{\gamma}\xi]} \quad (12.91)$$

Replacing $D(\theta)^{|x_n - \hat{x}_n|^2}$ with $M'_{z_n}(-\xi)$, the branch gains of Fig. 12.5 are now given by

$$a = \frac{I}{2} M'_{z_n}(-\xi)\bigg|_{|x_n - \hat{x}_n|^2 = 4} = \frac{I}{2}\left[\frac{1}{1 - 4[(2\bar{\gamma} + 1)\xi^2 - \bar{\gamma}\xi]}\right],$$

$$b = \frac{1}{2} M'_{z_n}(-\xi)\bigg|_{|x_n - \hat{x}_n|^2 = 2} = \frac{I}{2}\left[\frac{1}{1 - 2[(2\bar{\gamma} + 1)\xi^2 - \bar{\gamma}\xi]}\right],$$

$$c = \frac{1}{2} M'_{z_n}(-\xi)\bigg|_{|x_n - \hat{x}_n|^2 = 2} = \frac{1}{2}\left[\frac{1}{1 - 2[(2\bar{\gamma} + 1)\xi^2 - \bar{\gamma}\xi]}\right] \quad (12.92)$$

Thus, from (12.57), the transfer function is

$$T(\xi, I) = \frac{4ac}{1 - 2b} = \frac{I}{\{1 - 4[(2\bar{\gamma} + 1)\xi^2 - \bar{\gamma}\xi]\}\{1 - 2[(2\bar{\gamma} + 1)\xi^2 - \bar{\gamma}\xi] - I\}} \quad (12.93)$$

which from the TUB analogous to (12.52), namely,

$$P_b(E) \leq \frac{1}{2\pi j} \int_{\sigma' - j\infty}^{\sigma' + j\infty} \frac{1}{\xi} \frac{\partial}{\partial I} T(\xi, I)\bigg|_{I=1} d\xi \quad (12.94)$$

results in

$$P_b(E) \leq \frac{1}{2\pi j} \int_{\sigma' - j\infty}^{\sigma' + j\infty} \frac{1}{\xi}\left[\frac{1 - 2[(2\bar{\gamma} + 1)\xi^2 - \bar{\gamma}\xi]}{4\{1 - 4[(2\bar{\gamma} + 1)\xi^2 - \bar{\gamma}\xi]\}[(2\bar{\gamma} + 1)\xi^2 - \bar{\gamma}\xi]^2}\right] d\xi \quad (12.95)$$

For the case of fast Rayleigh fading, the MGF of (12.88) can be evaluated using a result in Ref. 23 which gives

$$M'_{z_n}(\xi) = \frac{1}{1 - |x_n - \hat{x}_n|^2[(2\bar{\gamma} + 1 + (1 - \rho^2)\bar{\gamma}^2)\xi^2 + \bar{\gamma}\rho\xi]} \quad (12.96)$$

where ρ is the correlation between the underlying complex Gaussian fading variables whose amplitudes are α_{n-1} and α_n. Note that for $\rho = 1$, (12.96) reduces to the result for slow fading given in (12.91), as it should. Analogous to (12.93), the transfer function is now

$$T(\xi, I) = \frac{I}{\{1 - 4[(2\bar{\gamma} + 1 + (1 - \rho^2)\bar{\gamma}^2)\xi^2 - \bar{\gamma}\rho\xi]\} \\ \times \{1 - 2[(2\bar{\gamma} + 1 + (1 - \rho^2)\bar{\gamma}^2)\xi^2 - \bar{\gamma}\rho\xi] - I\}} \quad (12.97)$$

which can then be used to obtain the TUB from (12.94), namely,

$$P_b(E) \leq \frac{1}{2\pi j} \int_{\sigma' - j\infty}^{\sigma' + j\infty} \frac{1}{\xi}\left[\frac{1 - 2[(2\bar{\gamma} + 1 + (1 - \rho^2)\bar{\gamma}^2)\xi^2 - \bar{\gamma}\rho\xi]}{4\{1 - 4[(2\bar{\gamma} + 1 + (1 - \rho^2)\bar{\gamma}^2)\xi^2 - \bar{\gamma}\rho\xi]\} \\ \times [(2\bar{\gamma} + 1 + (1 - \rho^2)\bar{\gamma}^2)\xi^2 - \bar{\gamma}\rho\xi]^2}\right] d\xi \quad (12.98)$$

12.3 NUMERICAL RESULTS: COMPARISON OF THE TRUE UPPER BOUNDS AND UNION–CHERNOFF BOUNDS

In this section we compare the various TUBs derived thus far with the corresponding union–Chernoff bounds obtained in Refs. 1, 2 and 23. For the purpose of illustration, we present numerical results for the rate $-\frac{1}{2}$ coded QPSK with two-state trellis example and only for the Rayleigh channel.

For coherent detection with known CSI, the TUB and union–Chernoff bounds are given in (12.61) and (12.62). Figure 12.7 is a plot of these two upper bounds. We observe a uniform superiority of about 1.5 dB for the TUB relative to the union–Chernoff bound.

For coherent detection with unknown CSI, the TUB is obtained by evaluating either (12.70) or (12.74). The union–Chernoff bound is obtained from Eqs. (58) and (60) of Ref. 1 and is given by

$$P_b(E) \leq \min_{\lambda \geq 0} \frac{\xi_1 \xi_2 D^{-24\lambda^2}}{(1 - \xi_2 D^{-8\lambda^2})^2} \tag{12.99}$$

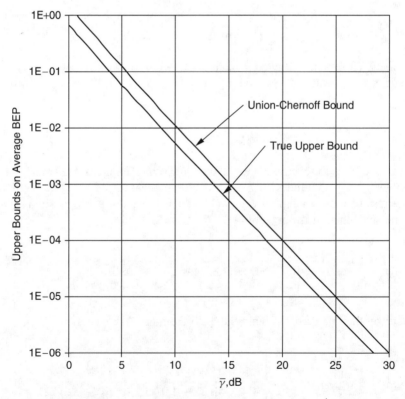

Figure 12.7. Upper bounds on the average bit error probability for rate $-\frac{1}{2}$ trellis-coded QPSK with two-state trellis: coherent detection with known channel state information, Rayleigh fading.

where

$$D \triangleq e^{-\bar{\gamma}/4}$$

$$\xi_1 = 1 - 4\sqrt{\pi}\lambda\bar{\gamma}\exp[(2\lambda\bar{\gamma})^2]Q(2\sqrt{2}\lambda\bar{\gamma}) \quad (12.100)$$

$$\xi_2 = 1 - 2\sqrt{\pi}\lambda\bar{\gamma}\exp[(\lambda\bar{\gamma})^2]Q(\sqrt{2}\lambda\bar{\gamma})$$

Superimposed on Fig. 12.8 are the TUB and union–Chernoff bound for the no CSI case as given above.

For differential detection with unknown CSI and slow fading, the TUB is obtained from (12.95), which can be evaluated using the Gauss–Chebyshev quadrature technique described in Appendix 9B.[16] In this regard, it is convenient to first to write the transfer function $T(\xi, I)$ of (12.93) in its infinite series form, namely,

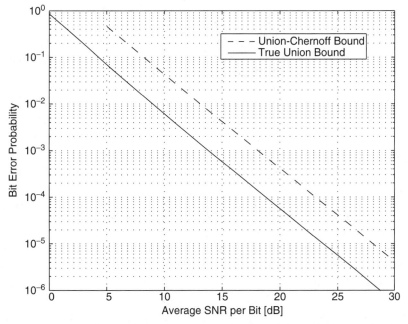

Figure 12.8. TUB and union–Chernoff bound on average BEP performance of coherent detection with unknown CSI in slow Rayleigh fading.

---

[16] For the particular integrand in (12.95), the best value of σ' (denoted by c in (9B.15)) to guarantee quick convergence can easily be evaluated as

$$\sigma' = \frac{\bar{\gamma}}{2(2\bar{\gamma}+1)}$$

$$T(\xi, I) = 4ac(1 + 2b + (2b)^2 + \cdots)$$

$$= \frac{I}{\{1 - 4[(2\overline{\gamma} + 1)\xi^2 - \overline{\gamma}\xi]\}\{1 - 2[(2\overline{\gamma} + 1)\xi^2 - \overline{\gamma}\xi]\}}$$

$$\times \left[1 + \frac{I}{1 - 2[(2\overline{\gamma} + 1)\xi^2 - \overline{\gamma}\xi]} + \frac{I^2}{\{1 - 2[(2\overline{\gamma} + 1)\xi^2 - \overline{\gamma}\xi]\}^2} + \cdots \right] \quad (12.101)$$

from which the derivative required in (12.94) has the corresponding infinite series form

$$\frac{\partial}{\partial I}T(\xi, I)\bigg|_{I=1} = \frac{1}{\{1 - 4[(2\overline{\gamma} + 1)\xi^2 - \overline{\gamma}\xi]\}\{1 - 2[(2\overline{\gamma} + 1)\xi^2 - \overline{\gamma}\xi]\}}$$

$$+ \frac{2}{\{1 - 4[(2\overline{\gamma} + 1)\xi^2 - \overline{\gamma}\xi]\}\{1 - 2[(2\overline{\gamma} + 1)\xi^2 - \overline{\gamma}\xi]\}^2}$$

$$+ \frac{3}{\{1 - 4[(2\overline{\gamma} + 1)\xi^2 - \overline{\gamma}\xi]\}\{1 - 2[(2\overline{\gamma} + 1)\xi^2 - \overline{\gamma}\xi]\}^3} + \cdots \quad (12.102)$$

Substituting (12.102) in (12.94) and applying the Gauss–Chebyshev technique of Appendix 9B term by term until additional terms produce a negligible change in the result is the most efficient method of evaluating (12.95). The union–Chernoff bound on average BEP for differential detection with unknown CSI and slow fading is given by

$$P_b(E) \leq \frac{\partial}{\partial I}T(\xi, I)\bigg|_{I=1, \xi=\overline{\gamma}/2(2\overline{\gamma}+1)} = \frac{4(1 + 2\overline{\gamma} + \overline{\gamma}^2/2)(2\overline{\gamma} + 1)^2}{\overline{\gamma}^4(1 + \overline{\gamma})^2} \quad (12.103)$$

which agrees with Eq. (40) of Ref. 2 after being specialized to the case of symmetric QPSK. Figure 12.9 is an illustration of the TUB and the union–Chernoff bound.

Finally for differential detection with unknown CSI and fast fading, the TUB is obtained from (12.98) which again is conveniently evaluated using the Gauss–Chebyshev quadrature technique.[17] Again, analogous to (12.102), an infinite series representation of the derivative of the transfer function in (12.97) is particularly helpful in efficiently carrying out the evaluation. The corresponding union–Chernoff bound is given by

$$P_b(E) \leq \frac{\partial}{\partial I}T(\xi, I)\bigg|_{I=1, \xi=\overline{\gamma}\rho/2[2\overline{\gamma}+1+(1-\rho^2)\overline{\gamma}^2]}$$

$$= \frac{1 + 2\zeta}{4\zeta^2(1 + 4\zeta)}, \quad \zeta \triangleq \frac{\rho^2\overline{\gamma}^2}{4[2\overline{\gamma} + 1 + (1 - \rho^2)\overline{\gamma}^2]} \quad (12.104)$$

---

[17] Here, the best value of σ' to guarantee quick convergence becomes

$$\sigma' = \frac{\overline{\gamma}\rho}{2[2\overline{\gamma} + 1 + (1 - \rho^2)\overline{\gamma}^2]}$$

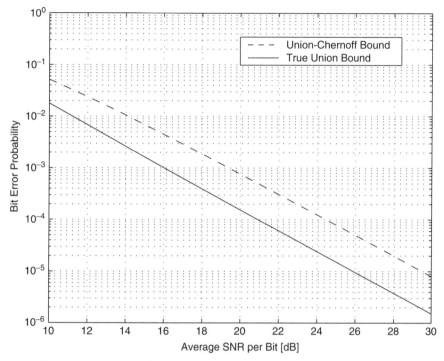

Figure 12.9. TUB and union–Chernoff bound on average BEP performance of differential detection with unknown CSI in slow Rayleigh fading.

which agrees with [23, Eqs. (34) and (35)] and furthermore reduces to (12.103) when $\rho = 1$ (slow fading). Note that in the limit of large $\overline{\gamma}$ and $\rho \neq 1$, the bound on BEP as given by (12.104) approaches a finite value given by

$$P_b(E) \leq \frac{2(1-\zeta^2)^2(2-\zeta^2)}{\zeta^4} \qquad (12.105)$$

which represents an *error floor* (i.e., regardless of how large we make the average SNR, the upper bound predicts a nonzero BEP).

To obtain numerical results for this case, we must apply an appropriate correlation model for the fading process. Mason [24] has tabulated the autocorrelation function (or equivalently, the power spectral density) for various types of fast fading processes of interest. These results were given in Table 2.1 where f_d denotes the Doppler spread and for convenience the variance of the fading process has been normalized to unity. Figure 12.10 is an illustration of the union–Chernoff bound and the TUB as computed from (12.104) and (12.98), respectively, for the land mobile channel with $f_d T_b$ as a parameter. The value of $f_d T_b = 0$ corresponds to the case of slow fading as per the results in Fig. 12.9.

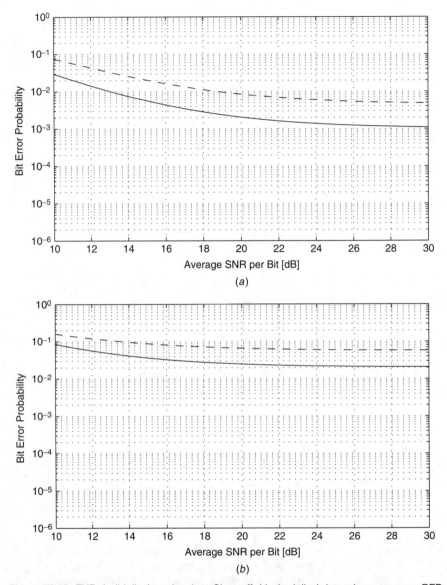

Figure 12.10. TUB (solid line) and union–Chernoff (dashed line) bound on average BEP performance of differential detection with unknown CSI in fast Rayleigh fading: land-mobile channel: (a) $f_d T_b = 0.05$; (b) $f_d T_b = 0.1$.

REFERENCES

1. D. Divsalar and M. K. Simon, "Trellis coded modulation for 4800–9600 bits/s transmission over a fading mobile satellite channel," *IEEE J. Sel. Areas Commun.*, vol. SAC-5, February 1987, pp. 162–175.

2. D. Divsalar and M. K. Simon, "The performance of trellis coded multilevel DPSK on a fading mobile satellite channel," *IEEE Trans. Veh. Technol.*, vol. VT-37, May 1988, pp. 78–91. Also see *ICC'87 Conf. Rec.*, Seattle, WA, June 7–10, 1987, pp. 21.2.1–21.2.7.
3. E. Biglieri, D. Divsalar, P. J. McLane, and M. K. Simon, *Introduction to Trellis Coded Modulation with Applications*. New York: Macmillan, 1990. Currently distributed by Prentice Hall, Upper Saddle River, NJ.
4. K. Y. Chan and A. Bateman, "The performance of reference based M-ary PSK with trellis coded modulation in Rayleigh fading," *IEEE Trans. Veh. Technol.*, vol. 41, May 1992, pp. 190–198.
5. S. B. Slimane and T. Le-Ngoc, "Tight bounds on the error probability of coded modulation schemes in Rayleigh fading channels," *IEEE Trans. Veh. Technol.*, vol. 44, February 1995, pp. 121–130.
6. A. J. Viterbi, "Convolutional codes and their performance in communication systems," *IEEE Trans. Commun. Technol.*, vol. COM-19, October 1971, pp. 751–772.
7. A. J. Viterbi and J. K. Omura, *Principles of Digital Communication and Coding*. New York: McGraw-Hill, 1979.
8. R. G. McKay, P. J. McLane, and E. Biglieri, "Error bounds for trellis coded MPSK on a fading mobile satellite channel," *IEEE Trans. Commun.*, vol. 39, December 1991, pp. 1750–1761.
9. J. K. Cavers and P. Ho, "Analysis of the error performance of trellis coded modulations in Rayleigh fading channels," *IEEE Trans. Commun.*, vol. 40, January 1992, pp. 74–80.
10. C. Tellambura, "Evaluation of the exact union bound for trellis coded modulations over fading channels," *IEEE Trans. Commun.*, vol. 44, December 1996, pp. 1693–1699.
11. J.-H. Kim, P. Ho, and J. K. Cavers, "A new analytical tool for evaluating the bit-error-rate of trellis coded modulation in Rician fading channel," *Proc. IEEE Veh. Technol. Conf. (VTC'97)*, vol. 3, Phoenix, AZ, May 1997, pp. 2012–2016.
12. J. K. Cavers, J.-H. Kim, and P. Ho, "Exact calculation of the union bound on performance of trellis-coded modulation in fading channels," *IEEE Trans. Commun.*, vol. COM-46, no. 5, May 1998, pp. 576–579. Also see *Proc. IEEE Int. Conf. Univ. Personal Commun. (ICUPC'96)*, vol. 2, Cambridge, MA, September 1996, pp. 875–880.
13. M. K. Simon and D. Divsalar, "Some new twists to problems involving the Gaussian probability integral," *IEEE Trans. Commun.*, vol. 46, February 1998, pp. 200–210.
14. G. Ungerboeck, "Channel coding with multilevel/phase signals," *IEEE Trans. Inf. Theory*, vol. IT-28, January 1982, pp. 55–67.
15. A. J. Viterbi, "Error bounds for convolutional codes and an asymptotically optimum decoding algorithm," *IEEE Trans. Inf. Theory*, vol. IT-13, April 1967, pp. 260–269.
16. E. K. Hall and S. G. Wilson, "Design and analysis of turbo codes on Rayleigh fading channels," *IEEE J. Sel. Areas Commun.*, vol. 16, February 1998, pp. 160–174.
17. E. Biglieri, C. Caire, G. Taricco, and J. Ventura-Traveset, "Simple method for evaluating error probabilities," *IEE Electron. Lett.*, vol. 32, February 1996, pp. 191–192.
18. E. Biglieri, C. Caire, G. Taricco, and J. Ventura-Traveset, "Computing error probabilities over fading channels: a unified approach," *Eur. Trans. Telecommun.*, vol. 9, February 1998, pp. 15–25.

19. D. Divsalar, "Performance of mismatched receivers on bandlimited channels," Ph.D. dissertation, University of California at Los Angeles, 1978.
20. M. Abramowitz and I. A. Stegun, *Handbook of Mathematical Functions with Formulas, Graphs, and Mathematical Tables*, 9th ed. New York: Dover Press, 1972.
21. D. A. Johnston and S. K. Jones, "Spectrally efficient communication via fading channels using coded multilevel DPSK," *IEEE Trans. Commun.*, vol. COM-29, no. 3, March 1981, pp. 276–284.
22. J. Proakis, *Digital Communications*, 3rd ed., New York: McGraw-Hill, 1995.
23. D. Divsalar and M. K. Simon, "Performance of trellis coded MDPSK on fast fading channels," *ICC'89 Conf. Rec.*, Boston, June 11–14, 1989, pp. 9.1.1–9.1.7.
24. L. J. Mason, "Error probability evaluation of systems employing differential detection in a Rician fast fading environment and Gaussian noise," *IEEE Trans. Commun.*, vol. COM-35, January 1987, pp. 39–46.

APPENDIX 12A: EVALUATION OF A MOMENT GENERATING FUNCTION ASSOCIATED WITH DIFFERENTIAL DETECTION OF *M*-PSK SEQUENCES

The decision variable associated with the pairwise error probability of differential detection of M-PSK sequences transmitted over a fading channel with no channel state information available to the receiver is a sum of RVs of the form[1]

$$z_n = \text{Re}\{w_n w_{n-1}^*(x_n - \hat{x}_n)^*\} \quad (12\text{A}.1)$$

where w_n and w_{n-1} are conditionally (on the fading) complex Gaussian random variables and $(x_n - \hat{x}_n)^*$ is a complex constant. Of interest in this appendix is the evaluation of the moment generating function (MGF) of z_n.

Recalling (12.78) and (12.79), then (12A.1) can be expressed as

$$z_n = (\alpha_n \sqrt{2E_s} v_{n-1} x_n + n_n)(\alpha_{n-1} \sqrt{2E_s} v_{n-1} + n_{n-1})^*(x_n - \hat{x}_n)^* \quad (12\text{A}.2)$$

Proakis [22, App. B] considers a quadratic form of complex Gaussian RVs X_n and Y_n, that is,

$$d_n = A|X_n|^2 + B|Y_n|^2 + CX_n Y_n^* + C^* X_n^* Y_n \quad (12\text{A}.3)$$

in which A, B, and C are constant weights. Our interest is in the case where $A = B = 0$ and X_n, Y_n are uncorrelated, whereupon (12A.3) becomes

$$d_n = 2\text{Re}\{C^* X_n^* Y_n\} \quad (12\text{A}.4)$$

---
[1] For simplicity of notation we shall omit the conditioning on **x** with the understanding that the RVs w_{n-1} and w_n are evaluated assuming that **x** is the transmitted vector.

APPENDIX 12A: EVALUATION OF A MOMENT GENERATING FUNCTION

whose MGF is given by Eq. (B-5) of Ref. 22, which for this special case becomes

$$M_{d_n}(s) = \frac{v^2}{-s^2 + v^2} \exp\left[\frac{v^2(\xi_{1n}s^2 + \xi_{2n}s)}{-s^2 + v^2}\right] \quad (12A.5)$$

with

$$\mu_{xx} = \frac{1}{2}E\{|X_n - \overline{X}_n|^2\}, \quad \mu_{yy} = \frac{1}{2}E\{|Y_n - \overline{Y}_n|^2\}$$

$$v = \sqrt{\frac{1}{4\mu_{xx}\mu_{yy}|C|^2}} \quad (12A.6)$$

$$\xi_{1n} = 2|C|^2(|\overline{X}_n|^2\mu_{yy} + |\overline{Y}_n|^2\mu_{xx}),$$

$$\xi_{2n} = C\overline{X}_n\overline{Y}_n^* + C^*\overline{X}_n^*\overline{Y}_n = 2\operatorname{Re}\{C^*\overline{X}_n^*\overline{Y}_n\}$$

Comparing (12A.1) with (12A.4), we draw the equivalences

$$X_n = w_{n-1}, \quad Y_n = w_n, \quad C = \frac{x_n - \hat{x}_n}{2} \quad (12A.7)$$

Recognizing the normalizations $|v_{k-1}| = |v_k| = |x_k| = 1$, the parameters in (12A.6) then become

$$|\overline{X}_n| = \alpha_{n-1}\sqrt{2E_s}, \quad |\overline{Y}_n| = \alpha_n\sqrt{2E_s}$$

$$\mu_{xx} = \frac{1}{2}E\{|n_{k-1}|^2\} = N_0, \quad \mu_{yy} = \frac{1}{2}E\{|n_n|^2\} = N_0,$$

$$v = \frac{1}{N_0|x_n - \hat{x}_n|} \quad (12A.8)$$

$$\xi_{1n} = N_0(\alpha_{n-1}^2 + \alpha_n^2)E_s|x_n - \hat{x}_n|^2,$$

$$\xi_{2n} = 2\alpha_{n-1}\alpha_n E_s \operatorname{Re}\{x_n(x_n - \hat{x}_n)^*\} = \alpha_{n-1}\alpha_n E_s|x_n - \hat{x}_n|^2$$

where the last equality in ξ_{2n} is obtained by using (12.11). Substituting (12A.8) in (12A.5) gives the conditional MGF of z_n as

$$M_{z_n}(s)|\alpha_{n-1}, \alpha_n = \frac{1}{1 - (sN_0|x_n - \hat{x}_n|)^2}$$

$$\times \exp\left\{\frac{(E_s/N_0)|x_n - \hat{x}_n|^2[(\alpha_{n-1}^2 + \alpha_n^2)s^2N_0^2 + \alpha_{n-1}\alpha_n sN_0]}{1 - (sN_0|x_n - \hat{x}_n|)^2}\right\} \quad (12A.9)$$

Since the unconditional PDF of z_n, namely, $p_{z_n}(z_n)$, is obtained by averaging the conditional PDF $p_{z_n|\alpha_n}(z_n|\alpha_n)$ over the PDF of α_n, the unconditional MGF is also the average of the conditional MGF over the PDF of α_n [see, e.g., (12.41) for the case of a Gaussian z_n]. Finally, then, the unconditional MGF of z_n is, from (12A.9)

$$M_{z_n}(s) = \frac{1}{1-(sN_0|x_n-\hat{x}_n|)^2}$$

$$\times \exp\left\{\frac{(E_s/N_0)|x_n-\hat{x}_n|^2[(\alpha_{n-1}^2+\alpha_n^2)s^2N_0^2+\alpha_{n-1}\alpha_n sN_0]}{1-(sN_0|x_n-\hat{x}_n|)^2}\right\}^{\alpha_{n-1},\alpha_n}$$

(12A.10)

INDEX

Additive white Gaussian noise (AWGN):
  bit error probability (BEP), 222, 511
  coded communication, 499, 511–512
  defined, 17
  fading channels and, 23, 27, 48, 57
  multicarrier DS-CDMA systems, 483
  multichannel receivers, 261, 313
  optimum receivers, detection of, 162–163, 181–182, 184, 186
  performance, detection methods:
    binary signaling, generic results, 218–219
    differentially coherent, 213–218
    ideal coherent, 194–205
    noncoherent, 209–210
    nonideal coherent, 206–209
    partially coherent, 210–213
  single carrier DS-CDMA systems, 476
  single channel reception, 193–219
Amount of fading (AF):
  defined, 18
  Hoyt, *see* Nakagami-q
  Nakagami-m, 23
  Nakagami-n, 21
  Nakagami-q, 21
  Rayleigh, 20
  Rice, *see* Nakagami-n
Amplitude modulation, multiple, *see* Multiple amplitude modulation (M-AM)
Amplitudes, fading channel (state of knowledge for optimum reception):
  known:
    known delays, unknown phases, 166–167
    known phases and delays, 159–163
  unknown:
    known delays and unknown phases, 168–188
    known phases and delays, 163–166
    unknown phases and delays, 188–191

Antenna array:
  multichannel receivers, 260, 324, 328–329
  optimum combining (OC), 437–438
Arbitrary two-dimensional signal constellation error probability integral, 116–117

Bayes rule, 158
Binary differential phase-shift-keying (BDPSK), *see* Differentially coherent detection, phase-shift-keying
Bhattacharyya parameter, 504, 513
Binary frequency-shift-keying (BFSK):
  defined, 44
  multichannel receivers:
    coherent, 343–344
    noncoherent, BDPSK and, 340–343
    noncoherent equal-gain combining
      bit error probability, average, 294–303
      characteristics of, generally, 290–291
      receiver structure, 291–294
  optimum receivers, detection of, 169, 171–172, 175–176
  selection combining, 340–343
  single channel receivers, 229–230
Binary phase-shift-keying (BPSK):
  average error probability performance, 111
  fading channels:
    optimum receivers for, 157
    single channel receivers, 230, 244
  ideal coherent detection, 37
  multichannel receivers:
    hybrid diversity schemes, 388–389, 400–401
    switched diversity, 317–318, 330, 357, 359, 369
  single carrier DS-CDMA systems, 475, 478
Bit error probability (BEP), average:
  characteristics of, 6–12
  coded communication, evaluation by transfer function bound, 510

535

536 INDEX

Bit error probability (BEP), average: (*Continued*)
  for decision statistic, quadratic form of complex Gaussian random variables, 421–426
  integrals involving:
    arbitrary two-dimensional signal constellation error probability integral, 116–117
    definite integrals associated with Rayleigh and Nakagami-m fading, 124–139
    Gaussian Q-function, 99–107, 118–121
    incomplete gamma function, 111–114
    Marcum Q-function, 107–111
    M-PSK error probability integral, 115–116, 121–124
  multicarrier DS-CDMA systems, 487–488
  multichannel receivers:
    M-ary orthogonal FSK, 304–311
    noncoherent equal-gain combining, 294–303
  optimum combining:
    MRC, comparison with, 467, 469
    Nakagami-m fading, 451–454
    Rayleigh fading, 442–448, 453
    Rician fading, 449–451, 453
    single interferer, 465–466
  optimum combining (OC) receivers, 441
  optimum receivers and, 169, 172, 179–180
  single channel receivers and, 214–215, 219, 221, 242
Bivariate Nakagami-m CDF, 149–150
Bivariate Rayleigh CDF, 144–146
Bivariate Rayleigh PDF, 142–145

Carrier synchronization loop, 51
Channel correlation models, *see* Correlation models, channel
Channel state information (CSI), 159, 261, 498, 520, 522, 524, 526–527
Chernoff bound, 71–72, 78–79, 216, 258
Classical functions, alternative representations:
  Gaussian Q-function, 70–74
  incomplete gamma function, 84
  Marcum Q-function, 74–89
  overview, 89–94
Co-channel interferers (CCI's), 437, 454, 462
Code-division multiple access (CDMA), 259, 263, 455. *See also* Direct-sequence code-division multiple access (DS-CDMA)
Coded communication:
  coherent detection:
    example of, 514–519
    pairwise error probability, evaluation of, 502–510
    system model, 499–502
    transfer function bound
      alternative formulation of, 513–514
      on average bit error probability, 510–513
  differentially coherent detection, 520–525
  overview, 497–499
  true upper bounds (TUB), 512–513, 517, 524
    comparison with union-Chernoff bounds, 526–530
Coherence bandwidth f_c, defined, 16
Coherence time T_c, defined, 16
Coherent equal gain combining, multichannel receivers:
  average output SNR, 279–281
  characteristics of, generally, 278–279
  error rate analysis:
    approximate, 288–289
    asymptotic, 289–290
    exact, 281–288
  exact error rate analysis:
    binary signals, 281–287
    M-PSK signals, extension to, 287–288
  outage probability performance, 313–314
  receiver structure, 279
Combined (time-shared) shadowed/unshadowed fading, 25–26
Combining, *see* Diversity combining
Communication, types of:
  differentially coherent detection:
    M-ary differential phase-shift-keying (M-DPSK), 59–65
      conventional detection, two-symbol observation, 60–63
      multiple symbol detection, 63–65
    $\pi/4$-differential QPSK, 65
  ideal coherent detection:
    M-ary frequency-shift keying (M-FSK), 43–44
    M-ary phase-shift-keyed (M-PSK), 35–42
    minimum shift-keying (MSK), 45–47
    multiple amplitude modulation (M-AM), 33
    multiple amplitude-shift-keying (M-ASK), 33
    overview, 31–32
    quadrature amplitude modulation (QAM), 34–35
    quadrature amplitude-shift-keying (QASK), 34–35
  noncoherent detection, 53–55
  nonideal coherent detection, 47–52
  partially coherent detection:

conventional detection, one-symbol
observation, 55–57
defined, 49
multiple symbol detection, 57–59
Composite log-normal shadowing/Nakagami-*m*
fading channel, 104–107
Composite multipath/shadowing, 24–25
Constant-envelope signal sets, 162–163, 166
Continuous phase frequency-shift-keying
(CPFSK), 45, 47
Correlation coefficient, multichannel receivers,
323
Correlation models, channel:
dual branch, nonidentically distributed
fading, 320–322
multiple branch, identically distributed
fading, 323–329
arbitrary correlation, 325–329
constant correlation, 323–324
exponential correlation, 324–325
numerical examples, 329–333
overview, 316–320
Cumulative distribution function (CDF):
correlative fading applications, 142–143
bivariate Rayleigh, 142–146
bivariate Nakagami-*m*, 149–150
for Gaussian random variable, 70
for maximum of two Nakagami-*m* random
variables, 149–152
for maximum of two Rayleigh random
variables, 146–149
multichannel receivers:
inversion of Laplace transform, numerical
techniques, 427–429

Deinterleaving, 498
Delays, fading channels (state of knowledge for
optimum reception):
known:
known amplitudes and unknown phases,
166–167
known amplitudes and phases, 159–163
known phases and unknown amplitudes,
163–166
unknown amplitudes and phases, 168–188
unknown:
unknown amplitudes and phases, 188–191
Differential decoding, 39, 47
Differential detection, 51
Differential detectors, 60
Differential encoding, 39, 59, 61
Differential quadrature phase-shift-keying
(DQPSK), *see* Differentially coherent
detection, phase-shift-keying

Differentially coherent detection:
in AWGN communication systems:
M-ary differential phase-shift-keying,
59–65
conventional detection, two-symbol
observation, 60–63
multiple symbol detection, 63–65
$\pi/4$-differential QPSK, 65
in fading communication systems
optimum receivers:
Nakagami-*m* fading, 186
overview, 181–183
Rayleigh fading, 183–185
multichannel receivers, fast fading:
noncoherent EGC over Rician fast
fading, 371–373
multichannel receivers, slow fading:
M-ary orthogonal FSK, 307, 309–310
noncoherent equal-gain combining:
average bit error probability,
294–303
characteristics of, generally,
290–291
receiver structure, 291–294
single channel receivers, fast fading:
M-ary differential phase-shift-keying,
248–251
single channel receivers, slow fading:
M-ary differential phase-shift-keying,
243–248
conventional detection, two-symbol
observation, 243–247
multiple symbol detection, 247–248
Differentially encoded *M*-ary
phase-shift-keying (*M*-DPSK), 202–204
Differential phase encoding, 184, 186
Direct-sequence code-division multiple access
(DS-CDMA):
channel model:
single carrier, 475–476
multicarrier, 481
numerical examples, 489–492
overview, 473–474
performance:
single carrier, 477–479
multicarrier, 483–492
receiver:
single carrier, 476
multicarrier, 481–482
transmitter:
single carrier, 474–475
multicarrier, 480–481
Dirichlet transformation, 398

538 INDEX

Diversity combining:
  complexity-performance tradeoffs, 264
  concept, 260
  hybrid schemes, *see* Hybrid diversity schemes, multichannel receivers
  mathematical modeling, 260–261
  pure schemes
    equal gain combining (EGC), *see* Equal gain combining, multichannel receivers
    maximal ratio combining (MRC), *see* Maximal ratio combining, multichannel receivers
    selection combining (SC), *see* Selection combining, multichannel receivers
    switch and stay combining (SSC), *see* Switch and stay combining, multichannel receivers
Doppler frequency shift, 370
Doppler spread f_d:
  defined, 16
  effect on error probability performance in fast fading:
    multichannel receivers, 373
    single-channel receivers, 250–251
  relation to correlation and spectral properties of fading process, 17

Encoder, defined, 501
Envelope fluctuations, in fading channels, 15–16, 503
Equal-gain combining (EGC):
  characteristics of, generally, 262–263
  coded communication, 507
  coherent, *see* Coherent equal gain combining
  defined, 11, 169
  noncoherent, *see* Noncoherent equal gain combining
  single carrier DS-CDMA systems, 478
Euler summation-based technique, 427–428

Fading channels. *See also specific types of fading channels*:
  characteristics of:
    envelope fluctuations, 15–16, 325
    fast fading, 16
    frequency-flat fading, 16–17
    frequency-selective fading, 16–17
    phase fluctuations, 15–16
    slow fading, 16
  coded communication:
    coherent detection, 499–519
    differentially coherent detection, 520–525
    overview, 497–499

  true upper bounds, compared with Union-Chernoff bounds, 526–530
  multichannel receivers, *see* Correlation models, channel
Nakagami-m, 104–107
Nakagami-n, 102
Nakagami-q, 101–102
optimum receivers for:
  known amplitudes and delays, unknown phases, 166–167
  known amplitudes, phases, and delays, 159–163
  known delays and unknown amplitudes and phases, 168–188
  known phases and delays, unknown amplitudes, 163–166
  unknown amplitudes, phases, and delays, 188–191
performance, detection:
  differentially coherent, 243–251
  ideal coherent, 220–234
  noncoherent detection, 239–241
  nonideal coherent, 234–239
  partially coherent, 242–243
single channel receivers:
  differentially coherent detection, 243–251
  ideal coherent detection, 220–234
  noncoherent detection, 239–241
  nonideal coherent detection, 234–239
  partially coherent detection, 242–243
Fast fading, defined, 16
Flat fading channels:
  frequency, 16–17
  modeling:
    combined (time-shared) shadowed/unshadowed fading, 25–26
    composite multipath/shadowing, 24–25
    log-normal shadowing, 23–24
    multipath fading, 18–23
Frequency nonselective fading, *see* Flat fading channels
Frequency selective fading:
  characteristics of, 16–17
  modeling, 26–28

Gamma/log-normal PDF, 25
Gauss quadrature integration:
  Gauss-Chebyshev, 9, 519, 428–429, 528
  Gauss-Hermite, 105, 230–231, 291, 519
Gaussian Q-function:
  alternative form, 71
  classical form, 70

integrals involving:
  composite log-normal
    shadowing/Nakagami-m fading
    channel, 104–107
  integer powers of, 118–121
  log-normal shadowing channel, 104–107
  Nakagami-m fading channel, 102–107
  Nakagami-n (Rice) fading channel, 102
  Nakagami-q (Hoyt) fading channel,
    101–102
  overview, 99–101
  Rayleigh fading channel, 101
  one-dimensional case, 70–72
  proofs of alternative form, 95–97
  two-dimensional case, 72–74
Generalized multilink fading channel, 260
Generalized selection combining (GSC):
  average error rate, 388–391
  average output SNR, 384, 386
  characteristics of, 5, 378–381
  Nakagami-m channels, performance over,
    391–403
  outage probability, 386–388
  statistics, 381–384
Generalized switch and stay combining
  (GSSC), *see* Generalized switched
  diversity
Generalized switched diversity, 403–411
Global System for Mobile (GSM), 455
Gray code bit-to-symbol mapping, 194

Hoyt fading channel:
  Gaussian Q-function and, 101–102
  Marcum Q-function and, 109
  multipath fading, 20–21
  single channel receivers, 241
Hybrid diversity schemes, multichannel
  receivers:
  generalized selection combining (GSC):
    average error rate, 388–391
    average output SNR, 384–386
    characteristics of, 378–381
    Nakagami-m channels, performance over,
      391–403
    outage probability, 386–388
    statistics, 381–384
  generalized switched diversity:
    average probability of error, 406–407
    characteristics of, 403–405
    output statistics, 406
  two-dimensional diversity schemes:
    numerical examples, 409–411
    performance analysis, 408–409

Ideal coherent detection:
  in communication systems:
    M-ary frequency-shift keying (M-FSK),
      43–44
    M-ary phase-shift-keyed (M-PSK), 35–42
    minimum-shift-keying (MSK), 45–47
    multiple amplitude modulation (M-AM),
      33
    multiple amplitude-shift-keying (M-ASK),
      33
    overview, 31–32
    quadrature amplitude modulation (QAM),
      34–35
    quadrature amplitude-shift-keying
      (QASK), 34–35
  multichannel receivers, performance over
    fading channels:
    coherent equal-gain combining, 278–290
    hybrid diversity schemes, 378–411
    maximal-ratio combining, 265–278,
      370–371
    selection combining, 333–347, 373–374
    switch and stay combining, 348–369,
      374–378
  single channel receivers, performance over
    AWGN channel:
    M-ary frequency-shift-keying, 204–205
    M-ary phase-shift-keying, 197–204
    minimum-shift-keying, 205
    multiple amplitude modulation, 194–195
    multiple amplitude-shift-keying, 194–195
    offset QPSK, 204
    $\pi/4$-QPSK, 202–204
    quadrature amplitude modulation,
      195–197
    quadrature amplitude-shift-keying,
      195–197
    staggered QPSK, *see* Offset QPSK
Interference, optimum combining receivers, *see*
  Optimum combining (OC) receivers
Interleaving, 498
Inverse Laplace transform method, for
  evaluating pairwise error probability
  508–509
Irreducible error probability, 250

Line-of-sight (LOS) path, multipath fading, 18,
  21–22. *See also* Rician fading
Log-normal shadowing:
  Gaussian Q-function and, 104–107
  Marcum Q-function and, 114
  overview, 23–24
Loop SNR, 51

540 INDEX

Marcum Q-function:
  bounds on, 86–90
  first-order:
    alternative forms, 76–81
    classical form, 74–75
  generalized (mth-order):
    alternative form, 82–86
    classical form, 81
  integrals involving:
    composite log-normal
      shadowing/Nakagami-m fading
      channel, 110–111
    log-normal shadowing channel,
      109–110
    Nakagami-m fading channel, 109
    Nakagami-n (Rice) fading channel, 109
    Nakagami-q (Hoyt) fading channel, 109
    overview, 107–108
    Rayleigh fading channel, 108
  multichannel receivers, 351, 421–426
  noncoherent equal-gain combining, 291
  Rice-Ie function, relations between, 257
  single channel receivers, 243
  Toronto function (incomplete), relations
    between, 75
M-ary differential phase-shift-keying
  (M-DPSK). See also Differentially
  coherent detection:
  characteristics of, 59–65, 93–94
  coded communication, 521–522
  multichannel receivers:
    fast fading, 371–373
    slow fading, 290–303
  optimum receivers, detection of, 181, 186
  single channel receivers:
    fast fading, 248–251
    generally, 213
    slow fading, 243–248
M-ary frequency-shift-keyed (M-FSK) signal:
  characteristics of, 43–44, 55, 111
  single channel receivers, 204–205, 229–234,
    240
  multichannel receivers:
    bit error probability, average, 304–309
    numerical examples, 309–311
M-ary phase-shift-keyed (M-PSK) signal:
  average symbol error rate, 271–272,
    390–391
  coded communication, 499, 532–534
  differentially coherent detection:
    multiple-symbol detection, 63–65
    overview, 59–60
    two-symbol observation, 60–64

differentially encoded:
  overview, 39–41
  single channel receivers, 202–204
  fast fading, 248–251
  ideal coherent detection, 35–39
  multichannel receivers, 271, 287–289, 357,
    390, 402–403
  multiple symbol detection, 217–218
  noncoherent detection, 53
  optimum combining, 448–449
  optimum receivers, detection of, 181–182,
    186
  partially coherent detection, 57
  single channel receivers, 197–204, 217–218,
    227, 248–251
Maximal ratio combining (MRC), multichannel
  receivers:
  characteristics of, generally, 5, 162, 265
  coded communication, 504
  multichannel receivers:
    diversity combining and, 262–263
    MGF-based approach, 267–275
    outdated/imperfect channel estimates,
      370–371
    outage probability performance, 315–316
    receiver structure, 265–267
    switched diversity, 340, 353, 360–361
  optimum combining, 452, 466–469
  SER expressions:
    asymptotic results, 275–278
    bounds, 275
  single carrier DS-CDMA systems, 478
Maximum delay spread T_{max}, 17
Maximum-likelihood (ML) decision rule, 158
Maximum-likelihood (ML) receiver, 501
Maximum-likelihood sequence estimator
  (MLSE), 502
Minimum-shift-keying (MSK):
  characteristics of, 45–47
  error probability performance of, in AWGN,
    205
  error probability performance of, in slow
    fading:
    single channel reception, 234
Moment generating function (MGF)-based
  approach, defined, 8
M-PSK error probability integral:
  integer powers of:
    overview, 121–122
    Rayleigh fading channel, 122–124
  Nakagami-m fading channel, 115–116
  overview, 115–116
  Rayleigh fading channel, 115
  single channel receivers, 223

INDEX   **541**

Multicarrier systems (MC-CDMA). *See also*
 Direct-sequence code-division multiple
 access (DS-CDMA):
 characteristics of, 479–480
 performance analysis:
  average bit error rate, 487–488
  conditional SNR, 483–487
  system and channel models:
   channel, 481
   notation, 482
   receiver, 481–482
   transmitter, 480–481
Multichannel receivers, performance of:
 bit error probability, 421–426
 coherent equal gain combining:
  approximate error rate analysis, 288–289
  asymptotic error rate analysis, 289–290
  average output SNR, 279–281
  characteristics of, generally, 278–279
  exact error rate analysis, 281–288
  receiver structure, 279
 diversity combining:
  complexity of, 264
  concept, 260
  mathematical modeling, 260–261
  techniques, survey of, 261–264
 fading correlation, impact of:
  correlated branches with nonidentical fading, 320–323
  identically distributed branches with constant correlation, 323–324
  identically distributed branches with exponential correlation, 324–325
  nonidentically distributed branches with arbitrary correlation, 325–329
  numerical examples, 329–333
  overview, 316–320
 hybrid diversity schemes:
  generalized selection combining (GSC), 378–403
  generalized switched diversity, 403–407
  two-dimensional diversity schemes, 408–411
 maximal-ratio combining:
  MGF-based approach, 268–275
  PDF-based approach, 267–268
  receiver structure, 265–267
  SER expressions, bounds and asymptotic, 275–278
 noncoherent equal-gain combining:
  DPSK, DQPSK, and BFSK, 290–303
  M-ary orthogonal FSK, 304–311

 numerical techniques for inversion of Laplace transform of cumulative distribution functions:
  Euler summation-based technique, 427–428
  Gauss-Chebyshev quadrature-based technique, 428–429
 outage probability performance:
  characteristics of, generally, 311–312
  coherent EGC, 313–314
  MRC and noncoherent EGC, 312–313
  numerical examples of, 314–315
 outdated or imperfect channel estimates, performance in presence of:
  MRC, 370–371
  noncoherent EGC over Rician fast fading, 371–373
  numerical results, 377–378
  selection combining, 373–374
  switched diversity, 374–477
 selection combining:
  average output SNR, 336–338
  average probability of error, 340–347
  characteristics of, generally, 333–335
  MGF of output SNR, 335–336
  outage probability, 338–340
 switched diversity:
  branch correlation, effect of, 366–369
  branch unbalance, effect of, 362–366
  performance of SSC over independent identically distributed branches, 348–362
Multipath fading:
 defined, 18
 Nakagami-m model, 22–23
 Nakagami-n (Rice) model, 21–22
 Nakagami-q (Hoyt) model, 20–21
 optimum receivers, detection of, 189
 Rayleigh model, 18, 20
Multiple access interference (MAI):
 single carrier DS-CDMA systems, 477
 multicarrier DS-CDMA systems, 483
Multiple amplitude modulation (M-AM):
 characteristics of, 33
 single channel receivers, 194–195, 220–221
 multichannel receivers, 272
Multiple amplitude-shift-keying (M-ASK), *see* Multiple amplitude modulation (M-AM)
Multiple interferers, optimum combining, 454–466
Multiple symbol detection, *see* Differentially coherent detection

542 INDEX

Nakagami-m fading channel:
  coded communication, 513
  definite integrals associated with, 124–139
  Gaussian Q-function and, 102–107, 119–121
  Marcum Q-function and, 109–113
  multicarrier systems, DS-CDMA, 488–492
  multichannel receivers, switched diversity:
    average output SNR, 352
    average probability of error, 359
    fading correlation, 318, 327, 331
    hybrid diversity schemes, 391–403
    noncoherent equal-gain combining, 308
    numerical examples, 314
    switched diversity, 349, 355
    optimum combining receivers, 451–454
    optimum receivers:
      characteristics of, 164–165
      detection by, 186–191
    single carrier systems, DS-CDMA, 478–479
    single channel receivers, 222, 246–247
Nakagami-n (Rice) fading channel, see Rician fading channel
Nakagami-q (Hoyt), see Hoyt fading channel
Noncentrality parameter, 74
Noncoherent combining loss, 172
Noncoherent detection:
  in communication systems, 53–55
  multichannel receivers, see Noncoherent equal-gain combining
  optimum receivers:
    Nakagami-m fading, 175–181
    Rayleigh fading, 168–175
  single channel receivers:
    AWGN channel performance, 209–210
    slow fading channel performance, 239–242
Noncoherent equal-gain combining, multichannel receivers:
  DPSK, DQPSK, and BFSK, 290–303
  M-ary orthogonal FSK, 304–311
  outage probability performance, 312–313
  outdated/imperfect channel estimates, 371–373
Nonideal coherent detection:
  in communication systems, 47–52
  single channel receivers:
    AWGN channel performance, 206–209
    slow fading channel performance, 234–239
Numerical techniques, inversion of Laplace transform of CDFs:
  Euler summation-based technique, 427–428
  Gauss-Chebyshev quadrature-based technique, 428–429

Offset QPSK (OQPSK)
  characteristics of, 41–42
  single channel receivers:
    AWGN channel performance, 204
    slow fading channel performance, 229
Optimum combining (OC):
  characteristics of, 437–438
  multiple interferers:
    number of interferers equal to or greater than number of array elements, 462–466
    number of interferers less than number of array elements, 456–462
  single interferer, 438–454
Optimum reception, performance evaluation:
  fading channels, optimum receivers for, 157–191
  ideal coherent detection, 36, 38–40, 44, 48, 50
  multichannel receivers, 259–433
  multiple symbol detection, 58
  for noncoherent detection, 54
  for partially coherent detection, 56
  single channel receivers, 193–258
Outage probability:
  as performance criterion, 5–6
  diversity combining, 141
  multichannel receivers:
    coherent EGC, 313–314
    MRC and noncoherent EGC, 312–313
    numerical examples of, 314–315
    performance, 311–312
    hybrid diversity schemes, 386–388
    switched diversity, 354–357
Outdated or imperfect channel estimates, performance in presence of:
  maximal-ratio combining, 370–371
  noncoherent EGC over Rician fast fading, 371–373
  numerical results, 377–378
  selection combining, 373–374
  switched diversity, 374–477

Pairwise error probability, coded communication:
  defined, 8, 497
  evaluation of:
    known channel state information, 503–506
    overview of, 502–503
    unknown channel state information, 506–510
Partial-band interference (PBI)
  multicarrier DS-CDMA systems, 485–487

Partially coherent detection:
  conventional detection, one-symbol observation, 55–57
  defined, 49
  M-ary phase-shift-keyed (M-PSK) signal, 57
  multiple symbol detection, 57–59
  single channel receivers:
    AWGN channel performance, 210–213
    slow fading channel performance, 242–243
Phase-locked loop (PLL), 51
Phase-shift-keying (PSK), *see* Binary phase-shift-keying
Phases, fading channels (state of knowledge for optimum reception):
  known:
    known amplitudes and delays, 159–163
    known phases, unknown amplitudes, 163–166
  unknown:
    known amplitudes and delays, 166–167
    known delays and unknown amplitudes, 168–188
    unknown amplitudes, phases, and delays, 188–191
$\pi/4$-QPSK, 202–204, 218
$\pi/4$-differential QPSK, detection of:
  communications systems, 65
  optimum receivers, detection of, 218
  single channel receivers, 251
Pilot tone-aided detection, 51
Power decay factor, 28
Precoded MSK, 47, 49–50
Probability density function (PDF):
  bivariate Rayleigh, 142–146
  composite gamma/log-normal, 25
  for maximum of two Nakagami-m random variables, 149–152
  for maximum of two Rayleigh random variables, 146–149
  Hoyt, *see* Nakagami-q
  Nakagami-m, 22–23
  Nakagami-n, 21–22
  Nakagami-q, 20–21
  Rayleigh, 18
  Rice, *see* Nakagami-n
  shadowed/unshadowed fading, 26

Quadrature phase-shift-keying (QPSK):
  characteristics of, 37–39
  error probability performance of, in AWGN, 197
  error probability performance of, in slow fading:
    single channel reception, 224

Quadrature amplitude modulation (QAM):
  characteristics of, 34–35
  error probability performance of, in AWGN, 195–197
  error probability performance of, in slow fading:
    single channel reception, 224
    multichannel reception, 272–275, 278, 361, 391, 404–405
Quadrature amplitude-shift-keying (QASK), *see* Quadrature amplitude modulation

RAKE receivers:
  fading channels, optimum receivers, 160–161
  multicarrier DS-CDMA systems, 479, 484
  multichannel, 263, 304, 314, 317, 330, 409
  optimum combining receivers, 439
  single carrier DS-CDMA systems, 478
Rayleigh fading channel:
  coded communication, 513
  evaluation of definite integrals associated with, 124–139
  Gaussian Q-function and, 101, 105–106, 118–119
  Marcum Q-function and, 108, 112
  M-PSK error probability integrals, integer powers of, 122–124
  multichannel receivers:
    hybrid diversity schemes, 381, 401
    outdated/imperfect channel estimates, 378
    switched diversity, 351–352, 354–355
    optimum combining (OC), average bit error probability, 442–448
  optimum receivers:
    characteristics of, 163–164
    detection by, 183–190
  single channel receivers, 223, 238, 245
Rician fading channel:
  characteristics of, 21–22, 102, 109, 113
  coded communication, 513
  fast, 248–249, 371–373
  Gaussian Q-function and, 102
  Marcum Q-function and, 109
  multichannel receivers, outdated/imperfect channel estimates, 371–373
  optimum combining receivers, 449–451
  single channel receivers, 223, 234, 239, 246, 248–249

Selection combining (SC):
  characteristics of, generally, 262–263
  defined, 141

Selection combining (SC): (*Continued*)
  multichannel receivers:
    average output SNR, 336–338
    average probability of error, 340–347
    characteristics of, generally, 333–335
    MGF of output SNR, 335–336
    outage probability, 338–340
    outdated/imperfect channel estimates, 373–374
    switched diversity, 360–361
Self-adaptive receivers, 159
Self-interference, 485
Shadowing:
  combined (time-shared) shadowed/unshadowed fading, 25–26
  composite multipath, 24–25
  log-normal, 23–24, 104–117, 109–111
Signal-to-noise ratio (SNR):
  as performance criterion, 4–5
  common fading channels, 19
  conditional, in multicarrier DS-CDMA systems, 483–484, 486
  flat fading channels, 17–18
  instantaneous, 99–100, 102, 107–109, 148, 150–151, 219
  multichannel receivers:
    coherent equal gain combining, 279–281
    diversity combining, 260, 264
    fading correlation, 323, 325, 327
    hybrid diversity schemes, 384–386
    M-ary orthogonal FSK, 304–307, 310
    noncoherent equal-gain combining, 300–301
    outage probability performance, 312–313, 331
    selection combining, 333–347
    switched diversity, 348–349, 352–354, 365–368, 376–377
  multipath fading, 20
  optimum receivers and, 164, 172
  shadowed/unshadowed fading, 26
Single carrier systems, DS-CDMA. *See also* Direct-sequence code-division multiple access (DS-CDMA):
  performance analysis:
    characteristics of, 477–478
    general case, 478
    Nakagami-m fading channels, application to, 478–479

system and channel models:
  channel model, 475–476
  receiver, 476
  transmitted signal, 474–475
Single channel reception:
  performance over AWGN channel, 193–219
  performance over fading channels, 219–251
Slow fading, defined, 16
Staggered QPSK (SQPSK),
  see Offset QPSK
Stein's unified analysis of error probability performance, 253–258
Switch and stay combining (SSC), *see* Switched diversity
Switched diversity:
  branch correlation, effect of, 366–369
  branch unbalance, effect of, 362–366
  generalized, hybrid diversity schemes:
    average probability of error, 406–407
    characteristics of, 403–405
    output statistics, 406
  multichannel receivers:
    branch correlation, effect of, 366–369
    branch unbalance, effect of, 362–366
    outdated/imperfect channel estimates, 374–377
  performance of SSC over independent identically distributed branches, 348–362
System performance measures:
  bit error probability (BEP), average, 6–12
  average signal-to-noise ratio (SNR), 4–5
  outage probability, 5–6

TDMA system, 455
Tikhonov distribution, 51
Toronto function (incomplete), relation to generalized Marcum Q-function, 81
Transfer function bound, *see* Coded communication, coherent detection
Trellis-coded modulation (TCM), 498, 511–512
True upper bounds (TUB),
  see Coded communication, coherent detection

Uniform error probability (UEP), 512
Union-Chernoff bound, 514, 526–530

Viterbi algorithm, 499